ANALYTICAL GROUNDWATER MECHANICS

Groundwater mechanics is the study of fluid flow in porous media. Focusing on applications and case studies, this book explains the basic principles of groundwater flow using mathematical expressions to describe a wide range of different aquifer configurations. Emphasis is placed throughout on the importance of developing simplified models that can be solved analytically to provide insight into complex groundwater flow scenarios and to allow better interpretation of the full numerical solution.

Focusing first on identifying the important features of a problem, the book explains how to translate practical questions into mathematical form and discusses the interpretation of the results. Illustrated with numerous real-world examples and graphical results, this is an ideal textbook for advanced undergraduate and graduate courses in earth science, geological engineering, and environmental engineering, as well as a useful reference for researchers and professionals.

OTTO D. L. STRACK is Professor of Civil Engineering and Geomechanics at the University of Minnesota. With almost 50 years' experience, he is an expert in groundwater mechanics with a particular interest in regional modeling of groundwater flow and transport, and he is the original developer of the analytic element method. He has received the Lifetime Achievement Award granted by the Minnesota Groundwater Association and is a correspondent of the Royal Dutch Academy of Sciences. He has published widely in the field, including the successful textbook *Groundwater Mechanics* (1989).

ANALYTICAL GROUNDWATER MECHANICS

OTTO D. L. STRACK
University of Minnesota

CAMBRIDGE
UNIVERSITY PRESS

CAMBRIDGE
UNIVERSITY PRESS

University Printing House, Cambridge CB2 8BS, United Kingdom

One Liberty Plaza, 20th Floor, New York, NY 10006, USA

477 Williamstown Road, Port Melbourne, VIC 3207, Australia

4843/24, 2nd Floor, Ansari Road, Daryaganj, Delhi – 110002, India

79 Anson Road, #06–04/06, Singapore 079906

Cambridge University Press is part of the University of Cambridge.

It furthers the University's mission by disseminating knowledge in the pursuit of education, learning, and research at the highest international levels of excellence.

www.cambridge.org
Information on this title: www.cambridge.org/9781107148833
DOI: 10.1017/9781316563144

First published 2017

Printed in the United States of America by Sheridan Books, Inc.

A catalogue record for this publication is available from the British Library.

Library of Congress Cataloging-in-Publication Data
Names: Strack, Otto D. L., 1943– author.
Title: Analytical groundwater mechanics / Otto D.L. Strack.
Description: New York, NY : Cambridge University Press, 2017. | Includes bibliographical references and index.
Identifiers: LCCN 2017032438 | ISBN 9781107148833 (Hardback : alk. paper)
Subjects: LCSH: Groundwater flow–Mathematical models.
Classification: LCC GB1197.7 .S765 2017 | DDC 551.49–dc23
LC record available at https://lccn.loc.gov/2017032438

ISBN 978-1-107-14883-3 Hardback

Contents

Preface

The subject matter covered in this text is the mathematical description of fluid flow through porous media. Some parts of this book have been taken from Strack [1989], but much of the material is newly written. The two primary objectives are as follows. The first is to instruct the reader in approximating groundwater flow problems in such a manner that they can be solved analytically. The analytic solution will help us to gain insight, prior to constructing a complete solution to the problem using some numerical method if a more elaborate model is required.

The second objective is to explain how to simplify a practical problem so that it is analytically tractable. It requires considerable skill and understanding to approximate an actual flow problem by a simpler one that can be solved analytically, yet provide insight into the essence of the original problem. Even if the problem cannot be handled adequately by simple means, and recourse to a numerical solution is necessary, the determination and interpretation of relatively crude approximate solutions often provides crucial insight. The understanding thus gained can be used with advantage in selecting and setting up a numerical model that may ultimately be used to solve the problem. Modern computational environments exist that are suitable for implementing analytical solutions with relative ease, and are capable of displaying the results in a variety of manners, often in graphical form. The availability of such environments greatly enhances the use of analytic solutions as compared with in the past.

In view of the primary two objectives, emphasis is placed on a detailed coverage of methods for solving a variety of problems, rather than on providing a catalogue of existing solutions. Application of complex variable methods greatly simplifies the method of solution of many groundwater flow problems. Although complex variable methods carry with them a certain level of intimidation for many, once understood, complex variables make major simplification possible, as compared with real variables. We introduce in this text complex variables using Wirtinger calculus (Wirtinger [1927]), which extends the use of complex variables to general two-dimensional problems. The implementation of complex variables in the majority of modern computational engines makes their use for obtaining analytic solutions attractive, especially in view of the primary objectives of this text.

The subject matter is organized in ten chapters. Each chapter covers one main topic. Chapter 1 is concerned with the basic concepts and the derivation of the governing equations. Chapter 2 deals with steady flow in a single aquifer. Steady shallow interface flow is the topic of Chapter 3, and two-dimensional flow in the vertical plane is covered in Chapter 4. Steady flow in leaky aquifers is covered in Chapter 5 and three-dimensional flow in Chapter 6. Transient flow is discussed in Chapter 7. A substantial part of the text, Chapter 8, is devoted to the use of complex variables for solving a variety of practical problems. It will become apparent to the reader that using complex variables instead of real ones leads to significant simplification, visible not only in the mathematical treatment, but also in the implementation in simple computer codes. An additional advantage of the use of complex variables is that for many of the problems flow nets can be easily produced, giving a graphical illustration of the flow patterns. The reader who completed the exercises will have produced a simple but flexible computer model capable of modeling features such as rivers, wells, lakes of various shapes, areas of recharge, and inhomogeneities in the hydraulic conductivity. Fluid particle paths and contaminant transport are covered in Chapter 9. An overview of numerical methods for solving groundwater flow problems is presented in Chapter 10; the reader is introduced to the basics of both finite difference and finite element methods. Applications as well as exercises are included in Chapter 2 through 10. There are two appendixes.

Many groundwater flow problems, in particular problems of regional aquifer flow, are not solved exactly in this text; approximate governing equations are derived for a variety of aquifer configurations. The mathematical formulations for these configurations are unified by the use of discharge potentials, expressed differently for each type of flow that corresponds to a particular aquifer configuration. The mathematical formulations thus become similar, and solutions and techniques are often transferable from one type of flow to another. An additional, but related, simplification of major practical interest is the application of the concept of vertically integrated flow, which results in accurate predictions of flow rates, while approximating the hydraulic heads.

Acknowledgments

The author is grateful to Dr. Erik Anderson, Dr. Henk M. Haitjema, and Dr. Kees A. Maas for their careful reading of portions of the text. They identified ambiguity in explanations, noted missing examples and inconsistencies, and suggested additions and deletions. The author is indebted to all three experts in the field for their contributions to this work.

The text could not have been created without the help of the author's wife, Andrine D. Strack, who unwaveringly lent her support to the author in his efforts. She created figures and identified, through multiple readings of the text, many places where the explanations lacked in clarity.

Work on this text proceeded over the past 10 years. The text was used in the classroom as it grew to completion. Many of the students attending the classes identified errors and inconsistencies, and the author is grateful for their comments.

1

Basic Equations

The flow of water in soil occurs through interconnected openings between the soil particles. The flow of water through the soil is erratic, and its velocity changes radically in space: the velocity is large in the small pores and small in the larger ones.

We do not need to determine the path that the water particles follow in their way through the soil for most engineering groundwater flow problems. It usually is sufficient to determine average velocities, average flow paths, the discharge flowing through a given area of soil, or the pressure distribution in the soil. We work, throughout this text, with averages and ignore the actual paths of flow. We use the term *rectilinear flow*, for example,when the average flow is in one direction.

The theory of groundwater flow is based on a law discovered by Henry Darcy [1856]. After the introduction of the basic concepts, we discuss the experiment performed by Darcy and present his law in its simplest form. We then present the generalized form of Darcy's law and the equation of continuity, and finish the chapter by combining these two equations into one governing equation for steady flow of a homogeneous fluid in a porous medium.

1.1 Basic Concepts

The basic quantities used to describe groundwater flow are velocity, discharge, pressure, and head. We discuss these quantities next.

1.1.1 The Specific Discharge

We define specific discharge as the volume of water flowing through a unit area of soil per unit time. The units of specific discharge are $[L^3/(L^2T)]$, or [L/T], and thus are the same as those of a velocity. Specific discharge sometimes is called discharge

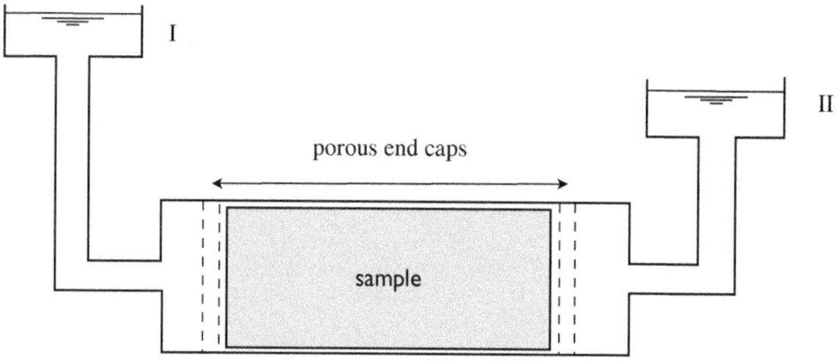

Figure 1.1 Schematic of Darcy's experiment.

velocity, but we use the term *specific discharge* to avoid confusion with a velocity. We represent the specific discharge by q.

The seepage velocity v is the average velocity at a point of the porous medium; it is the specific discharge divided by the area of voids present in a unit area of porous medium. If the porosity of the medium is n, then the area of voids per unit area is n and therefore

$$v = \frac{q}{n}. \tag{1.1}$$

Since n is always less than 1, v is always larger than q.

We illustrate the concepts of specific discharge and seepage velocity by considering water flowing through a cylindrical tube filled with sand, contained in the space between the end caps 1 and 2, as shown in Figure 1.1. This experimental setup is similar in principle to that used by Darcy to establish Darcy's law (Darcy [1856]). The cylinder is filled with water and is connected to two reservoirs, I and II, with different water levels. The water flows through the cylinder as a result of the difference in water levels. The level of reservoir II is controlled by overflow. We pour water into reservoir I in order to maintain its level. By measuring the rate at which water is poured into reservoir I, we determine the total amount, Q, of water flowing through the cylinder per unit time. The specific discharge is found by dividing Q by the cross-sectional area of the cylinder, A, i.e.,

$$q = \frac{Q}{A}. \tag{1.2}$$

We express the seepage velocity v with (1.1) as

$$v = \frac{q}{n} = \frac{Q}{nA}. \tag{1.3}$$

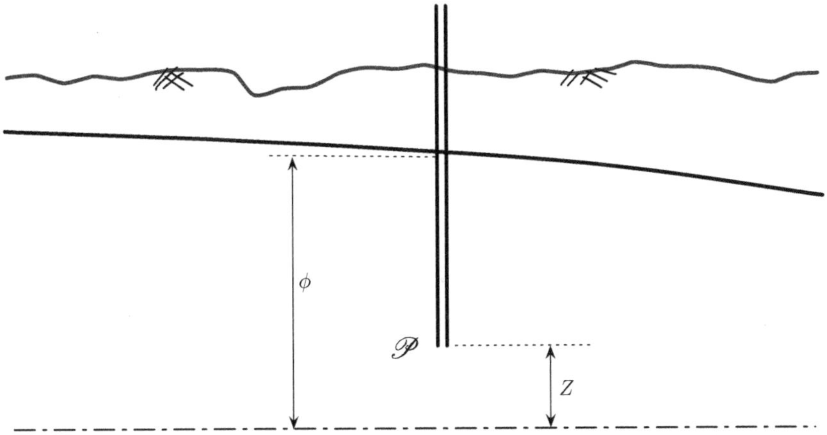

Figure 1.2 Definition of the head ϕ.

We obtain an expression for the time t for a water particle to travel through the sample from the seepage velocity v and the distance traveled, L:

$$vt = L \tag{1.4}$$

or

$$t = \frac{L}{v} = \frac{LnA}{Q}. \tag{1.5}$$

See (1.3).

1.1.2 Pressure and Head

The hydraulic head at a certain point $\mathscr{P}$ in a soil body is defined as the level to which the water rises in an open standpipe with its lower end at point $\mathscr{P}$ (see Figure 1.2). The *hydraulic head*, also simply called *head*,[1] is defined as a level and is measured with respect to a reference level or datum. We represent the hydraulic head by the letter ϕ. The units of ϕ are the units of length.

We find an expression for the pressure at point $\mathscr{P}$ from the weight of the water column above $\mathscr{P}$ in the standpipe. If the elevation of $\mathscr{P}$ above the reference level is Z [L], then the height of the water column above $\mathscr{P}$ is $\phi - Z$. Denoting the pressure as p [F/L^2], the density of water as ρ [M/L^3] and the acceleration of gravity as g [L/T^2], the pressure at $\mathscr{P}$ is

$$p = \rho g(\phi - Z). \tag{1.6}$$

[1] Other terms are in use as well, such as piezometric head and potentiometric head. We avoid using these terms; the word *piezometric* suggests that pressure is involved, and *potentiometric* suggests that the head is a potential, which it is not.

The elevation Z of point $\mathscr{P}$ above the reference level is known as the elevation head of point $\mathscr{P}$. The head, ϕ, can be expressed in terms of pressure, p, and elevation head, Z, by the use of (1.6) as follows:

$$\boxed{\phi = \frac{p}{\rho g} + Z} \tag{1.7}$$

The fraction $p/(\rho g)$, with the units of $[(F/L^2)/(F/L^3)] = [L]$ is the *pressure head.* We may express (1.7) as *(hydraulic) head equals pressure head plus elevation head.*

1.2 Darcy's Law

Darcy's law (Darcy [1856]) is an empirical relation for the specific discharge in terms of the head. The original form of this law is applicable to rectilinear flow of a homogeneous liquid only. A general form of Darcy's law exists; we present it after covering the case of rectilinear flow.

1.2.1 Rectilinear Flow

Darcy found that the amount of flow through a cylinder of sand of cross-sectional area A increases linearly with the difference in head at the ends of the sample (see Figure 1.3). The head at end cap 1 is ϕ_1. We see this from Figure 1.3: the pipe or hose connecting reservoir I to the sample can be viewed as a standpipe. Similarly, the head at end cap 2 is ϕ_2. Darcy's law for the experiment of Figure 1.3 is

$$Q = kA\frac{\phi_1 - \phi_2}{L}, \tag{1.8}$$

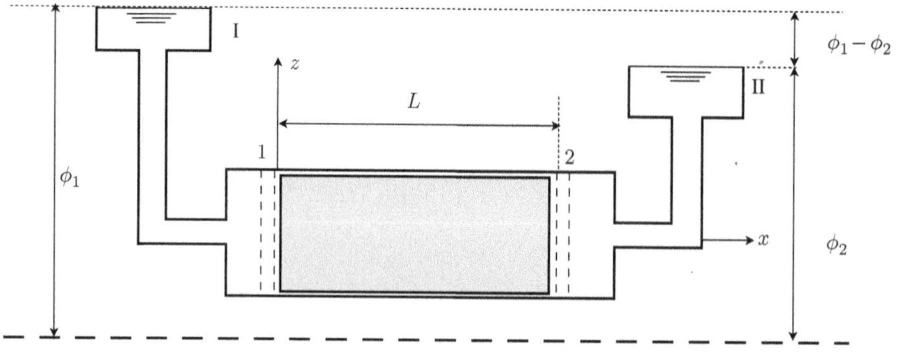

Figure 1.3 Darcy's experiment.

where Q is the discharge [L^3/T], A is the cross-sectional area of the sample [L^2], and L is the length of the sample [L]. The proportionality constant k is known as the hydraulic conductivity. It follows from (1.8) that the dimensions of k are those of a velocity, [L/T]. We sometimes use the term *resistance to flow*, borrowing this concept from electrokinetics. Resistance to flow is the inverse of hydraulic conductivity ($1/k$).

We write (1.8) in terms of the specific discharge q, with $q = Q/A$,

$$q = k\frac{\phi_1 - \phi_2}{L}. \tag{1.9}$$

If we measure the head at various points inside the sample of Figure 1.3, we find that it varies linearly over the sample. Choosing a coordinate system with the x-axis running along the axis of the sample with the origin at end cap 1, we obtain the following expression for ϕ:

$$\phi = \phi_1 + \frac{\phi_2 - \phi_1}{L}x. \tag{1.10}$$

This equation represents the straight line from $x = 0$, $\phi = \phi_1$, to $x = L$, $\phi = \phi_2$ in Figure 1.4, where the head is plotted as a function of position over the sample. It follows from (1.10) that

$$\frac{d\phi}{dx} = \frac{\phi_2 - \phi_1}{L}, \tag{1.11}$$

so that we may write (1.9) as

$$q_x = -k\frac{d\phi}{dx}. \tag{1.12}$$

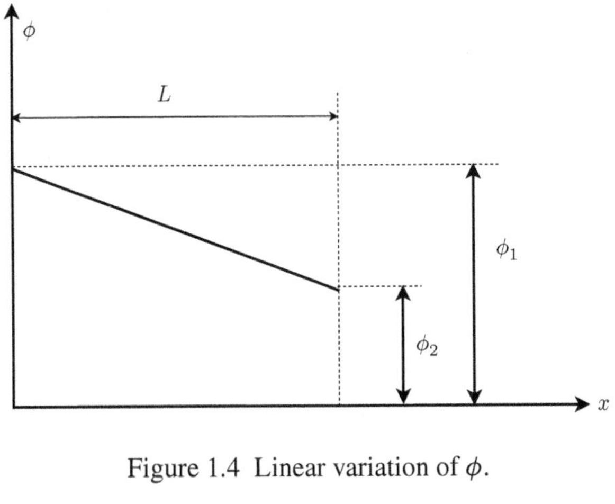

Figure 1.4 Linear variation of ϕ.

The index x in q_x is used to indicate that the specific discharge is in the x-direction. The derivative $d\phi/dx$ is known as the hydraulic gradient for flow in the x-direction.

1.2.2 Intrinsic Permeability

The hydraulic conductivity is a material constant, which depends on the properties of both the fluid and the soil. It is possible to define another constant, the (intrinsic) permeability κ, which depends only on the soil properties and is related to the hydraulic conductivity as

$$k = \frac{\kappa \rho g}{\mu}, \tag{1.13}$$

where μ is the dynamic viscosity [FT/L^2]. The dimension of κ is L^2. The intrinsic permeability is used primarily when the density or the viscosity of the fluid varies with position. In this text, however, only fluids with homogeneous properties are considered and therefore the classical hydraulic conductivity k is used. Values for k and κ are listed in Table 1.1 for some natural soils. An alternative unit for permeability is the Darcy, named after Henry Darcy. One Darcy, equal to 1000 μD (millidarcy), is equal to 0.9869233 μm^2.

1.2.3 Range of Validity of Darcy's Law

Darcy's law is restricted to specific discharges less than a certain critical value. The critical specific discharge depends on the grain size of the soil and the specific mass and the viscosity of the fluid. The criterion for assessing the validity of Darcy's

Table 1.1 *Permeabilities for some natural soils*

	k [m/s]	κ [m^2]
Clays	$< 10^{-9}$	$< 10^{-17}$
Sandy clays	$10^{-9} - 10^{-8}$	$10^{-16} - 10^{-15}$
Peat	$10^{-9} - 10^{-7}$	$10^{-16} - 10^{-14}$
Silt	$10^{-8} - 10^{-7}$	$10^{-15} - 10^{-14}$
Very fine sands	$10^{-6} - 10^{-5}$	$10^{-13} - 10^{-12}$
Fine sands	$10^{-5} - 10^{-4}$	$10^{-12} - 10^{-11}$
Coarse sands	$10^{-4} - 10^{-3}$	$10^{-11} - 10^{-10}$
Sand with gravel	$10^{-3} - 10^{-3}$	$10^{-10} - 10^{-9}$
Gravels	$> 10^{-2}$	$> 10^{-9}$

(Source: A. Verruijt, *Theory of Groundwater Flow*, © 1970, p. 10. Reprinted by permission of the author.)

law in a given case is expressed in terms of the Reynolds number Re, defined for groundwater flow as follows:

$$Re = \frac{D\rho}{\mu} q, \tag{1.14}$$

where D is the average grain diameter [L]. The Reynolds number is dimensionless. The range of validity of Darcy's law is defined by a relation obtained experimentally,

$$Re \leq 1. \tag{1.15}$$

If the Reynolds number is larger than 1, Darcy's law is not valid, and other, more complex, equations of motion must be used.

Darcy's law is valid for most cases of flow through soils. This is seen by substituting some average values for q, D, ρ, and μ in (1.14). The dynamic viscosity, μ, of water at a temperature of 10° C is about $1.3 * 10^{-3}$ N s/m^2 and ρ is about 10^3 kg/m^3. The average particle size of coarse sand is about $0.4 * 10^{-3}$ m in diameter. Substitution of these values in (1.14) yields

$$Re = (0.3 * 10^3) q. \tag{1.16}$$

This number is smaller than 1 if

$$q < 3.3 * 10^{-3} \text{ m/s} = 3.3 \text{ mm/s}. \tag{1.17}$$

This is a large value for the flow of groundwater. The hydraulic conductivity ranges from less than 10^{-9} m/s for clays to about 10^{-3} m/s for coarse sands. Furthermore, k is equal to the specific discharge occurring when the hydraulic gradient is 1, a large value. The specific discharge is the product of the hydraulic gradient and k, and is usually less than 10^{-3} m; Darcy's law indeed appears to be valid for most cases of flow through soils.

1.2.4 General Form of Darcy's Law

The flow is rarely rectilinear in practice, and neither the direction of flow nor the magnitude of the hydraulic gradient is known. The simple form (1.12) of Darcy's law, is not suitable for solving problems in practice; it is necessary to use a generalized form of Darcy's law, which gives a relation between the specific discharge vector and the hydraulic gradient. The direction of the specific discharge vector usually varies with position. The magnitude of this vector represents the amount of water flowing per unit time through a plane of unit area normal to the direction of flow. In three dimensions, the specific discharge vector has three components. With reference to a Cartesian coordinate system x, y, z, the three components of the specific discharge vector are represented as q_x, q_y, and q_z. The form of Darcy's law for three-dimensional flow through an isotropic porous medium is

$$q_x = -k\frac{\partial \phi}{\partial x}$$
$$q_y = -k\frac{\partial \phi}{\partial y} \tag{1.18}$$
$$q_z = -k\frac{\partial \phi}{\partial z}.$$

Because the three components of the specific discharge vector are proportional to minus the three components of the hydraulic gradient, with k as the proportionality factor, the specific discharge vector points in the direction opposite to the hydraulic gradient; groundwater flow occurs in the direction of decreasing head, hence the minus sign in Darcy's law. We sometimes represent the specific discharge vector with components (q_x, q_y, q_z) briefly as q_i. The index i then stands for x, y, or z. The partial derivatives $\partial\phi/\partial x$, $\partial\phi/\partial y$, and $\partial\phi/\partial z$ represent the three components of the hydraulic gradient. We may write the components of this vector as $\partial_x\phi, \partial_y\phi$, and $\partial_z\phi$, where the ∂ with the index stands for differentiation with respect to the coordinate represented by the index. The hydraulic gradient can then be written as

$$\partial_i\phi = \left[\frac{\partial \phi}{\partial x}, \frac{\partial \phi}{\partial y}, \frac{\partial \phi}{\partial z}\right]. \tag{1.19}$$

The notation with indices is known as the indicial notation or tensor notation, and has the advantage of compactness. The three equations (1.18), for example, can be written as one,

$$q_i = -k\partial_i\phi. \tag{1.20}$$

Darcy's law (1.18) may be written in terms of pressure by the use of (1.6),

$$q_x = -\frac{k}{\rho g}\frac{\partial p}{\partial x}$$
$$q_y = -\frac{k}{\rho g}\frac{\partial p}{\partial y} \tag{1.21}$$
$$q_z = -k - \frac{k}{\rho g}\frac{\partial p}{\partial z},$$

where the z-coordinate points vertically upward, so that $\partial Z/\partial z = 1$ (see Figure 1.2). Equations (1.18) and (1.21) are equivalent only if the density of the fluid is constant. Equation (1.18) is wrong in case ρ varies, as we demonstrate, following Verruijt [1970], by considering the case of groundwater of variable density at rest, so that

$q_x = q_y = q_z = 0$. We integrate (1.18) and use (1.7) to express ϕ in terms of the pressure,

$$\phi = \text{constant} = \frac{p}{\rho g} + Z. \tag{1.22}$$

Integration of (1.21) yields for this case

$$p = -\int_{Z_0}^{Z} \rho g dz, \tag{1.23}$$

where Z_0 is a reference level. Since (1.22) is not applicable to water of variable density at rest, and is obtained from (1.18), it follows that the latter equation cannot be used for cases of variable density, at least not with the definition (1.6) for ϕ. Equation (1.23), however, is correct and (1.21) is indeed valid for variable density.

Cases where variable density must be considered are not covered in this text, and we use Darcy's law in the form (1.18), with ϕ defined by (1.7).

1.2.5 Anisotropy

We assumed thus far that the hydraulic conductivity k is the same in all directions. In practice the soil often is layered; the hydraulic conductivity has different values in the directions parallel and normal to the layers. We call hydraulic conductivity anisotropic if its value depends on orientation. This is illustrated in Figure 1.5(a), where layers of sand are sandwiched between thin layers of clay. We consider the case in which there is no flow normal to the plane of drawing, the (x, y)-plane. We introduce Cartesian coordinates x^* and y^* such that the x^*-axis is parallel to the layers. It follows from Figure 1.5(b) that

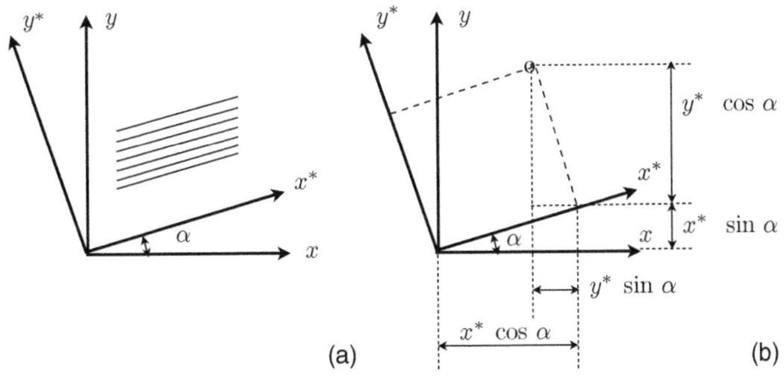

Figure 1.5 Anisotropic hydraulic conductivity.

$$x = x^* \cos\alpha - y^* \sin\alpha$$
$$y = x^* \sin\alpha + y^* \cos\alpha. \tag{1.24}$$

We write Darcy's law in terms of the x^*, y^*-coordinate system, denoting the components of the specific discharge vector in the x^* and y^* directions as q_x^* and q_y^*:

$$q_x^* = -k_1 \frac{\partial \phi}{\partial x^*}$$
$$q_y^* = -k_2 \frac{\partial \phi}{\partial y^*}, \tag{1.25}$$

where k_1 and k_2 represent the values of the hydraulic conductivity in the directions parallel and normal to the layers, respectively. These directions are called the principal directions, and k_1 and k_2 the principal values of the hydraulic conductivity.

We write Darcy's law in terms of vector components in the x- and y-directions. The expressions for q_x and q_y in terms of q_x^* and q_y^* are similar to (1.24):

$$q_x = q_x^* \cos\alpha - q_y^* \sin\alpha$$
$$q_y = q_x^* \sin\alpha + q_y^* \cos\alpha. \tag{1.26}$$

We obtain from (1.25)

$$q_x = -k_1 \frac{\partial \phi}{\partial x^*} \cos\alpha + k_2 \frac{\partial \phi}{\partial y^*} \sin\alpha$$
$$q_y = -k_1 \frac{\partial \phi}{\partial x^*} \sin\alpha - k_2 \frac{\partial \phi}{\partial y^*} \cos\alpha. \tag{1.27}$$

By application of the chain rule we find

$$\frac{\partial \phi}{\partial x^*} = \frac{\partial \phi}{\partial x}\frac{\partial x}{\partial x^*} + \frac{\partial \phi}{\partial y}\frac{\partial y}{\partial x^*} = +\frac{\partial \phi}{\partial x}\cos\alpha + \frac{\partial \phi}{\partial y}\sin\alpha$$
$$\frac{\partial \phi}{\partial y^*} = \frac{\partial \phi}{\partial x}\frac{\partial x}{\partial y^*} + \frac{\partial \phi}{\partial y}\frac{\partial y}{\partial y^*} = -\frac{\partial \phi}{\partial x}\sin\alpha + \frac{\partial \phi}{\partial y}\cos\alpha, \tag{1.28}$$

where the partial derivatives $\partial x/\partial x^*$, $\partial y/\partial x^*$, $\partial x/\partial y^*$, and $\partial y/\partial y^*$ are obtained by differentiating (1.24). Combining (1.27) and (1.28) we obtain Darcy's law for anisotropic hydraulic conductivity:

$$q_x = -k_{xx} \frac{\partial \phi}{\partial x} - k_{xy} \frac{\partial \phi}{\partial y}$$
$$q_y = -k_{yx} \frac{\partial \phi}{\partial x} - k_{yy} \frac{\partial \phi}{\partial y}, \tag{1.29}$$

where

$$
\begin{aligned}
k_{xx} &= k_1 \cos^2\alpha + k_2 \sin^2\alpha \\
k_{xy} &= k_{yx} = (k_1 - k_2)\sin\alpha\cos\alpha \\
k_{yy} &= k_1 \sin^2\alpha + k_2 \cos^2\alpha.
\end{aligned}
\tag{1.30}
$$

Note that each component of the specific discharge vector depends on both components of the hydraulic gradient. These two vectors therefore are not collinear, as opposed to the isotropic case. We see this also from (1.25), considering the case in which the hydraulic gradient is inclined at 45 degrees to the layers ($\partial\phi/\partial x^* = \partial\phi/\partial y^*$). Because k_1 is larger than k_2 the flow is inclined at an angle less than 45 degrees to the layers ($q_y^*/q_x^* < 1$).

For the general case of three-dimensional flow, Darcy's law (1.29) becomes

$$
\begin{aligned}
q_x &= -k_{xx}\frac{\partial\phi}{\partial x} - k_{xy}\frac{\partial\phi}{\partial y} - k_{xz}\frac{\partial\phi}{\partial z} \\
q_y &= -k_{yx}\frac{\partial\phi}{\partial x} - k_{yy}\frac{\partial\phi}{\partial y} - k_{yz}\frac{\partial\phi}{\partial z} \\
q_z &= -k_{zx}\frac{\partial\phi}{\partial x} - k_{zy}\frac{\partial\phi}{\partial y} - k_{zz}\frac{\partial\phi}{\partial z}.
\end{aligned}
\tag{1.31}
$$

The nine coefficients k_{ij} ($i = x, y, z$) are called the coefficients of a second rank tensor, the hydraulic conductivity tensor. Of these nine coefficients, only six are different; the tensor is symmetric:

$$
k_{xy} = k_{yx}, \qquad k_{xz} = k_{zx}, \qquad k_{yz} = k_{zy}. \tag{1.32}
$$

Consolidation of soils, which occurs primarily in the vertical direction, often causes the hydraulic conductivity to be less in the vertical direction than in the horizontal one. For such cases the two horizontal directions are principal directions of the hydraulic conductivity tensor, both equal to k, whereas the vertical component, also a principal value, is much less, k_{zz}. Darcy's law for such cases is:

$$
\begin{aligned}
q_x &= -k\frac{\partial\phi}{\partial x} \\
q_y &= -k\frac{\partial\phi}{\partial y} \\
q_z &= -k_{zz}\frac{\partial\phi}{\partial z}.
\end{aligned}
\tag{1.33}
$$

We consider problems with anisotropic hydraulic conductivity in this text, but, unless explicitly stated, assume that the hydraulic conductivity is isotropic.

1.3 Continuity of Flow

Darcy's law furnishes three equations of motion for the four unknowns q_x, q_y, q_z, and ϕ. A fourth equation is required for a complete description of groundwater flow. This equation is the mass balance equation, which for steady flow reduces to the equation of continuity of flow for points inside the porous medium. We consider continuity of flow by use of the concept of divergence.

1.3.1 Divergence

The divergence of the specific discharge vector is defined as the net release of fluid per unit volume and per unit time at a point in the porous medium. The concept of divergence is not limited to the specific discharge vector, but applies to the velocity field of any substance; it is useful in such considerations as mass balance. We use the concept of divergence in many places in this text.

We determine the divergence of the specific discharge vector for the cube shown in Figure 1.6. The sides of the cube are Δx, Δy, and Δz, and the center is at (x, y, z). The cube is so small that the variation of the magnitudes of the components of q_i along the sides can be neglected with respect to their value at the center of each side. The inflow through each side then is equal to the normal component of q_i multiplied by the area of the side. The net flow ΔQ [L^3/T] that leaves the cube is obtained from Figure 1.6, where only the components of q_i normal to the sides are shown:

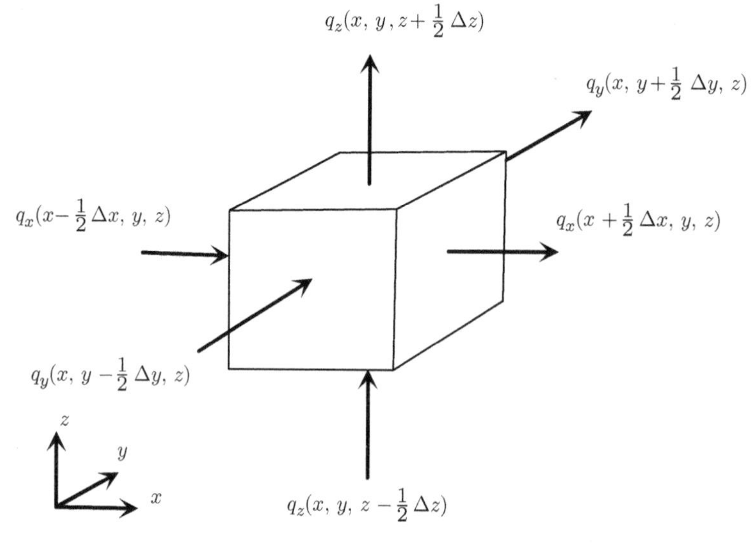

Figure 1.6 Continuity of flow.

$$\begin{aligned}\Delta Q = &\Delta y \Delta z[q_x(x+\tfrac{\Delta x}{2},y,z) - q_x(x-\tfrac{\Delta x}{2},y,z)] \\ &+ \Delta x \Delta z[q_y(x,y+\tfrac{\Delta y}{2},z) - q_y(x,y-\tfrac{\Delta y}{2},z)] \\ &+ \Delta x \Delta y[q_z(x,y,z+\tfrac{\Delta z}{2}) - q_z(x,y,z-\tfrac{\Delta z}{2})].\end{aligned} \tag{1.34}$$

Division by $\Delta x \Delta y \Delta z$ yields

$$\begin{aligned}&\frac{q_x(x+\frac{\Delta x}{2},y,z) - q_x(x-\frac{\Delta x}{2},y,z)}{\Delta x} \\ &+ \frac{q_y(x,y+\frac{\Delta y}{2},z) - q_y(x,y-\frac{\Delta y}{2},z)}{\Delta y} \\ &+ \frac{q_z(x,y,z+\frac{\Delta z}{2}) - q_z(x,y,z-\frac{\Delta z}{2})}{\Delta z} = \mathrm{div}(\vec{q}),\end{aligned} \tag{1.35}$$

where $\mathrm{div}(\vec{q})$ represents the divergence of the specific discharge vector. Passing to the limit for $\Delta x \to 0$, $\Delta y \to 0$, and $\Delta z \to 0$, the three terms in (1.35) become partial derivatives and we obtain

$$\boxed{\frac{\partial q_x}{\partial x} + \frac{\partial q_y}{\partial y} + \frac{\partial q_z}{\partial z} = \mathrm{div}(\vec{q})} \tag{1.36}$$

which is the expression for the divergence of the specific discharge vector.

The divergence of the specific discharge vector is zero for the case of steady flow, but also for the special case of transient flow where both the fluid and the porous medium are incompressible. The divergence of the specific discharge vector is non-zero only if both the flow is transient, and the porous medium and/or the fluid is compressible.

1.4 Optional Reading: The Curl of the Discharge Vector

The content of the remainder of this section is optional reading. The statement is often made that groundwater flow is irrotational. This statement is correct only for special cases and what follows addresses this issue.

1.4.1 Indicial Notation

In what follows we make use of indicial notation – compare (1.19) – and a notation known as the Einstein summation convention. We write the components of the specific discharge vector in indicial notation as q_i, where the index i represents either 1, 2, or 3, where 1 denotes the $x \equiv x_1$-direction, 2 the $y \equiv x_2$-direction, and 3 the $z \equiv x_3$-direction. The Einstein summation convention implies that summation is taken over every index that occurs twice in a single term. We write Darcy's law for flow in an anisotropic aquifer, (1.29), in compact form with indicial notation and Einstein summation convention as

$$q_i = -k_{ij}\frac{\partial \phi}{\partial x_j} = -k_{ij}\partial_j \phi. \tag{1.37}$$

Choosing i as 1, 2, and 3 gives the three equations for the three components of the specific discharge vector; the index j is to be summed over from 1 to 3.

We may also write the divergence of the specific discharge vector in compact form as

$$\frac{\partial q_i}{\partial x_i} = \partial_i q_i = div(\vec{q}). \tag{1.38}$$

1.4.2 The Curl of the Discharge Vector

The curl of a vector is defined as

$$C_i = \varepsilon_{ijk}\partial_j q_k = \vec{\nabla} \times \vec{q}. \tag{1.39}$$

The symbol ε_{ijk}, the Levi-Civita symbol, represents the alternating tensor of the third rank; the components of this tensor are zero if any two indices are equal, 1 if the indices ijk are an even permutation of 123, and -1 if the indices are an odd permutation of 123, or an even one of 132. For example, $\varepsilon_{133} = 0, \varepsilon_{312} = 1, \varepsilon_{321} = -1$.

A vector field that has a zero curl is said to be irrotational; otherwise the vector field is rotational. The vector field for groundwater flow is irrotational only if the hydraulic conductivity is both a constant and isotropic.

1.4.3 Variable Hydraulic Conductivity

If the hydraulic conductivity varies but is isotropic, the expression for the curl, C_l, of the specific discharge vector becomes, using indicial notation,

$$C_l = \varepsilon_{lmn}\partial_m q_n = -\varepsilon_{lmn}\partial_m(k\partial_n\phi) = -\varepsilon_{lmn}k\partial_m\partial_n\phi - \varepsilon_{lmn}\partial_m k\partial_n\phi. \tag{1.40}$$

The term $\varepsilon_{lmn}k\partial_m\partial_n\phi$ is zero. We can see this by considering the following identities

$$\varepsilon_{lmn}k\partial_m\partial_n\phi = \varepsilon_{lmn}k\partial_n\partial_m\phi = -\varepsilon_{lnm}k\partial_n\partial_m\phi = -\varepsilon_{lmn}k\partial_m\partial_n\phi = 0. \tag{1.41}$$

The first identity in (1.41) is true because the order of differentiation is immaterial for single-valued functions, and the hydraulic head is single-valued for physical reasons. The second identity follows from the definition of the alternating tensor. The third identity is simply a matter of renaming indices; this can be done because the indices are being summed over, and the symbol representing this sum is immaterial. The fourth and final identity follows because a quantity equal to minus itself must be zero. The expression for the curl, (1.40), thus reduces to

$$C_l = \varepsilon_{lmn}\partial_m q_n = -\varepsilon_{lmn}\partial_m k \partial_n \phi. \tag{1.42}$$

This expression is non-zero only if the gradients of hydraulic conductivity and hydraulic head are mutually orthogonal, or if the hydraulic conductivity is a constant.

1.4.4 Anistropy

Another case where the curl of the specific discharge vector is non-zero, i.e., where the flow is rotational, is if the hydraulic conductivity is anisotropic. In this case, Darcy's law is given by (1.37)

$$q_i = -k_{ij}\partial_j \phi. \tag{1.43}$$

The curl of the discharge vector is, assuming that k_{ij} is constant,

$$C_m = \varepsilon_{mni}\partial_n q_i = -\varepsilon_{mni}\partial_n[k_{ij}\partial_j \phi] = -\varepsilon_{mni}k_{ij}\partial_n\partial_j \phi. \tag{1.44}$$

This expression is zero only if $k_{ij} = k\delta_{ij}$, where δ_{ij} is the Kronecker delta, which is zero if $i \neq j$ and 1 if $i = j$. In this case, $k_{ij} = k\delta_{ij}$ corresponds to isotropic hydraulic conductivity, i.e., $k_{11} = k_{22}$ and $k_{12} = k_{21} = 0$. For the latter case, we obtain

$$C_m = -\varepsilon_{mni}k\delta_{ij}\partial_n\partial_j \phi = -\varepsilon_{mni}k\partial_n\partial_i \phi = 0, \tag{1.45}$$

where we set j equal to i, because for all other values the expression is zero. The final result is then zero, for the same reason as explained for (1.41). We conclude that the flow is rotational if the hydraulic conductivity is anisotropic, as asserted.

1.5 Fundamental Equations

Darcy's law and the continuity equation together provide four equations for the four unknown quantities q_x, q_y, q_z, and ϕ. We eliminate three of these four equations by substituting $-k\partial_i\phi$ for q_i in the continuity equation. This yields

$$\frac{\partial}{\partial x}\left[k\frac{\partial \phi}{\partial x}\right] + \frac{\partial}{\partial y}\left[k\frac{\partial \phi}{\partial y}\right] + \frac{\partial}{\partial z}\left[k\frac{\partial \phi}{\partial z}\right] = 0, \tag{1.46}$$

which is the governing equation for steady groundwater flow in an isotropic porous medium. If the hydraulic conductivity k is a constant, (1.46) reduces to

$$\boxed{\frac{\partial^2 \phi}{\partial x^2} + \frac{\partial^2 \phi}{\partial y^2} + \frac{\partial^2 \phi}{\partial z^2} = 0} \tag{1.47}$$

which is Laplace's equation in three dimensions. The fundamental equations for steady flow through isotropic porous media consist of either (1.46) or (1.47) and Darcy's law:

$$\boxed{\begin{aligned} q_x &= -k\frac{\partial \phi}{\partial x} \\ q_y &= -k\frac{\partial \phi}{\partial y} \\ q_z &= -k\frac{\partial \phi}{\partial z} \end{aligned}} \tag{1.48}$$

Solving problems of steady groundwater flow with constant hydraulic conductivity amounts to solving the differential equation of Laplace with the appropriate boundary conditions. Sometimes these conditions are given in terms of the head ϕ, as along the bank of a canal or river, and sometimes in terms of the discharge vector, as along a slurry wall where the component of q_i normal to the wall is zero.

Only few solutions to Laplace's equation in three dimensions exist. Fortunately, many practical groundwater flow problems either are two-dimensional or can be approximated by a two-dimensional analysis. We discuss such problems in the next chapter, along with approximate methods which make it possible to solve three-dimensional problems by a two-dimensional analysis. This two-dimensional analysis involves the vertically integrated specific discharge. The equations resulting from this vertical integration are not affected by the vertical anisotropy of a homogeneous aquifer; the rotational term in the equations does not affect the governing equations involving the vertically integrated flow.

2

Steady Flow in a Single Aquifer

Problems in the area of geomechanics have a special place in engineering; we deal with nature and are not in complete control of our designs, as is the case for man-made constructions. In the field of groundwater flow, we deal with the additional complication that the properties of the aquifers are hidden to us; we need to do tests to determine both the geometry and the relevant properties of the aquifer system.

Fortunately, compared with other fields of applied mechanics, steady flow of groundwater is governed by a comparatively simple differential equation. The difficulty in solving practical flow problems is the diversity of natural conditions encountered; to deal with this diversity, we categorize groundwater flow problems into types of flow according to the natural conditions under which the flow takes place. We define and discuss these types of flow in this chapter, treating one type of flow in each section.

Many of the natural types of flow are three-dimensional, but can be covered with good approximation by a two-dimensional analysis. We present the approximations for each type of flow, derive the basic equations, present some elementary methods of solution, and apply these to one or more practical problems. The scope of this chapter is limited to homogeneously permeable and isotropic porous media. Where possible, we introduce dependent variables for each type of flow in such a way that the various mathematical formulations are similar, if not identical.

The solution of practical problems may be viewed as a process that consists of three main steps:

1. Simplification of the real problem in such a way that it becomes mathematically tractable, while retaining the important characteristics of the problem
2. Formulation of the physical quantities in terms of abstract variables and functions used in the mathematical description of the problem
3. Determination of the mathematical solution and interpretation of the results in terms of the practical questions to be answered.

We apply this approach to practical problems in each section.

2.1 Horizontal Confined Flow

We call a flow horizontal if it takes place only in the horizontal plane; horizontal flow occurs if the vertical component of the specific discharge vector is zero throughout the flow region, i.e., if

$$q_z = 0. \tag{2.1}$$

Flow is called *confined* if the groundwater flows between two impermeable boundaries. These impermeable boundaries are known as the *confining boundaries*. If the confining boundaries are horizontal and the flow in the space between them is horizontal, we call the flow *horizontal confined flow*. The permeable layer between the confining layers is called a *water-bearing layer* or *aquifer*.

Horizontal confined flow either is one-dimensional or two-dimensional. In some cases, two-dimensional flow may be radial. We consider horizontal confined flow for cases of one-dimensional flow, radial flow, and two-dimensional flow below.

2.1.1 One-Dimensional Flow

An example of a case of one-dimensional flow is given in Figure 2.1. The flow occurs in an aquifer between two long, parallel, straight rivers. An (x,y,z)-coordinate system is selected as indicated in the figure. The x- and y-axes are normal and parallel to the rivers, respectively, and the z-axis points vertically upward. The origin is halfway between the rivers, the distance between the rivers is L, and the thickness of the aquifer is H. The heads in rivers $\mathscr{AB}$ $(x = -L/2)$ and $\mathscr{CD}$ $(x = L/2)$ are ϕ_1 and ϕ_2, respectively. It follows from symmetry that no flow occurs in the y-direction, so that, with (2.1),

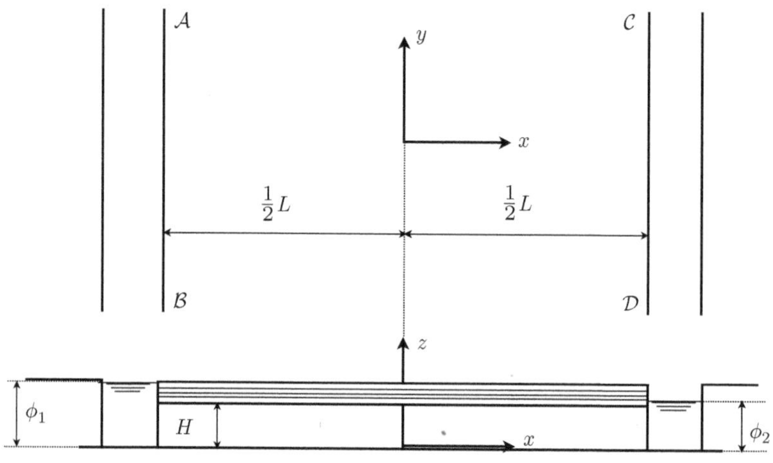

Figure 2.1 Horizontal confined flow between two parallel rivers.

$$q_y = q_z = 0. \tag{2.2}$$

It follows from (2.2) and Darcy's law, (1.18), that $\partial\phi/\partial y = \partial\phi/\partial z = 0$; ϕ does not vary in the y- and z-directions and is a function of x alone. Equation (1.47) therefore reduces to

$$\frac{d^2\phi}{dx^2} = 0 \tag{2.3}$$

and integration gives

$$\frac{d\phi}{dx} = A, \tag{2.4}$$

where A is a constant. Integration of (2.4) gives the general solution for one-dimensional flow:

$$\phi = Ax + B. \tag{2.5}$$

This solution covers all cases of one-dimensional flow. The particular solution that fits the flow case of Figure 2.1 is found by application of the boundary conditions, which we find from Figure 2.1:

$$\begin{aligned} x &= -\tfrac{1}{2}L, \qquad \phi = \phi_1 \\ x &= \tfrac{1}{2}L, \qquad \phi = \phi_2. \end{aligned} \tag{2.6}$$

The first boundary condition gives with (2.5)

$$\phi_1 = -\tfrac{1}{2}AL + B \tag{2.7}$$

and the second one yields

$$\phi_2 = \tfrac{1}{2}AL + B. \tag{2.8}$$

We solve (2.7) and (2.8) for A and B

$$A = -\frac{\phi_1 - \phi_2}{L}, \qquad B = \tfrac{1}{2}(\phi_1 + \phi_2). \tag{2.9}$$

Substitution of these values for A and B in (2.5) gives

$$\phi = -(\phi_1 - \phi_2)\frac{x}{L} + \tfrac{1}{2}(\phi_1 + \phi_2). \tag{2.10}$$

The specific discharge is found by the use of Darcy's law (1.18)

$$q_x = \frac{k(\phi_1 - \phi_2)}{L}, \qquad q_y = 0, \qquad q_z = 0. \tag{2.11}$$

It follows that the specific discharge does not vary with position, and increases linearly with both the difference in head at the two rivers and the coefficient of hydraulic conductivity. It decreases linearly with the distance, L, between the two rivers.

The discharge through the aquifer per unit length of river bank is found by multiplying the specific discharge q_x by the thickness of the aquifer, H. We write this discharge as Q_x, and find from (2.11)

$$Q_x = q_x H = \frac{kH(\phi_1 - \phi_2)}{L}. \tag{2.12}$$

2.1.2 Basic Equations

We usually are interested primarily in the discharge occurring over the thickness of the aquifer, rather than in the specific discharge. Accordingly, we introduce the discharge vector (Q_x, Q_y). This vector has one component, Q_x, for one-dimensional flow, defined by (2.12). The discharge vector has two components, Q_x and Q_y, for two-dimensional flow. Since the flow is horizontal, the hydraulic head is constant over the vertical, and thus also its horizontal gradient. The specific discharge is therefore constant over the vertical as well, and

$$\begin{aligned} Q_x &= Hq_x \\ Q_y &= Hq_y. \end{aligned} \tag{2.13}$$

The two components of the discharge vector (Q_x, Q_y) [L^2/T] can be expressed in terms of derivatives of the head ϕ by use of (2.13) and Darcy's law (1.18). This gives

$$\begin{aligned} Q_x &= Hq_x = H\left[-k\frac{\partial \phi}{\partial x}\right] \\ Q_y &= Hq_y = H\left[-k\frac{\partial \phi}{\partial y}\right]. \end{aligned} \tag{2.14}$$

Since both H and k are constants, (2.14) can be written as

$$\begin{aligned} Q_x &= -\frac{\partial [kH\phi]}{\partial x} \\ Q_y &= -\frac{\partial [kH\phi]}{\partial y}. \end{aligned} \tag{2.15}$$

These equations suggest the introduction of a new variable, Φ [L^3/T], defined as,

$$\Phi = kH\phi + C_c, \tag{2.16}$$

where C_c is an arbitrary constant and the index c stands for confined. Since $\partial C_c/\partial x = \partial C_c/\partial y = 0$, (2.15) can be written as

$$\begin{aligned} Q_x &= -\frac{\partial \Phi}{\partial x} \\ Q_y &= -\frac{\partial \Phi}{\partial y}. \end{aligned} \tag{2.17}$$

We refer to the function Φ as the *discharge potential for horizontal confined flow*, or briefly as the *potential* Φ. The term *potential* is derived from physics; a potential is a function that is linked to some vector field, in this case that of the discharge vector. The property that characterizes a potential is that its gradient equals the vector it defines. An example of a potential is the electric potential, whose gradient equals the electric current. Note that although the potential, according to this definition, should be $-\Phi$ (the discharge vector equals minus the gradient of Φ), we relax this definition somewhat and accept that the discharge equals minus the gradient of the potential.

Since $q_z = -k(\partial\phi/\partial z) = 0$ for all cases of horizontal confined flow, the governing equation becomes

$$\frac{\partial^2 \phi}{\partial x^2} + \frac{\partial^2 \phi}{\partial y^2} = 0. \tag{2.18}$$

Furthermore, it follows from (2.16) and (2.18) that

$$\frac{\partial^2}{\partial x^2}\left[\frac{\Phi - C_c}{kH}\right] + \frac{\partial^2}{\partial y^2}\left[\frac{\Phi - C_c}{kH}\right] = \frac{1}{kH}\left[\frac{\partial^2 \Phi}{\partial x^2} + \frac{\partial^2 \Phi}{\partial y^2}\right] = 0 \tag{2.19}$$

or

$$\frac{\partial^2 \Phi}{\partial x^2} + \frac{\partial^2 \Phi}{\partial y^2} = 0. \tag{2.20}$$

Problems of horizontal confined flow can be solved by determining a function Φ that fulfills Laplace's equation and satisfies the boundary conditions.

Equipotentials and Streamlines

Equipotentials are defined as curves of constant potential. A plot of equipotentials shows the variation of the potential throughout a flow domain in much the same way as the contour lines on a topographic map show the ground elevations. In addition, both the direction and the magnitude of the discharge vector can be derived from a plot of equipotentials, as is shown below. The discharge vector is directed normal to the equipotentials. We show this by introducing a local (s,n)-Cartesian coordinate system, where s is normal and n is tangent to the equipotential at a point $\mathscr{P}$; see Figure 2.2. We write (2.17) in terms of the coordinates s and n

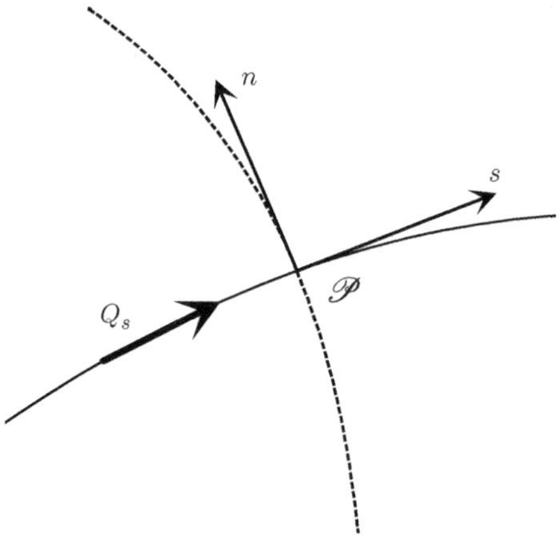

Figure 2.2 An equipotential and a streamline.

$$Q_s = -\frac{\partial \Phi}{\partial s}$$
$$Q_n = -\frac{\partial \Phi}{\partial n}. \tag{2.21}$$

The potential does not vary in the direction n, because n is tangent to the equipotential at $\mathscr{P}$. Hence, Q_n is zero so that the discharge vector indeed points in the direction s, which is normal to the equipotential. Because the flow is in the direction of decreasing potential and is normal to the equipotentials, groundwater flows in the direction of the steepest downward slope of the plot of equipotentials. *Discharge streamlines* are curves whose tangent at any point coincides with the discharge vector; see the solid curve in Figure 2.2. The word *discharge* in "discharge streamlines" is added to indicate that these streamlines correspond to the discharge vector, rather than to the specific discharge vector. When confusion is unlikely, however, we omit the word *discharge* and refer briefly to streamlines. Knowing that the streamlines are normal to the equipotentials, we can sketch them for any given plot of equipotentials. The magnitude of the discharge vector can be estimated from the distance between consecutive equipotentials, provided that the difference in values of Φ on such equipotentials is kept constant throughout the flow domain. If this difference is $\Delta\Phi$, and the distance between two consecutive equipotentials is Δs, then the first equation of (2.21) may be approximated as

$$Q_s \approx -\frac{\Delta \Phi}{\Delta s}. \tag{2.22}$$

Note that the dimensions of Q_s are [L^2/T], as it is the component of the discharge vector in the direction of flow (the other one is zero). Since $\Delta\Phi$ is constant, Q_s is inversely proportional to the distance between the equipotentials. It follows from the definition of the streamlines that no flow can occur across a streamline. Thus, the discharge flowing between two streamlines is constant, provided that no water enters the aquifer through the upper or lower boundaries. If the distance between two consecutive streamlines at a point $\mathscr{P}$ is Δn, then the discharge ΔQ [L^3/T] flowing between these streamlines is approximately $\Delta Q \approx Q_s \Delta n \approx -\Delta\Phi\Delta n/\Delta s$. If we draw the streamlines such that the distances Δn and Δs are approximately equal, ΔQ is about equal to $-\Delta\Phi$. The amount of flow between any pair of streamlines is then constant throughout the mesh of equipotentials and streamlines, which is called a flow net. For the simple case of one-dimensional flow shown in Figure 2.1, the equipotentials are straight lines that are parallel to the two rivers, and are all the same distance apart.

The use of the letter q for various types of discharge may be confusing; we summarize the usage adopted throughout this text as follows. We use the capital and lower cases of the letter q for discharge. The lower case is reserved for specific discharge, which has the dimension [L/T]. The upper case with index is reserved for the discharge vector (Q_x, Q_y), which has the dimensions [L^2/T]. The capital Q, without index, is reserved for discharge (per unit time), with dimensions [L^3/T].

Streamlines and the Stream Function

We obtain additional insight into most flow problems by determining streamlines in addition to equipotentials. We may create plots of equipotentials by programming the potential function and then apply some contouring routine to obtain the desired plot. A similar procedure might be used for obtaining plots of streamlines, provided that a function exists with the property that it is constant along streamlines. As we will see, this is possible only if the divergence of the discharge vector is zero, i.e., if there is no water entering or leaving the aquifer through the upper or lower boundaries. For such cases, we introduce the stream function Ψ, which is constant along streamlines. The gradient of $\Psi, \vec{\nabla}\Psi$, must be normal to the gradient of the potential because the streamlines are normal to the equipotentials. The gradient $\vec{\nabla}\Phi$ points in the direction of flow, whereas the gradient of the stream function, $\vec{\nabla}\Psi$, has no component in the direction of flow; it is constant along streamlines. The gradients $\vec{\nabla}\Phi$ and $\vec{\nabla}\Psi$ are shown in Figure 2.3.

The dot product of two orthogonal vectors is zero; the dot product equals the sum of the products of the corresponding components of the vectors. Since the gradients of the potential and the stream function are orthogonal we have

$$\frac{\partial\Phi}{\partial x}\frac{\partial\Psi}{\partial x} + \frac{\partial\Phi}{\partial y}\frac{\partial\Psi}{\partial y} = 0. \tag{2.23}$$

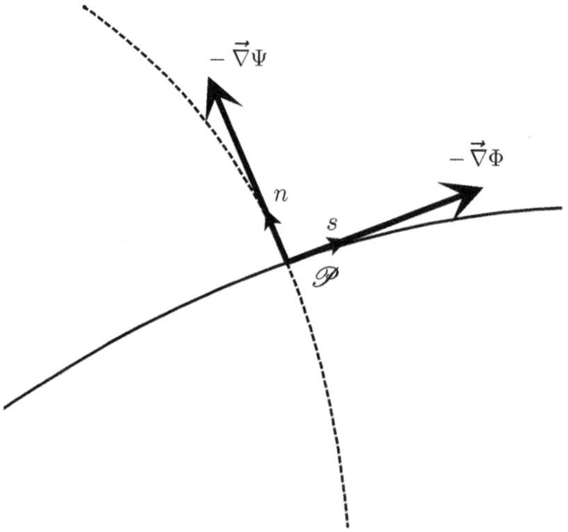

Figure 2.3 The gradients $-\vec{\nabla}\Phi$ and $-\vec{\nabla}\Psi$.

We satisfy this condition by choosing

$$\frac{\partial \Phi}{\partial x} = \frac{\partial \Psi}{\partial y} \tag{2.24}$$

$$\frac{\partial \Phi}{\partial y} = -\frac{\partial \Psi}{\partial x}. \tag{2.25}$$

The choice of the signs in the latter equations is arbitrary; we could have chosen the minus sign to occur in the first equation, rather than in the second; the signs as chosen correspond to an orientation of $\vec{\nabla}\Psi$ as shown in Figure 2.3. The equations (2.24) and (2.25) are the Cauchy-Riemann equations.

We noted that the stream function exists only if the divergence of the discharge vector is zero. We demonstrate that this is indeed the case by using the condition that the stream function be single-valued. Any single-valued function satisfies the condition that the order of differentiation is immaterial, i.e., that

$$\frac{\partial^2 \Psi}{\partial x \partial y} = \frac{\partial^2 \Psi}{\partial y \partial x}. \tag{2.26}$$

We differentiate (2.24) with respect to x and (2.25) with respect to y and add the results:

$$\frac{\partial^2 \Psi}{\partial x \partial y} - \frac{\partial^2 \Psi}{\partial y \partial x} = \frac{\partial^2 \Phi}{\partial x^2} + \frac{\partial^2 \Phi}{\partial y^2} = 0. \tag{2.27}$$

The condition that the stream function is single-valued requires that the potential satisfies the Laplace equation, i.e., that the divergence of the discharge vector is zero, as asserted.

We use equations (2.24) and (2.25) to determine the stream function from the discharge potentials for the various cases of flow that we consider in what follows.

Problem

2.1 Determine both $\Delta\Phi$ and Δs for the case of Figure 2.1, if there are to be four equipotentials, not counting the river banks. Draw the equipotentials, determine Q_s by the use of (2.22), and check this value against (2.12).

2.1.3 Radial Flow

Radial horizontal confined flow occurs when there is radial symmetry with respect to a vertical axis. An example of radial flow is the case of flow toward a well at the center of a circular island, as illustrated in Figure 2.4. We determine the potential for this flow case by applying Darcy's law and the equation of continuity separately, rather than solving the differential equation directly. Let the discharge of the well be Q [L^3/T], the radius of the well r_w, the radius of the island R, the head at the boundary ϕ_0, and the thickness of the aquifer H. It follows from the symmetry with respect to the axis of the well that the flow is radial; the discharge vectors at all points are directed toward the center of the island. Since the radial coordinate r points away from the center, the discharge vector Q_r points in the negative r direction, i.e.,

$$Q_r < 0. \tag{2.28}$$

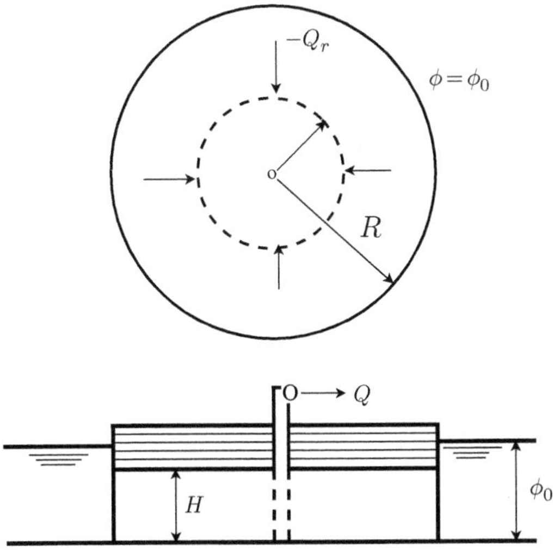

Figure 2.4 A well in a circular island.

It follows from continuity of flow that the same amount of water must pass through all cylinders with radius r ($r_w \leq r \leq R$) and height H. This amount equals the discharge Q of the well and may be expressed as follows, noting that $(-Q_r)$ is positive:

$$Q = 2\pi r(-Q_r). \tag{2.29}$$

It follows from Darcy's law (2.17) that

$$Q_r = -\frac{d\Phi}{dr} \tag{2.30}$$

and substitution in (2.29) gives:

$$Q = 2\pi r \frac{d\Phi}{dr} \tag{2.31}$$

or

$$\frac{d\Phi}{dr} = \frac{Q}{2\pi r}. \tag{2.32}$$

Integration yields

$$\Phi = \frac{Q}{2\pi} \ln r + C, \tag{2.33}$$

where C is a constant of integration. This equation represents the general solution for flow toward a well in an infinite confined aquifer. The value of the constant C for flow toward a well in a circular island is found from the boundary condition at $r = R$,

$$r = R, \qquad \phi = \phi_0. \tag{2.34}$$

Taking the arbitrary constant C_c in (2.16) as zero, the relation between potential and head becomes

$$\Phi = kH\phi \tag{2.35}$$

and the boundary condition may be expressed in terms of Φ as

$$r = R, \qquad \Phi = \Phi_0 = kH\phi_0. \tag{2.36}$$

The constant C in (2.33) is found on application of (2.36) to (2.33) and equals $\Phi_0 - Q/(2\pi)\ln R$, so that (2.33) becomes

$$\Phi = kH\phi = \frac{Q}{2\pi} \ln \frac{r}{R} + \Phi_0 \tag{2.37}$$

or, after division by kH, noticing that Φ_0 equals $kH\phi_0$,

$$\phi = \frac{Q}{2\pi kH} \ln \frac{r}{R} + \phi_0. \tag{2.38}$$

This equation is known as the Thiem equation (Thiem [1906]).

The Head at the Well

It may be necessary to calculate the head that must be maintained at the well to obtain a certain discharge of Q. If the radius of the well is r_w, this head, ϕ_w, is found from (2.38) by substituting r_w for r, which gives

$$\phi_w = \phi_0 + \frac{Q}{2\pi kH} \ln \frac{r_w}{R}. \tag{2.39}$$

Since the flow is confined, the water table at the well must be above the upper impervious boundary. Hence, ϕ_w must be greater than H:

$$\phi_w = \phi_0 + \frac{Q}{2\pi kH} \ln \frac{r_w}{R} \geq H. \tag{2.40}$$

If this condition is not fulfilled, the flow near the well is not confined and (2.38) is not applicable.

The Stream Function for a Well

We obtain an expression for the stream function for a well by integrating the Cauchy-Riemann equations (2.24) and (2.25). We rewrite these equations in radial form, to take advantage of the knowledge that the flow toward a well is radial. These equations are, in terms of radial coordinates r, θ:

$$\frac{\partial \Phi}{\partial r} = \frac{\partial \Psi}{r \partial \theta} \tag{2.41}$$

$$\frac{\partial \Phi}{r \partial \theta} = -\frac{\partial \Psi}{\partial r}. \tag{2.42}$$

The potential is independent of θ, i.e., $\partial \Phi / \partial \theta = 0$, so that the second equation implies that Ψ is independent of r. Integration of the first equation gives, with (2.37),

$$\frac{Q}{2\pi} \frac{1}{r} = \frac{1}{r} \frac{\partial \Psi}{\partial \theta}. \tag{2.43}$$

We integrate this, choosing the constant of integration as zero:

$$\Psi = \frac{Q}{2\pi} \theta. \tag{2.44}$$

The flow net for a single well in an infinite aquifer is shown in Figure 2.5. Recall that a flow net is obtained by plotting the contours of constant values of Φ and Ψ in a single plot with $\Delta\Phi = \Delta\Psi$. The values of the stream function are shown next to the corresponding streamline, and are obtained from (2.44). We see both from the figure and the expression for Ψ that the stream function is not single-valued, at least, not everywhere: along the negative x-axis, we obtain $\Psi = Q/2$ if we choose $\theta = \pi$, whereas we obtain $\Psi = -Q/2$ if we choose $\theta = -\pi$.

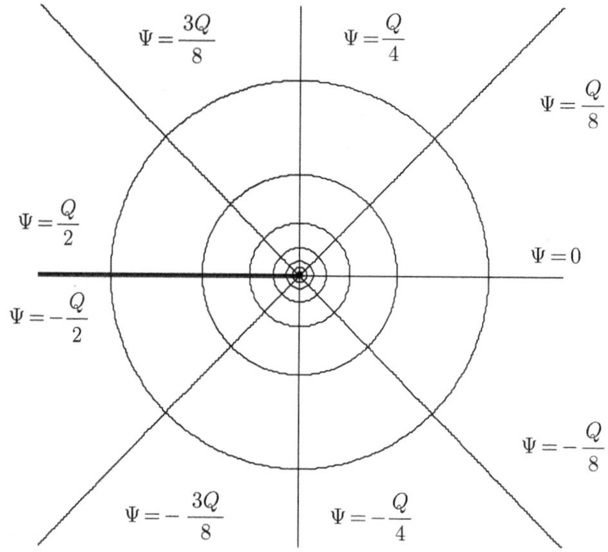

Figure 2.5 Equipotentials and streamlines for a well.

The reason for the dual values along a line that connects the well to infinity is that the condition for Ψ to be single-valued, i.e., the condition that the flow is divergence-free (no infiltration or extraction), is not satisfied at the well ($z=0$). We may view the discontinuity of Ψ, as a thought experiment, as follows. We consider the negative x-axis, called the *branch cut*, as a conduit between the well and infinity. We imagine that the water that flows from infinity to the well is transferred via the branch cut back to infinity, thereby satisfying the condition that there always must be continuity of flow. The jump in the stream function is indeed compatible with flow from the well through the branch cut to infinity. Clearly, this is a mathematical observation only, but explains why the branch cut is a mathematical necessity.

Problem

2.2 Prove that the equipotentials for the case of Figure 2.4 are circles. It is given that $r_w = 0.01 * R$. Draw nine equipotentials, not including the equipotentials $r = r_w$ and $r = R$, for a value of R of your choice. Do this by solving (2.40) for Q and substituting the resulting expression for Q in (2.37).

2.1.4 Two-Dimensional Flow

The differential equation that describes two-dimensional horizontal confined flow in the (x,y)-plane is Laplace's equation (2.20):

$$\frac{\partial^2 \Phi}{\partial x^2} + \frac{\partial^2 \Phi}{\partial y^2} = 0. \tag{2.45}$$

All solutions to problems of horizontal confined flow must satisfy this equation; the solution to a particular problem is selected by application of the boundary conditions.

A relatively straightforward method of solution is known as the *method of images* and is based on the *superposition principle*. We present this method below.

The Superposition Principle

The superposition principle states that if two different functions, say, $\overset{1}{\Phi}$ and $\overset{2}{\Phi}$, are solutions of Laplace's equation, then the function

$$\Phi(x,y) = C_1\overset{1}{\Phi}(x,y) + C_2\overset{2}{\Phi}(x,y) \tag{2.46}$$

also is a solution of Laplace's equation.

The proof of the superposition principle is as follows. Since $\overset{1}{\Phi}(x,y)$ and $\overset{2}{\Phi}(x,y)$ are solutions of the differential equation, we have

$$C_1\left[\frac{\partial^2\overset{1}{\Phi}}{\partial x^2} + \frac{\partial^2\overset{1}{\Phi}}{\partial y^2}\right] + C_2\left[\frac{\partial^2\overset{2}{\Phi}}{\partial x^2} + \frac{\partial^2\overset{2}{\Phi}}{\partial y^2}\right] = C_1 * 0 + C_2 * 0 = 0. \tag{2.47}$$

Since C_1 and C_2 are constants,

$$\begin{aligned}&\frac{\partial^2[C_1\overset{1}{\Phi}]}{\partial x^2} + \frac{\partial^2[C_2\overset{2}{\Phi}]}{\partial x^2} + \frac{\partial^2[C_1\overset{1}{\Phi}]}{\partial y^2} + \frac{\partial^2[C_2\overset{2}{\Phi}]}{\partial y^2}\\ &\quad = \frac{\partial^2[C_1\overset{1}{\Phi} + C_2\overset{2}{\Phi}]}{\partial x^2} + \frac{\partial^2[C_1\overset{1}{\Phi} + C_2\overset{2}{\Phi}]}{\partial y^2} = 0.\end{aligned} \tag{2.48}$$

It follows from (2.46) and (2.48) that

$$\frac{\partial^2\Phi}{\partial x^2} + \frac{\partial^2\Phi}{\partial y^2} = 0, \tag{2.49}$$

as asserted. Note that superposition of solutions is possible without violating the differential equation only if this equation is linear in Φ. Laplace's equation is only one of many linear differential equations for which superposition holds.

Analytic Elements

We present a number of expressions for both the discharge potential Φ and the stream function Ψ and use superposition to create a variety of solutions to practical problems. We begin by listing the solutions, which we call analytic elements, to be combined as needed. The first analytic element is the constant C; this is a trivial solution to the governing equations, but it is important to recognize that the constant is added to each solution just once.

The Constant The potential for a constant is

$$\Phi = C. \tag{2.50}$$

The second analytic element is the uniform flow field.

Uniform Flow If there is uniform flow with components Q_{x0} and Q_{y0}, then the potential is

$$\Phi = -Q_{x0}x - Q_{y0}y. \tag{2.51}$$

We verify that this potential indeed corresponds to uniform flow with components Q_{x0} and Q_{y0} by application of Darcy's law: $-\partial\Phi/\partial x = Q_{x0}$ and $-\partial\Phi/\partial y = Q_{y0}$. We obtain an expression for the stream function by integrating the Cauchy-Riemann equations (2.24) and (2.25), which gives for (2.24)

$$\frac{\partial\Psi}{\partial y} = \frac{\partial\Phi}{\partial x} = -Q_{x0} \rightarrow \Psi = -Q_{x0}y + f(x), \tag{2.52}$$

where $f(x)$ is an arbitrary function of x that can be added without violating the partial differential equation. We treat (2.25) in a similar manner

$$\frac{\partial\Psi}{\partial x} = -\frac{\partial\Phi}{\partial y} = Q_{y0} \rightarrow \Psi = Q_{y0}x + g(y), \tag{2.53}$$

where $g(y)$ is an arbitrary function of y. We satisfy both equations by choosing $f(x)$ in (2.52) as $Q_{y0}x$ and $g(y)$ as $-Q_{x0}y$. The result is

$$\Psi = -Q_{x0}y + Q_{y0}x. \tag{2.54}$$

A Well The expressions for the discharge potential and the stream function for a well at (x_w, y_w), with discharge Q and radius r_w, are

$$\Phi = \frac{Q}{2\pi}\ln\frac{r}{r_w} = \frac{Q}{4\pi}\ln\frac{r^2}{r_w^2} = \frac{Q}{4\pi}\ln\frac{(x-x_w)^2+(y-y_w)^2}{r_w^2} \tag{2.55}$$

$$\Psi = \frac{Q}{2\pi}\theta = \frac{Q}{2\pi}\arctan\frac{y-y_w}{x-x_w}. \tag{2.56}$$

Note that we added a constant to the discharge potential in such a way that it is zero at the well screen. This is not always necessary, but it is useful when implementing the potential for a well, which is fixed inside the well screen, in a computer program. Adding the constant such that the contribution of the well to the potential vanishes at the well, we can set the potential equal to zero when $r \le r_w$, thereby avoiding the singularity at $r = 0$, where the logarithm becomes $-\infty$.

2.1.5 Superposition

We take advantage of the principle of superposition by making a list of elementary solutions that are suitable for combination into discharge potentials that describe flow in settings that are far more complex than each of the individual parts could represent by itself. So far, we can identify the potential for constant head,

$$\Phi = C, \tag{2.57}$$

the potential and stream function for uniform flow with components Q_{x0} and Q_{y0},

$$\Phi = -Q_{x0}x - Q_{y0}y, \qquad \Psi = -Q_{x0}y + Q_{y0}x, \tag{2.58}$$

and the potential and stream function for a well of discharge Q at x_1, y_1,

$$\Phi = \frac{Q}{4\pi}\ln[(x-x_w)^2 + (y-y_w)^2], \qquad \Psi = \frac{Q}{2\pi}\arctan\frac{y-y_w}{x-x_w}. \tag{2.59}$$

We left the radius of the well out of the expression for the potential; it can be absorbed in the constant.

2.1.6 Application: A Well in a Field of Uniform Flow

We can now construct the potential for uniform flow with a well at the origin to determine the location of the point of zero flow rate, called the stagnation point. We do this by adding (2.57), (2.58) (with $Q_{y0} = 0$), and (2.59), which gives

$$\Phi = -Q_{x0}x + \frac{Q}{4\pi}\ln[x^2 + y^2] + C. \tag{2.60}$$

We obtain an expression for the stream function in a similar manner:

$$\Psi = -Q_{x0}y + \frac{Q}{4\pi}\arctan\frac{y}{x}. \tag{2.61}$$

We implement this function in a simple computer program and contour it to obtain the streamline pattern shown in Figure 2.6.

We need additional information to determine the constant C, such as an observed head in the aquifer at some point, either before or after the well began to operate. We do not need this constant for the present application since we are interested in determining the location of the stagnation point, which requires only the derivative

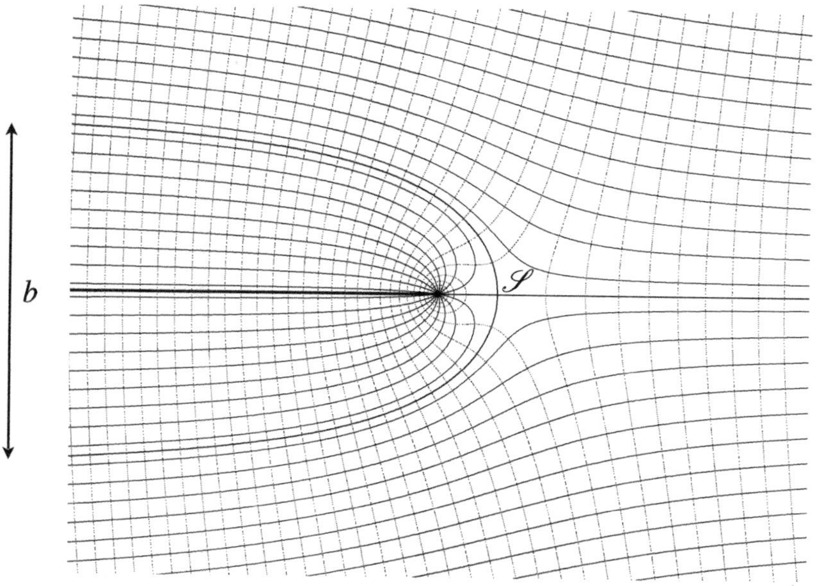

Figure 2.6 A well in a field of uniform flow.

of the potential. For that, we determine an expression for the discharge vector components:

$$Q_x = Q_{x0} - \frac{Q}{2\pi}\frac{x}{x^2+y^2} \tag{2.62}$$

$$Q_y = Q_{y0} - \frac{Q}{2\pi}\frac{y}{x^2+y^2}. \tag{2.63}$$

The component Q_y is zero for all $y = 0$; there is a stagnation point on the x-axis at $x = x_s$ where $Q_x = 0$,

$$Q_x(x_s, 0) = 0 = Q_{x0} - \frac{Q}{2\pi}\frac{1}{x_s}, \tag{2.64}$$

and solve this for x_s:

$$x_s = \frac{Q}{2\pi Q_{x0}}. \tag{2.65}$$

We express this in terms of the width of the uniform flow field captured by the well. We represent this width by b, shown in Figure 2.6, and express the discharge of the well as

$$Q = bQ_{x0} \tag{2.66}$$

so that (2.65) becomes

$$x_s = \frac{b}{2\pi}. \tag{2.67}$$

The streamlines through the stagnation point divide the flow into two zones; the flow within these streamlines is captured by the well, whereas the flow outside them passes by the well. We call these streamlines the *dividing streamlines.*

2.1.7 Application: A Well Capturing a Contaminated Zone

We consider the design of a well that must capture contaminated water flowing between two points $\mathscr{A}$ and $\mathscr{B}$ as shown in Figure 2.7. The well is to be placed at the centerline of the contaminated zone a distance d downstream from the front of the contaminated zone. We wish to determine the discharge of the well, such that precisely the contaminated groundwater is captured. The width of the contaminated zone is $2a$.

We compute the discharge Q of the well such that it captures the flow through the line $\mathscr{A}\mathscr{B}$; we do this by using the stream function, given by (2.61) of the previous application. The difference in value of the stream function between points $\mathscr{A}$ and $\mathscr{B}$ equals the flow between these points. In this case, however, we cannot simply compute the stream function at points $\mathscr{A}$ and $\mathscr{B}$ and subtract the values; there is a jump in the stream function along the negative x-axis. The stream function decreases from $\mathscr{A}$ to $\mathscr{B}$ because the potential increases in the negative x-direction and the potential and stream function form a right-hand coordinate system. Since the stream function jumps from $-Q/2$ to $Q/2$ crossing the cut in the positive y-direction, we have

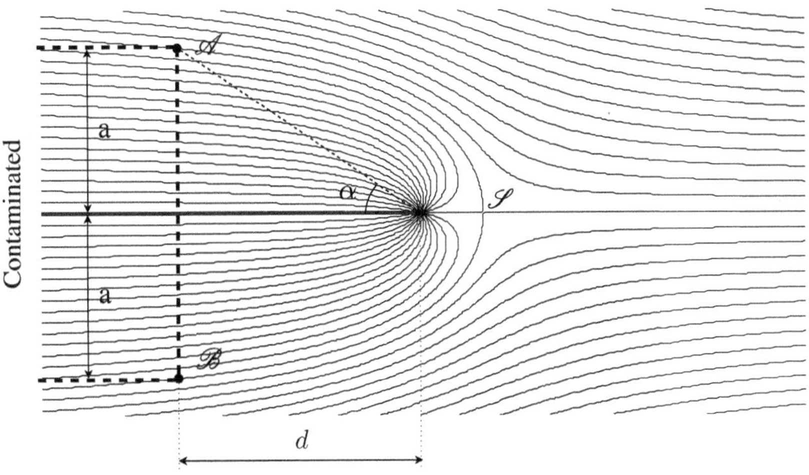

Figure 2.7 A well designed to capture a contaminated zone $\mathscr{A}\mathscr{B}$.

$$Q = \Psi_{\mathscr{B}} - \Psi_{\mathscr{A}} + Q. \tag{2.68}$$

The stream function at point $\mathscr{A}$ is found from (2.61) as

$$\Psi_{\mathscr{A}} = -Q_{x0}a + \frac{Q}{2\pi}\theta_{\mathscr{A}} = -Q_{x0}a + \frac{Q}{2\pi}(\pi - \alpha) = -Q_{x0}a + \frac{Q}{2\pi}\left(\pi - \arctan\frac{a}{d}\right). \tag{2.69}$$

We see from the figure that $\theta_{\mathscr{B}} = -\theta_{\mathscr{A}}$ so that

$$\Psi_{\mathscr{B}} = Q_{x0}a - \frac{Q}{2\pi}\left(\pi - \arctan\frac{a}{d}\right). \tag{2.70}$$

We use the expressions for $\Psi_{\mathscr{A}}$ and $\Psi_{\mathscr{B}}$ in (2.68):

$$0 = Q_{x0}a - \frac{Q}{2\pi}\left(\pi - \arctan\frac{a}{d}\right) - \left[-Q_{x0}a + \frac{Q}{2\pi}\left(\pi - \arctan\frac{a}{d}\right)\right] \tag{2.71}$$

or

$$Q\left(1 - \frac{1}{\pi}\arctan\frac{a}{d}\right) = 2Q_{x0}a. \tag{2.72}$$

We solve this for Q:

$$Q = \frac{2Q_{x0}a}{1 - 1/\pi\,\arctan(a/d)}. \tag{2.73}$$

Problem

2.3 Consider a confined aquifer of thickness H and hydraulic conductivity k. Hydraulic heads have been measured at three locations. You may assume that the flow is uniform. The values for H and k are H =10 m, k =10 m/d.

$$(x_1, y_1) = (0, 0), \qquad \phi_1 = 22 \text{ m}$$
$$(x_2, y_2) = (100, 100), \qquad \phi_2 = 21 \text{ m}$$
$$(x_3, y_3) = (200, -100), \qquad \phi_3 = 20.5 \text{ m}.$$

Questions:

1. Compute the two components of the uniform flow from your measurements. Hint: Write the expression for the discharge potential for uniform flow with components in both the x- and y-directions. Apply that equation to the three points with measured heads, and solve for your unknowns. Write expressions for the discharge vector components in symbolic form, and afterward substitute the given numbers in your equations for numerical evaluation.

2. Write the equation for the potential for a well of given discharge Q in this field of uniform flow. The well is placed at $x = -d = -1000$ m, $y = 0$, and you may assume that the head ϕ_3 will not be affected by the well (you need this assumption for computing the constant in your solution). Determine the maximum discharge of the well such that the flow near the well remains confined. The radius of the well is $r_w = 0.3$ m.
3. Determine the coordinates of the stagnation point. You may either orient your coordinate system in the direction of flow and use the results for a well in a field of uniform flow, or you may use the coordinates presented above (this will make the analysis more involved).

The Method of Images

The method of images is a combination of the superposition principle and symmetry. We consider the cases of horizontal confined flow illustrated in Figure 2.8(a) and (b). The dotted curves in the plan views are equipotentials and the solid curves are streamlines. Figure 2.8(a) represents the case of a discharge well and a recharge well in a confined aquifer of infinite extent. A discharge well is one that withdraws water from the aquifer. A recharge well pumps water into the aquifer. We use the

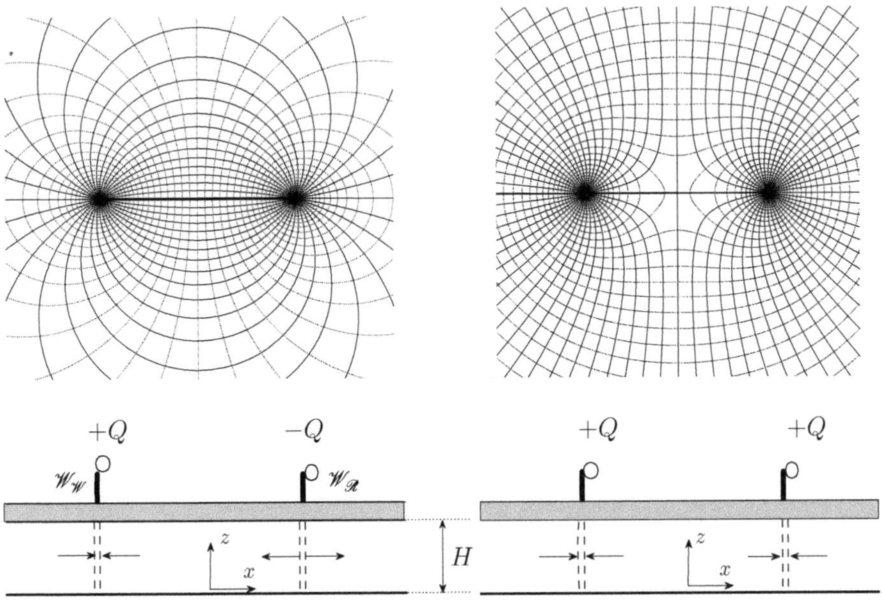

Figure 2.8 The method of images.

term *well* with reference to either a discharge or a recharge well. Both discharge and recharge may be represented as Q, where Q is positive for a discharge well and negative for a recharge well. Figure 2.8(b) represents the case of two discharge wells. We determine the potential as a function of position for each case, starting with Figure 2.8(a).

We choose the Cartesian (x,y,z)-coordinate system as shown in the figure, with the origin halfway between the wells and the x-axis passing through the recharge well. The distance between the wells is $2d$. The potential Φ for this flow problem is found by superposition of two potentials, $\overset{1}{\Phi}$ and $\overset{2}{\Phi}$. The function $\overset{1}{\Phi}$ is the potential for one well at $(-d,0)$ ($\mathscr{W}_{\mathscr{W}}$ in Figure 2.8(a)) in an infinite aquifer and is of the form (2.55)

$$\overset{1}{\Phi} = \frac{Q_1}{4\pi} \ln[(x+d)^2 + y^2], \tag{2.74}$$

where Q_1 is the discharge of the well.

The function $\overset{2}{\Phi}$ corresponds to one recharge well ($\mathscr{W}_{\mathscr{R}}$ in Figure 2.8(a)) in an infinite aquifer at $(d,0)$. An amount Q_2 is pumped into the aquifer at the recharge well, $\mathscr{W}_{\mathscr{R}}$. The potential $\overset{2}{\Phi}$ for a recharge well in an infinite aquifer is found from (2.55) by letting Q be negative, $Q = -Q_2$ $(Q_2 > 0)$,

$$\overset{2}{\Phi} = -\frac{Q_2}{4\pi} \ln[(x-d)^2 + y^2]. \tag{2.75}$$

We add the two potentials $\overset{1}{\Phi}$ and $\overset{2}{\Phi}$ and a constant C to obtain a new potential

$$\Phi = \frac{Q_1}{4\pi} \ln[(x+d)^2 + y^2] - \frac{Q_2}{4\pi} \ln[(x-d)^2 + y^2] + C \tag{2.76}$$

that represents a solution of Laplace's equation; it is the potential for flow from a recharge well $\mathscr{W}_{\mathscr{R}}$ toward a well $\mathscr{W}_{\mathscr{W}}$ (see Figure 2.8(a)).

For the special case of equal discharge and recharge,

$$Q_1 = Q_2 = Q, \tag{2.77}$$

(2.76) becomes, denoting the distances from the wells to an arbitrary point (x,y) as r_1 and r_2, and using that $\ln(r_1) - \ln(r_2) = \ln(r_1/r_2)$:

$$\Phi = \frac{Q}{2\pi} \ln\frac{r_1}{r_2} = \frac{Q}{4\pi} \ln\frac{r_1^2}{r_2^2} = \frac{Q}{4\pi} \ln\frac{(x+d)^2 + y^2}{(x-d)^2 + y^2} + C. \tag{2.78}$$

The equipotentials shown in Figure 2.8(a) correspond to this case. This solution has the property that the head is constant along the y-axis, which is the bisecting perpendicular of the line connecting the wells. Each point of the y-axis is at the

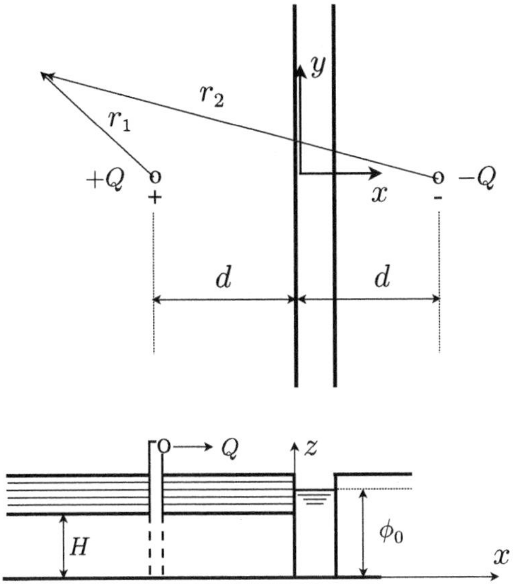

Figure 2.9 Flow toward a well near a river.

same distance from $\mathscr{W}_{\mathscr{W}}$ and $\mathscr{W}_{\mathscr{R}}$. Hence, for an arbitrary point of the y-axis we have that $r_1 = r_2 = r_0$, where r_0 is the distance from that point to either $\mathscr{W}_{\mathscr{R}}$ or $\mathscr{W}_{\mathscr{W}}$. Substitution of r_0 for both r_1 and r_2 in (2.78) yields

$$\Phi = \frac{Q}{2\pi} \ln \frac{r_0}{r_0} + C = C. \tag{2.79}$$

Hence, Φ is indeed constant along the y-axis. We see this also from the equipotentials, shown as the dotted contours in Figure 2.8(a).

It follows that the potential for flow from a recharge well toward a well can be used to solve the problem of a well near a straight river where the head is constant and equal to ϕ_0. This is illustrated in Figure 2.9. The boundary condition along the river bank is that ϕ is constant and equals ϕ_0. The value of x along the river bank is 0 and the distance from the well to the river bank is d. The function

$$\Phi = \frac{Q}{4\pi} \ln \frac{(x+d)^2 + y^2}{(x-d)^2 + y^2} + \Phi_0 \tag{2.80}$$

fulfills the boundary condition along the river bank, $x = 0$, and thus represents the discharge potential for the flow problem of Figure 2.9. We found the solution by use of the *image well* with discharge $-Q$ (a recharge well) at $x = d$, $y = 0$. The image well is not real but is used as a means to obtain the solution.

We next consider the case of two discharge wells shown in Figure 2.8(b). The potential for two wells of equal discharge Q and a distance $2d$ apart is obtained from (2.76) by replacing $-Q_2$ by $+Q$ and $+Q_1$ by $+Q$:

$$\Phi = \frac{Q}{4\pi}\ln\left[(x+d)^2+y^2\right] + \frac{Q}{4\pi}\ln\left[(x-d)^2+y^2\right] + C. \tag{2.81}$$

This solution has the property that no flow occurs across the y-axis, which is the perpendicular bisector of the line between the two wells in Figure 2.8(b). The y-axis is a streamline, which means that the specific discharge vector is directed along that line, i.e., $Q_x(0,y) = 0$. We verify this by computing the component Q_x of the discharge vector, which we find from (2.81) by use of Darcy's law, $Q_x = -\partial\Phi/\partial x$,

$$Q_x = -\frac{\partial \Phi}{\partial x} = -\frac{Q}{4\pi}\left[\frac{2(x+d)}{(x+d)^2+y^2} + \frac{2(x-d)}{(x-d)^2+y^2}\right], \tag{2.82}$$

which gives, for $x = 0$,

$$Q_x(0,y) = -\frac{Q}{4\pi}\left[\frac{2d}{d^2+y^2} - \frac{2d}{d^2+y^2}\right] = 0, \tag{2.83}$$

as asserted. The equipotentials in Figure 2.8(b) intersect the y-axis at right angles, because $Q_x(0,y)$ is zero and the flow is normal to the equipotentials. A more direct way to verify that the y-axis is a streamline is by examining the stream function along $x = 0$. We obtain an expression for the stream function for this case from (2.56):

$$\Psi = \frac{Q}{2\pi}(\theta_1+\theta_2) = \frac{Q}{2\pi}\left[\arctan\frac{y}{x+d} + \arctan\frac{y}{x-d}\right]. \tag{2.84}$$

Since $x = 0$ along the river bank, we have that $\theta_2 = \pi - \theta_1$, and $\theta_1 + \theta_2 = \pi$ for $y \geq 0$. The stream function is piecewise constant along the river bank; it is discontinuous across $y = 0, x \leq 0$, where θ_2 jumps from π just above the x-axis ($y = 0^+$), to $-\pi$ just below the x-axis ($y = 0^-$), so that $\theta_1 + \theta_2 = \pi$ for $y \geq 0$ and $\theta_1 + \theta_2 = -\pi$ for $y < 0$; (2.84) becomes

$$\begin{aligned} \Psi &= \frac{Q}{2\pi}\pi = \frac{Q}{2}, & y \geq 0 \\ \Psi &= -\frac{Q}{2\pi}\pi = -\frac{Q}{2}, & y < 0. \end{aligned} \tag{2.85}$$

The stream function is constant along streamlines; (2.85) is proof that the y-axis is a streamline.

We see that the solution for flow with two wells of equal discharge can be used to solve problems involving an impermeable boundary and wells as illustrated in Figure 2.10; the potential (2.81) has the property that no flow occurs across the y-axis (see Figure 2.10) and can be used to describe the problem of flow toward a well near an impervious boundary. The difference between the solutions to the problems of Figures 2.9 and 2.10 is the sign of the image well. The image well is

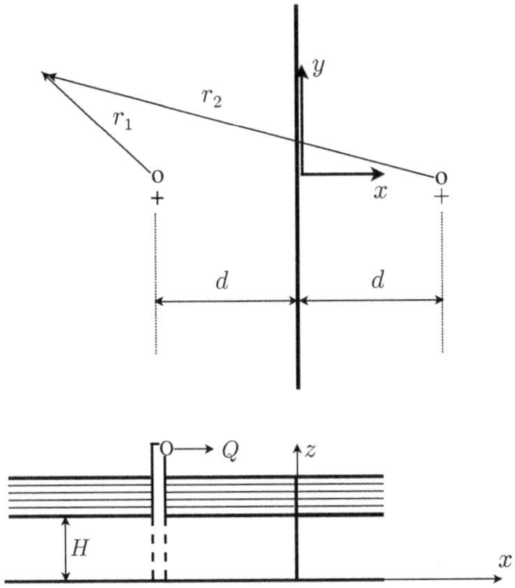

Figure 2.10 A well near impervious boundary.

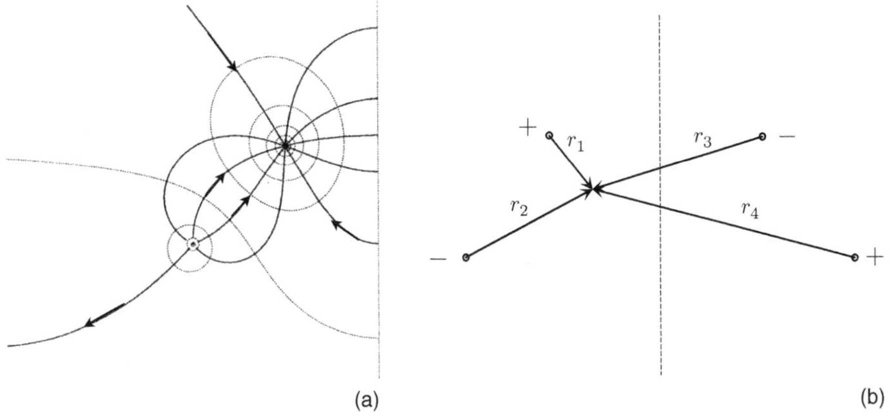

Figure 2.11 A well and a recharge well near a river.

a recharge well (indicated with a "−") in Figure 2.9 and a well (indicated with a "+") in Figure 2.10.

Examples of problems that can be solved by the use of the method of images are given in Figures 2.11(a) and 2.12(a). The image wells are indicated in Figures 2.11(b) and 2.12(b). The potentials are

$$\Phi = \frac{Q_1}{2\pi}\ln\frac{r_1}{r_3} - \frac{Q_2}{2\pi}\ln\frac{r_2}{r_4} + \Phi_0 \tag{2.86}$$

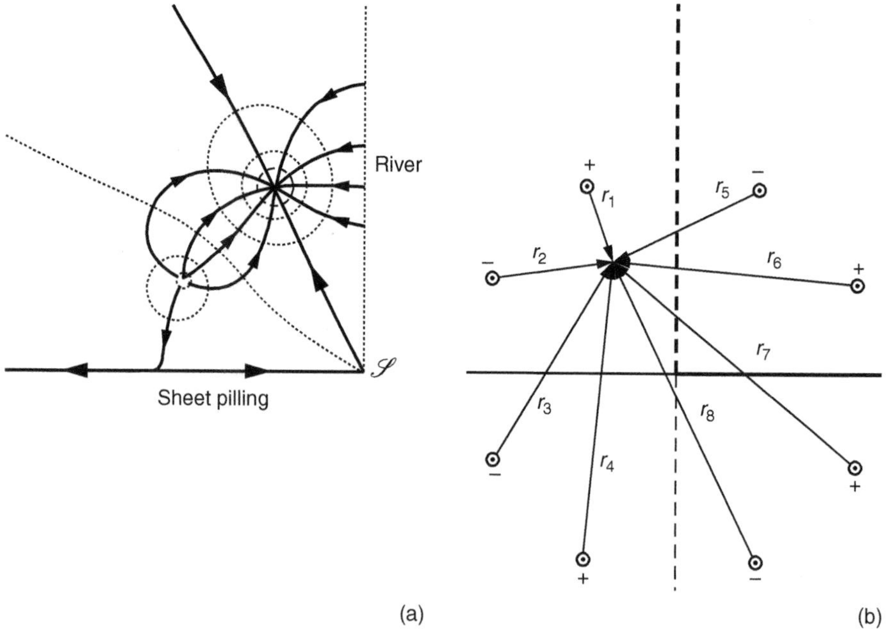

Figure 2.12 A well and a recharge well near a river and an impermeable wall.

and

$$\Phi = \frac{Q_1}{2\pi}\ln\frac{r_1 r_4}{r_5 r_8} - \frac{Q_2}{2\pi}\ln\frac{r_2 r_3}{r_6 r_7} + \Phi_0 \tag{2.87}$$

for the flow cases shown in Figures 2.11(a) and 2.12(a), respectively, which contain equipotentials (the dotted curves) and streamlines (the solid ones). The image wells are shown in Figures 2.11(b) and 2.12(b). Two equipotentials and two streamlines intersect at the point of intersection, $\mathscr{S}$, of the river and the impermeable wall: the equipotentials and streamlines are not orthogonal at this point. Point $\mathscr{S}$ is a stagnation point: the effects of all wells and image wells on the flow cancel there.

2.1.8 Partial Penetration

The boundaries discussed thus far, such as rivers and slurry walls, fully penetrate the aquifer; they are called *fully penetrating boundaries*. Boundaries that extend over part of the aquifer thickness are called *partially penetrating boundaries*. The flows in aquifers with partially penetrating boundaries have vertical components, which are large in the neighborhood of a partially penetrating boundary, but reduce rapidly with the distance from that boundary. As an illustration, equipotentials are shown in Figure 2.13 for the case of flow to a partially penetrating canal; the flow is symmetrical with respect to the canal, and only the left-hand side of a vertical

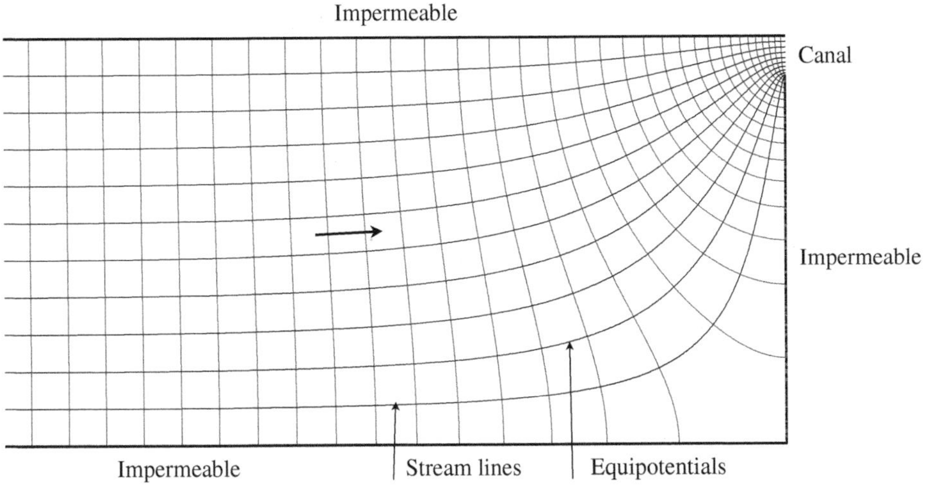

Figure 2.13 A partially penetrating river.

section of the aquifer through the canal is shown. The canal is long; the components of flow normal to the plane of the drawing can be neglected. The equipotentials are curved in the neighborhood of the canal, where the components of flow in the vertical direction are large and become nearly vertical at a distance of several times the aquifer thickness away from the canal. The application of an analysis of horizontal flow to such a case produces errors, as a result of neglecting the resistance to flow in the vertical direction (recall that resistance to flow is the inverse of hydraulic conductivity, $1/k$; a zero resistance implies an infinite hydraulic conductivity). The relative error decreases as the total resistance to horizontal flow increases, so that the ratio of the neglected resistance to the total resistance decreases. Thus, the errors become less important as the boundaries are further apart. The horizontal dimensions in real aquifers are often much larger than the vertical ones; for such cases the effect of partial penetration rarely need be considered. Wells also may be either fully penetrating or partially penetrating; a similar discussion as that given above applies to partially penetrating wells. We assume for the time being that all boundaries and wells in problems of horizontal flow are fully penetrating.

2.1.9 Hydraulic Head Contours

Hydraulic head contours are curves along which the head is constant. The head may be expressed linearly in terms of the potential; see (2.16):

$$\phi = \frac{\Phi - C_c}{kH}. \tag{2.88}$$

The value of ϕ associated with each contour is obtained from the corresponding value of Φ by the use of (2.88). Because ϕ is a linear function of Φ, the difference in value of ϕ, $\Delta\phi$, on two neighboring equipotentials is constant throughout the flow domain, provided that $\Delta\Phi$ is constant.

Problem

2.4 A system of four wells, located at the four corners of a building pit, will be used to reduce the pressures directly below the confining layer that separates the pit from a confined aquifer, to a level that is safe with regard to stability of the bottom of the pit. This level is given as $0.5 * 10^5$ N/m^2.

The thickness of the aquifer is $H = 25$ m and the hydraulic conductivity is $k = 10^{-5}$ m/s. The aquifer is bounded by a long, straight river (see Figure 2.14). The heads are measured with respect to the base of the aquifer, and the head along the entire river bank is 40 m. The soil above the upper confining bed is dry. The sides of the building pit are 40 m and its center is 80 m from the river bank. There is no flow in the aquifer before the excavation.

Questions:

1. Calculate the discharge, Q, of each well necessary to maintain the pressure at or below the safe level everywhere underneath the pit, assuming that all wells have the same discharge.
2. Calculate the two components of the discharge vector at the midpoint of the side of the pit closest to the canal.

Problem

2.5 A system of a well and a recharge well in a confined aquifer of thickness $H = 30$ m and $k = 10^{-4}$ m/s is used to produce groundwater for a cooling system. The wells are $2d = 100$ m apart, and the radius of each well is $r_w = 0.1$ m. The head in the aquifer is constant and equal to 35 m before the wells operate. The discharges of the well and the recharge well are $0.6 * 10^{-2}$ m^3/s and $-0.6 * 10^{-2}$ m^3/s, respectively.

Questions:

1. Calculate the largest distance from the midpoint between the wells at which the influence of the system on the head is 0.05 m.
2. Find the maximum discharge circulated by the system such that the aquifer remains confined at the well (neglect r_w with respect to d).

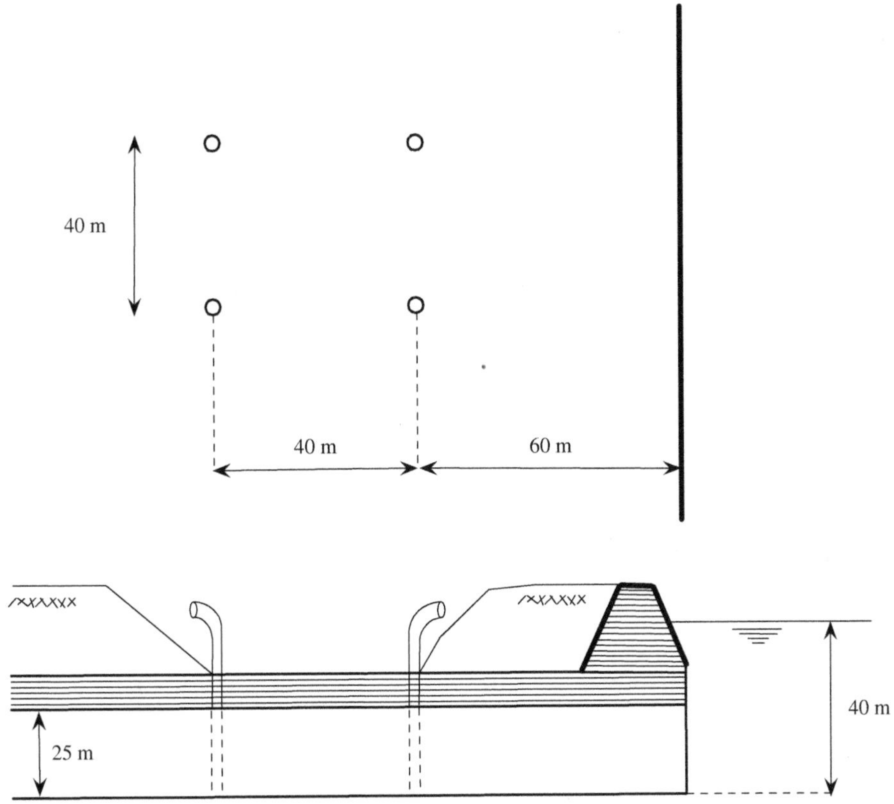

Figure 2.14 Dewatering of a building pit.

2.2 Shallow Unconfined Flow

An unconfined aquifer is one whose upper boundary is a water table. Problems with such a boundary are difficult to solve. We solve such problems with good approximation when the horizontal extent of the aquifer is much larger than the vertical one, so that the vertical head loss is small relative to the horizontal one. The approximation of taking the head to be constant in the vertical direction is due to Dupuit [1863] and Forchheimer [1914], and is known as the Dupuit-Forchheimer approximation. We view this approximation as corresponding to an anisotropic porous medium with an infinite vertical hydraulic conductivity (Strack [1984]). This view bears resemblance to the approach of Kirkham [1967], who viewed a Dupuit-Forchheimer medium as composed of slots filled with fluid and separated by thin semi-permeable walls, but rejected the analogy with an anisotropic porous medium.

A large class of groundwater flow problems can be solved with good results using the Dupuit-Forchheimer approximation. We consider such problems in this section and make the additional approximation that the pores of the soil beneath the groundwater table, called the phreatic surface, are completely filled with water, and the pores above it are completely filled with air. This approximation is justified for fairly permeable soils such as sand. The zone of the aquifer below the phreatic surface is the saturated zone.

We call a flow *shallow unconfined flow* if we neglect the vertical component of the gradient of the hydraulic head. The word *shallow* is added to avoid confusion; there are cases of unconfined flow where the resistance to flow in the vertical direction may not be neglected.

2.2.1 Basic Equations

We derived the basic equations for horizontal confined flow directly from the fundamental equations presented in Chapter 1; the general equations applied because no approximations were made. For the case of shallow unconfined flow, however, approximations are made; we must take these into account when deriving the basic equations: the equation of motion and the continuity equation. We introduce a discharge vector Q_i and a potential Φ for shallow unconfined flow in such a way that the basic equations become identical to those for horizontal confined flow.

The Dupuit-Forchheimer Approximation

The Dupuit-Forchheimer approximation is usually identified with horizontal flow, i.e., with neglecting the component q_z of the specific discharge vector. As we associate the Dupuit-Forchheimer approximation with an anisotropic porous medium with infinite vertical hydraulic conductivity, q_z is not zero, and can be computed on the basis of continuity of flow, as shown by Strack [1984]. The magnitude of the vertical component q_z is not associated with a variation of ϕ:

$$q_z = - \lim_{\substack{k_z \to \infty \\ \partial\phi/\partial z \to 0}} k_z \frac{\partial \phi}{\partial z} \neq 0. \tag{2.89}$$

The head is constant along vertical surfaces and the pressure distribution is hydrostatic. The Dupuit-Forchheimer approximation is applicable to aquifers bounded below by an impervious base, which we take for now as the reference level for the head; see Figure 2.15. If the elevation of the water table above the impervious base is h, the head at a point $\mathscr{P}$ of the phreatic surface may be expressed in terms of the pressure as follows (see (1.7)):

$$\phi = \frac{p}{\rho g} + Z = \frac{p}{\rho g} + h. \tag{2.90}$$

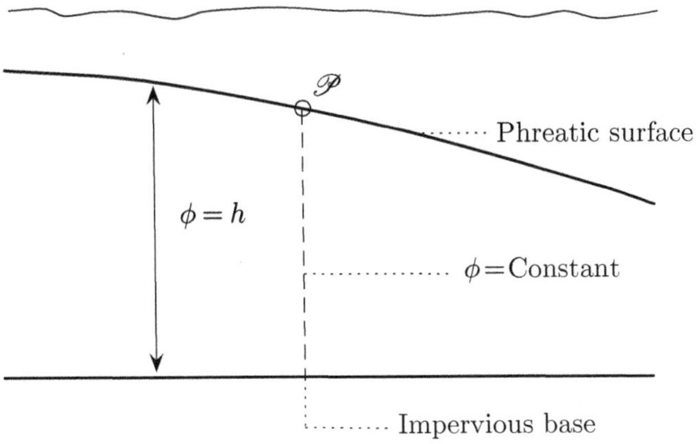

Figure 2.15 The Dupuit-Forchheimer approximation.

The pressure at the phreatic surface is equal to the atmospheric pressure. Taking the atmospheric pressure as zero, and with the aquifer base as the reference level for the head, (2.90) becomes

$$\phi = h. \tag{2.91}$$

The head along the phreatic surface is equal to the elevation of the water table above the base, h. Since ϕ is taken to be constant along vertical surfaces, ϕ equals h throughout the unconfined aquifer.

The Discharge Vector

We defined the discharge vector in Section 2.1 as the product of the specific discharge vector and the thickness of the aquifer H. For the case of unconfined flow, the aquifer thickness varies and equals h:

$$\begin{aligned} Q_x &= hq_x \\ Q_y &= hq_y. \end{aligned} \tag{2.92}$$

The physical interpretation for (Q_x, Q_y) is the same for shallow unconfined flow as for horizontal confined flow; the magnitude of (Q_x, Q_y) equals the amount of flow through a vertical section of unit width. We apply Darcy's law (1.18) to (2.92):

$$\begin{aligned} Q_x &= -kh\frac{\partial \phi}{\partial x} \\ Q_y &= -kh\frac{\partial \phi}{\partial y}. \end{aligned} \tag{2.93}$$

Since $h = \phi$, this becomes

$$
\begin{aligned}
Q_x &= -kh\frac{\partial h}{\partial x} \\
Q_y &= -kh\frac{\partial h}{\partial y}
\end{aligned}
\tag{2.94}
$$

or, since k is a constant,

$$
\begin{aligned}
Q_x &= -\frac{\partial\left[\frac{1}{2}kh^2\right]}{\partial x} \\
Q_y &= -\frac{\partial\left[\frac{1}{2}kh^2\right]}{\partial y}.
\end{aligned}
\tag{2.95}
$$

The Discharge Potential

We introduce the discharge potential Φ for shallow unconfined flow as

$$\Phi = \tfrac{1}{2}kh^2 + C_u, \tag{2.96}$$

where C_u is an arbitrary constant, and write Darcy's law, (2.95), for the discharge vector in terms of the discharge potential:

$$
\begin{aligned}
Q_x &= -\frac{\partial \Phi}{\partial x} \\
Q_y &= -\frac{\partial \Phi}{\partial y}.
\end{aligned}
\tag{2.97}
$$

These equations are the same as the ones for horizontal confined flow (see (2.17)).

If we do not take the base of the aquifer as the reference level for the hydraulic head ϕ, then

$$h = \phi - b \tag{2.98}$$

and (2.96) becomes

$$\boxed{\Phi = \frac{1}{2}k(\phi - b)^2 + C_u} \tag{2.99}$$

We will take C_u as zero and the base of the aquifer as the reference level for the head in this chapter, so that the discharge potential for unconfined flow reduces to

$$\Phi = \tfrac{1}{2}k\phi^2. \tag{2.100}$$

h is the *saturated thickness* of the aquifer, and kh the *transmissivity*, T:

$$T = kh. \tag{2.101}$$

Equipotentials and Hydraulic Head Contours

Equipotentials are defined in the same way as for horizontal confined flow; the concept of an equipotential is universal and is applicable to any potential function. Information regarding the discharge vector may be obtained from a plot of equipotentials as outlined in Section 2.1. Hydraulic head contours are obtained by letting the head ϕ be constant. The head ϕ is expressed in terms of Φ by the use of (2.96):

$$\boxed{\phi = \sqrt{\frac{2(\Phi - C_u)}{k}}} \qquad (2.102)$$

Because ϕ is constant whenever Φ is constant, each equipotential is a hydraulic head contour. A plot of equipotentials obtained for constant $\Delta\Phi$ may be viewed as a plot of hydraulic head contours, with the value of ϕ on each contour obtained from the value of Φ by the use of (2.102). However, ϕ is not a linear function of Φ, and therefore $\Delta\phi$, the difference in value of ϕ on two consecutive equipotentials, is not constant. A plot of hydraulic head contours with constant $\Delta\phi$ must be constructed separately from a plot of equipotentials. Equipotentials serve to gain insight in the magnitude and direction of the discharge vector. Hydraulic head contours are used as a graphical representation of the head ϕ.

Problem

2.6 Demonstrate that the magnitude of the specific discharge vector is approximately inversely proportional to the distance between two neighboring hydraulic head contours (compare (2.22)).

The Continuity Equation

The divergence of the discharge vector must be zero in order that the condition of continuity of flow is satisfied. The discharge vector is a two-dimensional quantity, and the divergence of a vector in three dimensions, (1.36), reduces to:

$$\frac{\partial Q_x}{\partial x} + \frac{\partial Q_y}{\partial y} = \operatorname{div}(\vec{Q}) = 0. \qquad (2.103)$$

This equation is the continuity equation for shallow unconfined flow without rainfall or evaporation.

2.2.2 Differential Equation for the Discharge Potential

Substitution of expressions (2.97) for Q_x and Q_y in (2.103) yields

$$\frac{\partial^2 \Phi}{\partial x^2} + \frac{\partial^2 \Phi}{\partial y^2} = 0. \qquad (2.104)$$

We see from (2.104) and (2.20) that the governing equations for horizontal confined and shallow unconfined flow are the same; we treat these flows as mathematically the same. We achieved this by the use of the discharge potential Φ, introduced by Girinskii [1946a] in a somewhat different form.

The difference between horizontal confined flow and shallow unconfined flow lies in the expressions for Φ. We defined Φ for horizontal confined flow as a linear function of ϕ:

$$\Phi = kH\phi + C_c \tag{2.105}$$

and for shallow unconfined flow, Φ is given by a quadratic function:

$$\Phi = \tfrac{1}{2}k\phi^2. \tag{2.106}$$

Plots of equipotentials can be viewed as plots of hydraulic head contours for cases of horizontal confined flow. For cases of shallow unconfined flow, however, the plots of equipotentials and hydraulic head contours differ in the spacing between the contours.

We solve problems of unconfined flow in three steps: first, we express the boundary conditions in terms of the potential; second, we solve the differential equation with these boundary conditions; and third, we obtain the head from the potential by the use of (2.102). Applications of such solutions are given in the following.

2.2.3 Planar Flow

We consider the case of no flow in the y-direction; this does not imply that the flow is one-dimensional as for horizontal confined flow. There must be a vertical component of flow, because the water table changes in elevation as a result of the variation in hydraulic head. That we do not concern ourselves at present with the vertical component of flow does not make the flow one-dimensional, but only our description of it. We refer to such cases of flow as *planar flow*.

We consider an unconfined aquifer bounded by two long, parallel rivers. A cross section through the aquifer is given in Figure 2.16. The heads along the river banks are ϕ_1 and ϕ_2 and the distance between the banks is L. An (x, y, z)-coordinate system is chosen as indicated in the figure. The y-coordinate is parallel to the rivers and the origin is midway between the two river banks. The boundary conditions in terms of the head ϕ are

$$\begin{aligned} x &= -\tfrac{1}{2}L, \qquad \phi = \phi_1 \\ x &= \tfrac{1}{2}L, \qquad \phi = \phi_2 \end{aligned} \tag{2.107}$$

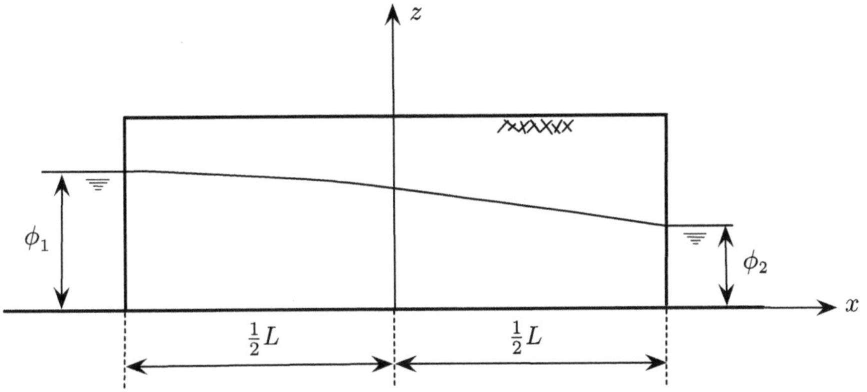

Figure 2.16 One-dimensional flow in an unconfined aquifer.

and may be expressed in terms of the potential Φ by the use of (2.106):

$$\begin{aligned} x = -\tfrac{1}{2}L, \qquad & \Phi = \Phi_1 = \tfrac{1}{2}k\phi_1^2 \\ x = \tfrac{1}{2}L, \qquad & \Phi = \Phi_2 = \tfrac{1}{2}k\phi_2^2. \end{aligned} \tag{2.108}$$

For one-dimensional flow, Φ is a function of x alone, and the differential equation (2.104) reduces to $d^2\Phi/dx^2 = 0$, with the general solution:

$$\Phi = Ax + B. \tag{2.109}$$

Application of the boundary conditions (2.108) yields

$$\begin{aligned} \Phi_1 &= \tfrac{1}{2}k\phi_1^2 = -\tfrac{1}{2}AL + B \\ \Phi_2 &= \tfrac{1}{2}k\phi_2^2 = \tfrac{1}{2}AL + B. \end{aligned} \tag{2.110}$$

Hence,

$$A = \frac{\Phi_2 - \Phi_1}{L}, \qquad B = \tfrac{1}{2}(\Phi_1 + \Phi_2), \tag{2.111}$$

and the expression (2.109) for the potential becomes

$$\Phi = -(\Phi_1 - \Phi_2)\frac{x}{L} + \tfrac{1}{2}(\Phi_1 + \Phi_2). \tag{2.112}$$

The discharge Q_x is found from $Q_x = -\partial\Phi/\partial x$:

$$Q_x = \frac{\Phi_1 - \Phi_2}{L}. \tag{2.113}$$

Substitution of the expressions (2.108) for Φ_1 and Φ_2 gives

$$Q_x = \frac{k(\phi_1^2 - \phi_2^2)}{2L}. \tag{2.114}$$

This formula for the flow through an unconfined aquifer is due to Dupuit. It was shown by Charny [1951] that the Dupuit formula is exact, notwithstanding the simplifying approximations used to derive it.

An expression for the head ϕ is found from (2.102) and (2.112):

$$\phi = \sqrt{\tfrac{1}{2}(\phi_1^2 + \phi_2^2) - \frac{\phi_1^2 - \phi_2^2}{L}x}, \qquad C_u = 0. \tag{2.115}$$

This equation represents a parabola, known as the Dupuit parabola.

Problem

2.7 Draw four hydraulic head contours for the case of Figure 2.16 if $\phi_1 = 20$ m, $\phi_2 = 10$ m, and $L = 1000$ m.

2.2.4 Shallow Unconfined Radial Flow to a Well

The results obtained in Section 2.1 for confined flow can be applied directly to unconfined flow, because the governing equations for confined and unconfined flow in terms of the discharge potentials are identical. The potential for flow to a well in a circular island, illustrated in Figure 2.17, is given by the Thiem formula, written in terms of the discharge potential, (2.37),

$$\Phi = \frac{Q}{2\pi} \ln \frac{r}{R} + \Phi_0, \tag{2.116}$$

where the potential is now defined in terms of ϕ by (2.106) rather than (2.105), so that the expression in terms of ϕ becomes

$$\tfrac{1}{2}k\phi^2 = \frac{Q}{2\pi} \ln \frac{r}{R} + \tfrac{1}{2}k\phi_0^2 \tag{2.117}$$

or

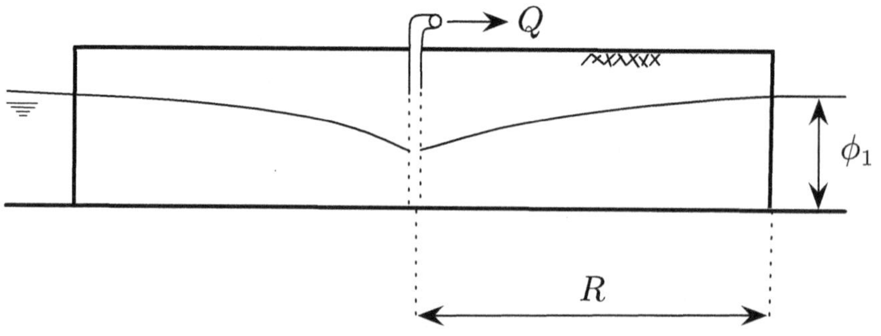

Figure 2.17 Unconfined flow toward a well.

$$\phi = \sqrt{\frac{Q}{\pi k} \ln \frac{r}{R} + \phi_0^2}. \tag{2.118}$$

Although the potentials for radial confined and radial unconfined flow are the same, the expressions for ϕ, (2.38) and (2.118), are quite different.

The superposition principle applies to the potential, but not to the head; the differential equation is linear in terms of Φ, but not in terms of ϕ. Superposition of two solutions of the form (2.116) therefore is allowed, but superposition of two expressions of the form (2.118) is wrong.

2.2.5 Application: Computing the Maximum Discharge of a Well Near a Long River

A well is planned in an aquifer bounded by a long river. We wish to determine the maximum discharge of the well such that no river water is captured. We also must verify that the well can actually pump the computed discharge i.e., the potential at the well screen must be greater than or equal to zero. We represent the first of these two conditions mathematically as

$$Q_x(0, y) \geq 0. \tag{2.119}$$

This condition implies that there is outflow all along the river bank, so that no river water enters the aquifer anywhere; the well captures only water flowing in the aquifer toward the well and the river.

The second condition implies that the head inside the well screen must be greater than or equal to zero. Note that in reality the head directly outside the well screen is larger than the head inside the well screen; the well screen is a seepage face, where the same conditions apply as along a bluff, provided that the pressure in the well screen above the water level is atmospheric; see Figure 2.18. We are neglecting the effect of the resistance to flow imposed by the well screen in this analysis. The maximum discharge of a well at (x_w, y_w) of radius r_w is found from

$$\Phi(x_w + r_w, y_w) \geq 0. \tag{2.120}$$

Note that we neglect the small variation in head along the well screen due to the uniform flow or, in general, effects other than by the well.

The head at a distance L from the river bank is measured prior to operation of the well and is ϕ_1. The well will be placed at a distance d from the river bank. We choose the (x, y, z)-coordinate system as indicated in Figure 2.19, with the x-axis pointing normal to the river away from the aquifer, and the y-axis along the river bank. We determine first the potential Φ for the uniform flow in the x-direction

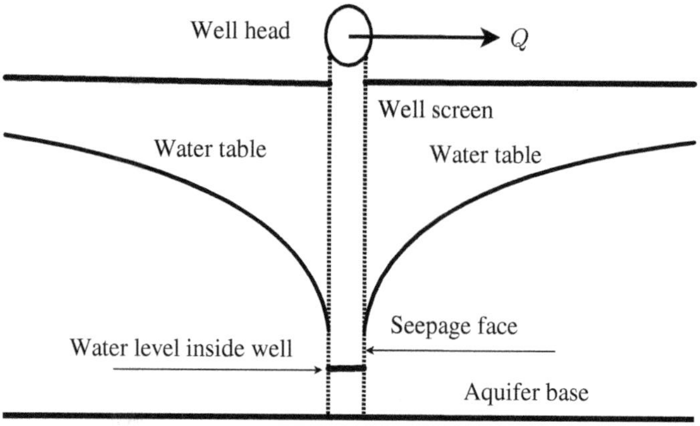

Figure 2.18 The seepage face along a well screen.

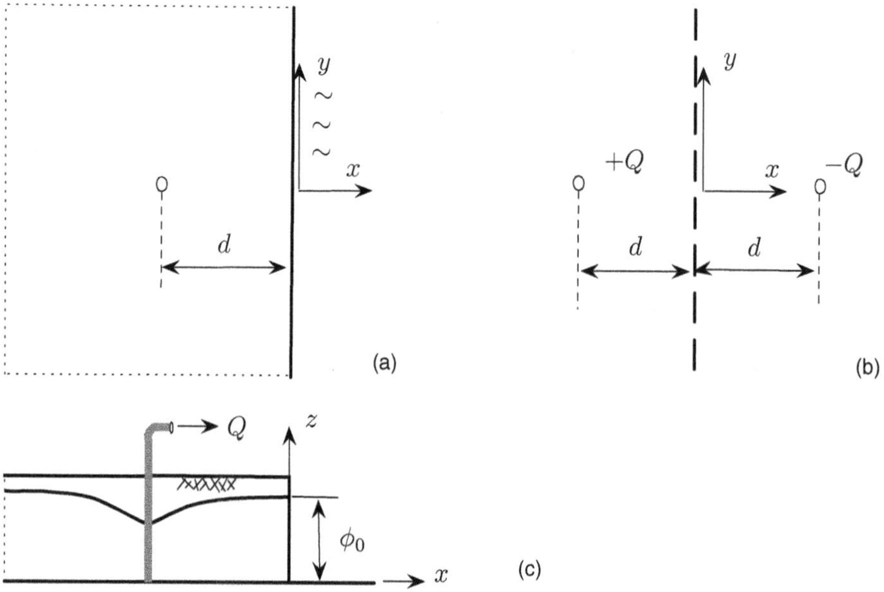

Figure 2.19 A well near a river.

applicable before the well operates. The boundary conditions are

$$\begin{aligned} x &= 0, & \phi &= \phi_0, & \Phi &= \Phi_0 = \tfrac{1}{2}k\phi_0^2 \\ x &= -L, & \phi &= \phi_1, & \Phi &= \Phi_1 = \tfrac{1}{2}k\phi_1^2. \end{aligned} \qquad (2.121)$$

We apply the Dupuit formula to determine the discharge Q_{xo} flowing toward the river before the well operates:

$$Q_{x0} = \frac{\Phi_1 - \Phi_0}{L}. \tag{2.122}$$

The derivative of the potential must be $-Q_{x0}$ and the potential is equal to Φ_0 along the river bank, i.e.,

$$\Phi = -Q_{x0}x + \Phi_0. \tag{2.123}$$

We next consider a well drawing an amount Q from the aquifer and $Q_{x0} = 0$. We consider the flow as steady, which implies that the well has been operating sufficiently long that changes over time have become negligible. We obtain an expression for the potential Φ by the method of images; the image well is a recharge well at a distance d from the river bank (see Figure 2.19(b)). The corresponding potential is

$$\Phi = \frac{Q}{4\pi}\ln\left[(x+d)^2+y^2\right] - \frac{Q}{4\pi}\ln\left[(x-d)^2+y^2\right] + C. \tag{2.124}$$

We obtain an expression for the complete solution by assuming that the uniform flow, in reality caused by effects far from the area of interest but left out of the analysis, remains unchanged by the pumping. Note that the head at $(-L,0)$ will change due to the influence of the well. We obtain the potential for the combined flow, i.e., for flow toward the river with the well operating by adding the potentials (2.123) and (2.124)

$$\Phi = -Q_{x0}x + \Phi_0 + \frac{Q}{4\pi}\ln\left[(x+d)^2+y^2\right] - \frac{Q}{4\pi}\ln\left[(x-d)^2+y^2\right] + C. \tag{2.125}$$

Note that we always must add a constant when superimposing solutions to ensure that all boundary conditions are met. The potential (2.125) must fulfill the boundary condition along the river; Φ must be equal to Φ_0 for $x = 0$. It follows from (2.125) that

$$x = 0, \qquad \Phi = \Phi_0 + C. \tag{2.126}$$

In order that Φ be Φ_0 for $x = 0$, the constant C must vanish,

$$C = 0, \tag{2.127}$$

so that (2.125) becomes

$$\begin{aligned}\Phi &= -Q_{x0}x + \frac{Q}{4\pi}\ln\left[(x+d)^2+y^2\right] - \frac{Q}{4\pi}\ln\left[(x-d)^2+y^2\right] + \Phi_0 \\ &= -Q_{x0}x + \frac{Q}{4\pi}\ln\frac{(x+d)^2+y^2}{(x-d)^2+y^2} + \Phi_0.\end{aligned} \tag{2.128}$$

The argument of the logarithm in the final expression approaches unity near infinity; the potential (2.128) far away reduces to the one for the flow prior to

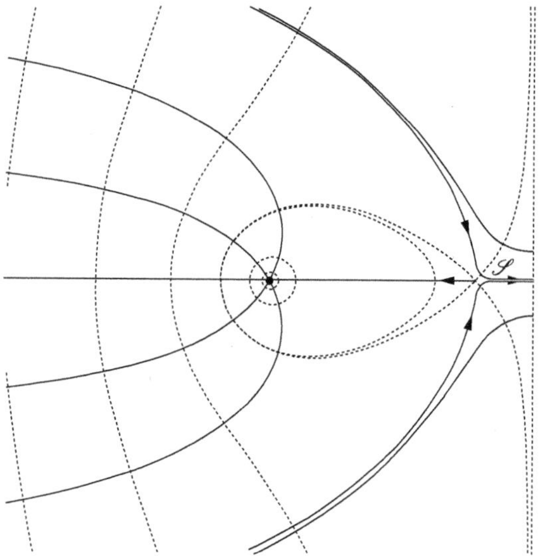

Figure 2.20 Equipotentials for flow toward a well near a river.

operating the well; the influence of the well is noticeable only in a limited area. A plot of equipotentials and streamlines is given in Figure 2.20 as an illustration. The lowest value of Φ ($\Phi_{\min}$) occurs at the well screen. The curves are obtained by letting Φ be $\Phi_{\min} + j\Delta\Phi$, $j = 0, 1, 2, \ldots, n-1$, where n is the total number of equipotentials. An exception is the special equipotential adjacent to the river, which intersects itself at point $\mathscr{S}$ in Figure 2.20. The flow stagnates at this point of intersection, where both Q_x and Q_y are equal to zero: this point is a stagnation point. Along the x-axis, the flow to the right of $\mathscr{S}$ is to the river, and the flow to the left of $\mathscr{S}$ is to the well. The streamline through the stagnation point is called the *dividing streamline*; the flow to the one side of the dividing streamline is to the well, whereas the flow to the other side is to the river.

The objective of the analysis is to determine the maximum discharge of the well such that no river water enters the aquifer, expressed by (2.119). We differentiate the potential (2.128) partially with respect to x:

$$\frac{\partial \Phi}{\partial x} = -Q_{x0} + \frac{Q}{4\pi}\left[\frac{2(x+d)}{(x+d)^2 + y^2} - \frac{2(x-d)}{(x-d)^2 + y^2}\right]. \tag{2.129}$$

We apply the condition (2.119):

$$Q_x = -\frac{\partial \Phi}{\partial x} = Q_{x0} - \frac{Q}{4\pi}\frac{4d}{d^2 + y^2} \geq 0 \tag{2.130}$$

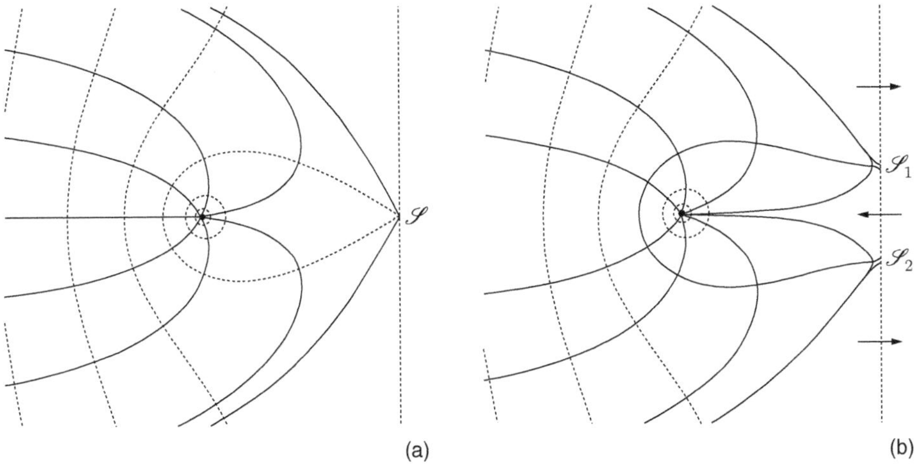

Figure 2.21 Streamlines and equipotentials for (a) the critical case ($Q = \pi d Q_{x0}$) and (b) the case that $\pi d Q_{x0}/Q = 0.95$.

or

$$\frac{Q}{4\pi}\frac{4d}{y^2+d^2} \leq Q_{x0}. \tag{2.131}$$

The expression to the left of the $\leq$ sign has its maximum value for $y=0$. Thus, condition (2.131) will be fulfilled along the entire river bank, i.e., for all values of y, if

$$\frac{Q}{\pi}\frac{1}{d} \leq Q_{x0} \tag{2.132}$$

or

$$Q \leq \pi d Q_{x0}. \tag{2.133}$$

This equation gives the maximum discharge of the well [L^3/T] in terms of Q_{x0} [L^2/T] and d [L]. This maximum discharge increases linearly with the distance between well and river bank.

When $Q = \pi d Q_{x0}$, the critical case, Q_x is positive along the entire river bank, except at the origin, where $Q_x = 0$. The river bank is an equipotential and therefore Q_y is zero there. Thus, both Q_x and Q_y are zero at $x = 0, y = 0$, for the critical case; the origin is a stagnation point. This is illustrated in Figure 2.21(a), where the equipotentials and streamlines are shown for the critical case. If the ratio $\pi d Q_{x0}/Q$ is greater than one, the stagnation point lies to the left of the river bank and outflow occurs along the entire bank. The equipotentials shown in Figure 2.20 apply to this case; $\pi d Q_{x0}/Q = 1.05$. If the ratio $\pi d Q_{x0}/Q$ is less than one, inflow occurs along a portion of the river bank, centered at the origin. This is shown in Figure 2.21(b), which applies when $\pi d Q_{x0}/Q = 0.95$. The section of inflow is between the two stagnation points labeled $\mathscr{S}_1$ and $\mathscr{S}_2$.

We must verify that the well indeed can produce the desired discharge, i.e., that (2.120) is satisfied. We set $x = -d + r_w, y = 0$ in (2.128) and neglect r_w relative to d:

$$\Phi_w = Q_{x0}d + \frac{Q}{4\pi}\ln(r_w^2) - \frac{Q}{4\pi}\ln(4d^2) + \Phi_0 \geq 0 \tag{2.134}$$

or

$$\frac{Q}{2\pi}\ln\frac{2d}{r_w} \leq Q_{x0}d + \Phi_0, \tag{2.135}$$

so that the maximum discharge is

$$Q_{\max} = \frac{2\pi(Q_{x0}d + \Phi_0)}{\ln(2d/r_w)}. \tag{2.136}$$

Problem

2.8 Determine the y-coordinates of the two stagnation points in Figure 2.21(b).

Problem

2.9 Consider the problem of one-dimensional shallow unconfined flow through a dam with vertical faces (see Figure 2.22). The dam is divided into two zones, consisting of different materials that have hydraulic conductivities k_1 and k_2. The lengths of the zones are L_1 and L_2, and the heads along the faces are ϕ_1 and ϕ_2.

Questions:

1. Express Q_x in terms of the parameters of the problem.
2. Calculate Q_x if $k_1 = 10$ m/d, $k_2 = 10^{-6}$ m/d, $L_1 = 100$ m, $L_2 = 100$ m, $\phi_1 = 20$ m, and $\phi_2 = 10$ m.

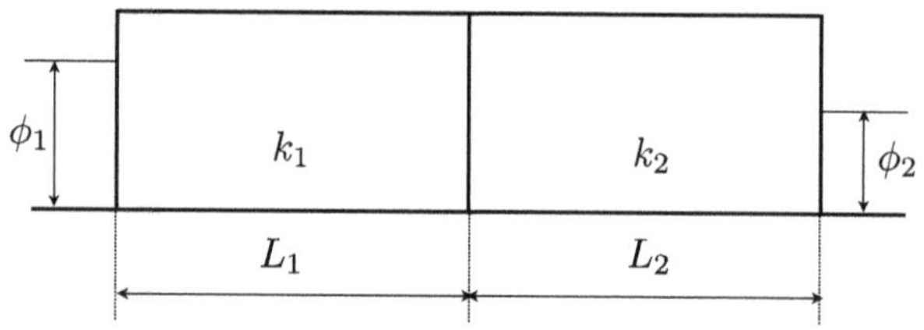

Figure 2.22 Flow through an inhomogeneous dam.

Problem

2.10 A dewatering system consisting of four wells will be used to obtain dry working conditions in a building pit in an unconfined aquifer of hydraulic conductivity $k_1 = 10$ m/d. The aquifer is bounded by a long straight river and there is no flow before the wells operate. The head along the river bank is constant, and is $\phi_0 = 25$ m above the base of the aquifer. The building pit is square and the bottom of the pit will be 20 m above the aquifer base. The sides of the pit are 50 m in length, and the center of the pit is 100 m away from the river bank.

Questions:

1. Calculate the discharge Q per well that is necessary to obtain dry working conditions.
2. Answer question 1 if it is given that there is a uniform flow of discharge $Q_{x0} = 0.04$ m^2/d in the x-direction before the wells operate.
3. Verify that the wells can pump the calculated discharges by checking that the heads at the wells are above the base of the aquifer. The radii of all wells are 0.25 m.

Problem

2.11 Consider the problem of one-dimensional shallow unconfined flow through a dam with vertical faces (see Figure 2.22). The dam is divided into two zones, consisting of different materials which have hydraulic conductivities k_1 and k_2. The lengths of the zones are both L and the head along the upstream face is ϕ_1. The head along the downstream face is below the base of the aquifer. There is no infiltration from rainfall ($N = 0$).

Questions:

1. Express Q_x in terms of the parameters of the problem.
2. Calculate Q_x if $k_1 = 10$ m/d and $k_2 = 0.2$ m/d, $L_1 = 100$ m, and $\phi_1 = 20$ m.

2.3 Combined Confined and Unconfined Flow

Combined shallow confined and unconfined flow occurs in horizontally confined aquifers if the head is less than the aquifer thickness H in some zones and greater than H in others. Such unconfined zones may occur around wells or near boundaries. Problems of combined shallow confined and unconfined flow can be solved in a way suggested by Girinskii [1946a], as explained in the following.

2.3.1 Method of Solution

We refer to the boundaries between confined and unconfined zones as the *interzonal boundaries* and measure ϕ with respect to the base of the aquifer. Along the interzonal boundaries the head equals the thickness, H, of the aquifer.

We use a single potential Φ throughout the flow region, given by (2.105) in the confined zones

$$\Phi = kH\phi + C_c, \qquad \phi \geq H, \tag{2.137}$$

and by (2.106) in the unconfined zones

$$\Phi = \tfrac{1}{2}k\phi^2, \qquad \phi < H. \tag{2.138}$$

We choose the constant C_c such that the potential is continuous across the interzonal boundary; (2.137) and (2.138) must yield the same value when $\phi = H$:

$$kH^2 + C_c = \tfrac{1}{2}kH^2. \tag{2.139}$$

We solve for C_c:

$$C_c = -\tfrac{1}{2}kH^2. \tag{2.140}$$

The single potential Φ is now defined, depending on the magnitude of ϕ, as follows:

$$\boxed{\Phi = kH\phi - \tfrac{1}{2}kH^2, \qquad \phi \geq H} \tag{2.141}$$

$$\boxed{\Phi = \tfrac{1}{2}k\phi^2, \qquad \phi < H} \tag{2.142}$$

The interzonal boundary is defined by the constant value of the discharge potential that corresponds to $\phi = H$:

$$\boxed{\Phi_{\text{interzonal}} = \tfrac{1}{2}kH^2} \tag{2.143}$$

The discharge vector has the same physical meaning for horizontal confined flow as it has for shallow unconfined flow: it points in the direction of flow and its magnitude equals the amount of water flowing through a cross section of unit width that extends over the saturated thickness of the aquifer. By continuity of flow, the component of the discharge vector normal to the interzonal boundary, Q_n, must be the same on either side of that boundary (see Figure 2.23). Furthermore, the interzonal boundary is an equipotential, $\Phi = \frac{1}{2}kH^2$, and therefore the tangential component of the discharge vector along the interzonal boundary is zero. The discharge vector is thus continuous across the interzonal boundary; there exists a single discharge vector throughout the aquifer:

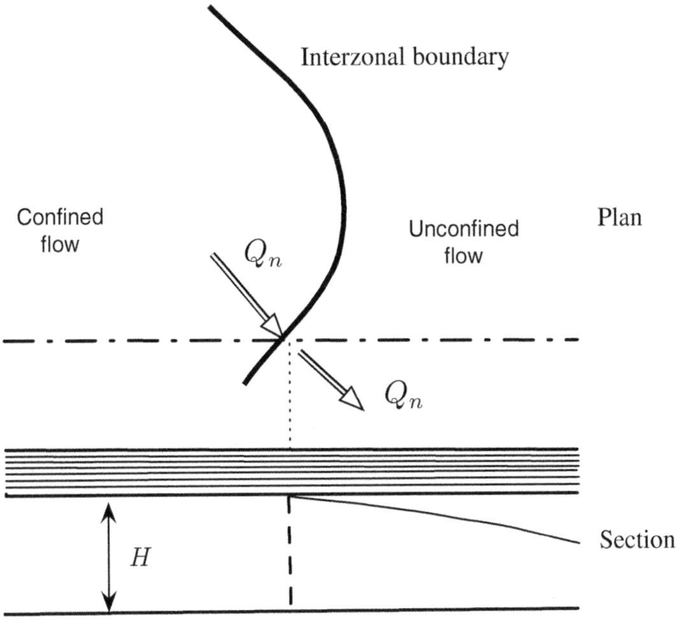

Figure 2.23 Continuity of flow across the interzonal boundary.

$$Q_x = -\frac{\partial \Phi}{\partial x} \tag{2.144}$$

$$Q_y = -\frac{\partial \Phi}{\partial y}. \tag{2.145}$$

The continuity equation (2.103); must be satisfied in the entire flow domain, including the interzonal boundary,

$$\frac{\partial Q_x}{\partial x} + \frac{\partial Q_y}{\partial y} = 0, \tag{2.146}$$

and (2.144) through (2.146) yield Laplace's equation for the potential Φ,

$$\frac{\partial^2 \Phi}{\partial x^2} + \frac{\partial^2 \Phi}{\partial y^2} = 0. \tag{2.147}$$

It follows that problems of combined confined/unconfined flow can be solved without taking the location of the interzonal boundary into account; if so desired, it can be calculated after determining Φ as a function of position. A choice must be made between (2.141) and (2.142) when writing the boundary conditions in terms of Φ; (2.141) applies along boundary segments where the flow is confined, and (2.142) must be used where the flow is unconfined. When calculating the head in order to obtain hydraulic head contours, either (2.141) or (2.142) must be chosen according to the value of Φ:

$$\boxed{\phi = \frac{\Phi + \frac{1}{2}kH^2}{kH}, \qquad \Phi \geq \tfrac{1}{2}kH^2} \tag{2.148}$$

$$\boxed{\phi = \sqrt{\frac{2\Phi}{k}}, \qquad \Phi < \tfrac{1}{2}kH^2} \tag{2.149}$$

Some examples of the application of this approach follow.

2.3.2 Application: One-Dimensional Flow

Consider the problem of one-dimensional flow illustrated in Figure 2.24. The aquifer is confined, of length L, and of thickness H. An (x, y, z)-coordinate system is chosen as shown in the figure. The boundary conditions are

$$x = -\tfrac{1}{2}L, \qquad \phi = \phi_1 > H \tag{2.150}$$

and

$$x = \tfrac{1}{2}L, \qquad \phi = \phi_2 < H. \tag{2.151}$$

The aquifer is confined at $x = -L/2$ and unconfined at $x = L/2$, and the boundary values of Φ are obtained by the use of (2.141) and (2.142), respectively:

$$x = -\tfrac{1}{2}L, \qquad \Phi = \Phi_1 = kH\phi_1 - \tfrac{1}{2}kH^2 \tag{2.152}$$

and

$$x = \tfrac{1}{2}L, \qquad \Phi = \Phi_2 = \tfrac{1}{2}k\phi_2^2. \tag{2.153}$$

The expression for the potential is identical to that obtained for the problem of Figure 2.16 (see (2.112)),

$$\Phi = -(\Phi_1 - \Phi_2)\frac{x}{L} + \tfrac{1}{2}(\Phi_1 + \Phi_2), \tag{2.154}$$

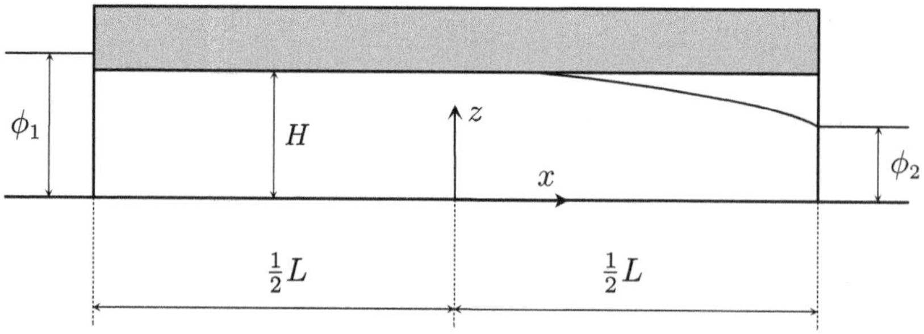

Figure 2.24 One-dimensional combined confined/unconfined flow.

but the expressions for Φ_1 and Φ_2 in terms of the head are different from those applying to (2.112). The expression for Q_x becomes

$$Q_x = \frac{\Phi_1 - \Phi_2}{L} = \frac{kH\phi_1 - \frac{1}{2}kH^2 - \frac{1}{2}k\phi_2^2}{L}. \tag{2.155}$$

The location of the interzonal boundary, $x = x_b$, is obtained from (2.154) by letting $\Phi = \frac{1}{2}kH^2$, which gives with (2.152) and (2.153),

$$\frac{x_b}{L} = \frac{\frac{1}{2}(kH\phi_1 - \frac{3}{2}kH^2 - \frac{1}{2}k\phi_2^2)}{kH\phi_1 - \frac{1}{2}kH^2 - \frac{1}{2}k\phi_2^2}. \tag{2.156}$$

Note that $x_b/L = -\frac{1}{2}$ when $\phi_1 = H$, and $x_b/L = \frac{1}{2}$ when $\phi_2 = H$.

Problem

2.12 Determine the potential as a function of x for the confined and unconfined zones in Figure 2.24 separately. Join the two solutions by requiring that there be continuity of flow across the interzonal boundary. Express both the discharge vector and x_b in terms of the parameters of the problem and compare these results with (2.155) and (2.156).

Problem

2.13 For the case of radial flow toward a well, draw nine hydraulic head contours between the concentric circles $r = r_w$ and $r = R$, if $r_w = 0.01R$, $\phi_0 = 1.2H$, and $\phi_w = 0.2H$.

2.3.3 Application: Dewatering Problem

We apply the technique outlined in this section to a dewatering problem where a building pit is dug through the upper confining bed of an aquifer. A case of a circular building pit, to be dewatered by means of six wells, is illustrated in Figure 2.25. An (x, y, z)-Cartesian coordinate system is chosen with the origin at the center of the pit as shown in the figure. The wells are at a distance B from the center of the pit and are spaced uniformly along the circle $x^2 + y^2 = B^2$. The bottom of the pit is at an elevation H_0 above the base of the aquifer, which has thickness H, with $H_0 < H$. There is no flow before the wells operate. The wells all have a discharge Q, which we determine from the condition that the head inside a cylinder of radius B around the center of the pit is less than or equal to H_0:

$$0 \leq x^2 + y^2 \leq B^2, \qquad \phi < H_0. \tag{2.157}$$

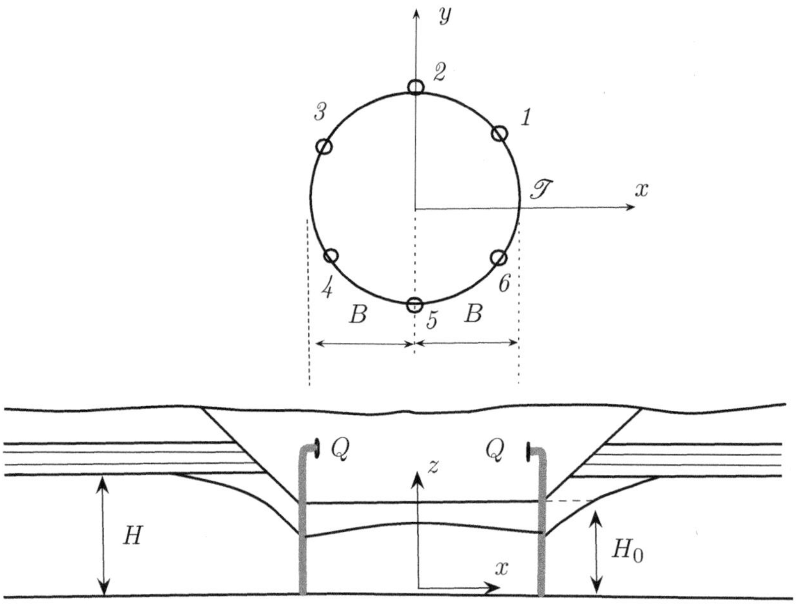

Figure 2.25 A circular building pit dug into a confined aquifer.

The head at a point $\mathscr{P}$ with coordinates $x_p = R, R \gg B$, $y_p = 0$, is assumed to be unaffected by the wells and is ϕ_0, where $\phi_0 > H$. The potential for this flow problem has the form

$$\Phi = \frac{Q}{2\pi}\sum_{j=1}^{6} \ln\left[r_j(x,y)\right] + C, \tag{2.158}$$

where $r_j(x,y)$ is the local radial coordinate emanating from well j, which is centered at (x_j, y_j), i.e.,

$$r_j^2(x,y) = (x - x_j)^2 + (y - y_j)^2. \tag{2.159}$$

The coordinates of the six wells are obtained from Figure 2.25:

$$x_1 = x_6 = -x_3 = -x_4 = B\cos\frac{\pi}{6} \tag{2.160}$$

$$y_1 = y_3 = -y_4 = -y_6 = \tfrac{1}{2}B \tag{2.161}$$

$$x_2 = x_5 = 0 \tag{2.162}$$

$$y_2 = -y_5 = B. \tag{2.163}$$

The condition at point $\mathscr{P}$ is

$$x = R, \qquad y = 0, \qquad \phi = \phi_0 > H, \qquad \Phi = \Phi_0 = kH\phi_0 - \tfrac{1}{2}kH^2, \tag{2.164}$$

and application to (2.158) yields

$$C = \Phi_0 - \frac{Q}{2\pi}\sum_{j=1}^{6}\ln r_j(R,0), \tag{2.165}$$

so that (2.158) becomes

$$\Phi = \frac{Q}{2\pi}\sum_{j=1}^{6}\ln\frac{r_j(x,y)}{r_j(R,0)} + \Phi_0. \tag{2.166}$$

The groundwater table along the boundary of the pit will be highest at points midway between the wells, such as point $\mathscr{T}$ in Figure 2.25. Water flows from these points to the center of the pit and from there to the wells; the head is higher at point $\mathscr{T}$ than anywhere inside the cylinder of radius B. Condition (2.157) will thus be met if the head at point $\mathscr{T}$ is equal to H_0. The flow is unconfined at $\mathscr{T}$ so that the potential there equals $\frac{1}{2}kH_0^2$. Substitution of B for x, zero for y, and $\frac{1}{2}kH_0^2$ for Φ in (2.166) yields, with (2.164),

$$\tfrac{1}{2}kH_0^2 = \frac{Q}{2\pi}\sum_{j=1}^{6}\ln\frac{r_j(B,0)}{r_j(R,0)} + kH\phi_0 - \tfrac{1}{2}kH^2. \tag{2.167}$$

We solve the latter equation for Q:

$$Q = 2\pi\frac{\frac{1}{2}kH_0^2 - kH\phi_0 + \frac{1}{2}kH^2}{\sum_{j=1}^{6}\ln\frac{r_j(B,0)}{r_j(R,0)}}. \tag{2.168}$$

The interzonal boundary $\Phi = \frac{1}{2}kH^2$ and the equipotential $\Phi = \frac{1}{2}kH_0^2$ are shown in Figure 2.26(a) for $B/R = 0.1$, $B/H = 10$, $H_0/H = 0.7$, and $\phi_0/H = 1.4$. The corresponding value for $Q/(kH^2)$ obtained from (2.168) is 0.31. A cross-sectional view along the x-axis is shown in Figure 2.26(b).

Problem

2.14 A dewatering system consisting of four wells will be used to obtain dry working conditions in a building pit in a confined aquifer of hydraulic conductivity $k = 10^{-5}$ m/s and thickness $H = 20$ m. The aquifer is bounded by a long straight river and there is no flow before the wells operate. The head along the river bank is constant and is $\phi_0 = 25$ m above the base of the aquifer. The building pit is square and the bottom of the pit will be 10 m above the aquifer base. The sides of the pit are 50 m in length, and the center of the pit is 100 m away from the river bank.

Use four wells to dewater the pit, placing one well at each corner. Assume that all wells pump the same discharge.

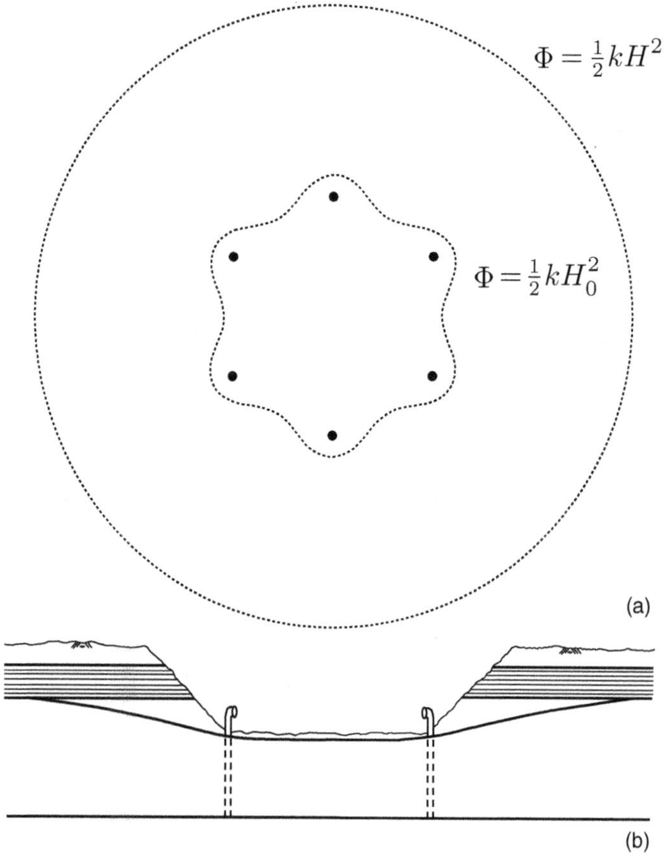

Figure 2.26 The equipotentials $\Phi = \frac{1}{2}kH^2$ and $\Phi = \frac{1}{2}kH_0^2$ (a) and sectional view (b).

Questions:

1. Determine the magnitude of the discharge Q of each well, necessary to obtain dry working conditions in the pit.
2. Demonstrate that indeed dry working conditions are obtained throughout the pit.
3. Determine the head at the screen of each of the four wells if it is given that the wells have a radius of 0.3 m.

2.4 Vertically Integrated Flow

Application of the Dupuit-Forchheimer approximation leads to a discharge potential for shallow unconfined flow. The characteristic property of the discharge potential is that it minus its gradient equals the discharge vector. We next consider the flow three-dimensional and integrate the two horizontal components q_x and q_y vertically, and construct an exact expression for the discharge potential to compare this with the approximate one introduced earlier in this section. We refer to this treatment of flow as *vertically integrated flow*. We will see that under certain, rather general, conditions the two potentials are equal along certain common boundaries. This implies that the discharges obtained from the Dupuit-Forchheimer approximation are not affected by the approximation; the same values would be obtained using a fully three-dimensional model.

Charny [1951] demonstrated that the discharge potential used in the Dupuit-Forchheimer approximation yields the exact expression for the vertically integrated discharge, both for a dam with vertical faces and for a well in an unconfined aquifer. Youngs [1966] demonstrated that the vertically integrated discharge vector can be written in terms of a discharge potential for a wide range of problems. An example of unconfined flow through a dam with vertical faces is shown in Figure 2.27.

The following analysis shows that the negative of the gradient of the exact discharge potential equals the vertically integrated flow, represented by the discharge vector. We further show that the boundary values of this exact discharge potential can be computed for many practical problems; these boundary values are equal to those obtained with the Dupuit-Forchheimer approximation for the following *vertical* boundaries:

- Boundaries with given head
- River banks, including bluffs where the water table is below the base of the aquifer
- Boundaries with given discharge
- Impermeable boundaries
- Well screens

2.4.1 Comprehensive Discharge and Comprehensive Discharge Potential

We define the comprehensive discharge vector as the vertically integrated specific discharge vector and write the expression for its components Q_i, $i = 1,2$ [L^2/T] in terms of the specific discharge vector components $q_i, (i = 1,2)$ [L/T] by the use of Darcy's law

$$Q_i = \int_0^h q_i dz = -\int_0^h k \frac{\partial \phi(x,y,z)}{\partial x_i} dz, \qquad i = 1,2, \quad x_1 \equiv x, \quad x_2 \equiv y, \tag{2.169}$$

where $\phi(x,y,z)$ is the hydraulic head, and h is the elevation of the phreatic surface above the base. We apply Leibniz's rule for differentiation of an integral with variable upper bound:

$$I = \frac{\partial}{\partial x_i}\int_0^h k\phi(x,y,z)dz$$

$$= \int_0^h k\frac{\partial\phi(x,y,z)}{\partial x_i}dz + k\phi(x,y,h)\frac{\partial h}{\partial x_i}. \quad (2.170)$$

We use this expression in (2.169):

$$Q_i = -\frac{\partial}{\partial x_i}\int_0^h k\phi(x,y,z)dz + k\phi(x,y,h)\frac{\partial h}{\partial x_i}. \quad (2.171)$$

We set $\phi(x,y,h) = h$ and rewrite the second term:

$$Q_i = -\frac{\partial}{\partial x_i}\int_0^h k\phi(x,y,z)dz + \frac{\partial}{\partial x_i}\left[\tfrac{1}{2}kh^2\right]. \quad (2.172)$$

We can express Q_i as the gradient of a single function,

$$Q_i = -\frac{\partial\Phi}{\partial x_i}, \quad (2.173)$$

so that a comprehensive potential Φ indeed exists and is given by

$$\Phi = \int_0^h \phi(x,y,z)dz - \tfrac{1}{2}kh^2. \quad (2.174)$$

We write this in a slightly different form,

$$\Phi = kh\left[\frac{1}{h}\int_0^h \phi(x,y,z)dz - \tfrac{1}{2}h\right] = kh\left(\tilde{\phi} - \tilde{h}\right), \quad (2.175)$$

where $\tilde{\phi}$ is the average hydraulic head, and $\tilde{h}$ the elevation halfway between the base of the aquifer and the water table (see Strack et al. [2005]).

The pressure and hydraulic head are related as

$$\phi = \frac{p}{\rho g} + Z \rightarrow \frac{p}{\rho g} = \phi - Z, \quad (2.176)$$

where Z is the elevation of the point considered. Since $\frac{1}{2}h$ is a point midway between the base of the aquifer and the phreatic surface, it represents the average value of the elevation Z on a vertical saturated section. We may therefore rewrite (2.175) as

$$\Phi = kh(\tilde{P} + \tilde{h} - \tilde{h}) = kh\tilde{P}, \tag{2.177}$$

where $\tilde{P}$ is the average pressure head [L]; the pressure head P is defined as

$$P = \frac{p}{\rho g}. \tag{2.178}$$

We may write the discharge potential in two equivalent forms:

$$\Phi = kh(\tilde{\phi} - \tilde{h}) = kh\tilde{P}. \tag{2.179}$$

The expressions (2.175) and (2.177) are valid for both confined and unconfined flow. For the former case, the expressions reduce to

$$\Phi = kH\bar{\phi} - \tfrac{1}{2}kH^2 = kH\tilde{P}. \tag{2.180}$$

We introduce the transmissivity, T, defined as

$$T = kH, \tag{2.181}$$

so that we may write expression (2.180) as

$$\Phi = T(\tilde{\phi} - \tfrac{1}{2}H) = T\tilde{P}. \tag{2.182}$$

2.4.2 Boundary Values

We can compute the value for the discharge potential precisely if the boundary is vertical and the head along the entire boundary is known, e.g., the upstream boundary with given head ϕ_1 in Figure 2.27. The value of the discharge potential along the downstream face is more difficult to determine. The average value of the head is

$$h\bar{\phi} = \int_0^h \phi dz = \int_0^{\phi_2} \phi_2 dz + \int_{\phi_2}^h z dz = \phi_2^2 + \tfrac{1}{2}\left(h^2 - \phi_2^2\right) = \tfrac{1}{2}\phi_2^2 + \tfrac{1}{2}h^2. \tag{2.183}$$

We substitute this expression for $\bar{\phi}$ in (2.174):

$$\Phi_2 = k\left(\tfrac{1}{2}\phi_2^2 + \tfrac{1}{2}h^2 - \tfrac{1}{2}h^2\right) = \tfrac{1}{2}k\phi_2^2. \tag{2.184}$$

We see that the value for the discharge potential is identical to that obtained using the Dupuit-Forchheimer approximation. For the case of a bluff, shown in Figure 2.28 where the entire face is exposed to air, the potential reduces to zero.

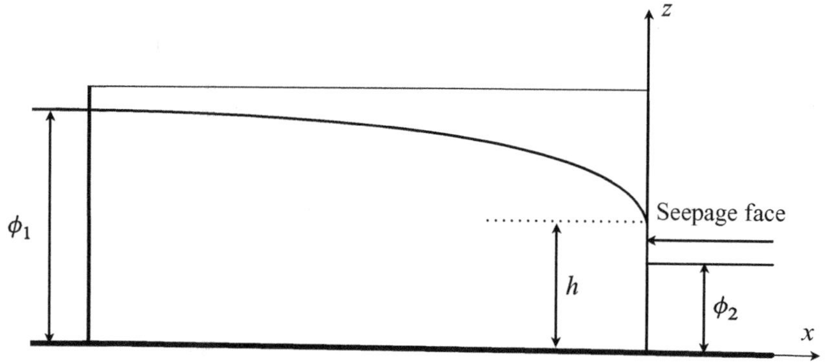

Figure 2.27 A dam with vertical faces and a seepage face downstream.

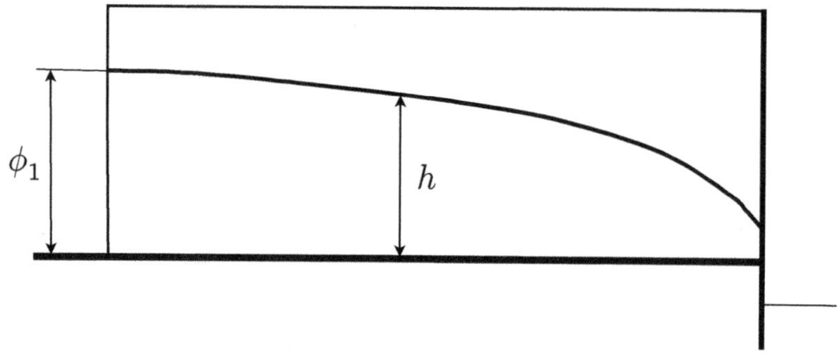

Figure 2.28 Flow toward a bluff.

2.4.3 Differential Equation

The discharge potential for vertically integrated flow satisfies the same partial differential equation as the one obtained using the Dupuit-Forchheimer approximation. This is so because the relation between discharge potential and the discharge vector is the same, whether the potential is obtained with or without the Dupuit-Forchheimer approximation. We obtain the discharge potential as a function of position by solving its partial differential equation, Laplace's equation:

$$\frac{\partial^2 \Phi}{\partial x^2} + \frac{\partial^2 \Phi}{\partial y^2} = 0. \tag{2.185}$$

We determine Φ as a function of x and y by solving Laplace's equation with the proper boundary conditions. Since the discharge potential as well as its gradient can be determined exactly for vertical boundaries with given discharge, pressure, or head, the resulting values computed for the discharge vector are not affected by the Dupuit-Forchheimer approximation for this class of boundary conditions.

2.4.4 Anisotropy

The expression for the discharge potential remains the same if the hydraulic conductivity is anisotropic with horizontal and vertical principal axes and with the two horizontal components equal to k. In that case Darcy's law for the specific discharge vector in three dimensions becomes

$$\begin{aligned} q_x &= -k\frac{\partial \phi}{\partial x} \\ q_y &= -k\frac{\partial \phi}{\partial y} \\ q_z &= -k_{zz}\frac{\partial \phi}{\partial z}, \end{aligned} \tag{2.186}$$

where k_{zz} is the vertical component of the hydraulic conductivity tensor. The discharge potential is obtained by integrating the two horizontal components of the specific discharge vector. Since neither of these components contains k_{zz}, the expression for the discharge potential remains valid, even if the aquifer is much less conductive in the vertical direction than in the horizontal one.

Problem

2.15 Demonstrate by differentiation of (2.174) that the gradient of the discharge potential indeed equals expressions (2.169).

2.5 Shallow Unconfined Flow with Rainfall

Water may percolate downward through the soil above the phreatic surface of an unconfined aquifer. This water may infiltrate, for example, as the result of rainfall or artificial infiltration, or may infiltrate through the bottom of a creek or pond above the phreatic surface. In the first case the infiltration occurs over the entire aquifer or a large part of it. In the latter two cases the infiltration is local. We use the term *recharge* to represent the various sources of infiltration into the saturated zone of the aquifer.

We begin by deriving the basic equations for shallow unconfined flow with rainfall, and afterward discuss both uniform and local recharge for one-dimensional, radial, and two-dimensional shallow unconfined flow along with some applications.

2.5.1 Basic Equations

We represent the recharge rate as N [L/T], which is the amount of water per unit area that enters the aquifer per unit time. The resistance to flow in the vertical direction is neglected in accordance with the Dupuit-Forchheimer approximation,

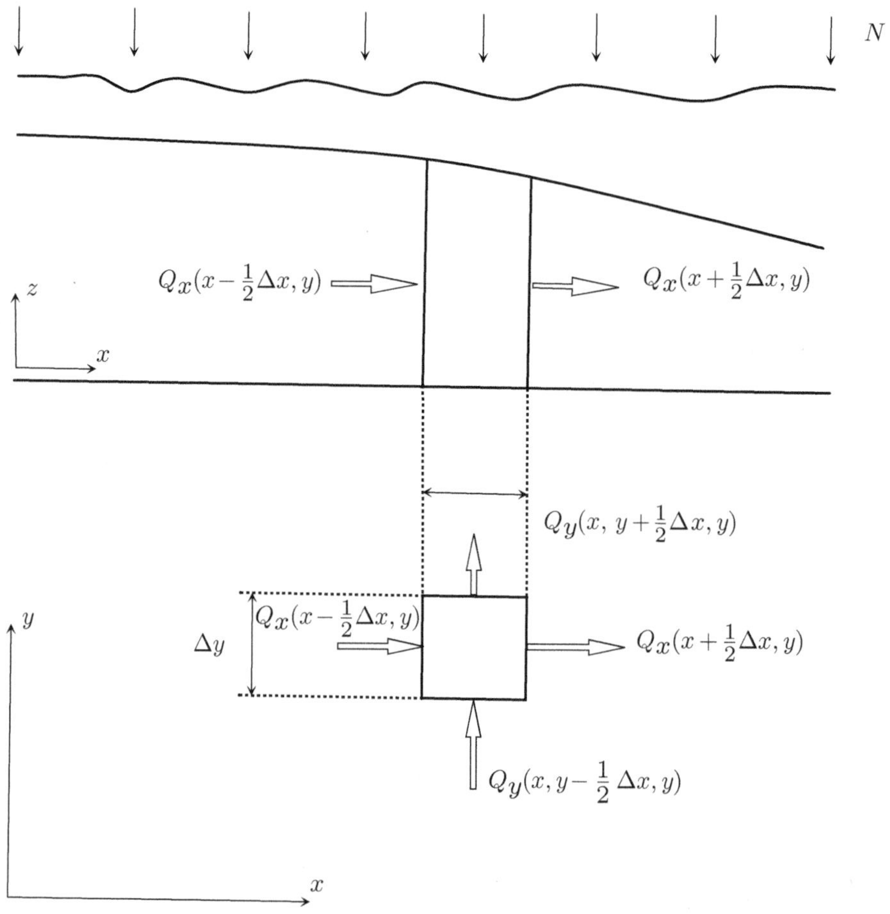

Figure 2.29 Unconfined flow with rainfall.

and the recharge is modeled as taking place without any vertical change in head. We do not consider the flow through the unsaturated zone of the infiltrated water on its way into the aquifer.

The condition that the inflow into the aquifer per unit area and unit time is equal to the outflow implies that the divergence of the discharge vector must be equal to the rate of infiltration, so that (see Figure 2.29):

$$\boxed{\frac{\partial Q_x}{\partial x} + \frac{\partial Q_y}{\partial y} = N} \tag{2.187}$$

The expression for the potential in terms of the head, (2.142), remains unchanged, and the discharge vector equals minus the gradient of Φ. Hence, the differential equation for the potential becomes the Poisson equation:

$$\boxed{\frac{\partial^2 \Phi}{\partial x^2} + \frac{\partial^2 \Phi}{\partial y^2} = \nabla^2 \Phi = -N} \tag{2.188}$$

where the operator ∇^2 stands for $\partial^2/\partial x^2 + \partial^2/\partial y^2$. The infiltration rate N may be a function of position, but we begin by considering uniform recharge.

2.5.2 Planar Flow

For planar flow, with no flow occurring in the y-direction, (2.188) becomes

$$\frac{d^2 \Phi}{dx^2} = -N \tag{2.189}$$

with the general solution

$$\Phi = -\tfrac{1}{2} N x^2 - Q_{x0} x + B, \tag{2.190}$$

where Q_{x0} represents the uniform flow component that would occur in the absence of rainfall. As an example, consider the case of flow between two long and straight parallel rivers, illustrated in Figure 2.30. The distance between the rivers is L. The origin of an (x, y, z)-coordinate system is chosen to be halfway between the rivers, with the x-axis normal to the river banks. The boundary conditions are

$$x = -\tfrac{1}{2} L, \qquad \phi = \phi_1, \qquad \Phi = \Phi_1 = \tfrac{1}{2} k \phi_1^2 \tag{2.191}$$

and

$$x = \tfrac{1}{2} L, \qquad \phi = \phi_2, \qquad \Phi = \Phi_2 = \tfrac{1}{2} k \phi_2^2. \tag{2.192}$$

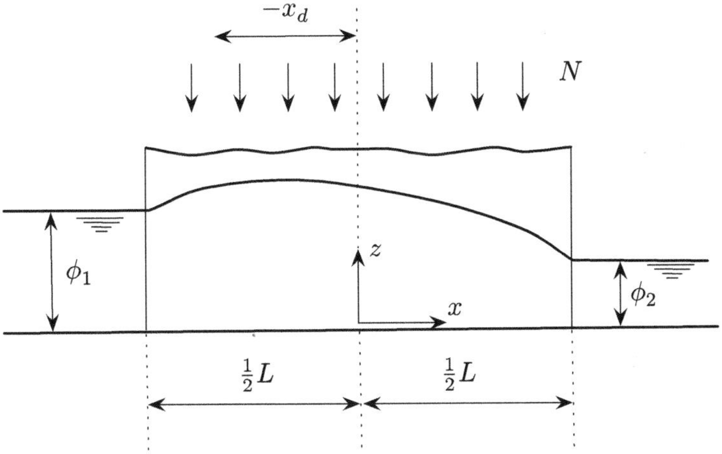

Figure 2.30 One-dimensional unconfined flow with rainfall.

Application of the boundary conditions (2.191) and (2.192) to the general solution (2.190) yields

$$\Phi = -\tfrac{1}{2}N\left[x^2 - (\tfrac{1}{2}L)^2\right] - \frac{\Phi_1 - \Phi_2}{L}x + \tfrac{1}{2}(\Phi_1 + \Phi_2). \qquad (2.193)$$

We differentiate (2.193) to obtain an expression for the discharge vector component Q_x:

$$Q_x = -\frac{d\Phi}{dx} = Nx + \frac{\Phi_1 - \Phi_2}{L}. \qquad (2.194)$$

The location of the divide, $x = x_d$, the point where Φ has a maximum and $d\Phi/dx = 0$, is found by setting Q_x equal to zero:

$$Nx_d = -\frac{\Phi_1 - \Phi_2}{L}, \qquad -\tfrac{1}{2}L \le x_d \le \tfrac{1}{2}L. \qquad (2.195)$$

The divide is at the center of the aquifer when $\Phi_1 = \Phi_2$, as expected. There will not always be a divide: if the value obtained for x_d from (2.195) is larger than $\frac{1}{2}L$ or less than $-\frac{1}{2}L$, there is no divide between the two boundaries, and the flow is in one direction throughout the aquifer. If x_d is greater than $\frac{1}{2}L$, all flow is in the negative x-direction and all rainwater flows to the boundary at $x = -\frac{1}{2}L$. There is a divide if $-\frac{1}{2}L < x_d < \frac{1}{2}L$; the amount of rainwater flowing per unit width of the aquifer to the boundary at $x = \frac{1}{2}L$ is $(\frac{1}{2}L - x_d)N$, whereas the remaining amount, $(\frac{1}{2}L + x_d)N$, flows toward the boundary at $x = -\frac{1}{2}L$.

Problem

2.16 Consider the problem of flow through an aquifer bounded by two long parallel rivers and of differing hydraulic conductivities in two zones, as shown in Figure 2.31. Zone 1 is in $-L \le x \le 0$ and zone 2 in $0 \le x \le L$. The hydraulic conductivity values in zones 1 and 2 are k_1 and k_2, respectively. The head at $x = -L$ is ϕ_1 and the river level along $x = L$ is below the base of the aquifer. There is infiltration at rate N over the entire aquifer.

Questions:

1. Express Q_x in terms of the parameters of the problem.
2. Calculate Q_x at $x = 0$ if $k_1 = 10$ m/d and $k_2 = 0.2$ m/d, $L_1 = 100$ m, and $\phi_1 = 20$ m.

2.5.3 Estimating Hydraulic Conductivity

Questions concerning groundwater flow can generally be answered only if the hydraulic conductivity is known or, at least, if we can estimate its value. The values

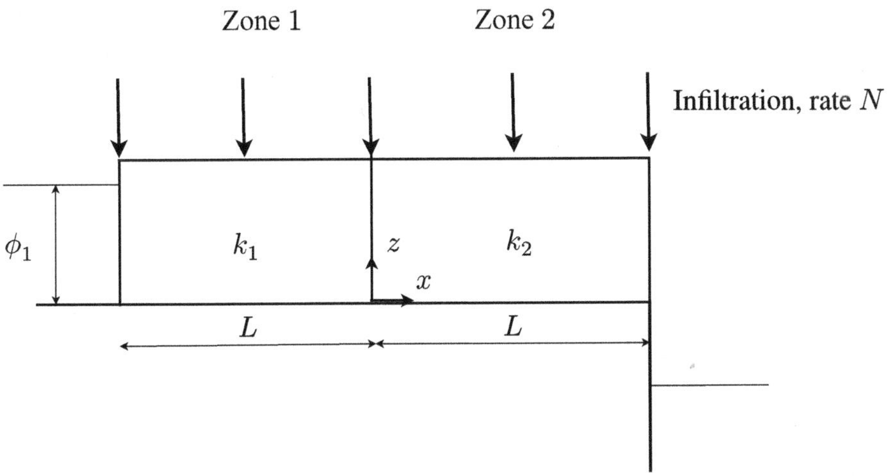

Figure 2.31 Flow through an inhomogeneous dam bounded by a bluff.

of hydraulic conductivities in the field vary enormously, not only from area to area, but also locally. Variations of an order of magnitude or more are quite common. What we often are interested in is an average value of the hydraulic conductivity, which we can use to answer modeling questions concerned with flow over relatively large areas. Expression (2.193) for the discharge potential makes it possible to get some insight, first into what is required to estimate the average hydraulic conductivity and, second, to find ways to make such an estimate. Equation (2.193) does not involve the hydraulic conductivity, because it is an expression for the potential, rather than the head, and we must recognize that the values of the potential are linearly dependent on the hydraulic conductivity. Expressing the potential in terms of head, we obtain from (2.193)

$$\tfrac{1}{2}k\phi^2 = -\tfrac{1}{2}N\left[x^2 - (\tfrac{1}{2}L)^2\right] - \tfrac{1}{2}k\frac{\phi_1^2 - \phi_2^2}{L}x + \tfrac{1}{4}k(\phi_1^2 + \phi_2^2). \tag{2.196}$$

We multiply both sides of the equation by $2/k$:

$$\phi^2 = -\tfrac{1}{2}\frac{N}{k}[x^2 - (\tfrac{1}{2}L)^2] - \tfrac{1}{2}\frac{\phi_1^2 - \phi_2^2}{L}x + \tfrac{1}{4}(\phi_1^2 + \phi_2^2). \tag{2.197}$$

We draw two important observations from this expression. The first is that if there is no rainfall, i.e., if $N = 0$, the hydraulic conductivity cannot be computed, even if the head is known at all points between the rivers. This follows from the equation

$$\phi^2 = -\tfrac{1}{2}\frac{\phi_1^2 - \phi_2^2}{L}x + \tfrac{1}{4}(\phi_1^2 + \phi_2^2), \tag{2.198}$$

which does not contain the hydraulic conductivity at all. Computation of the hydraulic conductivity for this case requires that discharge between the rivers be known. The second observation is that only the ratio N/k is involved in the expression for head; only this ratio can be computed from heads observed in the aquifer. Thus, if the recharge from rainfall is known, then the hydraulic conductivity can be computed (or, realistically, its average value estimated). Most groundwater modeling projects therefore should begin by obtaining as many data on discharges as possible, for example, recharge from rainfall, evaporation from swamps, and discharges of wells.

2.5.4 Superposition

The differential equation for shallow unconfined flow with rainfall is the Poisson equation (2.188),

$$\nabla^2\Phi = -N. \tag{2.199}$$

Because the Poisson equation is linear in terms of the potential, superposition of potentials is possible without violating the differential equation. Let $\overset{1}{\Phi}$ be a solution to the Poisson equation,

$$\nabla^2\overset{1}{\Phi} = -N, \tag{2.200}$$

and let $\overset{2}{\Phi}$ be a solution to the Laplace equation,

$$\nabla^2\overset{2}{\Phi} = 0. \tag{2.201}$$

The sum, $\Phi = \overset{1}{\Phi} + \overset{2}{\Phi}$, fulfills the Poisson equation

$$\nabla^2\Phi = \nabla^2(\overset{1}{\Phi} + \overset{2}{\Phi}) = \nabla^2\overset{1}{\Phi} + \nabla^2\overset{2}{\Phi} = -N. \tag{2.202}$$

Superposition is useful for solving problems of unconfined flow with rainfall. We solve these problems in two steps: first, a particular solution to the Poisson equation is selected without considering boundary conditions, and second, a solution to Laplace's equation is determined such that the sum of the two solutions meets the boundary conditions.

2.5.5 Application: A Well Near a River in an Aquifer with Recharge from Rainfall

We consider the problem of a well near a river in an unconfined aquifer with hydraulic conductivity k and recharge from rainfall at rate N. As before, we determine the maximum discharge Q of a well at $x = -d, y = 0$, such that no river water is drawn by the well. The river is long and we choose the y-axis along the river bank, with the x-axis pointing out of the aquifer. The head at a point $x = -L,\ y = 0$, is measured to be ϕ_1 and the head along the river is ϕ_0. The potential that incorporates rainfall and is constant along the river is given by (2.190). We add the potential for a well at $x = -d, y = 0$, and its image

$$\Phi = -\tfrac{1}{2}Nx^2 - Q_{x0}x + B + \frac{Q}{4\pi}\ln[(x+d)^2 + y^2] - \frac{Q}{4\pi}\ln[(x-d)^2 + y^2]. \quad (2.203)$$

We apply the boundary condition along the river and set x equal to zero,

$$\Phi_0 = \tfrac{1}{2}k\phi_0^2 = B, \quad (2.204)$$

so that the expression for the potential becomes

$$\Phi = -\tfrac{1}{2}Nx^2 - Q_{x0}x + \frac{Q}{4\pi}\ln[(x+d)^2 + y^2] - \frac{Q}{4\pi}\ln[(x-d)^2 + y^2] + \Phi_0. \quad (2.205)$$

We determine the constant Q_{x0} by using that the measurement of the observation well at $x = -L, y = 0$, was taken prior to operating the well, so that

$$\Phi_1 = \tfrac{1}{2}k\phi_1^2 = -\tfrac{1}{2}NL^2 + Q_{x0}L + \Phi_0. \quad (2.206)$$

We solve this for Q_{x0} and obtain

$$Q_{x0} = \tfrac{1}{2}k\frac{\phi_1^2 - \phi_0^2}{L} + \tfrac{1}{2}NL. \quad (2.207)$$

Note that the value of Q_{x0} depends on the infiltration rate N.

The maximum discharge for which no river water enters the aquifer occurs when the flow rate at $x = y = 0$ is zero. We differentiate expression (2.205) with respect to x for $y = 0$ and obtain the following expression for $Q_x(x,0)$:

$$Q_x = Nx + Q_{x0} - \frac{Q}{2\pi}\frac{1}{x+d} + \frac{Q}{2\pi}\frac{1}{x-d}. \quad (2.208)$$

We set this equal to zero at $x = 0$ and obtain an expression for the maximum discharge of the well, $Q_{\max}$:

$$Q_{x0} - \frac{Q_{\max}}{\pi d} = 0. \quad (2.209)$$

Surprisingly, we obtain the same expression for the maximum discharge, $Q_{\max}$, as for the case without infiltration from rainfall:

$$Q_{\max} = \pi d Q_{x0}. \quad (2.210)$$

We must remember that the expression for Q_{x0} involves the recharge rate N; it affects the result indirectly.

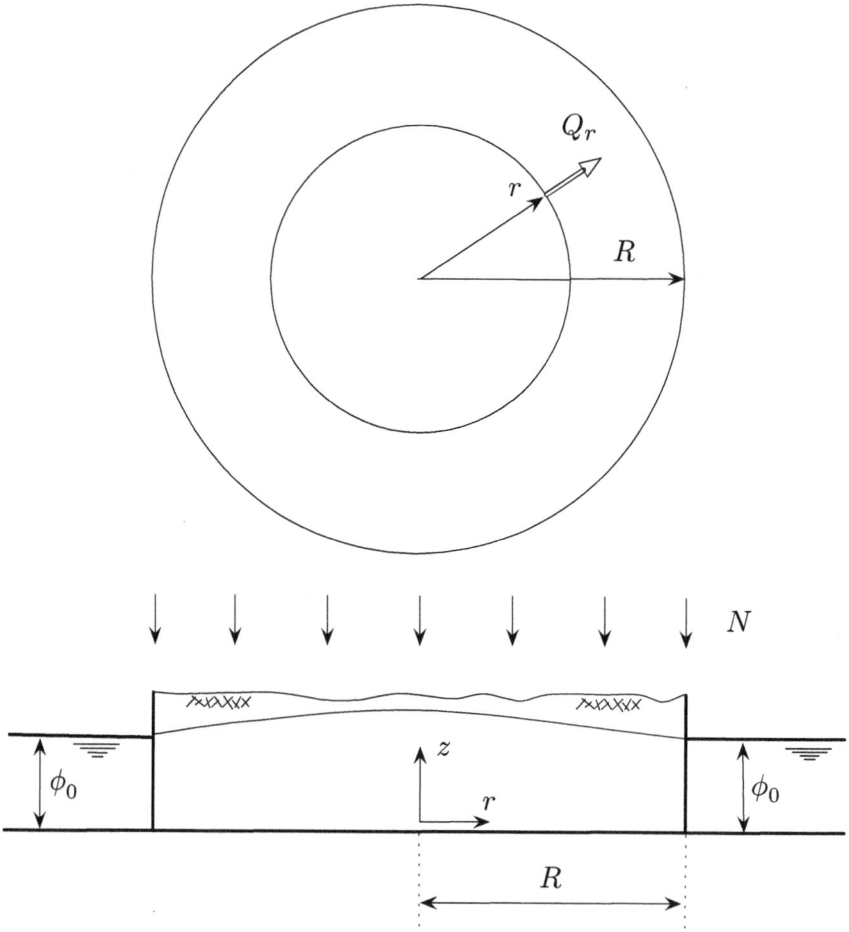

Figure 2.32 Radial shallow unconfined flow with rainfall.

2.5.6 Radial Flow

We obtain an expression for the potential for radial flow with rainfall by applying both Darcy's law and the condition of continuity of flow (see Figure 2.32). We consider the case of rain falling on a circular island of radius R. There is no well in the island and all of the infiltrated rainwater flows to the boundary. We use the radial coordinate r pointing away from the center of the island and apply the condition that the head at $r = R$ is ϕ_0. Hence,

$$r = R, \qquad \phi = \phi_0, \qquad \Phi = \Phi_0 = \tfrac{1}{2}k\phi_0^2. \tag{2.211}$$

The amount of rainwater infiltrating inside a cylinder of radius r equals $N\pi r^2$. By continuity of flow, this amount of water must flow through the wall of the cylinder.

Since the flow is radial, the component Q_r of the discharge vector is constant along the cylinder wall; the total flow through this wall equals $2\pi r Q_r$. The continuity equation is

$$2\pi r Q_r = N\pi r^2 \tag{2.212}$$

or

$$Q_r = \tfrac{1}{2}Nr. \tag{2.213}$$

Application of Darcy's law yields

$$Q_r = -\frac{d\Phi}{dr} = \tfrac{1}{2}Nr, \tag{2.214}$$

so that

$$\boxed{\Phi = -\tfrac{1}{4}Nr^2 + C} \tag{2.215}$$

where C is a constant of integration. Application of the boundary condition (2.211) to (2.215) yields $C = \frac{1}{4}NR^2 + \Phi_0$ so that (2.215) becomes

$$\Phi = -\tfrac{1}{4}N(r^2 - R^2) + \Phi_0. \tag{2.216}$$

2.5.7 Radial Flow to a Well

The problem of radial unconfined flow with a well at the center of a circular island is illustrated in Figure 2.33. The discharge of the well is Q, the radius of the island is R, and the head at $r = R$ is ϕ_0. The boundary condition at $r = R$ is given by (2.211). The particular solution of the Poisson equation is (2.216). We add the potential for unconfined flow to a well, (2.116), to (2.216):

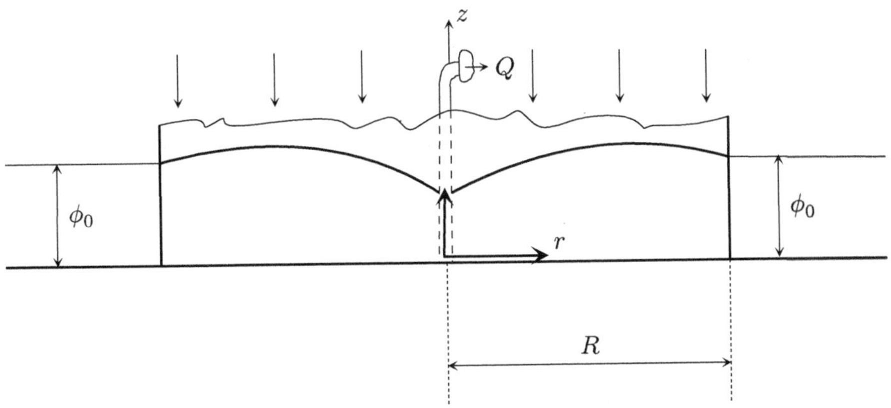

Figure 2.33 Radial flow to a well.

$$\Phi = -\tfrac{1}{4}N(r^2 - R^2) + \frac{Q}{2\pi}\ln\frac{r}{R} + C. \tag{2.217}$$

We introduced a new constant C and omitted Φ_0 in (2.116) and (2.216); we must introduce a new constant after superimposing individual solutions and determine it from the boundary condition. Application of the boundary condition (2.211) to (2.217) yields $C = \Phi_0$, and (2.217) becomes

$$\Phi = -\tfrac{1}{4}N(r^2 - R^2) + \frac{Q}{2\pi}\ln\frac{r}{R} + \Phi_0. \tag{2.218}$$

We obtain the expression for the component Q_r of the discharge vector from (2.218) by differentiation,

$$Q_r = -\frac{d\Phi}{dr} = \tfrac{1}{2}Nr - \frac{Q}{2\pi r}. \tag{2.219}$$

The divide is at $r = r_d$, where $Q_r = 0$, i.e.,

$$\tfrac{1}{2}Nr_d = \frac{Q}{2\pi r_d} \tag{2.220}$$

or

$$r_d = \sqrt{\frac{Q}{\pi N}}, \qquad r_d \le R. \tag{2.221}$$

As in the case of one-dimensional flow, there may not be a divide in the aquifer; there is a divide only when $r_d < R$.

2.5.8 Rainfall on an Island with an Elliptical Boundary

Circular islands rarely occur in reality; a solution whereby the boundary of the island is an ellipse would be more useful. Consider the following expression for the potential:

$$\Phi = -\tfrac{1}{2}\frac{N}{\frac{1}{a^2} + \frac{1}{b^2}}\left[\left(\frac{x}{a}\right)^2 + \left(\frac{y}{b}\right)^2 - 1\right] + \Phi_0. \tag{2.222}$$

This potential fulfills the differential equation

$$\nabla^2\Phi = -N \tag{2.223}$$

and is equal to Φ_0 along the boundary of an ellipse with principal axes of lengths a and b:

$$\left(\frac{x}{a}\right)^2 + \left(\frac{y}{b}\right)^2 = 1, \qquad \Phi = \Phi_0. \tag{2.224}$$

In view of these properties, the potential (2.222) describes the flow in an elliptic island.

We may generalize the potential (2.222) for the case when the principal axis of length a makes an angle α with the x-axis. We introduce an (x^*, y^*)-Cartesian coordinate system with its origin at the center of the ellipse, with the x^*-axis coinciding with the major axis of the ellipse; the potential then becomes:

$$\Phi = -\frac{1}{2}\frac{N}{\frac{1}{a^2}+\frac{1}{b^2}}\left[\left(\frac{x^*}{a}\right)^2+\left(\frac{y^*}{b}\right)^2-1\right]+\Phi_0. \tag{2.225}$$

We express the coordinates x^* and y^* in terms of x and y, using Figure 1.5, and with the origin of the (x^*, y^*)-coordinate system at $x = x_c$, $y = y_c$:

$$\begin{aligned} x^* &= (x-x_c)\cos\alpha + (y-y_c)\sin\alpha \\ y^* &= -(x-x_c)\sin\alpha + (y-y_c)\cos\alpha. \end{aligned} \tag{2.226}$$

The potential (2.225) may be written alternatively in a form that applies if either a or b is zero

$$\Phi = -\frac{1}{2}\frac{N}{a^2+b^2}\left[b^2(x^*)^2+a^2(y^*)^2-a^2b^2\right]+\Phi_0. \tag{2.227}$$

We write this in terms of x and y by the use of (2.226):

$$\begin{aligned} \Phi = -\frac{1}{2}\frac{N}{a^2+b^2}\{ & b^2\left[(x-x_c)^2\cos^2\alpha+2(x-x_c)(y-y_c)\sin\alpha\cos\alpha\right. \\ & \left.+(y-y_c)^2\sin^2\alpha\right] \\ & +a^2\left[(x-x_c)^2\sin^2\alpha-2(x-x_c)(y-y_c)\sin\alpha\cos\alpha\right. \\ & \left.+(y-y_c)^2\cos^2\alpha\right]-a^2b^2\}+\Phi_0. \end{aligned} \tag{2.228}$$

We collect terms and simplify

$$\begin{aligned} \Phi = -\frac{1}{2}\frac{N}{a^2+b^2} & \left[(a^2\sin^2\alpha+b^2\cos^2\alpha)(x-x_c)^2\right. \\ & -2(a^2-b^2)(x-x_c)(y-y_c)\sin\alpha\cos\alpha \\ & \left.+(a^2\cos^2\alpha+b^2\sin^2\alpha)(y-y_c)^2-a^2b^2\right]+\Phi_0. \end{aligned} \tag{2.229}$$

This function covers all cases of uniform infiltration presented so far: if $a = 0$ and $\alpha = 0$, then (2.229) reduces to

$$\Phi = -\tfrac{1}{2}N(x-x_c)^2+\Phi_0 = -\tfrac{1}{2}Nx^2+Nx_cx-\tfrac{1}{2}Nx_c^2+\Phi_0, \tag{2.230}$$

which is equivalent to (2.190) with Nx_c replacing $-Q_{x0}$ and $-\frac{1}{2}Nx_c^2+\Phi_0$ replacing B. If $a = b = R$, then (2.229) reduces to the potential (2.216) for radial flow with rainfall. We differentiate this function with respect to x and y to obtain the components Q_x and Q_y of the discharge vector:

$$Q_x = -\frac{\partial\Phi}{\partial x} = \frac{N}{a^2+b^2}\big[(a^2\sin^2\alpha + b^2\cos^2\alpha)(x-x_c) \\ - (a^2-b^2)(y-y_c)\sin\alpha\cos\alpha\big]$$

$$Q_y = -\frac{\partial\Phi}{\partial y} = \frac{N}{a^2+b^2}\big[-(a^2-b^2)(x-x_c)\sin\alpha\cos\alpha \\ + (a^2\cos^2\alpha + b^2\sin^2\alpha)(y-y_c)\big]. \tag{2.231}$$

2.6 Unconfined Flow with Local Infiltration

When the bottom of a ditch or a pond is not in contact with the phreatic surface, or in the case of irrigation, water leaks down to the phreatic surface and joins the flow in the aquifer. We will not model the unsaturated flow in the zone between the bottom of the ditch or pond and the phreatic surface, but approximate the flow in the unsaturated zone as vertical and constant in time. We first consider infiltration through the bottom of a long ditch for cases of one-dimensional flow, and second the infiltration through the bottom of a circular pond.

2.6.1 One-Dimensional Flow

We consider infiltration or extraction over strips or ditches, followed by the special case of a ditch of negligible width.

2.6.2 Influence Function for Infiltration below a Strip; the Ditch Function

We create a function that can be used to simulate infiltration at a unit rate [L/T] over a strip of width b bounded by the lines $x = x_1$ and $x = x_2$. We construct this function in such a manner that it is internally consistent, i.e., that both the function itself and its derivative are continuous throughout the domain. We further choose this function so that it is symmetrical with respect to the center of the ditch and that it is zero along the boundaries of the ditch; we call this function the *ditch function.* Note that these choices are arbitrary; we add other solutions to meet the boundary conditions of the problem we are solving. We demonstrate in what follows that the function, which we call $G_d(x;x_1,x_2)$, indeed meets the conditions we have chosen. Note that we introduced the notation for this function, with its parameters in parentheses, in order to facilitate implementation in a computer program. We separate the variable x from the parameters x_1 and x_2 by a semicolon. The function $G_d(x;x_1,x_2)$ is

$$
\begin{aligned}
G_d &= \tfrac{1}{2}b(x-x_1), && x < x_1 \\
G_d &= -\tfrac{1}{2}(x-x_1)(x-x_2), && x_1 \le x \le x_2 \\
G_d &= -\tfrac{1}{2}b(x-x_2), && x_2 < x.
\end{aligned}
\tag{2.232}
$$

We refer to the function G_d as the influence function for a strip or ditch; it satisfies the differential equation $d^2G_d/dx^2 = -1$ in the area of unit infiltration, and the differential equation $d^2G_d/dx^2 = 0$ elsewhere. Furthermore, G_d is zero at $x = x_1$ and $x = x_2$, when approached from both sides, and is thus continuous. The same holds for the derivative, which is equal to $\frac{1}{2}b$ at $x = x_1$ and equal to $-\frac{1}{2}b$ at $x = x_2$. Note that the derivative is zero for $x = \frac{1}{2}(x_1 + x_2)$, i.e., at the center of the strip.

2.6.3 Infiltration along a Line

If the strip is replaced by a line at $x = x_0$, and there is a unit infiltration rate [L^2/T] per unit length of line, then the function, which we will call $G_l(x;x_0)$, reduces to

$$
\begin{aligned}
G_l &= \tfrac{1}{2}(x-x_0), && x < x_0 \\
G_l &= -\tfrac{1}{2}(x-x_0), && x_0 < x.
\end{aligned}
\tag{2.233}
$$

Note that half of the infiltrated water flows to the left, and half to the right.

2.6.4 Application: Infiltration along a Strip between Two Parallel Rivers

We consider the case of infiltration through the bottom of a long ditch of width b, which is parallel to two long rivers that bound an unconfined aquifer. There is no rainfall; a Cartesian coordinate system is chosen as shown in Figure 2.34; the distance between the rivers is L; and the heads along the two rivers are ϕ_1 and ϕ_2:

$$
\begin{aligned}
x &= -\tfrac{1}{2}L, && \phi = \phi_1, && \Phi = \Phi_1 = \tfrac{1}{2}k\phi_1^2 \\
x &= \tfrac{1}{2}L, && \phi = \phi_2, && \Phi = \Phi_2 = \tfrac{1}{2}k\phi_2^2.
\end{aligned}
\tag{2.234}
$$

The rate of infiltration through the bottom of the ditch is N_1 [m/s]. The coordinates of the boundaries of the ditch are x_1 and x_2 (see Figure 2.34), so that

$$
x_2 - x_1 = b. \tag{2.235}
$$

We add a linear function to the influence function

$$
\Phi = Ax + B + N_1 G_d(x;x_1,x_2). \tag{2.236}
$$

The boundary condition at $x = -L/2$ is that $\Phi = \Phi_1 = \frac{1}{2}k\phi_1^2$, i.e.,

$$
\Phi_1 = -\tfrac{1}{2}AL + B + N_1 G_d(-L/2;x_1,x_2) = -\tfrac{1}{2}AL + B + \tfrac{1}{2}N_1 b\left(-\tfrac{1}{2}L - x_1\right). \tag{2.237}
$$

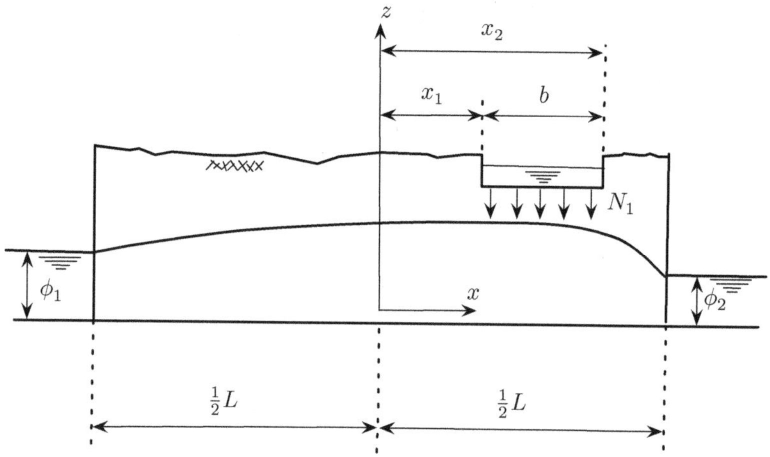

Figure 2.34 Local infiltration for one-dimensional flow. Note: $\xi_1 = x_1; \xi_2 = x_2$.

We apply the boundary condition at $x = L/2$, using the proper representation for $N_1 G_d$, in (2.232), and obtain, with $\Phi_2 = \frac{1}{2}k\phi_2^2$:

$$\Phi_2 = \tfrac{1}{2}AL + B - \tfrac{1}{2}N_1 b(\tfrac{1}{2}L - x_2). \tag{2.238}$$

We add (2.237) to (2.238),

$$\Phi_1 + \Phi_2 = 2B - \tfrac{1}{2}N_1 b(L + x_1 - x_2) = 2B - \tfrac{1}{2}N_1 b(L - b) \tag{2.239}$$

and solve for B:

$$B = \tfrac{1}{2}(\Phi_1 + \Phi_2) + \tfrac{1}{4}N_1 b(L - b). \tag{2.240}$$

We subtract (2.238) from (2.237):

$$\Phi_1 - \Phi_2 = -AL + \tfrac{1}{2}N_1 b(-\tfrac{1}{2}L - x_1 + \tfrac{1}{2}L - x_2) = -AL - \tfrac{1}{2}N_1 b(x_1 + x_2) \tag{2.241}$$

and solve for A:

$$A = \frac{\Phi_2 - \Phi_1}{L} - \tfrac{1}{2}N_1 b\frac{(x_1 + x_2)}{L}. \tag{2.242}$$

Note that when infiltration occurs over the entire area, i.e., $x_1 = -L/2$ and $x_2 = L/2$, the solution reduces to the case of uniform infiltration.

We obtain an expression for the discharge potential by substituting the expressions for A and B in (2.236):

$$\Phi = -\frac{\Phi_1 - \Phi_2}{L}x + \tfrac{1}{2}(\Phi_1 + \Phi_2) + N_1\left[G_d(x; x_1, x_2) - \tfrac{1}{2}b(x_1 + x_2)\frac{x}{L} + \tfrac{1}{4}b(L - b)\right]. \tag{2.243}$$

We differentiate the potential to obtain the discharge Q_x:

$$Q_x = \left[\Phi_1 - \Phi_2 + \tfrac{1}{2}N_1 b(x_1 + x_2)\right]\frac{1}{L} - N_1\frac{dG_d}{dx}. \tag{2.244}$$

We can determine the amount of infiltrated water that leaves the left boundary by computing Q_x at $x = x_1$:

$$Q_x(x_1) = \left[\Phi_1 - \Phi_2 + \tfrac{1}{2}N_1 b(x_1 + x_2)\right]\frac{1}{L} - \tfrac{1}{2}N_1 b. \tag{2.245}$$

The amount of infiltrated water that leaves the right boundary is:

$$Q_x(x_2) = \left[\Phi_1 - \Phi_2 + \tfrac{1}{2}N_1 b(x_1 + x_2)\right]\frac{1}{L} + \tfrac{1}{2}N_1 b. \tag{2.246}$$

Divide.

We find the divide and the maximum elevation of the water table by setting Q_x equal to zero:

$$0 = \left[\Phi_1 - \Phi_2 + \tfrac{1}{2}N_1 b(x_1 + x_2)\right]\frac{1}{L} - N_1\frac{dG_d}{dx}. \tag{2.247}$$

The only possible maximum value is below the area of infiltration, i.e., $G_d = -\frac{1}{2}(x - x_1)(x - x_2)$, so that

$$0 = \left[\Phi_1 - \Phi_2 + \tfrac{1}{2}N_1 b(x_1 + x_2)\right]\frac{1}{L} + N_1[x_s - \tfrac{1}{2}(x_1 + x_2)], \tag{2.248}$$

where x_s represents the location of the stagnation point. We solve for x_s:

$$x_s = -\frac{1}{N_1 L}\left[\Phi_1 - \Phi_2 + \tfrac{1}{2}N_1 b(x_1 + x_2)\right] + \tfrac{1}{2}(x_1 + x_2). \tag{2.249}$$

A divide exists only if x_s is between x_1 and x_2, so that the condition for a divide to exist is:

$$x_1 \le -\frac{1}{N_1 L}\left[\Phi_1 - \Phi_2 + \tfrac{1}{2}N_1 b(x_1 + x_2)\right] + \tfrac{1}{2}(x_1 + x_2) \le x_2. \tag{2.250}$$

2.6.5 Superposition of Strips of Infiltration

We may use the function G_d to model infiltration through n strips, located at $x_{2j-1} \le x \le x_{2j}, j = 1, 2, \ldots, n$, where the infiltration rate from the jth strip is N_j. We obtain the potential for this case by superposition:

$$\Phi = Ax + B + \sum_{j=1}^{n} N_j G_d(x; x_{2j-1}, x_{2j}), \tag{2.251}$$

where the constants A and B are determined from the boundary conditions.

Problem

2.17 Establish, for the flow problem of Figure 2.34, the conditions for which:

1. All water infiltrating through the bottom of the ditch leaves the aquifer at $x = -\frac{1}{2}L$.
2. All water infiltrating through the bottom of the ditch leaves the aquifer at $x = \frac{1}{2}L$.
3. Some of the water infiltrating through the bottom of the ditch leaves the aquifer at $x = -\frac{1}{2}L$, and some at $x = \frac{1}{2}L$. Determine the fraction of the total amount that leaves the aquifer at $x = -\frac{1}{2}L$.

Problem

2.18 Two ditches of width b are used for infiltration into an unconfined aquifer. The ditches are long and parallel to the two boundaries of the aquifer; the flow is in the (x,z)-plane. The distance from the center of each ditch to the nearest boundary is $d + b/2$; the width of the ditches is b; the distance between the boundaries of the aquifer is L; and the heads at the boundaries are ϕ_1 and ϕ_2. The infiltration rate through the bottom of the one ditch is N_1 and of the other is N_2. The hydraulic conductivity is k.

Questions:

1. Determine expressions for the ratios N_1/k and N_2/k so that the phreatic surface between the ditches is horizontal and a distance H above the aquifer base.
2. Determine the conditions that H, ϕ_1, ϕ_2, b, and d must satisfy in order that the infiltration rates N_1 and N_2 are less than k. (Uniform infiltration at a rate larger than k is physically impossible.)

2.6.6 Influence Function for a Circular Pond; the Pond Function

We determine an influence function for modeling infiltration through circular ponds, the *pond function*, in much the same manner as the ditch for planar flow. We derive the function for an infinite aquifer and choose it such that it vanishes at the boundary of the pond. We represent this function as $G_p(x,y;x_1,y_1,R)$, where the index p stands for pond, where x_1 and y_1 are the coordinates of the center of the pond, and R is its radius. The influence function applies to a unit infiltration rate through the bottom of the pond. Hence, G_p must fulfill the following differential equations:

$$0 \le r \le R, \qquad \nabla^2 G_p = -1 \tag{2.252}$$

$$R \le r < \infty, \qquad \nabla^2 G_p = 0, \tag{2.253}$$

where

$$r = \sqrt{(x-x_1)^2 + (y-y_1)^2}. \tag{2.254}$$

We obtain the expression for G_p valid below the pond from (2.216) with $N = 1$ and $\Phi_0 = 0$:

$$0 \le r \le R, \qquad G_p = -\tfrac{1}{4}(r^2 - R^2). \tag{2.255}$$

The discharge at $r = R$ is directed normal to the boundary of the pond because the flow is radial, and the total discharge flowing out of the pond of unit infiltration rate is Q, with

$$Q = \pi R^2. \tag{2.256}$$

Continuity of flow through the wall of the cylinder of radius R about the center of the pond requires that the infiltrated water enters the area outside this cylinder radially. Furthermore, G_p must fulfill Laplace's equation outside the pond, and the potential for a recharge well with $Q = -\pi R^2$ meets the conditions:

$$R \le r < \infty, \qquad G_p = -\frac{\pi R^2}{2\pi}\ln\frac{r}{R} = -\frac{R^2}{2}\ln\frac{r}{R}. \tag{2.257}$$

Expressions (2.255) and (2.257) both vanish at $r = R$, so that the potential is continuous at $r = R$. The function G_p is:

$$0 \le r \le R, \qquad G_p(x,y;x_1,y_1,R) = -\tfrac{1}{4}[(x-x_1)^2 + (y-y_1)^2 - R^2] \tag{2.258}$$

$$R \le r < \infty, \qquad G_p(x,y;x_1,y_1,R) = -\frac{R^2}{4}\ln\frac{(x-x_1)^2 + (y-y_1)^2}{R^2}. \tag{2.259}$$

2.6.7 Application: A Circular Pond of Infiltration in a Field of Uniform Flow

We consider the case of an infiltration pond in a field of uniform flow of discharge Q_{x0}. Far away from the pond, the flow is uniform in the x-direction and has a discharge Q_{x0}. We henceforth refer to flow fields near infinity as *far fields*. We examine under what circumstances stagnation points exist, and what their locations are. The potential for uniform flow with a pond of radius R and centered at $x = -d, y = 0$, has the form

$$\Phi = -Q_{x0}x + N_1 G_p(x,y;-d,0,R). \tag{2.260}$$

We determine the location of a stagnation point by differentiating the potential with respect to x and y, and then setting these derivatives to zero. By symmetry, Q_y is zero along the x-axis; we do not need to compute Q_y and set it to zero, because this is enforced by symmetry. We differentiate (2.260) with respect to x for $y=0$ and obtain

$$\frac{\partial \Phi}{\partial x} = -Q_{x0} + N_1 \frac{\partial G_p(x,0;-d,0,R)}{\partial x}. \tag{2.261}$$

The expression for the derivative of the function G_p depends on whether the stagnation point is underneath the pond or not. We first consider that the stagnation point is underneath the pond. In that case, the expression for the partial derivative of G_p with respect to x along $y=0$ is

$$\frac{\partial G_p}{\partial x} = -\tfrac{1}{2}(x+d), \qquad |x+d| \leq R. \tag{2.262}$$

We set the discharge component Q_x equal to zero at the stagnation point, denote the x-coordinate of the stagnation point by x_s, and obtain

$$Q_x(x_s,0) = Q_{x0} + \tfrac{1}{2}N_1(x_s+d) = 0, \qquad |x_s+d| \leq R. \tag{2.263}$$

We solve for x_s and note that the stagnation point is underneath the pond

$$x_s + d = -\frac{2Q_{x0}}{N_1}, \qquad |x_s+d| = 2\left|\frac{Q_{x0}}{N_1}\right| \leq R. \tag{2.264}$$

We rewrite the inequality in dimensionless form

$$x_s + d = -\frac{2Q_{x0}}{N_1}, \qquad \left|\frac{2Q_{x0}}{N_1 R}\right| \leq 1. \tag{2.265}$$

Note that x_s+d is negative for positive Q_{x0}/N_1; i.e., the stagnation point is upstream from the center of the pond.

We consider next that the stagnation point is not underneath the pond. In that case, the derivative of G_p with respect to x along $y=0$ is obtained from (2.259) to be

$$\frac{\partial G_p(x,0;-d,0,R)}{\partial x} = -\frac{R^2}{2}\frac{1}{x+d}. \tag{2.266}$$

Again, we set the discharge component Q_x equal to zero at $x=x_s$ and $y=0$:

$$Q_x(x_s,0) = Q_{x0} + \frac{N_1 R^2}{2(x_s+d)} = 0, \qquad |x_s+d| \geq R. \tag{2.267}$$

We solve this for x_s+d:

$$x_s + d = -\frac{N_1 R^2}{2Q_{x0}}, \qquad |x_s+d| = \left|\frac{N_1 R^2}{2Q_{x0}}\right| \geq R, \tag{2.268}$$

or, rewriting the inequality in dimensionless form,

$$x_s + d = -\frac{N_1 R^2}{2Q_{x0}}, \qquad \left|\frac{N_1 R}{2Q_{x0}}\right| \geq 1. \tag{2.269}$$

Note that when (2.265) is satisfied, i.e., if there is a stagnation point underneath the pond, there will also be one outside the pond, since (2.269) will be satisfied also. Thus, there are two possibilities. If $|2Q_{x0}/(N_1 R)| < 1$, there will be two stagnation points, one below the pond and one outside the area of the pond (it will be upstream from the pond if N_1/Q_{x0} is positive). Conversely, if $|2Q_{x0}/(N_1 R)| \geq 1$, there will be no stagnation point at all, with the exception of the case in which $2Q_{x0}/(N_1 R) = 1$, in which case the two stagnation points coincide; there is a single stagnation point on the boundary of the pond on the upstream side.

2.6.8 Application: A Circular Pond of Infiltration in a Field of Uniform Flow with a Well

It is sometimes of interest to capture water infiltrated over a circular area, for example if irrigation via a center pivot system for agricultural purposes carries contaminants. For such cases we place a well downstream to capture the infiltrated water; we are interested to delineate the contaminated zone, bounded by the dividing streamlines and extending to the stagnation point downstream from the well. We determine the location of the stagnation point, which bounds the area of infiltrated water on the downstream side.

The potential for a circular center pivot system at $x = -d, y = 0$, of infiltration rate N_1 with a well of discharge Q at $x = d, y = 0$, is

$$\Phi = -Q_{x0}x + N_1 G_p(x,y;-d,0,R) + \frac{Q}{4\pi}\ln[(x-d)^2 + y^2] + C. \tag{2.270}$$

We differentiate the potential with respect to x:

$$\frac{\partial \Phi}{\partial x} = -Q_{x0} + N_1 \frac{\partial G_p(x,y;-d,0,R)}{\partial x} + \frac{Q}{4\pi}\frac{2(x-d)}{(x-d)^2 + y^2}. \tag{2.271}$$

The expression for the discharge along the x-axis is

$$Q_x(x,0) = -\frac{\partial \Phi}{\partial x} = Q_{x0} - N_1 \frac{\partial G_p(x,0;-d,0,R)}{\partial x} - \frac{Q}{2\pi}\frac{1}{x-d}. \tag{2.272}$$

The purpose of the well is to capture all of the infiltrated water, i.e.,

$$Q = \pi N_1 R^2. \tag{2.273}$$

Stagnation Point below the Pond

We first assume that there is a stagnation point below the pond; in that case the derivative of G_p is given by (2.262), and we set the expression for the discharge at $x = x_s, y = 0$, equal to zero:

$$Q_x(x_s, 0) = Q_{x0} + \tfrac{1}{2}N_1(x_s + d) - \tfrac{1}{2}N_1 R^2 \frac{1}{x_s - d} = 0. \tag{2.274}$$

We multiply both sides by $2(x_s - d)/N_1$:

$$x_s^2 + \frac{2Q_{x0}}{N_1}x_s - \left[d^2 + R^2 + \frac{2Q_{x0}}{N_1}d\right] = 0. \tag{2.275}$$

This quadratic equation has the following two roots:

$$(x_s)_{1,2} = -Q_{x0}/N_1 \pm \sqrt{(Q_{x0}/N_1)^2 + 2(Q_{x0}/N_1)d + d^2 + R^2} \tag{2.276}$$

or

$$(x_s)_{1,2} = -Q_{x0}/N_1 \pm \sqrt{(Q_{x0}/N_1 + d)^2 + R^2}. \tag{2.277}$$

The root with the plus sign never corresponds to a point underneath the pond; we demonstrate this by writing the condition for the stagnation point to be below the pond as

$$(x_s)_2 = -Q_{x0}/N_1 + \sqrt{(Q_{x0}/N_1 + d)^2 + R^2} \le -d + R \tag{2.278}$$

or

$$\sqrt{(Q_{x0}/N_1 + d)^2 + R^2} \le Q_{x0}/N_1 - d + R. \tag{2.279}$$

This condition cannot be fulfilled. We consider the negative root next, i.e., the root that would correspond to a stagnation point upstream from the center of the pond:

$$x_s = -Q_{x0}/N_1 - \sqrt{(Q_{x0}/N_1 + d)^2 + R^2}. \tag{2.280}$$

The stagnation point will fall in the area below the pond if

$$-R \le x_s + d \le R \tag{2.281}$$

or

$$-R \le -Q_{x0}/N_1 + d - \sqrt{(Q_{x0}/N_1 + d)^2 + R^2} \tag{2.282}$$

and

$$-Q_{x0}/N_1 + d - \sqrt{(Q_{x0}/N_1 + d)^2 + R^2} \le R. \tag{2.283}$$

It can be verified that the second equation is always satisfied, provided that $d > R$ and that $Q_{x0}/N_1 > 0$. We use the equals sign in the first equation, and obtain, after adding $Q_{x0}/N_1 - d$ to both sides and squaring,

$$(Q_{x0}/N_1 - d)^2 - 2R(Q_{x0}/N_1 - d) + R^2 = (Q_{x0}/N_1 + d)^2 + R^2 \tag{2.284}$$

or

$$-4Q_{x0}d/N_1 - 2RQ_{x0}/N_1 + 2Rd = 0. \tag{2.285}$$

We solve this equation for Q_{x0}/N_1 and obtain

$$\frac{Q_{x0}}{N_1} = \frac{Rd}{R+2d}. \tag{2.286}$$

A stagnation point below the pond will occur if the uniform flow divided by the infiltration rate is less than the critical value given by (2.286), i.e.,

$$\frac{Q_{x0}}{N_1} \leq \frac{Rd}{R+2d}. \tag{2.287}$$

Stagnation Point Outside the Pond

If the stagnation point falls outside the area below the pond, the derivative of the function G_p is, for $y = 0$:

$$N_1 \frac{\partial G_p(x,y;-d,0,R)}{\partial x} = -\tfrac{1}{2}N_1R^2\frac{1}{x+d}. \tag{2.288}$$

The expression for the discharge $Q_x(x,0)$ now becomes

$$Q_x(x,0) = Q_{x0} + \tfrac{1}{2}N_1R^2\frac{1}{x+d} - \tfrac{1}{2}N_1R^2\frac{1}{x-d}. \tag{2.289}$$

We set this expression equal to zero for $x = x_s$:

$$Q_x(x_s,0) = Q_{x0} + \tfrac{1}{2}N_1R^2\frac{1}{x_s+d} - \tfrac{1}{2}N_1R^2\frac{1}{x_s-d} = 0. \tag{2.290}$$

We multiply both sides by $x_s^2 - d^2$:

$$Q_{x0}(x_s^2 - d^2) + \tfrac{1}{2}N_1R^2(x_s - d - x_s - d) = 0 \tag{2.291}$$

or

$$Q_{x0}x_s^2 = Q_{x0}d^2 + N_1dR^2. \tag{2.292}$$

We divide both sides by Q_{x0} and solve for x_s:

$$(x_s)_{1,2} = \pm\sqrt{\frac{N_1R^2d}{Q_{x0}} + d^2}. \tag{2.293}$$

We consider that the stagnation point is upstream from the center of the pond ($x_s < 0$), and we obtain for the quantity $-x_s - d$

$$-x_s - d = -d + \sqrt{\frac{N_1 R^2 d}{Q_{x0}} + d^2}. \tag{2.294}$$

The expression to the right of the equals sign is always positive, that is, the stagnation point is upstream from the pond, as assumed. The modulus of $-x_s - d$ must be greater than or equal to R in order that the stagnation point indeed is outside the area below the pond. Thus, the condition for a stagnation point to occur is

$$-d + \sqrt{\frac{N_1 R^2 d}{Q_{x0}} + d^2} \geq R. \tag{2.295}$$

We add d to both sides of the equation and square both sides, to obtain

$$\frac{N_1 R^2 d}{Q_{x0}} + d^2 \geq (R+d)^2 \tag{2.296}$$

or

$$\frac{N_1 R^2 d}{Q_{x0}} \geq R^2 + 2Rd, \tag{2.297}$$

which yields, after division by $R^2 d$:

$$\frac{N_1}{Q_{x0}} \geq \frac{R+2d}{Rd}. \tag{2.298}$$

We rewrite this equation as

$$\frac{Q_{x0}}{N_1} \leq \frac{Rd}{R+2d}, \tag{2.299}$$

which corresponds exactly to the condition that a stagnation point exists below the pond, (2.287).

The positive root of (2.293) is

$$x_s = \sqrt{\frac{N_1 R^2 d}{Q_{x0}} + d^2}. \tag{2.300}$$

This stagnation point is always downstream from the well, since $|x_s| > d$ and d must be greater than R for the well to be outside the infiltration area. Thus, there will always be a stagnation point downstream from the well. If condition (2.299) is satisfied, then there exist an additional two stagnation points: one below the pond, and one outside the area below the pond. When the equals sign prevails in (2.299), the two stagnation points coincide, and lie exactly at $x = -d - R$. If (2.299) is not satisfied, only a single stagnation point exists, downstream from the

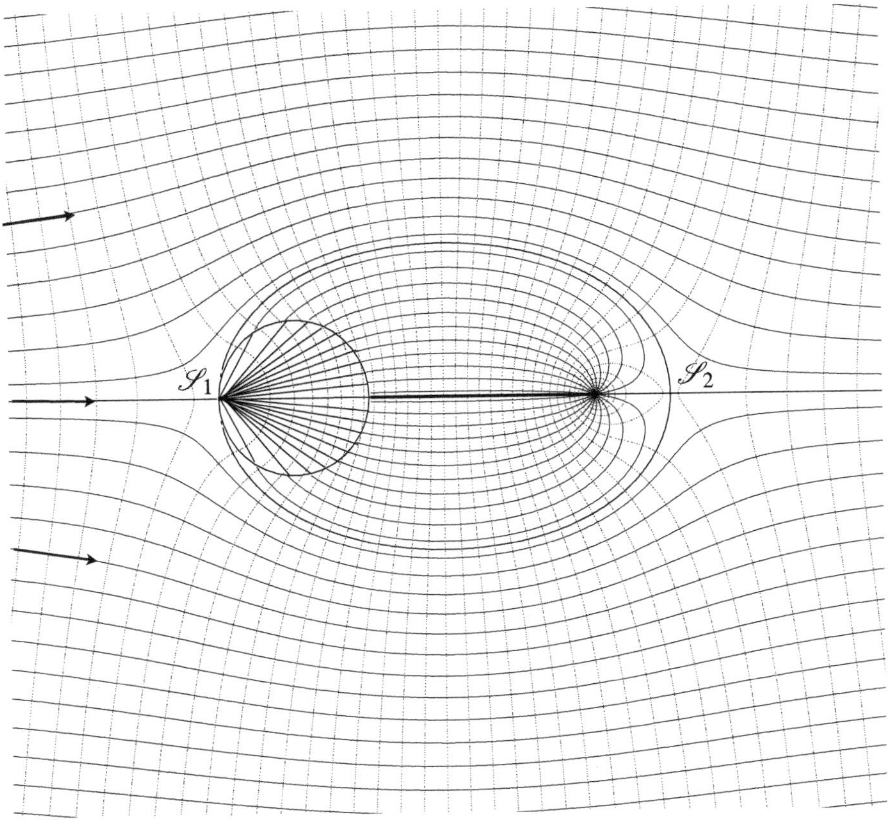

Figure 2.35 A pond with a well downstream, which captures all of the infiltrated water; the curve through the two stagnation points, $\mathscr{S}_1$ and $\mathscr{S}_2$, the streamline $\Psi = 0$, bounds the plume.

well. As an example, a flow net is shown in Figure 2.35 for a stagnation point just upstream from the pond, i.e., (2.299) applies. The data are $Q_{x0}/(kH) = 0.4$, $R/d = 0.5$, and $N_1 = Q_{x0}(R+2d)/(Rd)$. Since (2.299) applies, the expression for x_s reduces to

$$x_s = \sqrt{\frac{R+2d}{Rd}R^2d + d^2} = \sqrt{R^2 + 2dR + d^2} = R + d. \tag{2.301}$$

The two stagnation points, the one upstream from the center of the pond and the one downstream from the well, are at the same distance from the origin.

2.6.9 Application: A Circular Pond in a Field of Uniform Flow near a River

As a second application, we solve the problem of flow from a circular pond of radius R and infiltration rate N_1 with its center at a distance d from the bank of a

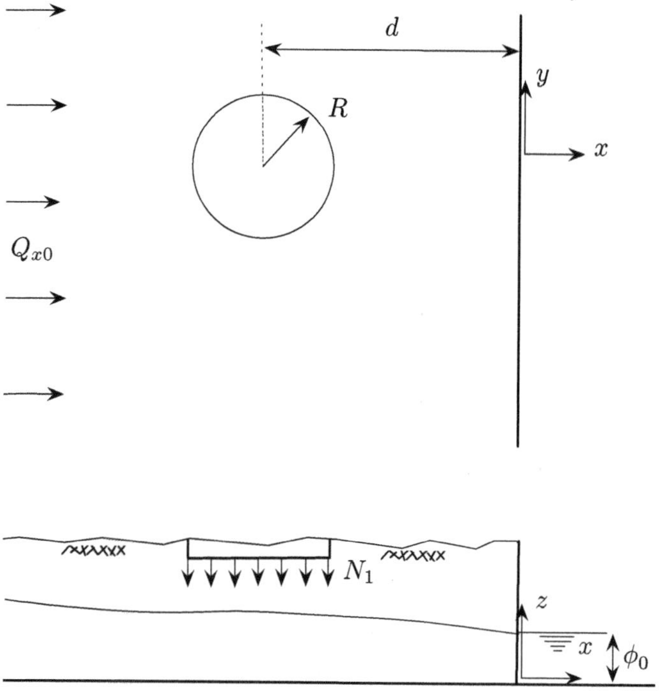

Figure 2.36 Flow from a circular pond toward a river.

long river, shown in Figure 2.36. The head along the river bank is ϕ_0. An (x,y,z)-coordinate system is chosen as indicated, with the y-axis along the river bank, the x-axis pointing away from the aquifer, and the negative x-axis passing through the center of the pond.

The function $G_p(x,y,x_1,y_1,R)$ for this case becomes

$$0 \le r \le R, \qquad G_p(x,y;-d,0,R) = -\tfrac{1}{4}\left[(x+d)^2 + y^2 - R^2\right] \tag{2.302}$$

$$R \le r < \infty, \qquad G_p(x,y;-d,0,R) = -\frac{R^2}{4}\ln\frac{(x+d)^2+y^2}{R^2}, \tag{2.303}$$

where

$$r^2 = (x+d)^2 + y^2. \tag{2.304}$$

The potential due to the flow from the pond and valid in the area outside the pond equals that of a recharge well of strength $\pi N_1 R^2$. We meet the boundary condition along the river bank,

$$x = 0, \qquad \phi = \phi_0, \qquad \Phi = \Phi_0 = \tfrac{1}{2}k\phi_0^2, \tag{2.305}$$

by placing an image pond of discharge $\pi N_1 R^2$ at $x = d, y = 0$. Since the pond will not intersect the river bank, all points in the aquifer will be in the area outside the image discharge pond, which can thus always be represented by an image discharge well for points in the aquifer. The complete complex potential for this problem is

$$\Phi = -Q_{x0}x + N_1 G_p(x,y;-d,0,R) + \frac{N_1 R^2}{4} \ln \frac{(x-d)^2 + y^2}{R^2} + C. \tag{2.306}$$

The constant C is found by applying the boundary condition (2.305) to (2.306). Since $r > R$ along the river bank, (2.303) represents G_p there and we obtain

$$\Phi_0 = \frac{-N_1 R^2}{4} \ln \frac{d^2 + y^2}{R^2} + \frac{N_1 R^2}{4} \ln \frac{d^2 + y^2}{R^2} + C, \tag{2.307}$$

so that

$$C = \Phi_0. \tag{2.308}$$

We substitute this expression for C in the expression for Φ, (2.306)

$$\Phi = -Q_{x0}x + N_1 G_p(x,y;-d,0,R) + \frac{N_1 R^2}{4} \ln \frac{(x-d)^2 + y^2}{R^2} + \Phi_0, \tag{2.309}$$

where G_p is defined by (2.302) and (2.303).

Stagnation Points

The case of the pond near the river is mathematically equivalent to that of a pond and a discharge well that captures all of its water; the line between the pond and the well in Figure 2.35 that bisects the line connecting the center of the pond and the well, the y-axis, can be interpreted as a river. Thus, the results obtained for that case apply to the case of a pond near a river as well.

Problem

2.19 Infiltration occurs at a rate N_1 through the bottom of a circular pond of radius R in an infinite aquifer. If the pond were not present, there would be no flow. A well is drilled through the bottom of the pond and screened in the aquifer. Determine the location and the discharge of this well such that exactly all infiltrated water is captured. Determine an expression for the potential as a function of position.

Problem

2.20 Write a computer program to verify your answer to Problem 2.17. Choose k =10 m/day, ϕ_1= 15 m, $\phi_2 = 20$ m, $b = 40$ m, and $L = 1000$ m. Use your program to plot the phreatic surface for all three cases considered.

Problem

2.21 Write routines for the potential for a pond at an arbitrary location and unit infiltration rate, a well, and uniform flow. Write a routine capable of computing the discharge vector and hydraulic head for the case of a pond in a field of uniform flow. Choose $k = 10$ m/d, $Q_{x0} = 0.4$ m^2/d, $R = 100$ m, and the head at $x = L$, $L = 1000$ m, is $\phi_0 = 25$ m. Use the formulas presented in the text for the coordinates of the stagnation point to choose N_1 such that there is a single stagnation point, that there are two stagnation points, and that there is no stagnation point. Check your code by verifying that the discharge is indeed zero at the stagnation point(s) when they occur. Produce flow nets for all three cases.

Problem

2.22 Use the computer model you developed for problem 2.20 to simulate the case of flow of a pond in a field of uniform flow with a well. Use your model to verify the solution presented in the text for this case; choose the parameters as given for Problem 2.20 and answer the same questions for this case.

2.7 Vertically Varying Hydraulic Conductivity

We consider aquifer systems that may be viewed as a stack of aquifers of different hydraulic conductivities, sometimes separated by layers of very low hydraulic conductivity. We call the separating strata *aquicludes* if they are impermeable, and *leaky layers* or *aquitards* if their hydraulic conductivity is not zero, but much less than the hydraulic conductivities of the aquifers. If the vertical variations in the hydraulic conductivity are relatively small, and the resistance to flow in the vertical direction may be neglected, the head will vary little in the vertical direction. Such problems may be solved by the use of potentials first introduced by Girinskii [1946a]. Problems where different heads must be assigned to each aquifer in the system will be discussed in Chapter 5, Section 5.2.

Using discharge potentials for stratified aquifers simplifies the governing equations to such an extent that the equations for the potentials and discharges in terms of position are not affected by the stratification. The geometry of the strata and their hydrological properties enter only via boundary conditions and the equations that relate the discharge potential to the hydraulic head. We present discharge potentials in a form somewhat different from that introduced by Girinskii, emphasizing continuity of the potentials across interzonal boundaries. We then present the potential as introduced by Girinskii, along with a generalization due to Youngs [1971].

Strack and Ausk [2015] present discharge potentials for vertically integrated flow in stratified coastal aquifers, and demonstrate that the results for the vertically integrated flow are exact, provided that the discharge potentials are continuous throughout the domain, and the boundaries are vertical. Strack [2017] developed exact expressions for the discharge potentials valid for flow in aquifer systems. We present a simplified approach here, but it should be borne in mind that the discharges computed are accurate, regardless of the contrast in hydraulic conductivities, provided that the conditions outlined in this section are met. The hydraulic head in the strata will be realistic only if the contrast in hydraulic conductivity is not too large.

2.7.1 Continuous Discharge Potentials

The two components of the discharge vector are defined by the relation:

$$
\begin{aligned}
Q_x &= \int_{Z_b}^{Z_t} q_x(x,y,z)dz \\
Q_y &= \int_{Z_b}^{Z_t} q_y(x,y,z)dz,
\end{aligned}
\qquad (2.310)
$$

where Z_b and Z_t represent the lower and upper boundaries of the saturated zone of flowing groundwater in the aquifer. Note that equations (2.310) are valid for three-dimensional flow. We take the hydraulic conductivity k to be a function only of the coordinate in vertical direction, z:

$$
k = k(z). \qquad (2.311)
$$

We use Darcy's law, and write (2.310) as

$$
\begin{aligned}
Q_x &= -\int_{Z_b}^{Z_t} k(z)\frac{\partial\phi}{\partial x}dz \\
Q_y &= -\int_{Z_b}^{Z_t} k(z)\frac{\partial\phi}{\partial y}dz.
\end{aligned}
\qquad (2.312)
$$

According to the Dupuit-Forchheimer approximation, ϕ is independent of z, so that:

$$
\begin{aligned}
Q_x &= -\frac{\partial\phi}{\partial x}\int_{Z_b}^{Z_t} k(z)dz \\
Q_y &= -\frac{\partial\phi}{\partial y}\int_{Z_b}^{Z_t} k(z)dz.
\end{aligned}
\qquad (2.313)
$$

We assume that Z_t and Z_b are functions of ϕ alone:

$$\begin{aligned} Z_t &= Z_t(\phi) \\ Z_b &= Z_b(\phi). \end{aligned} \tag{2.314}$$

All types of flow discussed so far satisfy this condition. We define the transmissivity, T, of the aquifer as

$$\boxed{T = \int_{Z_b}^{Z_t} k(z)dz = T(\phi)} \tag{2.315}$$

which is a function of ϕ alone, as follows from (2.314). We now write the expressions (2.313) for the components of the discharge vector as

$$\begin{aligned} Q_x &= -T(\phi)\frac{\partial \phi}{\partial x} = -\frac{\partial \Phi}{\partial x} \\ Q_y &= -T(\phi)\frac{\partial \phi}{\partial y} = -\frac{\partial \Phi}{\partial y}, \end{aligned} \tag{2.316}$$

where the discharge potential Φ is

$$\boxed{\Phi = \int T(\phi)d\phi + C} \tag{2.317}$$

The constant of integration, C, must be chosen such that the potential is continuous across interzonal boundaries. Since the gradient of the discharge potential is continuous throughout the flow domain, the discharge vector is continuous also, including across the interzonal boundaries.

The components of the specific discharge vector are functions of z because the hydraulic conductivity is a function of z:

$$\begin{aligned} q_x &= -k(z)\frac{\partial \phi}{\partial x} \\ q_y &= -k(z)\frac{\partial \phi}{\partial y}. \end{aligned} \tag{2.318}$$

We express these components in terms of the components of the discharge vector by the use of (2.316):

$$\begin{aligned} q_x &= \frac{k(z)}{T}Q_x \\ q_y &= \frac{k(z)}{T}Q_y. \end{aligned} \tag{2.319}$$

We obtain all discharge potentials presented so far by the use of (2.317). For example, if the flow is unconfined with a base at $Z_b = 0$ and the hydraulic conductivity is uniform, then

$$T(\phi) = \int_0^{\phi} k d\phi = k\phi, \qquad (2.320)$$

so that (2.317) yields:

$$\Phi = \int k\phi d\phi + C = \tfrac{1}{2}k\phi^2 + C. \qquad (2.321)$$

2.7.2 Flow in an Aquifer with an Arbitrary Number of Strata

We present an approach to include the effect of strata on the flow in an approximate fashion. The analysis is valid only if the differences in the hydraulic conductivities are small enough that variations of the head in the vertical direction can be neglected. We consider an aquifer between two parallel impervious horizontal boundaries, but allow for combined confined/unconfined flow in the uppermost stratum.

We number the strata as $1, 2, \ldots, n$, where stratum 1 is the bottom stratum, represent the base of stratum j as b_j, and denote the base of the upper confining bed as b_{n+1}. The thickness, H_n, of stratum j is

$$H_j = b_{j+1} - b_j, \qquad j = 1, 2, \ldots, n. \qquad (2.322)$$

The hydraulic conductivity is uniform within each stratum, and is represented for stratum j as k_j. If stratum j is fully saturated, then its transmissivity is T_j, with

$$T_j = k_j H_j = k_j(b_{j+1} - b_j). \qquad (2.323)$$

The head at any point of the aquifer may vary between the base elevation (b_1) and values larger than the elevation of the upper confining bed (b_{n+1}). If ϕ is equal to some value between b_m and b_{m+1} ($m = 1, 2, \ldots, n$), then the strata $m+1, m+2, \ldots, n$ do not contribute to the transmissivity. Hence, if T represents the total transmissivity of the aquifer, then

$$T = \sum_{j=1}^{m-1} T_j + k_m(\phi - b_m), \qquad b_m \leq \phi \leq b_{m+1}. \qquad (2.324)$$

We adopt the convention that summations are not to be carried out (i.e., are set to zero) if the upper bound ($m-1$ in this case) is less than the lower one. We express the potential Φ in terms of the head ϕ by the use of (2.317) and (2.324) as follows:

$$\Phi = \int \left[\sum_{j=1}^{m-1} T_j + k_m(\phi - b_m) \right] d\phi + C_m, \qquad b_m \le \phi \le b_{m+1}. \tag{2.325}$$

We integrate this:

$$\Phi = \sum_{j=1}^{m-1} T_j\phi + \tfrac{1}{2}k_m(\phi - b_m)^2 + C_m, \qquad b_m \le \phi \le b_{m+1}. \tag{2.326}$$

where the index m in the constant C_m denotes that the aquifer is saturated up to a level somewhere in stratum m.

We choose the constants C_m $(m = 1, 2, \ldots, n)$ such that the potential varies continuously if the head varies continuously between all possible levels in the aquifer. We accomplish this by chosing C_m such that (2.326) becomes:

$$\Phi = \sum_{j=1}^{m-1} \left[T_j(\phi - b_j) - \tfrac{1}{2}k_j(b_{j+1} - b_j)^2 \right] + \tfrac{1}{2}k_m(\phi - b_m)^2,$$

$$b_m \le \phi \le b_{m+1} \quad m = 2, 3, \ldots, n \tag{2.327}$$

$$\Phi = \tfrac{1}{2}k_1(\phi - b_1)^2, \qquad b_1 \le \phi \le b_2 \quad m = 1. \tag{2.328}$$

We demonstrate that the potential (2.327) indeed varies continuously across the interzonal boundary where the groundwater table intersects the base of stratum m. We determine the potential for $b_{m-1} \le \phi \le b_m$ $(m > 1)$, and show that the resulting expression is equal to (2.327) for the limiting case that $\phi = b_m$:

$$\Phi = \sum_{j=1}^{m-2} \left[T_j(\phi - b_j) - \tfrac{1}{2}k_j(b_{j+1} - b_j)^2 \right] + \tfrac{1}{2}k_{m-1}(\phi - b_{m-1})^2,$$

$$b_{m-1} \le \phi \le b_m, \qquad m > 1. \tag{2.329}$$

We set $\phi = b_m$ in (2.327) and (2.328),

$$\Phi_m = \sum_{j=1}^{m-1} \left[T_j(b_m - b_j) - \tfrac{1}{2}k_j(b_{j+1} - b_j)^2 \right], \qquad m > 1$$

$$\Phi_1 = 0, \tag{2.330}$$

where Φ_m represents the value of the potential for $\phi = b_m$, and do the same with (2.329):

$$\Phi_m = \sum_{j=1}^{m-2} \left[T_j(b_m - b_j) - \tfrac{1}{2}k_j(b_{j+1} - b_j)^2 \right] + \tfrac{1}{2}k_{m-1}(b_m - b_{m-1})^2, \quad m > 1. \tag{2.331}$$

The difference between the latter two expressions for Φ_m is

$$T_{m-1}(b_m - b_{m-1}) - \tfrac{1}{2}k_{m-1}(b_m - b_{m-1})^2 - \tfrac{1}{2}k_{m-1}(b_m - b_{m-1})^2 = 0, \qquad (2.332)$$

which is zero, since $T_{m-1} = k_{m-1}(b_m - b_{m-1})$; the potential varies continuously across the interzonal boundary where the phreatic surface passes through the base of stratum m, as asserted. Since this is true for any m between 1 and n, the potential (2.327) is valid for any position of the phreatic surface below the upper boundary.

If the aquifer is entirely confined, then the potential becomes

$$\boxed{\Phi = \sum_{j=1}^{n} \left[T_j(\phi - b_j) - \tfrac{1}{2}k_j(b_{j+1} - b_j)^2 \right], \qquad b_{n+1} \le \phi} \qquad (2.333)$$

We can use the potential given by either (2.327) or (2.333) to model flow in a stratified system; the potential obeys either the Laplace or the Poisson equation, depending on whether or not there is infiltration. We solve the governing equation for the potential in the usual manner, but the boundary conditions must be expressed in terms of the potential.

Once the potential is known as a function of position, we invert either (2.327) or (2.333) to compute the head ϕ. The flow is confined in all strata if $\Phi > \Phi_{n+1}$; we obtain Φ_{n+1} from (2.333):

$$\Phi_{n+1} = \sum_{j=1}^{n} \left[T_j(b_{n+1} - b_j) - \tfrac{1}{2}k_j(b_{j+1} - b_j)^2 \right]. \qquad (2.334)$$

We obtain an expression for the head in terms of the potential from (2.333):

$$\phi = \frac{\Phi + \sum_{j=1}^{n} \left[T_j b_j + \tfrac{1}{2}k_j(b_{j+1} - b_j)^2 \right]}{\overset{n}{T}}, \qquad \Phi > \Phi_{n+1}, \qquad (2.335)$$

where

$$\overset{n}{T} = \sum_{j=1}^{n} T_j. \qquad (2.336)$$

We write (2.335) alternatively:

$$\phi = \frac{\Phi - \sum_{j=1}^{n} \left[T_j(b_{n+1} - b_j) - \tfrac{1}{2}k_j(b_{j+1} - b_j)^2 \right]}{\overset{n}{T}} + \frac{\sum_{j=1}^{n} T_j b_{n+1}}{\overset{n}{T}}, \qquad \Phi > \Phi_{n+1} \qquad (2.337)$$

or

$$\boxed{\phi = b_{n+1} + \frac{\Phi - \Phi_{n+1}}{\overset{n}{T}}, \qquad \Phi > \Phi_{n+1}} \qquad (2.338)$$

where we used (2.334) and (2.336).

If the potential has some value between Φ_m and Φ_{m+1}, then (2.327) applies, which we write as:

$$\Phi = \sum_{j=1}^{m-1} \left\{ T_j[(\phi - b_m) + (b_m - b_j)] - \tfrac{1}{2}k_j(b_{j+1} - b_j)^2 \right\} + \tfrac{1}{2}k_m(\phi - b_m)^2,$$

$$\Phi_m \le \Phi \le \Phi_{m+1}, \quad m > 1, \tag{2.339}$$

or, with (2.330),

$$\boxed{\Phi = \sum_{j=1}^{m-1} T_j(\phi - b_m) + \Phi_m + \tfrac{1}{2}k_m(\phi - b_m)^2, \qquad \Phi_m \le \Phi \le \Phi_{m+1}} \tag{2.340}$$

We write this as a quadratic equation for $(\phi - b_m)$:

$$\tfrac{1}{2}k_m(\phi - b_m)^2 + \sum_{j=1}^{m-1} T_j(\phi - b_m) - (\Phi - \Phi_m) = 0, \qquad \Phi_m \le \Phi \le \Phi_{m+1}. \tag{2.341}$$

We solve this for $(\phi - b_m)$ in terms of Φ:

$$(\phi - b_m)_{1,2} = \frac{-\overset{m-1}{T} \pm \sqrt{\left(\overset{m-1}{T}\right)^2 + 2k_m(\Phi - \Phi_m)}}{k_m}, \qquad \Phi_m \le \Phi \le \Phi_{m+1}, \tag{2.342}$$

where $\overset{m-1}{T}$ is defined by (2.336) (with $m-1$ replacing n). Since $\phi - b_m$ is positive, the plus sign applies, so that

$$\boxed{\phi = b_m - \frac{\overset{m-1}{T}}{k_m} + \sqrt{\left(\frac{\overset{m-1}{T}}{k_m}\right)^2 + \frac{2(\Phi - \Phi_m)}{k_m}}, \qquad \Phi_m \le \Phi \le \Phi_{m+1}} \tag{2.343}$$

We solve problems in stratified aquifers in steps: first, we express boundary conditions in terms of the potential, Φ. Second, we solve the boundary-value problem in terms of Φ, which satisfies the Poisson equation $\nabla^2\Phi = -N$ where N represents recharge. Finally, we solve for the head ϕ, using (2.343).

2.7.3 Girinskii Potentials

The potentials introduced by Girinskii are rather similar to the potential (2.317); they differ in form and do not contain the additive constant, but otherwise render the same result.

For the derivation of Girinskii potentials use is made of Leibnitz's rule for differentiation of a definite integral (see, for example, Abramowitz and Stegun [1965]):

$$\frac{\partial}{\partial x}\int_{Z_b}^{Z_t} F(x,y,z)dz = \int_{Z_b}^{Z_t} \frac{\partial}{\partial x}F(x,y,z)dz + F(x,y,Z_t)\frac{\partial Z_t}{\partial x} - F(x,y,Z_b)\frac{\partial Z_b}{\partial x}. \tag{2.344}$$

Since $k(z)$ is independent of x and y, the integrands in (2.312) may be interpreted as the derivatives $\partial F/\partial x$ and $\partial F/\partial y$, and the application of Leibnitz's rule to (2.312) yields

$$Q_x = -\frac{\partial}{\partial x}\int_{Z_b}^{Z_t} k(z)\phi dz + k(Z_t)\phi_t\frac{\partial Z_t}{\partial x} - k(Z_b)\phi_b\frac{\partial Z_b}{\partial x}$$

$$Q_y = -\frac{\partial}{\partial y}\int_{Z_b}^{Z_t} k(z)\phi dz + k(Z_t)\phi_t\frac{\partial Z_t}{\partial y} - k(Z_b)\phi_b\frac{\partial Z_b}{\partial y}, \tag{2.345}$$

where

$$\phi_t = \phi(x,y,Z_t), \qquad \phi_b = \phi(x,y,Z_b). \tag{2.346}$$

Note that these expressions are valid for three-dimensional flow. Girinskii potentials are obtained by adopting the Dupuit-Forchheimer approximation. If the flow is unconfined, then $Z_t = h$, $Z_b = 0$, and $\phi = \phi_t = h$ so that (2.345) may be written as

$$Q_x = -\frac{\partial}{\partial x}\int_0^{\phi} k(z)\phi dz + k(\phi)\phi\frac{\partial \phi}{\partial x} = -\frac{\partial}{\partial x}\int_0^{\phi} k(z)(\phi - z)dz$$

$$Q_y = -\frac{\partial}{\partial y}\int_0^{\phi} k(z)\phi dz + k(\phi)\phi\frac{\partial \phi}{\partial y} = -\frac{\partial}{\partial y}\int_0^{\phi} k(z)(\phi - z)dz. \tag{2.347}$$

Note that Leibniz's rule is applied in reverse order to write the second term in each equation as the derivative of an integral. The Girinskii potential thus becomes

$$\Phi = \int_0^{h} k(z)(\phi - z)dz. \tag{2.348}$$

If k is constant, the familiar potential for unconfined flow is obtained:

$$\Phi = k\phi^2 - \tfrac{1}{2}k\phi^2 = \tfrac{1}{2}k\phi^2. \tag{2.349}$$

Equation (2.348) is applicable to cases of confined flow provided that the upper limit in the integral is set equal to the aquifer thickness H. That this is true does not follow from (2.345) directly: setting $Z_t = H$ and $Z_b = 0$ yields only the first term of the integrand in (2.348). However, if $h = H$ the second term in this integrand yields a constant on integration and thus may be included. This constant, fortuitously, makes the potential continuous with (2.348) across the interzonal boundary. Setting $h = H$ in (2.348) we obtain for $k(z) = k =$ constant:

$$\Phi = kH\phi - \tfrac{1}{2}kH^2. \tag{2.350}$$

Girinskii also presented a potential for interface flow (Girinskii [1946b]) but this function is not continuous with (2.349) across the interzonal boundary with an unconfined zone. Strack [1976] presented a continuous discharge potential for interface flow.

Youngs [1971] did not adopt the Dupuit-Forchheimer approximation, and expressed ϕ in (2.345) in terms of the pressure p by the use of the relation $\phi = p/(\rho g) + z$. This yields the following for Q_x:

$$\begin{aligned} Q_x = &-\frac{\partial}{\partial x}\int_{Z_b}^{Z_t} k(z)\frac{p}{\rho g}dz - \frac{\partial}{\partial x}\int_{Z_b}^{Z_t} zk(z)dz \\ &+ k(Z_t)\left(\frac{p_t}{\rho g} + Z_t\right)\frac{\partial Z_t}{\partial x} - k(Z_b)\left(\frac{p_b}{\rho g} + Z_b\right)\frac{\partial Z_b}{\partial x}, \end{aligned} \tag{2.351}$$

where

$$p_t = p(x, y, Z_t), \qquad p_b = p(x, y, Z_b). \tag{2.352}$$

Application of Leibniz's rule to the second integral in (2.351) yields after rearrangement:

$$Q_x = -\frac{\partial}{\partial x}\int_{Z_b}^{Z_t} k(z)\frac{p}{\rho g}dz + k(Z_t)\frac{p_t}{\rho g}\frac{\partial Z_t}{\partial x} - k(Z_b)\frac{p_b}{\rho g}\frac{\partial Z_b}{\partial x}. \tag{2.353}$$

If the flow is unconfined, $p_t = 0$ and we obtain

$$Q_x = -\frac{\partial}{\partial x}\int_{Z_b}^{Z_t} k(z)\frac{p}{\rho g}dz - k(Z_b)\frac{p_b}{\rho g}\frac{\partial Z_b}{\partial x}. \tag{2.354}$$

A similar expression results for Q_y. Youngs applied (2.354) together with the condition of continuity of flow to establish an approximate method for modeling flow in aquifers with lower boundaries of any shape.

Of particular interest is that (2.354) is exact: no approximations have been made. For the case of interface flow, p_b may be expressed in terms of Z_b, and the right-hand side of (2.354) may be written as the derivative of a potential. For one-dimensional flow, Q_x is a constant, equal to the difference in values of the potential at the boundaries. If these boundaries are vertical and the pressures along them are known, then the exact value of Q_x can be computed. In this way Youngs demonstrated that (3.56) is exact.

Problem

2.23 Solve the problem of one-dimensional horizontal confined flow in an aquifer system consisting of two strata, each of thickness $\frac{1}{2}H$. The boundary conditions are

$$x = -\tfrac{1}{2}L, \quad \phi = \phi_1, \qquad x = \tfrac{1}{2}L, \quad \phi = \phi_2.$$

The hydraulic conductivities in the upper and lower strata are k_1 and k_2, respectively. The conditions along any interface between two strata of different hydraulic conductivities are that there is continuity of both flow and head across the interface. Solve the problem in each stratum separately and demonstrate that the conditions across the interface are met. Determine the discharge through each stratum and the total discharge through the system. Obtain a second solution to the problem by integrating the potential given in integral form by (2.317); compare the two solutions. Is the answer approximate or exact?

3

Steady Interface Flow

We refer to a flow as steady interface flow when the aquifer contains two fluids that are separated by an interface rather than a transition zone, and when one of the fluids is at rest. Coastal aquifers may be modeled with good approximation by assuming interface flow, where the freshwater flows over saltwater at rest. We call the flow *shallow interface flow* if the aquifer is sufficiently shallow with respect to its extent so that the resistance to flow in the vertical direction may be neglected; we omit the word "steady" as this chapter is limited to steady flow.

3.1 Condition along the Interface

A cross section through an aquifer with freshwater flowing over saltwater at rest is shown in Figure 3.1, which is an approximation of reality. There actually exists both a seepage face and an outflow face; groundwater exits the aquifer above sea level through the seepage face, and below sea level through the outflow face. The actual conditions along the coast are illustrated in Figure 3.2.

The aquifer is bounded above by a phreatic surface and below by an interface that separates the freshwater from saltwater at rest. The distances between these surfaces and sea level are denoted as h_f and h_s, where the indices f and s stand for fresh and salt. All heads are measured from a reference level, which lies a distance H_s below sea level. The head in the freshwater, ϕ, is expressed in terms of the pressure p_f by the use of (1.7) as

$$\phi = \frac{p_f}{g\rho_f} + Z_f, \tag{3.1}$$

where ρ_f is the density of the freshwater and the elevation head Z_f is measured from the reference level. The head ϕ_s in the saltwater is written similarly in terms of the pressure p_s in the saltwater,

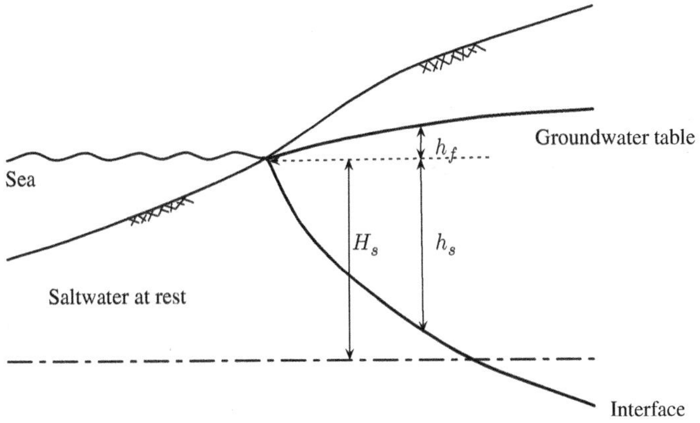

Figure 3.1 The Ghyben-Herzberg equation.

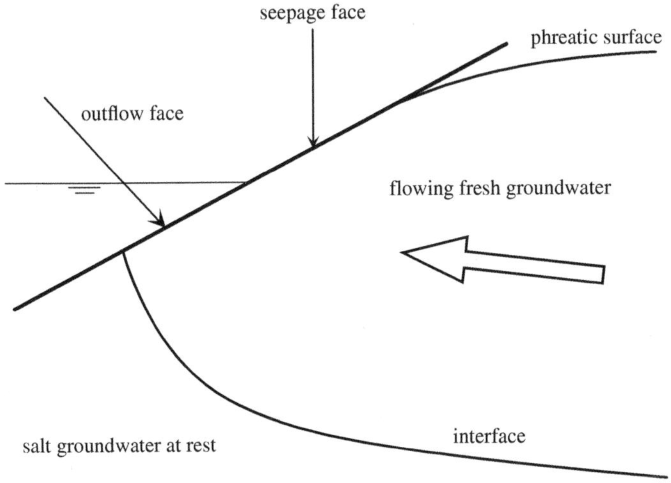

Figure 3.2 Actual conditions near the coast, showing the phreatic surface, the interface, the seepage face, and the outflow face.

$$\phi_s = \frac{p_s}{g\rho_s} + Z_s, \tag{3.2}$$

where ρ_s is the density of the saltwater and Z_s the elevation head in the saltwater. The pressure must be single-valued at all points of the interface, such as point $\mathscr{P}$ in Figure 3.1, where Z_f and Z_s in the expressions for the elevation head in the freshwater and saltwater zones are equal to Z, since the two zones meet at the interface. This gives, with (3.1) and (3.2),

$$g\rho_f(\phi - Z) = g\rho_s(\phi_s - Z). \tag{3.3}$$

The head in the saltwater zone is H_s, so that $\phi_s = H_s$. We solve for the elevation of the interface, Z, and obtain

$$\boxed{Z = \frac{\rho_s}{\rho_s - \rho_f} H_s - \frac{\rho_f}{\rho_s - \rho_f} \phi} \tag{3.4}$$

This equation is valid for saltwater at rest and an interface separating the flowing freshwater from the saltwater. If we adopt the Dupuit-Forchheimer approximation, then the hydraulic head contours are vertical surfaces, $Z = H_s - h_s$, and $\phi = H_s + h_f$ (see Figure 3.1), and we obtain from (3.4) after some manipulation:

$$h_f = h_s \frac{\rho_s - \rho_f}{\rho_f}. \tag{3.5}$$

This equation is known as the Ghyben-Herzberg equation and was derived independently by Badon-Ghyben and Drabbe [1888–1889] and Herzberg [1901]. It is of interest to note that DuCommun [1828] should really be credited with the discovery of the Ghyben-Herzberg equation (Carlston [1963]). It may be noted that the Ghyben-Herzberg equation is equally valid when the upper boundary of the aquifer is a horizontal impermeable boundary rather than a phreatic surface. In that case, h_f represents the hydraulic head with respect to sea level.

3.2 Discharge Potentials

We have seen that the application of discharge potentials to confined flow, unconfined flow, semiconfined flow, and flow in stratified aquifers leads to simplification of the mathematical formulations and results in accurate computation of discharges. Discharge potentials also are beneficial in working with interface flow, and we introduce them in what follows. The discharge potentials for interface flow were introduced by Strack [1976], and are obtained as the ones given in Section 2.3 for horizontal confined and shallow unconfined flow. We denote the thickness of the freshwater zone in the aquifer as h and limit the analysis to problems of flow where h is a linear function of the head ϕ, i.e.,

$$\boxed{h = \alpha\phi + \beta} \tag{3.6}$$

where α and β are constants that depend on the type of flow. The components of the discharge vector, Q_x and Q_y, are equal to h times the components q_x and q_y of the specific discharge vector (compare (2.92)), and we obtain with Darcy's law:

$$\begin{aligned} Q_x &= hq_x = -kh\frac{\partial\phi}{\partial x} \\ Q_y &= hq_y = -kh\frac{\partial\phi}{\partial y}. \end{aligned} \tag{3.7}$$

We let α in (3.6) be unequal to zero, substitute (3.6) for h in (3.7), and integrate:

$$Q_x = -k(\alpha\phi+\beta)\frac{\partial\phi}{\partial x} = -\frac{\partial}{\partial x}\left[\tfrac{1}{2}k\alpha\left(\phi+\frac{\beta}{\alpha}\right)^2+C\right], \qquad \alpha \neq 0$$
$$Q_y = -k(\alpha\phi+\beta)\frac{\partial\phi}{\partial y} = -\frac{\partial}{\partial y}\left[\tfrac{1}{2}k\alpha\left(\phi+\frac{\beta}{\alpha}\right)^2+C\right], \qquad \alpha \neq 0, \tag{3.8}$$

where C is a constant of integration and the hydraulic conductivity k is constant. The potential Φ for $\alpha \neq 0$ is

$$\boxed{\Phi = \tfrac{1}{2}k\alpha\left(\phi+\frac{\beta}{\alpha}\right)^2+C, \qquad \alpha \neq 0} \tag{3.9}$$

We obtain the potential (2.138) for shallow unconfined flow from (3.9) by setting $\phi = h$, or $\alpha = 1$, $\beta = 0$, which yields $\Phi = \frac{1}{2}k\phi^2 + C$. If $\alpha = 0$, (3.6) and (3.7) yield

$$Q_x = \beta q_x = -k\beta\frac{\partial\phi}{\partial x} = -\frac{\partial}{\partial x}(k\beta\phi+C)$$
$$Q_y = \beta q_y = -k\beta\frac{\partial\phi}{\partial y} = -\frac{\partial}{\partial y}(k\beta\phi+C), \tag{3.10}$$

which gives the following expression for the potential Φ:

$$\boxed{\Phi = k\beta\phi + C \qquad (\alpha = 0)} \tag{3.11}$$

We obtain the expression for the potential for horizontal confined flow from (3.11) by setting $\beta = H$. For all cases, the discharge vector equals the gradient of the potential:

$$Q_x = -\frac{\partial\Phi}{\partial x}$$
$$Q_y = -\frac{\partial\Phi}{\partial y}. \tag{3.12}$$

3.3 Shallow Confined Interface Flow

We refer to the flow as shallow confined interface flow when the aquifer is bounded above by a horizontal impermeable boundary and below by an interface (see Figure 3.3). If the upper boundary of the aquifer is a distance H above the reference level, we express the aquifer thickness h as

$$h = H - Z. \tag{3.13}$$

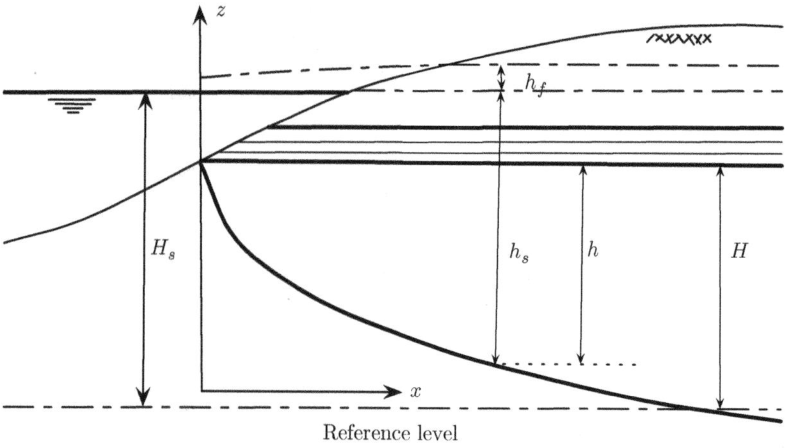

Figure 3.3 Shallow confined interface flow.

We use (3.4) to express Z in terms of ϕ and H:

$$h = \frac{\rho_f}{\rho_s - \rho_f}\phi + H - \frac{\rho_s}{\rho_s - \rho_f}H_s. \tag{3.14}$$

The constants α and β in (3.6) are

$$\alpha = \frac{\rho_f}{\rho_s - \rho_f}, \qquad \beta = -\frac{\rho_s}{\rho_s - \rho_f}H_s + H. \tag{3.15}$$

We obtain the potential for shallow confined interface flow from (3.9) and (3.15) as

$$\Phi = \tfrac{1}{2}k\frac{\rho_f}{\rho_s - \rho_f}\left[\phi - \frac{\rho_s}{\rho_f}H_s + \frac{\rho_s - \rho_f}{\rho_f}H\right]^2 + C_{ci}, \tag{3.16}$$

where C_{ci} replaces the constant C; the index ci stands for confined interface flow.

The condition of continuity is independent of the type of flow and is the same as the one for shallow unconfined flow given in Section 2.2,

$$\frac{\partial Q_x}{\partial x} + \frac{\partial Q_y}{\partial y} = 0. \tag{3.17}$$

The differential equation for the potential is

$$\nabla^2\Phi = 0. \tag{3.18}$$

The governing equations in terms of the potential are the same as those treated in Sections 2.1 through 2.3; the difference between the various types of flows appears only in the definition of the potentials in terms of head.

3.3.1 Application: Uniform Flow to the Coast

We consider the case of uniform flow toward the coast as an application. We choose an (x, y, z)-coordinate system as shown in Figure 3.3, with the x-axis pointing inland from the coast, the y-axis along the coast, and the z-axis vertically upward. The head at a distance L from the coast is ϕ_1. The flow is one-dimensional in the negative x-direction and the corresponding solution to (3.18) is

$$\Phi = Ax + B. \tag{3.19}$$

The boundary condition along the coast is obtained in terms of ϕ as follows. The pressure p_f at a point on the coast and just below the confining layer equals the weight of the column of *saltwater* above that point, which equals $g\rho_s(H_s - H)$ (see Figure 3.3). The elevation of the confining layer above the reference level is H, and we apply (3.1) to express the head ϕ_0 along the coast as

$$x = 0: \qquad \phi = \phi_0 = \frac{p_f}{g\rho_f} + Z = \frac{\rho_s}{\rho_f}(H_s - H) + H = \frac{\rho_s}{\rho_f}H_s - \frac{\rho_s - \rho_f}{\rho_f}H. \tag{3.20}$$

We find the corresponding value for the potential from (3.16),

$$x = 0: \qquad \Phi = \Phi_0 = C_{ci}, \tag{3.21}$$

and express $\Phi = \Phi_1$ at $x = L$ in terms of ϕ,

$$x = L: \qquad \Phi = \Phi_1 = \tfrac{1}{2}k\frac{\rho_f}{\rho_s - \rho_f}\left[\phi_1 - \frac{\rho_s}{\rho_f}H_s + \frac{\rho_s - \rho_f}{\rho_f}H\right]^2 + C_{ci}. \tag{3.22}$$

We apply these boundary conditions to (3.19):

$$\Phi = \frac{\Phi_1 - \Phi_0}{L}x + \Phi_0, \tag{3.23}$$

and express the head ϕ as a function of x using (3.23) with the definition (3.16) for Φ:

$$\phi = \frac{\rho_s}{\rho_f}H_s - \frac{\rho_s - \rho_f}{\rho_f}H + \sqrt{\frac{2}{k}\frac{\rho_s - \rho_f}{\rho_f}\left[\frac{\Phi_1 - \Phi_0}{L}x + \Phi_0 - C_{ci}\right]}. \tag{3.24}$$

Problem

3.1
1. Demonstrate that the values of ϕ computed by the use of expression (3.24) are not affected by the choice of C_{ci}.
2. Plot nine points of the interface between $x = L = 10H$ and $x = 0$ if $\rho_s = 1025$ kg/m^3, $\rho_f = 1000$ kg/m^3, $H_s = 1.1H$, and $\phi_1 = 1.3H$.
3. Why is it not necessary to know the value of k for the latter exercise?
4. What is h along the coast? Do you believe that this value is realistic? If not, why would the model predict an unrealistic value?

3.4 Shallow Confined Interface Flow/Confined Flow

We consider an aquifer with a horizontal impervious base, which we choose as the reference level, as shown in Figure 3.4. Proper choice of the constant C_{ci} in the definition of the potential for shallow confined interface flow enables us to make the transition from confined interface flow to confined flow. We do this in much the same way as in Section 2.3, where we combined unconfined flow with confined flow in a single aquifer. We represent the zones of confined interface flow and of confined flow as zones 1 and 2, respectively; see Figure 3.4. We represent the potential in the confined zone by (2.141):

$$\text{zone 2:} \quad \Phi = kH\phi - \tfrac{1}{2}kH^2. \tag{3.25}$$

The interzonal boundary is a vertical surface through the tip of the saltwater wedge, where $h = H$. We obtain the corresponding value for ϕ from (3.14),

$$\phi_t = \frac{\rho_s}{\rho_f} H_s, \tag{3.26}$$

where the subscript t refers to tip. We require that the potential, which is defined by (3.16) in zone 1 and by (3.25) in zone 2, is single-valued across the tip of the saltwater wedge; substitution of ϕ_t for ϕ in both (3.16) and (3.25) must give the same value:

$$\tfrac{1}{2}k\frac{\rho_f}{\rho_s - \rho_f}\left[\frac{\rho_s}{\rho_f}H_s - \frac{\rho_s}{\rho_f}H_s + \frac{\rho_s - \rho_f}{\rho_f}H\right]^2 + C_{ci} = kH\frac{\rho_s}{\rho_f}H_s - \tfrac{1}{2}kH^2 \tag{3.27}$$

so that

$$C_{ci} = k\frac{\rho_s}{\rho_f}HH_s - \tfrac{1}{2}k\frac{\rho_s}{\rho_f}H^2 \tag{3.28}$$

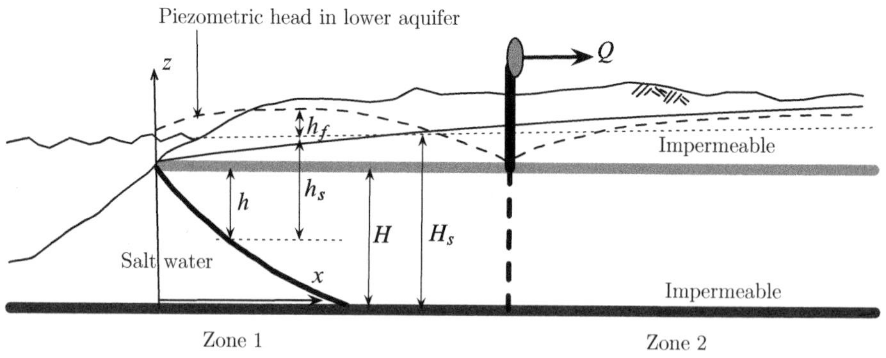

Figure 3.4 Combined confined flow/confined interface flow. (Source: Adapted from Strack [1976].)

and the potential in zone 1 becomes, with (3.16),

$$\text{zone 1:}\quad \Phi = \tfrac{1}{2}k\frac{\rho_f}{\rho_s-\rho_f}\left[\phi - \frac{\rho_s}{\rho_f}H_s + \frac{\rho_s-\rho_f}{\rho_f}H\right]^2 + k\frac{\rho_s}{\rho_f}HH_s - \tfrac{1}{2}k\frac{\rho_s}{\rho_f}H^2. \quad (3.29)$$

Problem

3.2 Consider the problem solved in Section 3.3.1, but now with an impermeable base at a distance H_s below sea level. Determine the position of the tip of the saltwater wedge in terms of x.

3.4.1 Application: Flow to a Well in a Coastal Aquifer

We next consider the case of flow to a well in the confined coastal aquifer shown in Figure 3.4. The well is at a distance d from the coast, and we determine the maximum discharge of the well such that no saltwater is pumped. A plan view of the aquifer cross section through the well is shown in Figure 3.5. The flow without the well is represented by (3.23). We define Q_{x0} as the discharge flowing in the x-direction when the well is not present and rewrite (3.23) as

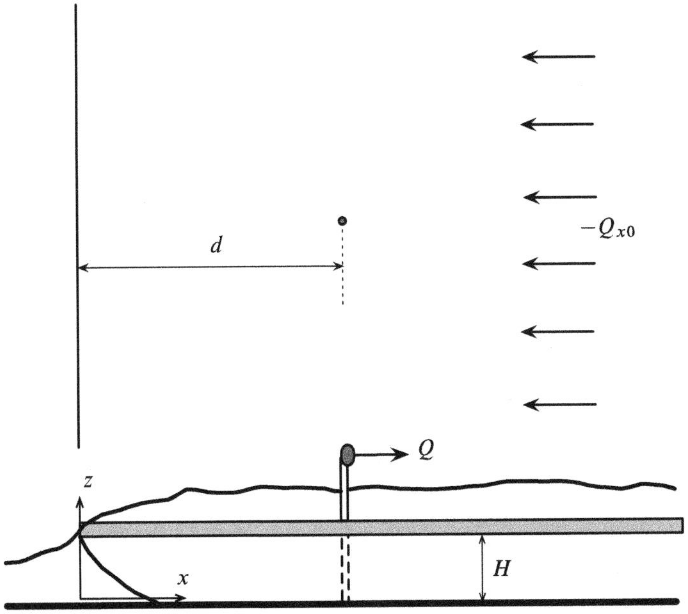

Figure 3.5 Confined flow toward a well in a coastal aquifer.

$$\Phi = -Q_{x0}x + \Phi_0. \tag{3.30}$$

Note that Q_{x0} is negative because the flow is in the negative x-direction to the coast. We obtain an expression for Φ_0 from (3.21) with C_{ci} given by (3.28):

$$\Phi_0 = k\frac{\rho_s}{\rho_f}HH_s - \tfrac{1}{2}k\frac{\rho_s}{\rho_f}H^2. \tag{3.31}$$

We use the method of images to obtain an expression for the potential for flow with a well. The potential for a well of discharge Q at $x = d$, $y = 0$, and an image recharge well at $x = -d$, $y = 0$, gives $\Phi = 0$ along the coast. Addition of the latter potential to (3.30) gives the function that meets the boundary conditions:

$$\Phi = -Q_{x0}x + \frac{Q}{4\pi}\ln\frac{(x-d)^2+y^2}{(x+d)^2+y^2} + \Phi_0. \tag{3.32}$$

The equation for the tip of the wedge of saltwater is found from (3.32) by setting Φ equal to the constant value that the potential assumes along the tip, which we obtain by substituting expression (3.26), ϕ_t, for ϕ in either (3.25) or (3.29):

$$\Phi_t = k\frac{\rho_s}{\rho_f}HH_s - \tfrac{1}{2}kH^2, \tag{3.33}$$

and the equation for the location of the tip of the wedge becomes, using (3.31) for Φ_0,

$$k\frac{\rho_s}{\rho_f}HH_s - \tfrac{1}{2}kH^2 = -Q_{x0}x + \frac{Q}{4\pi}\ln\frac{(x-d)^2+y^2}{(x+d)^2+y^2} + k\frac{\rho_s}{\rho_f}HH_s - \tfrac{1}{2}k\frac{\rho_s}{\rho_f}H^2. \tag{3.34}$$

We write this in terms of two dimensionless variables, λ_c and μ, by dividing both sides by $-\frac{1}{2}Q_{x0}d$ as follows:

$$\lambda_c = 2\frac{x}{d} + \frac{\mu}{2\pi}\ln\frac{(\frac{x}{d}-1)^2+(\frac{y}{d})^2}{(\frac{x}{d}+1)^2+(\frac{y}{d})^2}, \tag{3.35}$$

where

$$\lambda_c = -\frac{kH^2}{dQ_{x0}}\frac{\rho_s-\rho_f}{\rho_f}, \qquad \mu = -\frac{Q}{dQ_{x0}}. \tag{3.36}$$

Note that Q_{x0} is negative because the uniform flow is in the negative x-direction; μ is positive. If Q is gradually increased, while keeping all other flow parameters constant, the tip will tend to move inland and to the well. This is illustrated in Figure 3.6 where the tip of the wedge is represented in the dimensionless $(x/d, y/d)$-plane for $\lambda_c = 0.5$; only the part of the tip for $y/d \geq 0$ is shown. The curve

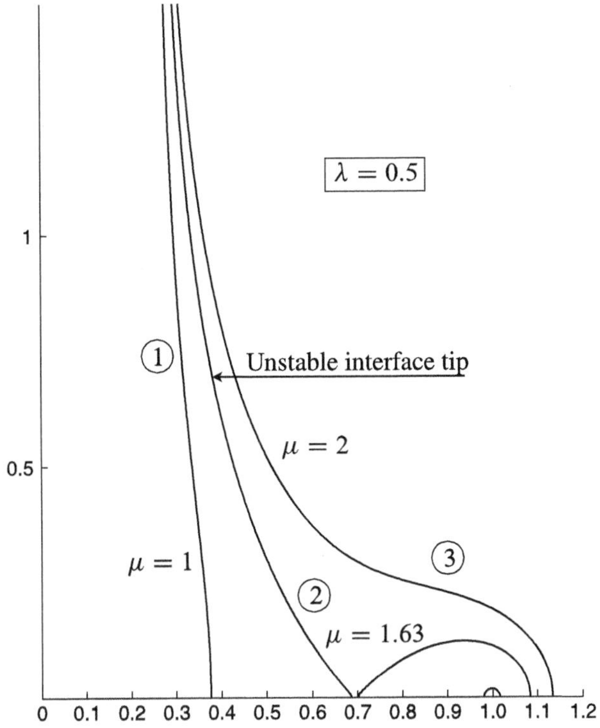

Figure 3.6 Locations of the tip for various values of μ.

for $\mu = 2$ intersects the x-axis inland from the well; the well is pumping saltwater, and the assumption that the saltwater is at rest is violated. The curve labeled $\mu = 1.63$ applies when the saltwater is on the verge of starting to flow. We see this in Figure 3.7, where the equipotentials are shown that apply to the unstable case. The curve through point $\mathscr{S}$ represents the tip of the saltwater wedge. Although the values of Φ along the equipotentials are greater than Φ_t in most of the aquifer inland from the tip, there exists an area where the values of Φ are less than Φ_t. Point $\mathscr{S}$ is a stagnation point, and although the head at $\mathscr{S}$ is just high enough to keep the tip at point $\mathscr{S}$, any reduction in head at $\mathscr{S}$ will cause the saltwater to enter the area of low heads: the situation is unstable.

Unstable conditions correspond to the stagnation point $\mathscr{S}$ coinciding with the tip of the saltwater wedge. These conditions are described mathematically by requiring that the coordinates (x_s, y_s) of the stagnation point fulfill equation (3.35) for the tip. We obtain equations for the coordinates of the stagnation point by setting the expressions for Q_x and Q_y, obtained by differentiation of (3.32), equal to zero:

$$Q_x = -\frac{\partial \Phi}{\partial x} = Q_{x0} - \frac{Q}{2\pi}\left[\frac{x_s - d}{(x_s - d)^2 + y_s^2} - \frac{x_s + d}{(x_s + d)^2 + y_s^2}\right] = 0 \qquad (3.37)$$

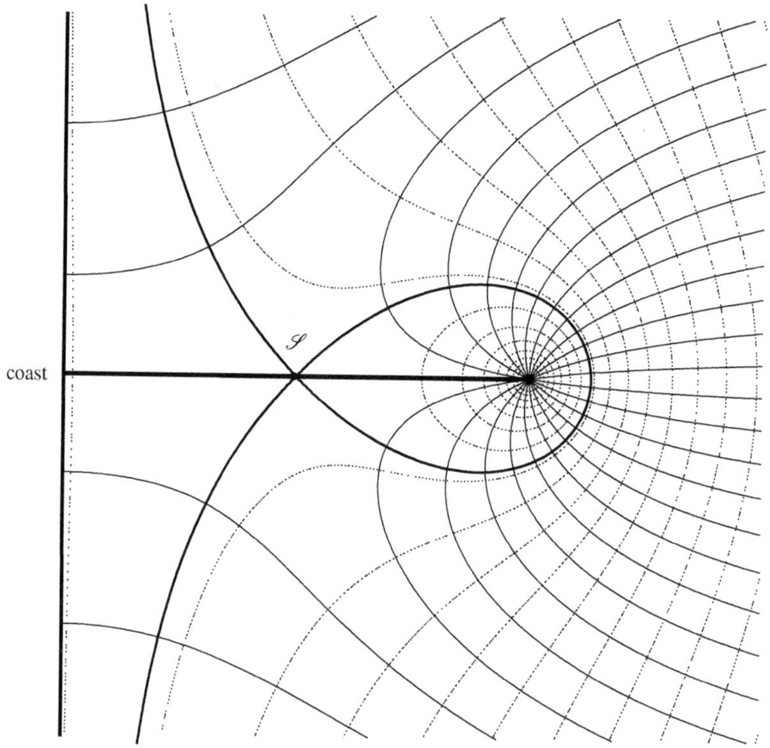

Figure 3.7 Equipotentials for the unstable case.

and

$$Q_y = -\frac{\partial \Phi}{\partial y} = -\frac{Q}{2\pi}\left[\frac{y_s}{(x_s-d)^2+y_s^2} - \frac{y_s}{(x_s+d)^2+y_s^2}\right] = 0. \tag{3.38}$$

The only roots of (3.38) occur for either $y_s = 0$ or $x_s = 0$. The case $x_s = 0$ can be disregarded, however, because for $x_s = 0$ the stagnation point(s) are on the coast, which means that saltwater is entering the aquifer: this violates the assumption that saltwater is at rest. Thus, $y_s = 0$, and (3.37) becomes

$$Q_{x0} - \frac{Q}{2\pi}\left[\frac{1}{x_s-d} - \frac{1}{x_s+d}\right] = Q_{x0} - \frac{Q}{2\pi}\left[\frac{2d}{x_s^2-d^2}\right] = 0, \qquad y_s = 0. \tag{3.39}$$

We solve this equation for x_s and obtain, writing $-Q/(Q_{x0}d)$ as μ,

$$\frac{x_s}{d} = +\sqrt{1-\frac{\mu}{\pi}}, \qquad y_s = 0, \tag{3.40}$$

where the plus sign applies because x_s is known to be positive. Substitution of this for x_s and y_s in (3.35) gives the following condition for instability to occur:

$$\lambda_c = 2\sqrt{1 - \frac{\mu}{\pi}} + \frac{\mu}{\pi} \ln \frac{1 - \sqrt{(1 - \mu/\pi)}}{1 + \sqrt{(1 - \mu/\pi)}}. \tag{3.41}$$

This equation represents a curve, shown in the lower part of Figure 3.8, where λ_c is plotted versus μ. This curve divides the plane into two domains: points above the curve correspond to two-fluid flow, and points below it to one-fluid flow. The points

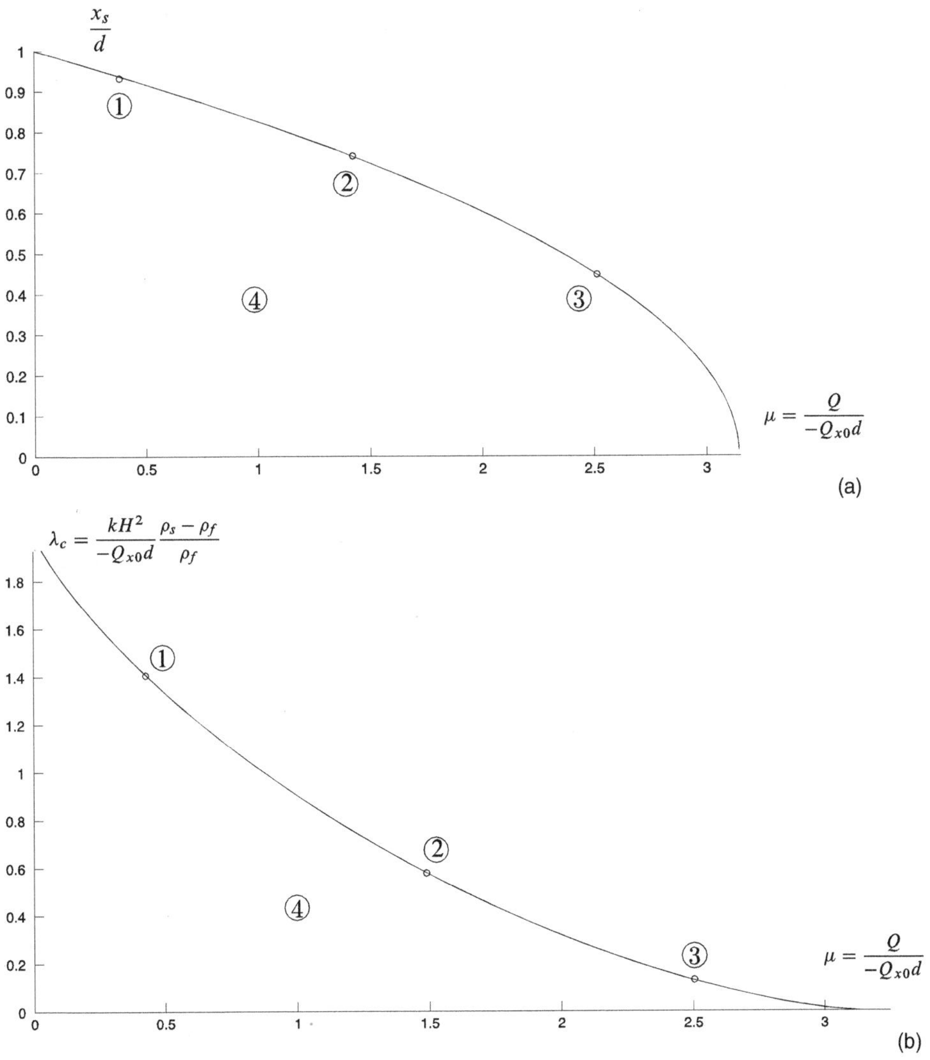

Figure 3.8 Flow parameters for an unstable situation. (Source: Adapted from Strack [1976].)

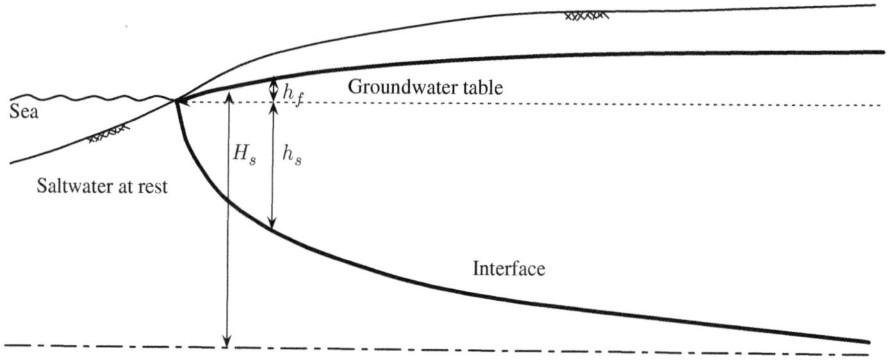

Figure 3.9 Shallow unconfined interface flow.

in the plot labeled as (1), (2), and (3) correspond to the curves labeled equally in Figure 3.6. Equation (3.40) is represented graphically in Figure 3.8(a).

3.5 Shallow Unconfined Interface Flow

A coastal unconfined aquifer is shown in Figure 3.9. The saturated thickness h now may be expressed in terms of Z and the head ϕ as

$$h = \phi - Z. \tag{3.42}$$

We use (3.4) for the elevation of the interface and obtain

$$h = \phi + \frac{\rho_f}{\rho_s - \rho_f}\phi - \frac{\rho_s}{\rho_s - \rho_f}H_s = \frac{\rho_s}{\rho_s - \rho_f}(\phi - H_s) \tag{3.43}$$

so that the constants α and β in (3.6) become

$$\alpha = \frac{\rho_s}{\rho_s - \rho_f}, \qquad \beta = -\frac{\rho_s}{\rho_s - \rho_f}H_s. \tag{3.44}$$

We express the potential Φ in terms of ϕ with (3.9) as

$$\Phi = \tfrac{1}{2}k\frac{\rho_s}{\rho_s - \rho_f}(\phi - H_s)^2 + C_{ui}. \tag{3.45}$$

Problem

3.3 1. Determine the head as a function of position for the problem described in Section 3.3.1, but with a phreatic surface replacing the upper confining boundary.

2. Plot nine points of the interface between $x = L = 10H_s$ and $x = 0$ if $\rho_s = 1025$ kg/m^3, $\rho_f = 1000$ kg/m^3, and $\phi_1 = 1.1H_s$.

3.6 Unconfined Interface Flow/Confined Flow

Two types of flow will occur in a coastal aquifer if the lower boundary is an impermeable base; the flow will be shallow unconfined interface flow near the coast and shallow unconfined flow at some distance away from the coast. We solve these problems again by the use of a single potential Φ. We choose the constant C_{ui} (3.45) such that Φ is continuous across the interzonal boundary, i.e., at the tip of the wedge, where $\phi = \rho_s H_s/\rho_f$. We apply this condition of continuity to (2.138) and (3.45),

$$\tfrac{1}{2}k\left[\frac{\rho_s}{\rho_f}H_s\right]^2 = \tfrac{1}{2}k\frac{\rho_s}{\rho_s-\rho_f}\left[\frac{\rho_s-\rho_f}{\rho_f}H_s\right]^2 + C_{ui} \tag{3.46}$$

or

$$C_{ui} = \tfrac{1}{2}k\frac{\rho_s}{\rho_f}H_s^2. \tag{3.47}$$

The expressions for the potential become

$$\Phi = \tfrac{1}{2}k\frac{\rho_s}{\rho_s-\rho_f}(\phi - H_s)^2 + \tfrac{1}{2}k\frac{\rho_s}{\rho_f}H_s^2, \qquad \phi \le \frac{\rho_s}{\rho_f}H_s \tag{3.48}$$

and

$$\Phi = \tfrac{1}{2}k\phi^2, \qquad \phi \ge \frac{\rho_s}{\rho_f}H_s. \tag{3.49}$$

3.6.1 Application: A Well in an Unconfined Coastal Aquifer

We solve the problem illustrated in Figure 3.5 for an aquifer with a phreatic surface as the upper boundary. The expression for the potential as a function of position is as before, i.e., (3.32). However, the expression for the potential along the coast, Φ_0, is different. We see from Figure 3.9 that the head along the coast is H_s, so that we obtain the potential Φ_0 from (3.48) as

$$\Phi_0 = \tfrac{1}{2}k\frac{\rho_s}{\rho_f}H_s^2. \tag{3.50}$$

We find an expression for the potential along the tip of the wedge by setting ϕ equal to $\rho_s H_s/\rho_f$ in either (3.48) or (3.49):

$$\Phi_t = \tfrac{1}{2}k\left[\frac{\rho_s}{\rho_f}H_s\right]^2. \tag{3.51}$$

The equation for the tip of the wedge is obtained by setting Φ equal to Φ_t in (3.32), which yields, with (3.50):

$$\tfrac{1}{2}k\left[\frac{\rho_s}{\rho_f}H_s\right]^2 = -Q_{x0}x + \frac{Q}{4\pi}\ln\frac{(x-d)^2+y^2}{(x+d)^2+y^2} + \tfrac{1}{2}k\frac{\rho_s}{\rho_f}H_s^2. \tag{3.52}$$

We write this equation in terms of dimensionless variables by dividing both sides by $-\frac{1}{2}Q_{x0}d$, yielding

$$\lambda_u = 2\frac{x}{d} + \frac{\mu}{2\pi}\ln\frac{(\frac{x}{d}-1)^2+(\frac{y}{d})^2}{(\frac{x}{d}+1)^2+(\frac{y}{d})^2}, \tag{3.53}$$

where μ is given by (3.36) and λ_u is defined as

$$\lambda_u = -\frac{kH_s^2}{dQ_{x0}}\frac{\rho_s}{\rho_f}\frac{\rho_s-\rho_f}{\rho_f}. \tag{3.54}$$

The present case of an aquifer bounded above by a phreatic surface differs from the confined case only in the expressions for Φ_0 and λ_u or λ_c; compare (3.53) and (3.35). The results represented in Figure 3.8 may therefore be applied to the present case by replacing λ_c by λ_u.

We next examine the location of the most inland point of the wedge as a function of the parameters μ and λ_u or λ_c. It appears from Figure 3.6 that the most inland point lies on the x-axis. We denote the x-coordinate of this point as L_c and L_u for the confined and unconfined cases, respectively. Expressions for L_c and L_u are obtained by setting y equal to zero and x equal to L_c in (3.35), and to L_u in (3.53). The resulting equations are represented graphically in Figure 3.10.

The case of uniform combined shallow unconfined interface flow and unconfined flow toward the coast without a well was solved by Henry [1959] without making use of discharge potentials. The discharge Q_{x0} for that case may be expressed as

$$Q_{x0} = -\frac{\Phi_1-\Phi_0}{L}, \tag{3.55}$$

where Φ_1 is the potential at some distance away from the coast and Φ_0 the potential along the coast. If $\phi_1 > \rho_s H_s/\rho_f$ and $\phi_0 = H_s$ represent the corresponding values of the head, then (3.55) may be written as follows (see (3.48) and (3.50)):

$$Q_{x0} = -\frac{\frac{1}{2}k\left[\phi_1^2 - \frac{\rho_s}{\rho_f}H_s^2\right]}{L}. \tag{3.56}$$

Youngs [1971] showed that (3.56) gives the exact value of the discharge (i.e., the discharge obtained without making the Dupuit-Forchheimer approximation) if the boundary at $x = L$ is an equipotential and if the coast is vertical. Strack and Ausk [2015] extended Young's solution to a stratified aquifer.

Problem

3.4 Consider the problem shown in Figure 3.5, but now with the well inactive, the upper impermeable layer elevated above sea level ($H > H_s$), and assuming that $\phi_1 > H$.

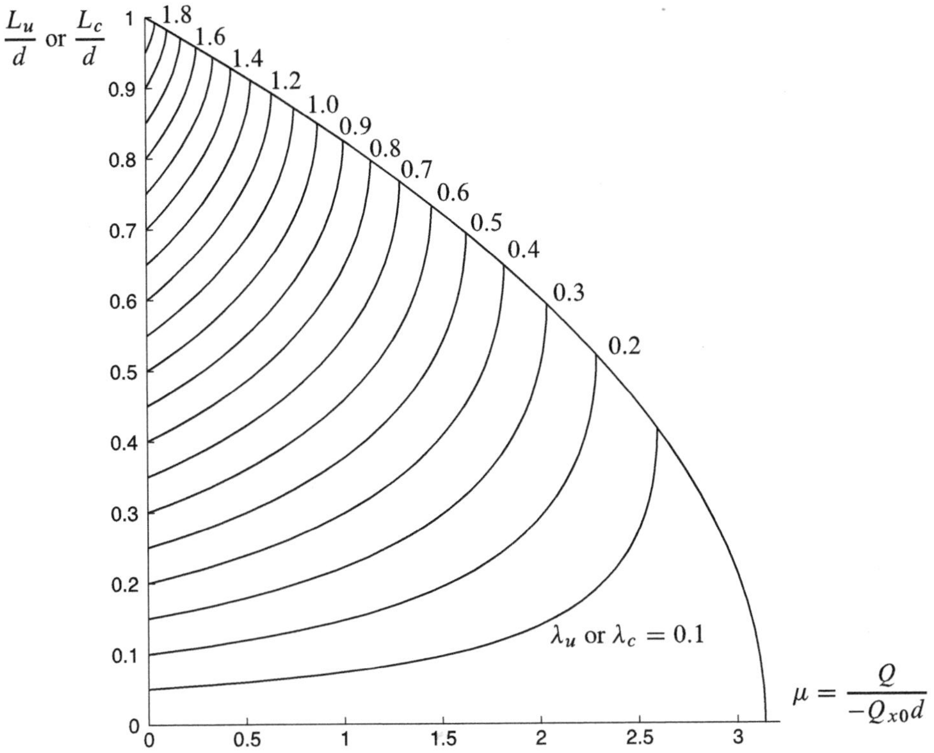

Figure 3.10 L_c/d and L_u/d as functions of λ_c, λ_u, and μ. (Source: Adapted from Strack [1976].)

Questions:

1. List three types of flow that occur in the aquifer if $H > \rho_s H_s/\rho_f$.
2. Select the constants in the various expressions for the potential such that it is continuous throughout the aquifer for $H > \rho_s H_s/\rho_f$. Determine the location of the tip of the saltwater wedge if $L = 10H$, $\rho_s = 1025$ kg/m^3, $\rho_f = 1000$ kg/m^3, $H_s = 0.8H$, $\phi_1 = 1.1H$. Plot nine points of the interface between the coast and the tip of the wedge.
3. List the three types of flow that occur in the aquifer if $H < \rho_s H_s/\rho_f$. Repeat question 2 for $H < \rho_s H_s/\rho_f$, and $H_s = 0.98H$ rather than $0.8H$.

3.7 Shallow Unconfined Interface Flow with Rainfall

Shallow unconfined interface flow with rainfall may be treated in the same way as unconfined flow without an interface. This is possible because the basic equations in terms of the discharge vectors and the potentials are identical for these types of flow.

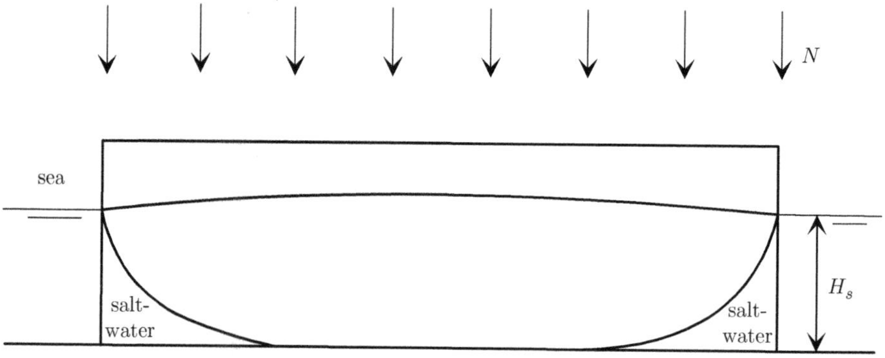

Figure 3.11 A circular island in the sea.

3.7.1 The Interface in a Circular Island

Consider the problem of shallow unconfined interface flow combined with unconfined flow in the circular island of radius R illustrated in Figure 3.11. The potential may be written in the form

$$\Phi = -\tfrac{1}{4}N(r^2 - R^2) + \Phi_0, \tag{3.57}$$

where the potential is defined according to the type of flow as:

$$\begin{aligned} \Phi &= \tfrac{1}{2}k\frac{\rho_s}{\rho_s - \rho_f}[\phi - H_s]^2 + \tfrac{1}{2}k\frac{\rho_s}{\rho_f}H_s^2, \qquad && \phi \le \frac{\rho_s}{\rho_f}H_s \\ \Phi &= \tfrac{1}{2}k\phi^2, && \phi \ge \frac{\rho_s}{\rho_f}H_s \end{aligned} \tag{3.58}$$

(see (3.48) and (3.49)). The value for Φ_0 is given by (3.50). The tip of the saltwater wedge occurs when $\Phi = \Phi_t$ (see (3.51)), and the corresponding radius r_t is found from (3.57) as

$$r_t = R\sqrt{1 - \frac{2k}{N}\left[\frac{H_s}{R}\right]^2 \frac{\rho_s}{\rho_f}\frac{\rho_s - \rho_f}{\rho_f}}. \tag{3.59}$$

Note that the interface lies entirely above the base of the aquifer when the root is imaginary.

Problem

3.5 Determine the potential for the problem of Figure 3.11, but with a well of discharge Q at the center of the island. Find the equation that holds for the unstable condition of the interface.

Problem

3.6 Determine an expression for the location of the tip of the saltwater wedge for rainfall on an island with an elliptic boundary, using (2.222).

4

Two-Dimensional Flow in the Vertical Plane

Two-dimensional flow in the vertical plane occurs if there exists a horizontal direction in which there is no flow. As an example, a case of flow underneath an impermeable dam is shown in Figure 4.1. The flow occurs in the aquifer underneath the dam and is driven by the difference in head along the bottoms of the two lakes that the dam separates. Few cases of true two-dimensional flow in the vertical plane occur in reality, but a two-dimensional analysis often gives a good approximation. If the dam in Figure 4.1 is long with respect to its width, the results obtained from a two-dimensional analysis will be accurate near the central portion of the dam and erroneous near the ends.

The boundary of the flow domain often consists of several segments such as the two horizontal streamlines and the two equipotentials in Figure 4.1. The elementary methods of solution discussed thus far are applicable only to problems with at most two orthogonal infinite impermeable boundaries or equipotentials as boundaries. We focus our attention in this section on deriving the basic equations and give an application of the method of images. We further discuss an approximate method, known as the graphical method, and apply it to a few problems involving dams. For all problems of two-dimensional flow in the vertical plane, we use (x, y, z)-Cartesian coordinates with the z-axis normal to the plane of flow, the x- and y-axes in the plane of flow, and the y-axis pointing vertically upward (see Figure 4.1).

4.1 Basic Equations

We obtain the basic equations directly from Darcy's law and the continuity equation presented in Chapter 1. There is no flow in the z-direction, so that

$$q_z = 0. \tag{4.1}$$

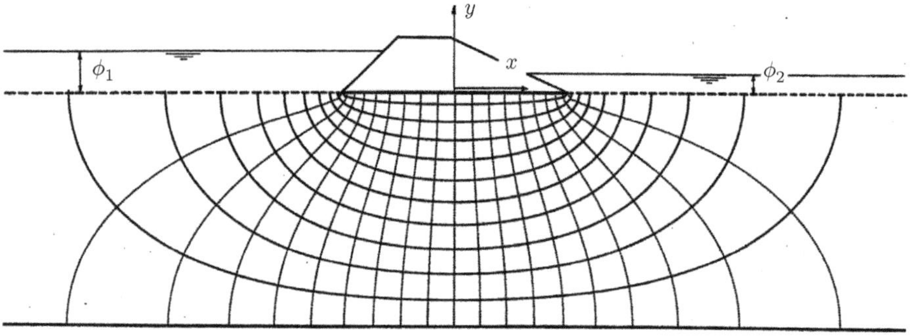

Figure 4.1 Flow underneath a dam.

All discharges are calculated over a slice of aquifer of width B normal to the plane of flow. The components Q_x and Q_y of the discharge vector are defined accordingly,

$$\begin{aligned} Q_x &= Bq_x \\ Q_y &= Bq_y. \end{aligned} \tag{4.2}$$

Darcy's law (1.48) gives with (4.2)

$$\begin{aligned} Q_x &= -\frac{\partial(kB\phi)}{\partial x} \\ Q_y &= -\frac{\partial(kB\phi)}{\partial y} \end{aligned} \tag{4.3}$$

and we define the discharge potential Φ for two-dimensional flow in the vertical plane as

$$\Phi = kB\phi. \tag{4.4}$$

The components Q_x and Q_y may be written in terms of Φ as

$$\begin{aligned} Q_x &= -\frac{\partial\Phi}{\partial x} \\ Q_y &= -\frac{\partial\Phi}{\partial y}. \end{aligned} \tag{4.5}$$

With (4.1) the continuity equation (1.36) becomes

$$\frac{\partial q_x}{\partial x} + \frac{\partial q_y}{\partial y} = 0 \tag{4.6}$$

or, after multiplication by the constant B,

$$\frac{\partial Q_x}{\partial x} + \frac{\partial Q_y}{\partial y} = 0. \tag{4.7}$$

Combination of (4.5) and (4.7) yields Laplace's equation for the potential Φ,

$$\boxed{\nabla^2 \Phi = 0} \tag{4.8}$$

The basic equations (4.5) and (4.8) are formally equal to those for the various types of shallow flow without leakage and infiltration. The elementary solutions and techniques discussed so far for the latter types of flow can therefore be applied to two-dimensional flow in the vertical plane.

4.2 Horizontal Drains

A horizontal drain is a drainpipe with a horizontal axis. If the head inside the drain is kept at a fixed level by pumping, flow to the drain results. A case of flow toward a horizontal drain is illustrated in Figure 4.2. The drain lies in a deep aquifer, which is bounded above by an impermeable layer.

Flow to a single drain of infinite length in an infinite aquifer is radial, and the potential that fulfills Laplace's equation for radial flow is

$$\Phi = \frac{Q}{2\pi} \ln r + C, \tag{4.9}$$

where Q is the discharge flowing into the drain per length B. The aquifer in Figure 4.2 is semi-infinite; there is an impermeable boundary along $y = 0$. If the center of the drain is on the y-axis at a depth d below the origin, then an image drain of discharge Q must be placed at $x = 0,\ y = d$, in order to meet the boundary condition that the x-axis is a streamline. The potential is

$$\Phi = \frac{Q}{4\pi}\{\ln[x^2 + (y+d)^2] + \ln[x^2 + (y-d)^2]\} + C, \tag{4.10}$$

where C is a constant. If we add a uniform flow of discharge Q_{x0} parallel to the boundary, a term $-Q_{x0}x$ must be added to (4.10):

$$\Phi = -Q_{x0}x + \frac{Q}{4\pi}\{\ln[x^2 + (y+d)^2] + \ln[x^2 + (y-d)^2]\} + C. \tag{4.11}$$

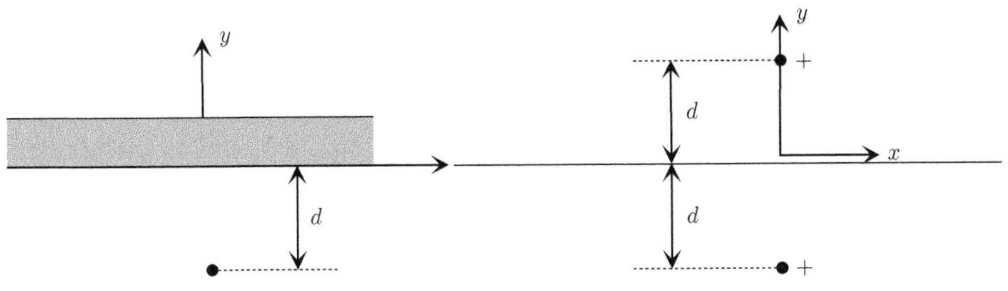

Figure 4.2 Flow toward a horizontal drain.

We evaluate C by requiring that the head be equal to some given value ϕ_0 at a point $\mathscr{P}$ with coordinates $x = x_0$, $y = y_0$:

$$\begin{aligned} x &= x_0, \quad & y &= y_0, \quad & \phi &= \phi_0 \\ x &= x_0, \quad & y &= y_0, \quad & \Phi &= \Phi_0 = kB\phi_0. \end{aligned} \tag{4.12}$$

We apply boundary condition (4.12) to the potential (4.11),

$$\Phi = -Q_{x0}(x - x_0) + \frac{Q}{4\pi} \ln \frac{[x^2 + (y+d)^2][x^2 + (y-d)^2]}{[x_0^2 + (y_0 + d)^2][x_0^2 + (y_0 - d)^2]} + \Phi_0, \tag{4.13}$$

and obtain an expression for the head in the aquifer by dividing (4.13) by kB.

4.3 The Graphical Method

The graphical method is simple and requires little time and equipment to obtain a good approximation for many problems. It is used mostly to determine pressure distributions in embankments and dams, behind earth-retaining structures, or for estimating seepage through dams and underneath hydraulic structures. The method is based on the properties of equipotentials and streamlines (see Section 2.1). Equipotentials are curves along which the potential is constant. Streamlines are curves defined by requiring that the discharge vector is everywhere tangent to the streamline. Since the flow is normal to the equipotentials, as shown in Section 2.1, the streamlines and equipotentials are mutually orthogonal.

4.3.1 Flow Nets

A flow net is a graphical representation of equipotentials and streamlines in a flow region. An example of a flow net is shown in Figure 4.3. The intervals between consecutive streamlines and equipotentials are arbitrary but should be taken small enough so that the sides of the rectangles of the net are approximately straight. Consider the rectangle $\mathscr{A}\mathscr{B}\mathscr{F}\mathscr{E}$ with sides Δs and Δn in Figure 4.3. The directions s and n are along the streamlines and equipotentials as shown in the figure. The magnitude of s at an arbitrary point $\mathscr{P}$ of a streamline equals the arc length of the section of that streamline between $\mathscr{P}$ and the equipotential $\mathscr{A}\mathscr{D}$. The arc lengths of sections of the equipotentials are expressed similarly in terms of n. We represent the flow through side $\mathscr{A}\mathscr{B}$ as ΔQ. Assuming that Q_s varies only slightly along $\mathscr{A}\mathscr{B}$, ΔQ will be equal to the value of Q_s at a point halfway between $\mathscr{A}$ and $\mathscr{B}$, multiplied by Δn, i.e.,

$$\Delta Q \approx Q_s \Delta n. \tag{4.14}$$

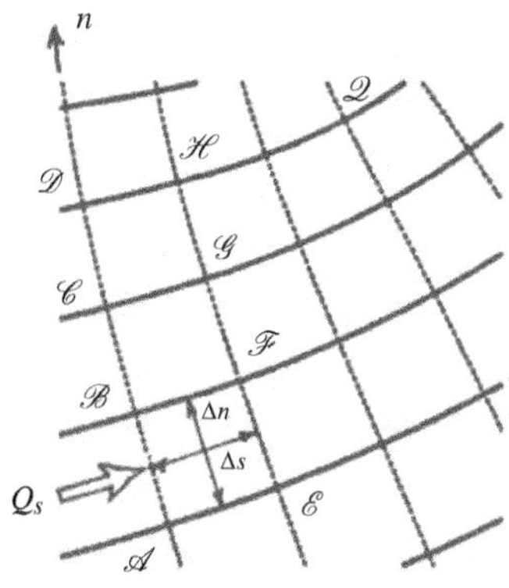

Figure 4.3 A flow net.

Furthermore, $Q_s = -d\Phi/ds$ by Darcy's law or, approximately,

$$\Delta Q \approx -\frac{\Delta\Phi}{\Delta s}\Delta n, \tag{4.15}$$

where $\Delta\Phi$ is the difference in value of Φ on the equipotentials $\mathscr{E}\mathscr{F}$ and $\mathscr{A}\mathscr{B}$, i.e.,

$$\Delta\Phi = \Phi_{\mathscr{E}\mathscr{F}} - \Phi_{\mathscr{A}\mathscr{B}}. \tag{4.16}$$

Since Φ decreases in the direction of flow, $\Phi_{\mathscr{E}\mathscr{F}}$ is less than $\Phi_{\mathscr{A}\mathscr{B}}$ so that $\Delta\Phi$ is negative and we write (4.15) as

$$\Delta Q \approx \frac{|\Delta\Phi|}{\Delta s}\Delta n. \tag{4.17}$$

We draw the flow net such that the rectangles are approximately square: $\Delta s \approx \Delta n$, and (4.17) becomes

$$\Delta Q \approx |\Delta\Phi|. \tag{4.18}$$

No water can cross the streamlines: ΔQ is a constant inside the channel between the streamlines $\mathscr{B}\mathscr{F}$ and $\mathscr{A}\mathscr{E}$ by continuity,

$$\Delta Q = |\Delta\Phi| = \text{constant}. \tag{4.19}$$

Hence, the drop in potential, $|\Delta\Phi|$, is the same for each pair of consecutive equipotentials: if the potential is known at the bordering equipotentials of the flow net, $|\Delta\Phi|$ can be found. If Φ along the equipotentials $\mathscr{A}\mathscr{D}$ and $\mathscr{P}\mathscr{Q}$ in Figure 4.3 is known, then

$$|\Delta\Phi| = \frac{\Phi_{\mathscr{A}\mathscr{D}} - \Phi_{\mathscr{P}\mathscr{Q}}}{m_e}, \tag{4.20}$$

where m_e, the number of drops in potential, is 3 in the case of Figure 4.3. Since $|\Delta\Phi|$ is a constant throughout the flow net, ΔQ is the same for each pair of consecutive streamlines, by (4.18). We compute the total discharge through the flow net by multiplying ΔQ by the number of channels between the boundary streamlines, m_s. Thus, by (4.18) and (4.20),

$$Q = m_s \Delta Q \approx m_s |\Delta\Phi| = \frac{m_s}{m_e}\left(\Phi_{\mathcal{AD}} - \Phi_{\mathcal{PQ}}\right). \tag{4.21}$$

The channels between the streamlines are called flow channels. The channels between the equipotentials are called equipotential channels.

We draw the following conclusion: if a flow net is a mesh of approximate squares, then the discharge flowing through is approximated by (4.21). The numbers m_s and m_e represent the numbers of flow channels and equipotential channels, respectively. The value of Φ on any equipotential can be determined, and the values of Φ at points in between equipotentials may be computed either by interpolation or by locally refining the mesh. We find the pressure from the head using the relation $p = \rho g[\phi - Z]$ where $\phi = \Phi/(kB)$.

A solution obtained by drawing a flow net becomes more accurate as the squares are made smaller. Some applications of the graphical technique follow.

4.3.2 Applications of the Graphical Method

A good flow net fulfills the following conditions:

1. Equipotentials and streamlines are everywhere orthogonal.
2. Equipotentials and streamlines form approximate squares.
3. The flow net satisfies the boundary conditions for the problem.

The flow net for flow underneath a dam with a cutoff wall is shown in Figure 4.4. The boundary conditions are that $\mathcal{SPT}$ and $\mathcal{VW}$ are streamlines and that $\mathcal{RS}$ and $\mathcal{TU}$ are equipotentials; the streamlines of the net must intersect the boundaries $\mathcal{RS}$ and $\mathcal{TU}$ at right angles and the equipotentials must be perpendicular to $\mathcal{VW}$ and $\mathcal{SPT}$. The flow net is obtained by a trial-and-error procedure. One approach is to draw the streamlines first; in the case of Figure 4.4 from $\mathcal{TU}$ to $\mathcal{RS}$. The equipotentials are drawn next so as to obtain a pattern of approximate squares. The first estimate of the streamlines usually appears inaccurate after drawing the equipotentials; redrawing of the streamlines and equipotentials is necessary until the flow net is good.

The number of flow channels need not be an integer. The lowest curved streamline in Figure 4.4 was drawn in order to improve the flow net. The width of the channel enclosed between that streamline and the boundary streamline $\mathcal{VW}$ is

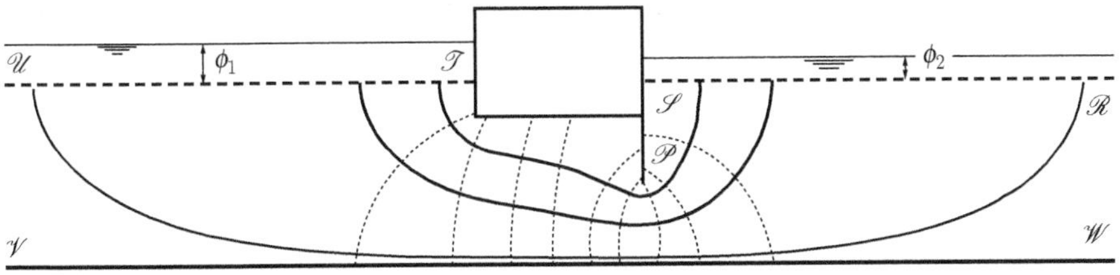

Figure 4.4 Flow underneath a dam with a cutoff wall. (Source: A. Verruijt, *Theory of Groundwater Flow*, © 1970, p. 147. Reprinted by permission of the author.)

about one-tenth of the width of a flow channel. The total number of flow channels between $\mathscr{RS}$ and $\mathscr{VW}$ is therefore 3.1 (see Verruijt [1970]).

The discharge flowing underneath the dam from $\mathscr{TU}$ to $\mathscr{RS}$ is found as follows. The difference in potential between $\mathscr{TU}$ and $\mathscr{RS}$ is

$$\Phi_1 - \Phi_2 = kB(\phi_1 - \phi_2). \tag{4.22}$$

There are $m_e = 10$ equipotential channels between $\mathscr{TU}$ and $\mathscr{RS}$, so that

$$|\Delta\Phi| = \frac{\Phi_1 - \Phi_2}{m_e} = \frac{k(\phi_1 - \phi_2)B}{10}. \tag{4.23}$$

Since $\Delta Q \approx |\Delta\Phi|$ and since there are $m_s = 3.1$ flow channels, the total discharge flowing underneath the dam is approximately given by

$$Q \approx m_s \Delta Q \approx m_s |\Delta\Phi| = 3.1 \frac{kB(\phi_1 - \phi_2)}{10}. \tag{4.24}$$

If the difference in the heads between $\mathscr{TU}$ and $\mathscr{RS}$ is

$$\phi_1 - \phi_2 = 2.0 \text{ m} \tag{4.25}$$

and if the hydraulic conductivity is 10^{-4} m/s,

$$k = 10^{-4} \text{ m/s} = 8.64 \text{ m/d}, \tag{4.26}$$

then the seepage underneath the dam per B meters of length is found from (4.24) to be

$$Q = 3.1 \frac{10^{-4} * 2}{10} B = 6.2 * 10^{-5} * B \text{ m}^3/\text{s} = 5.4 * B \text{ m}^3/\text{d}. \tag{4.27}$$

It may be noted that the scale of the drawing has no influence on the discharge, which solely depends on the geometry of the flow domain and the difference $\Phi_1 - \Phi_2$. A lengthening of the sheet piling $\mathscr{SP}$, however, will result in an increase of the number of equipotential channels; the discharge will decrease accordingly (see Verruijt [1970]).

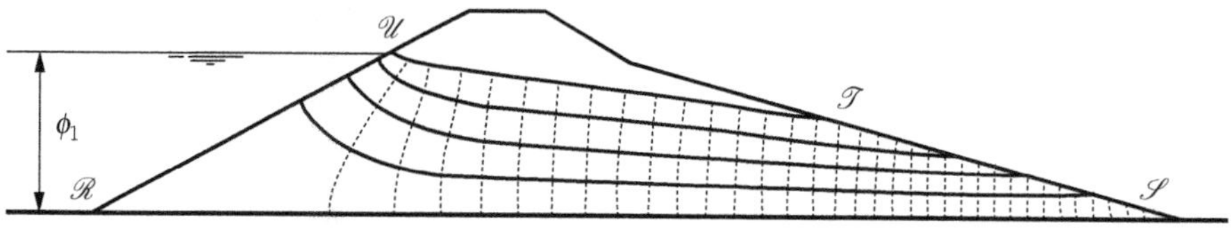

Figure 4.5 Flow through a dam with a phreatic surface. (Source: A. Verruijt, *Theory of Groundwater Flow*, © 1970, p. 147. Reprinted by permission of the author.)

Another flow net is shown in Figure 4.5. It represents the flow through a dam. The upper boundary $\mathcal{TU}$ of the flow net is a phreatic surface. It is assumed that the pores of the soil beneath the phreatic surface are completely filled with water whereas the pores above it are filled with air; the phreatic surface is a streamline. The phreatic surface represents a special kind of boundary condition: neither its form nor its position is known beforehand; it is called a free surface or a free boundary. The boundary conditions along this free surface are that (1) it is a streamline and (2) the pressure along it is equal to the atmospheric pressure, which usually is taken as zero. As follows from (1.7) the head ϕ along the free surface equals its elevation,

$$\phi = \frac{p}{\rho g} + Y = Y, \tag{4.28}$$

where Y is now used to represent the elevation head rather than Z. Since $\Phi = kB\phi$, it follows that

$$\Phi = kBY \tag{4.29}$$

along the free surface. This implies that the vertical distance, ΔY, between the points of intersection of two consecutive equipotentials with the free surface is the same along the entire free surface, as indicated in Figure 4.6. The free surface is drawn by trial and error: its form and position are estimated, the flow net is drawn with the free surface as the boundary streamline, and it is determined whether the boundary condition (4.29) is met. If this condition is not met, the free surface is modified and the flow net redrawn. This procedure is repeated until the flow net is good.

The boundary segment $\mathcal{ST}$ of Figure 4.5 is called a seepage face; the water there leaves the dam, and the pressure is zero. Hence, the boundary condition along the seepage face is given by (4.29); the intersection points of consecutive equipotentials with the seepage face must be the same vertical distance apart along the entire seepage face.

Point $\mathcal{T}$ in Figure 4.5 is the intersection point of the free surface with the seepage face. The free surface is tangent to the seepage face at point $\mathcal{T}$ (see, e.g., Strack

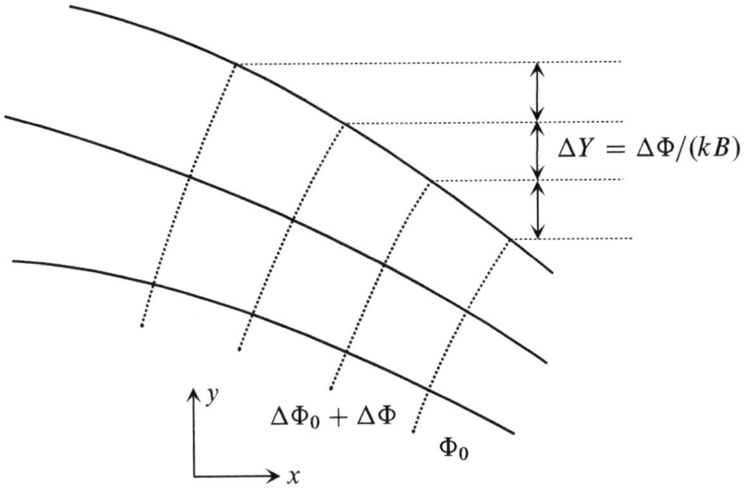

Figure 4.6 The boundary condition along the phreatic surface.

[1989]). The other boundary conditions for the case of Figure 4.5 are that $\mathscr{U}\mathscr{R}$ is an equipotential ($\Phi = kB\phi_1 = \Phi_1$) and that $\mathscr{R}\mathscr{S}$ is a streamline.

It requires several trials to obtain a good flow net involving a free surface such as the one represented in Figure 4.5. However, when the flow net is drawn carefully, the result will be quite accurate.

The total discharge flowing per unit width through the dam of Figure 4.5 is found as follows. The number of equipotential channels is 35. The potential decreases from $kB\phi_1$ along $\mathscr{R}\mathscr{S}$ to 0 at point $\mathscr{S}$, so that

$$|\Delta\Phi| = \frac{kB\phi_1}{35}. \tag{4.30}$$

Since there are four flow channels, the total discharge, Q, per unit length of dam is

$$Q = m_s \Delta Q = 4\Delta Q \approx 4|\Delta\Phi| = \frac{4kB\phi_1}{35}. \tag{4.31}$$

If ϕ_1 is 10 meters and k is again 10^{-4} m/s or 8.64 m/d, and if $B = 1$ m,

$$\phi_1 = 10 \text{ m}, \qquad k = 10^{-4} \text{ m/s} = 8.64 \text{ m/d}, \qquad B = 1 \text{ m}, \tag{4.32}$$

then Q becomes

$$Q = \frac{4}{35}10^{-3} \text{ m}^3/\text{s} = 1.14 * 10^{-4} \text{ m}^3/\text{s} \approx 10 \text{ m}^3/\text{d}. \tag{4.33}$$

Problem

4.1 A long waterway will be dug into a thick horizontal impervious clay layer resting on a layer of fine sand of thickness H (see Figure 4.7). The layer

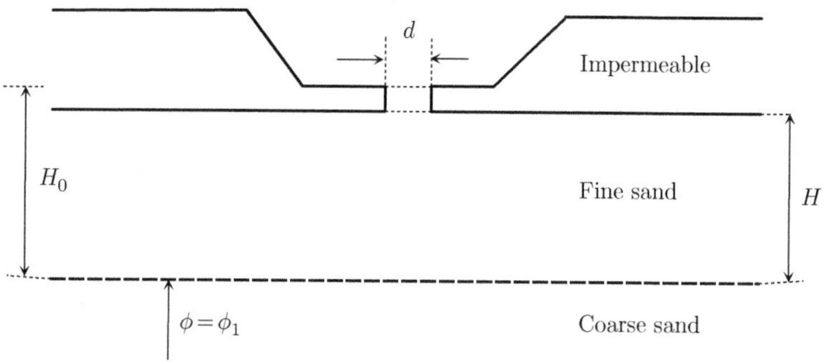

Figure 4.7 Flow toward a sand drain.

of fine sand is bounded below by coarse sand. The head in the coarse sand is constant and equal to ϕ_1 ($\phi_1 > H$). A long sand drain of width d runs along the center of the waterway and is used for pressure relief during construction. The head underneath the trench at the interface between clay and fine sand must be kept below level H_0. The flow is two-dimensional.

Question:

Determine the maximum allowable head that must be maintained at the bottom of the sand drain. (Note that the resistance to flow in the sand drain can be neglected and that it is assumed that any head between ϕ_1 and H can be maintained at the bottom of the sand drain.) Copy the figure and draw your flow net in the figure using the graphical method.

4.4 Anisotropy

The graphical method is not directly applicable to problems with anisotropic hydraulic conductivity, because there does not exist a potential: if x^* and y^* are Cartesian coordinates with the x^*-axis parallel to the major principal direction of hydraulic conductivity tensor, then Darcy's law is given by (1.25):

$$q_x^* = -k_1 \frac{\partial \phi}{\partial x^*}$$
$$q_y^* = -k_2 \frac{\partial \phi}{\partial y^*}. \qquad (4.34)$$

Indeed, we cannot define a single potential Φ since the coefficients in the two preceding equations are different.

It is possible, however, to transform the flow domain to another domain with coordinates X, Y such that a potential Φ may be defined in that domain. Let the transformation be given by

$$X = x^* \tag{4.35}$$

$$Y = \beta y^*. \tag{4.36}$$

We choose the constant β such that the head ϕ satisfies Laplace's equation in the transformed domain. We use (4.34) with the continuity equation (1.36), written in terms of the (x^*, y^*)-coordinate system,

$$\frac{\partial q_x^*}{\partial x^*} + \frac{\partial q_y^*}{\partial y^*} = -k_1 \frac{\partial^2 \phi}{\partial (x^*)^2} - k_2 \frac{\partial^2 \phi}{\partial (y^*)^2} = 0. \tag{4.37}$$

With (4.35) this becomes

$$-k_1 \frac{\partial^2 \phi}{\partial X^2} - k_2 \beta^2 \frac{\partial^2 \phi}{\partial Y^2} = 0. \tag{4.38}$$

In order to obtain Laplace's equation the constant β must be equal to

$$\beta = \sqrt{\frac{k_1}{k_2}}. \tag{4.39}$$

Since the head fulfills Laplace's equation in terms of X and Y, we can solve the problem in the transformed domain. We cannot compute flow rates from such a solution without first determining the isotropic hydraulic conductivity $\underset{i}{k}$ in the transformed domain. We do this by considering a case of one-dimensional flow in the x^*-direction as follows (see Figure 4.8). Darcy's law in the transformed domain is

$$\begin{aligned} q_X &= -\underset{i}{k} \frac{\partial \phi}{\partial X} \\ q_Y &= -\underset{i}{k} \frac{\partial \phi}{\partial Y}. \end{aligned} \tag{4.40}$$

The flow net in the (x^*, y^*)-plane is such that a mesh of squares is obtained on transformation to the (X, Y)-plane. We denote the difference in head between consecutive potentials in either plane as $\Delta\phi$ and express the discharge ΔQ flowing between two streamlines as

$$\Delta Q = q_x^* \Delta y^* = -k_1 \frac{\Delta \phi}{\Delta x^*} \Delta y^* = -k_1 \frac{\Delta \phi}{\Delta X} \frac{\Delta Y}{\beta}, \tag{4.41}$$

where we used (4.35) and (4.36). It follows from (4.40) that

$$\Delta Q = q_X \Delta Y = -\underset{i}{k} \frac{\Delta \phi}{\Delta X} \Delta Y. \tag{4.42}$$

The above two expressions for ΔQ must be equal, so that

$$\underset{i}{k} = \frac{k_1}{\beta} = \sqrt{k_1 k_2}. \tag{4.43}$$

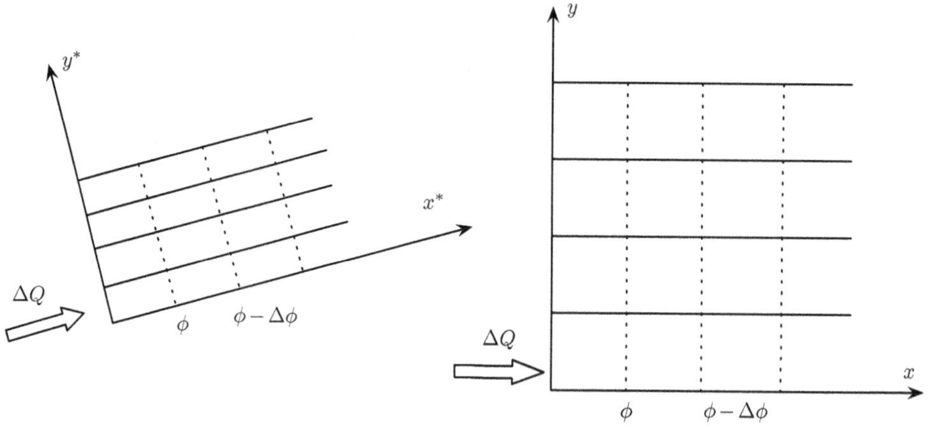

Figure 4.8 Transformation of a flow net.

We write Darcy's law for the discharge vector components Q_X and Q_Y in terms of a potential Φ defined in the isotropic domain:

$$
\begin{aligned}
Q_X &= -\underset{i}{k}B\frac{\partial \phi}{\partial X} = -\frac{\partial \Phi}{\partial X} \\
Q_Y &= -\underset{i}{k}B\frac{\partial \phi}{\partial Y} = -\frac{\partial \Phi}{\partial Y},
\end{aligned}
\tag{4.44}
$$

where

$$\Phi = \underset{i}{k}B\phi = B\phi\sqrt{k_1k_2}. \tag{4.45}$$

4.4.1 Application: Flow through an Anisotropic Aquifer Underneath a Dam with a Cutoff Wall

As an application, we reconsider the case of flow underneath a dam with a cutoff wall. The thickness of the aquifer and the depth of the cutoff wall are half of what they are for the case of Figure 4.4. The principal directions of the hydraulic conductivity are horizontal and vertical; $k_1 = 10^{-4}$ m/s, and $k_2 = k_1/4$:

$$k_2 = k_1/4 = 0.25 * 10^{-4} \text{ m/s}. \tag{4.46}$$

The difference in head across the dam is $\phi_1 - \phi_2$, with

$$\phi_1 - \phi_2 = 2 \text{ m}. \tag{4.47}$$

The transformation formulas become, with $x^* = x$ and $y^* = y$:

$$
\begin{aligned}
X &= x \\
Y &= \beta y = \sqrt{\frac{k_1}{k_2}}y = 2y.
\end{aligned}
\tag{4.48}
$$

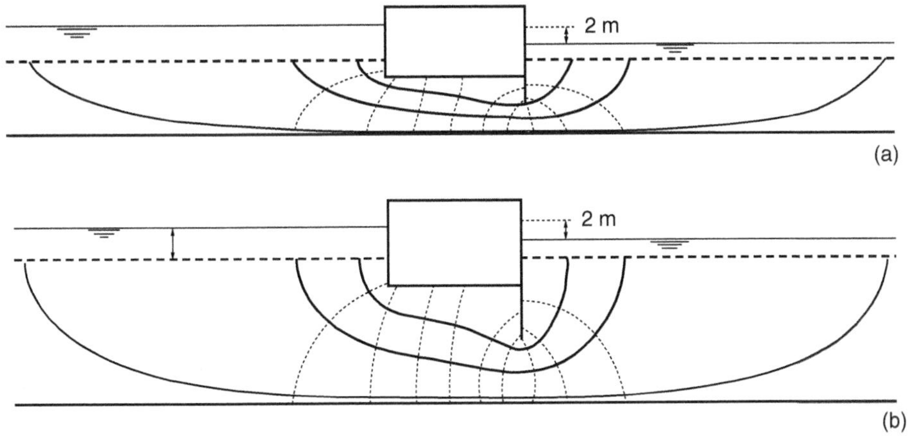

Figure 4.9 Flow in an anisotropic aquifer underneath a dam.

We obtain the boundary of the transformed flow domain, shown in Figure 4.9(b), from the flow domain in Figure 4.9(a) by multiplying all vertical coordinates by 2. The flow net in the isotropic domain is identical to that shown in Figure 4.4, and is reproduced in Figure 4.9(b). This flow net is transformed by the use of (4.48), which results in the flow net in the anisotropic domain shown in Figure 4.9(a).

We express the potential Φ in terms of ϕ with (4.45) and (4.46):

$$\Phi = \underset{i}{k} B\phi = B\sqrt{\tfrac{1}{4}k_1^2}\phi = \tfrac{1}{2}Bk_1\phi = 0.5 * 10^{-4} B\phi \text{ m}^3/\text{s}. \tag{4.49}$$

We compute the difference in value of Φ across the flow domain from (4.47) and (4.49):

$$\Phi_1 - \Phi_2 = 10^{-4} B \text{ m}^3/\text{s}. \tag{4.50}$$

The number of equipotential channels (m_e) is 10, so that

$$|\Delta\Phi| = \frac{\Phi_1 - \Phi_2}{m_e} B = 10^{-5} B \text{ m}^3/\text{s}. \tag{4.51}$$

There are 3.1 flow channels (compare (4.24)); the total discharge Q underneath the dam equals

$$Q = 3.1\,\Delta Q = 3.1\,|\Delta\Phi| = 3.1 * 10^{-5} B \text{ m}^3/\text{s}. \tag{4.52}$$

It is usually difficult to estimate the effect of anisotropy; this effect depends on the direction of flow. The value of k_2 will, for example, have little effect if the flow is primarily in the x^*-direction.

5

Steady Flow in Leaky Aquifer Systems

The confining beds of a shallow confined aquifer are never truly impermeable; we indicate by the term "confined" that the leakage through the confining beds is negligibly small. If the leakage cannot be neglected, the aquifer is referred to as a semiconfined aquifer and the leaking confining bed as a leaky layer, a semipermeable layer, or an aquitard. We use the term *shallow leaky aquifer flow* whenever the aquifer is sufficiently shallow that the resistance to flow in the vertical direction may be neglected.

An important property of leaky systems is that there exists an equilibrium condition if there are no features in the system, wells, for example, that induce leakage. We discuss in this chapter how to describe the leakage induced by features in systems of two aquifers, separated by a single leaky layer.

5.1 Shallow Semiconfined Flow

The simplest case of shallow leaky aquifer flow occurs if the head is constant above the upper semipermeable layer, and the lower boundary of the aquifer is impermeable. Such a case applies if there is little or no flow in the aquifer above the semipermeable layer. We refer to this type of flow as semiconfined flow. We derive the basic equations for this type of flow and consider several examples. The equilibrium condition in a semiconfined aquifer, i.e., in a system without features that induce leakage, is that the head in the lower aquifer equals the constant head in the upper aquifer.

5.1.1 Basic Equations

A vertical section through a semiconfined aquifer is shown in Figure 5.1; the hydraulic conductivity and thickness of the aquifer are k and H. We label all

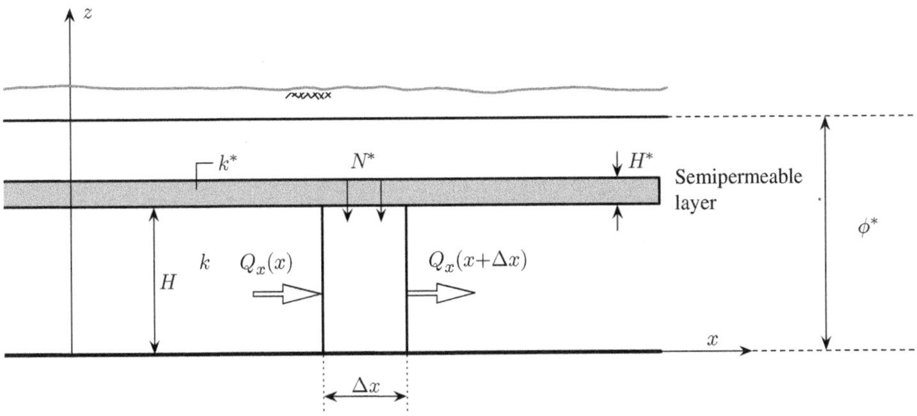

Figure 5.1 A semiconfined aquifer.

quantities associated with the semipermeable layer with an asterisk: the hydraulic conductivity and thickness of the semipermeable layer are k^* and H^*. The constant head in the aquifer above the leaky layer is ϕ^*. All heads are measured with respect to the impermeable base of the aquifer.

The hydraulic conductivity of the leaky layer is much less (at least a factor of 10) than the hydraulic conductivity k of the aquifer,

$$k^* \ll k. \tag{5.1}$$

The pressure, and therefore the head, is continuous across the interface between the aquifer and the leaky layer; the horizontal components of the gradient of ϕ, $(\partial\phi/\partial x,\ \partial\phi/\partial y)$, are continuous across the interface. This implies, with Darcy's law, that the horizontal component of flow in the leaky layer is a factor k^*/k less than that in the aquifer. Conversely, we expect a considerable difference between the heads ϕ and ϕ^*, and therefore a vertical component of flow that is much larger than the horizontal one, inside the leaky layer. This is the basis for neglecting the horizontal component of flow in the leaky layer with respect to the vertical one.

We view a semiconfined aquifer as a confined one with a variable infiltration rate N^* through the upper confining bed, as shown in Figure 5.1, and use the potential for shallow confined flow introduced in Section 2.1,

$$\Phi = kH\phi + C_{sc}, \tag{5.2}$$

where C_{sc} is a constant. The discharge vector equals the negative of the gradient of the potential, as usual,

$$Q_x = -\frac{\partial\Phi}{\partial x}, \qquad Q_y = -\frac{\partial\Phi}{\partial y}. \tag{5.3}$$

The differential equation for this potential is the same as that for unconfined flow with infiltration rate N^* (see (2.188) on page 71):

$$\nabla^2\Phi = -N^*. \tag{5.4}$$

By Darcy's law, the leakage N^*, which is positive downward, is equal to

$$N^* = k^*\frac{\phi^* - \phi}{H^*}. \tag{5.5}$$

We multiply both the numerator and denominator by kH:

$$N^* = -\frac{kH(\phi - \phi^*)}{kHH^*/k^*}. \tag{5.6}$$

We introduce the leakage factor Λ, which has the dimension of length and is defined as

$$\Lambda = \sqrt{kHH^*/k^*} = \sqrt{kHc}, \tag{5.7}$$

where c is the resistance of the leaky layer,

$$c = H^*/k^*, \tag{5.8}$$

and has the dimension of time.

We choose the constant C_{sc} in (5.2) as $-kH\phi^*$, i.e.,

$$\boxed{\Phi = kH(\phi - \phi^*)} \tag{5.9}$$

so that (5.6) becomes, with (5.7)

$$N^* = -\frac{\Phi}{\Lambda^2}. \tag{5.10}$$

The differential equation (5.4) in terms of the potential is

$$\boxed{\nabla^2\Phi = \frac{\Phi}{\Lambda^2}} \tag{5.11}$$

This equation is the *modified Helmholtz equation*.

5.1.2 General Solution for Planar Flow

For the case of planar flow, the differential equation (5.11) reduces to

$$\frac{d^2\Phi}{dx^2} = \frac{\Phi}{\Lambda^2}. \tag{5.12}$$

We obtain the general solution to this equation by positing a function of the form

$$\Phi = Ce^{\alpha x}, \tag{5.13}$$

which yields

$$C\alpha^2 e^{\alpha x} = \frac{C}{\Lambda^2} e^{\alpha x} \tag{5.14}$$

so that

$$\alpha = \pm 1/\Lambda. \tag{5.15}$$

The general solution to (5.12) therefore is

$$\Phi = Ae^{x/\Lambda} + Be^{-x/\Lambda}, \tag{5.16}$$

where the two constants A and B are determined from the boundary conditions.

Sometimes it is advantageous to write the general solution in terms of combinations of the exponential functions known as the hyperbolic cosine and sine, defined as

$$\cosh\frac{x}{\Lambda} = \tfrac{1}{2}(\mathrm{e}^{x/\Lambda} + \mathrm{e}^{-x/\Lambda}) \tag{5.17}$$

and

$$\sinh\frac{x}{\Lambda} = \tfrac{1}{2}(\mathrm{e}^{x/\Lambda} - \mathrm{e}^{-x/\Lambda}). \tag{5.18}$$

The first equation is symmetrical with respect to the origin, whereas the second one is anti-symmetrical. The general solution can be written as a combination of the hyperbolic sine and cosine:

$$\Phi = A\cosh\frac{x}{\Lambda} + B\sinh\frac{x}{\Lambda}. \tag{5.19}$$

Both (5.16) and (5.19) represent general solutions to the governing differential equation, but in different forms. Most problems suggest through their setting which form of the general solution is to be preferred, but we choose either one or the other. For that reason, using the same names for the constants should not cause confusion. As a general rule, (5.19) is advantageous if the problem can be separated in symmetrical and anti-symmetrical parts, which is not the case for problems where the domain is semi-infinite.

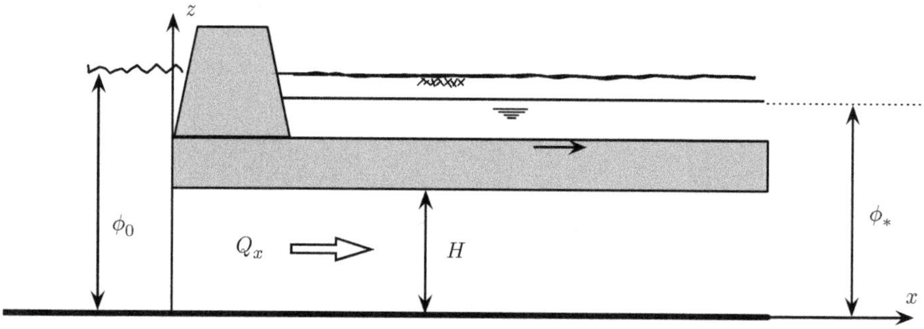

Figure 5.2 Leakage into an aquifer bounded by a long river.

5.1.3 Application: Leakage into an Aquifer Bounded by a Long River

As an application, we consider the problem of leakage into a semiconfined aquifer bounded by a long river, sketched in Figure 5.2. The boundary conditions are that the head approaches ϕ^* far away from the river and equals ϕ_0 along the river bank. We choose the y-axis along the river bank and the x-axis as shown in the figure, and the boundary conditions are

$$\begin{aligned} x = 0: \quad & \phi = \phi_0 \\ x \to \infty: \quad & \phi \to \phi^*. \end{aligned} \tag{5.20}$$

We express the potential by (5.9) and write these conditions as:

$$\begin{aligned} x = 0: \quad & \Phi = \Phi_0 = kH(\phi_0 - \phi^*) \\ x \to \infty: \quad & \Phi \to 0. \end{aligned} \tag{5.21}$$

The aquifer is semi-infinite, so that separation into symmetric and anti-symmetric parts is not useful; we choose the general solution (5.16),

$$\Phi = Ae^{x/\Lambda} + Be^{-x/\Lambda}. \tag{5.22}$$

The second condition in (5.21) can be satisfied only if A is zero,

$$A = 0, \tag{5.23}$$

and application of the first one gives

$$\Phi_0 = B, \tag{5.24}$$

so that the solution becomes

$$\Phi = \Phi_0 e^{-x/\Lambda} = kH(\phi_0 - \phi^*)e^{-x/\Lambda}. \tag{5.25}$$

The total leakage into the aquifer is, by continuity of flow, equal to the discharge leaving the aquifer at the river bank. It follows from Darcy's law (5.3) with (5.25) that

$$Q_x = -\frac{\partial \Phi}{\partial x} = \frac{\Phi_0}{\Lambda} e^{-x/\Lambda}, \tag{5.26}$$

which becomes at $x = 0$:

$$Q_x(0) = \frac{\Phi_0}{\Lambda} = \frac{kH(\phi_0 - \phi^*)}{\Lambda}. \tag{5.27}$$

The leakage is proportional to the difference in head and inversely proportional to the leakage factor.

Problem

5.1 A semiconfined aquifer is bounded by two infinite rivers that are a distance L apart. The y-axis is midway between the rivers, and the heads along the river banks at $x = -L/2$ and $x = L/2$ are ϕ_1 and ϕ_2, respectively. The head above the leaky layer is ϕ^*. Use the symbols for the aquifer parameters introduced in the text.

Questions:

1. Determine the head as a function of position.
2. Determine the total leakage into the aquifer in terms of the parameters of the problem.

Problem

5.2 A long straight dam separates two lakes as shown in Figure 5.3; the levels in the lakes are ϕ_1^* and ϕ_2^*. A leaky layer forms the bottoms of the lakes, which may be approximated as being semi-infinite, and separates the lakes from an aquifer of thickness H. The hydraulic conductivity and thickness

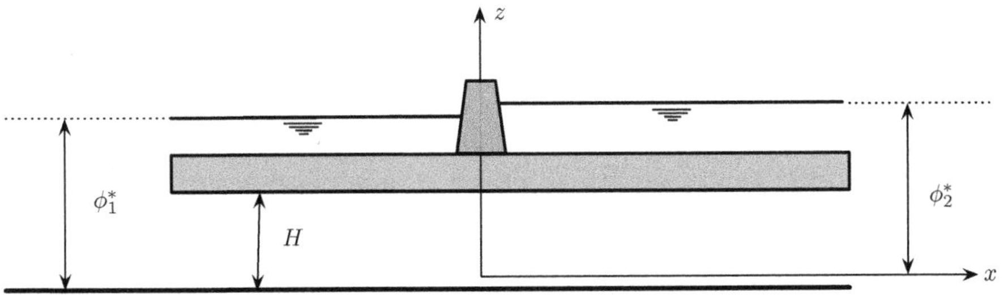

Figure 5.3 Leakage underneath a dam separating two lakes.

of the leaky layer are k^* and H^*. The centerline of the dam falls along the y-axis, and the width of the dam may be neglected.

Questions:

1. Find the head in the aquifer as a function of position.
2. Derive an expression for the flow underneath the dam per unit length of the dam.

Problem

5.3 Determine the flow underneath the dam of Figure 5.4 if an impervious sheet is placed over the interval $0 \le x \le L$ as shown in the figure.

5.1.4 Application: Leakage into a River

We consider the case of surface water/groundwater interaction between a confined aquifer of thickness H and hydraulic conductivity k and a river of width $2b$ that is separated from the aquifer by a leaky bottom of resistivity c. We choose the origin

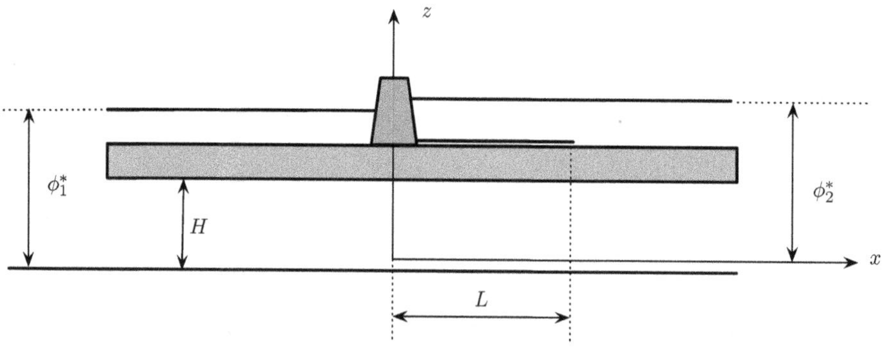

Figure 5.4 Leakage underneath a dam with an impervious sheet.

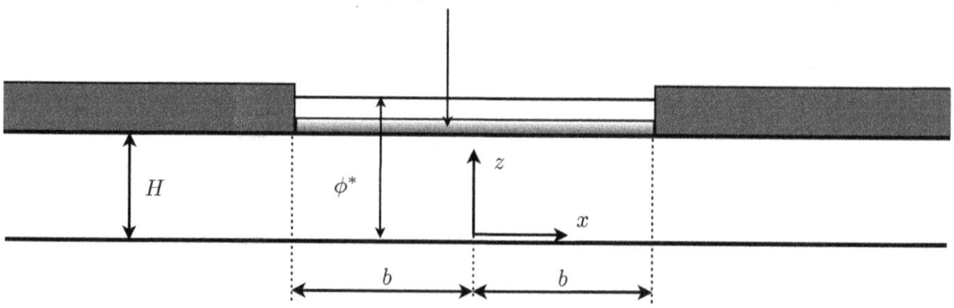

Figure 5.5 Cross sectional view of the aquifer.

of a Cartesian (x, y, z)-coordinate system such that the y-axis coincides with the centerline of the river and the z-axis points vertically upward; see Figure 5.5.

We determine the discharge Q flowing into the river per unit width of river as a function of the aquifer parameters and the heads ϕ_L and ϕ_R, given at $x = -L$ and at $x = L$, respectively; the boundary conditions are

$$x = -L: \quad \phi = \phi_L \tag{5.28}$$

$$x = L: \quad \phi = \phi_R. \tag{5.29}$$

The flow in the portion of aquifer below the river is semiconfined, with the discharge potential Φ defined as

$$\Phi = kH(\phi - \phi^*), \tag{5.30}$$

where Λ is the leakage factor, with

$$\Lambda = \sqrt{kHc}. \tag{5.31}$$

We label the zone of semiconfined flow as zone II, and the zones of confined flow as zones I ($x < -b$) and III ($x > b$). We apply definition (5.30) for the potential also in zones I and III; we are free to add an arbitrary constant to the potentials for confined and unconfined flow. This choice renders a potential that is continuous at $x = \pm b$.

We separate the solution to this problem into two parts: a symmetrical one, represented by a potential $\underset{s}{\Phi}$, and an anti-symmetrical one, represented by a potential $\underset{a}{\Phi}$:

$$\Phi = \underset{s}{\Phi} + \underset{a}{\Phi} \tag{5.32}$$

and we split the boundary conditions accordingly:

$$x = -L: \quad \underset{s}{\Phi} = \tfrac{1}{2}(\Phi_L + \Phi_R) = \Phi_0, \quad \underset{a}{\Phi} = \tfrac{1}{2}(\Phi_L - \Phi_R) = \Phi_1 \tag{5.33}$$

$$x = L: \quad \underset{s}{\Phi} = \tfrac{1}{2}(\Phi_L + \Phi_R) = \Phi_0, \quad \underset{a}{\Phi} = -\tfrac{1}{2}(\Phi_L - \Phi_R) = -\Phi_1. \tag{5.34}$$

The discharge potential must satisfy the boundary conditions

$$x = -L: \quad \Phi = \Phi_L = kH(\phi_L - \phi^*) \tag{5.35}$$

$$x = L: \quad \Phi = \Phi_R = kH(\phi_R - \phi^*). \tag{5.36}$$

As we separate the solution into symmetrical and anti-symmetrical components, we write the general solution as (5.19):

$$\Phi = A\cosh\frac{x}{\Lambda} + B\sinh\frac{x}{\Lambda}. \tag{5.37}$$

We use two other hyperbolic functions, the hyperbolic tangent,

$$\tanh\frac{x}{\Lambda} = \frac{\sinh(x/\Lambda)}{\cosh(x/\Lambda)}, \tag{5.38}$$

and the hyperbolic cotangent,

$$\coth\frac{x}{\Lambda} = \frac{\cosh(x/\Lambda)}{\sinh(x/\Lambda)}. \tag{5.39}$$

The Symmetrical Case

We represent the potential in zone I for the symmetrical case as

$$\underset{s}{\Phi} = -\underset{s}{Q_{x0}}x + \underset{s}{C}, \qquad x < -b, \tag{5.40}$$

and apply (5.33) to obtain $\underset{s}{C} = -\underset{s}{Q_{x0}}L + \Phi_0$, so that (5.40) becomes

$$\underset{s}{\Phi} = -\underset{s}{Q_{x0}}(x+L) + \Phi_0. \tag{5.41}$$

The potential $\underset{s}{\Phi}$ in zone II is symmetrical with respect to $x = 0$, and thus cannot contain the hyperbolic sine, i.e.,

$$\underset{s}{\Phi} = A\cosh\frac{x}{\Lambda}, \qquad |x| \le b. \tag{5.42}$$

Application of the condition of continuity of head at $x = -b$ yields

$$-\underset{s}{Q_{x0}}(L-b) + \Phi_0 = A\cosh\frac{b}{\Lambda} \tag{5.43}$$

and the condition of continuity of flow requires that

$$\underset{s}{Q_{x0}} = \frac{A}{\Lambda}\sinh\frac{b}{\Lambda}. \tag{5.44}$$

Division of both sides of (5.43) by the corresponding sides of (5.44) gives

$$\frac{-\underset{s}{Q_{x0}}(L-b) + \Phi_0}{\underset{s}{Q_{x0}}} = -(L-b) + \frac{\Phi_0}{\underset{s}{Q_{x0}}} = \Lambda\coth\frac{b}{\Lambda} \tag{5.45}$$

or

$$\underset{s}{Q_{x0}} = \frac{\Phi_0}{L-b+\Lambda\coth\frac{b}{\Lambda}}. \tag{5.46}$$

We obtain an expression for the constant A from (5.44):

$$A = \frac{\Lambda \underset{s}{Q_{x0}}}{\sinh\frac{b}{\Lambda}} \tag{5.47}$$

with $\underset{s}{Q_{x0}}$ given by (5.46). We observe from (5.46) that the discharge into the lake from one side is equal to

$$\underset{s}{Q_{x0}} = \frac{kH(\phi_0 - \phi^*)}{L - b + \Lambda \coth(b/\Lambda)}. \tag{5.48}$$

We divide both numerator and denominator by Λ:

$$\underset{s}{Q_{x0}} = \frac{\frac{kH}{\Lambda}(\phi_0 - \phi^*)}{\frac{L-b}{\Lambda} + \coth(b/\Lambda)}. \tag{5.49}$$

For fixed values of Λ, the discharge into the lake ($\underset{s}{Q_{x0}}$ is half this discharge) is proportional to a factor ν, defined as

$$\nu = \frac{kH}{\Lambda} = \frac{kH}{\sqrt{kHc}} = \sqrt{\frac{kH}{c}} \qquad [\mathrm{L/T}]. \tag{5.50}$$

We may write (5.49) in terms of the factor ν, which gives

$$\underset{s}{Q_{x0}} = \frac{\nu(\phi_0 - \phi^*)}{\frac{L-b}{\Lambda} + \coth\frac{b}{\Lambda}}. \tag{5.51}$$

The distribution of leakage is governed by the leakage factor Λ, which we refer to as the *space-leakage factor*, whereas the discharge is governed by yet another factor, which we call the *flux-leakage factor* ν. It appears that if we change both the transmissivity of the aquifer, kH, and the resistance c in such a manner that the product (i.e., Λ) remains constant, the leakage into the lake increases with ν. For example, if we increase kH by a factor of 2, and decrease c by a factor of 2, then Λ remains constant, but ν increases by a factor of 2; i.e., the leakage into the lake will double, whereas the distribution of the leakage over the lake bottom remains unchanged.

The Equivalent Length This same discharge would be obtained if the entire aquifer were confined and the head ϕ^* were applied at a distance

$$l^* = \Lambda \coth\frac{b}{\Lambda} \tag{5.52}$$

from the lake shore. We call l^* the equivalent length (Verruijt [1970]); it represents the length of aquifer that provides the same resistance to flow as the leaky bottom. When Λ is small relative to b, i.e., when

$$\frac{b}{\Lambda} \gg 1, \tag{5.53}$$

then $\coth(b/\Lambda) \sim 1$ and l^* reduces to Λ:

$$l^* \sim \Lambda, \qquad \frac{b}{\Lambda} \gg 1. \tag{5.54}$$

Substitution of expression (5.47) for A in expression (5.42) for $\underset{s}{\Phi}$ yields

$$\underset{s}{\Phi} = \frac{\Lambda \underset{s}{Q_{x0}}}{\sinh(b/\Lambda)} \cosh\frac{x}{\Lambda}. \tag{5.55}$$

An expression for the leakage $\underset{s}{\gamma}$ is obtained from (5.55) by division by Λ^2

$$\underset{s}{\gamma} = \frac{\underset{s}{\phi} - \phi^*}{c} = \frac{kH(\underset{s}{\phi} - \phi^*)}{kHc} = \frac{\underset{s}{\Phi}}{\Lambda^2} = \frac{\underset{s}{Q_{x0}}}{\Lambda \sinh(b/\Lambda)} \cosh\frac{x}{\Lambda}. \tag{5.56}$$

When Λ is large relative to b, the leakage varies relatively little with x for $|x| \le b$. When Λ is small as compared with b, however, γ varies significantly across the river bottom. If x/Λ is large, and x is positive, then $\cosh\frac{x}{\Lambda} \sim \frac{1}{2}e^{x/\Lambda}$, and $\sinh\frac{b}{\Lambda} \sim \frac{1}{2}e^{b/\Lambda}$, so that (5.56) may be approximated as follows:

$$\underset{s}{\gamma} \sim \frac{\underset{s}{Q_{x0}}}{\Lambda} e^{(x-b)/\Lambda}. \tag{5.57}$$

Considering the portion $0 \ge x \ge b$ of the river bottom, we observe that

$$\underset{s}{\gamma}(0) \sim \frac{\underset{s}{Q_{x0}}}{\Lambda} e^{-b/\Lambda} \tag{5.58}$$

and the maximum value occurs at the river bank, $x = b$,

$$\underset{s}{\gamma}(b) \sim \frac{\underset{s}{Q_{x0}}}{\Lambda}. \tag{5.59}$$

As Λ decreases, $\underset{s}{\gamma}(b)$ increases, but $\underset{s}{\gamma}(0)$ decreases (note that $e^{-b/\Lambda}/\Lambda$ decreases for decreasing Λ and approaches zero as Λ approaches zero); the leakage decreases rapidly with distance from the river bank.

For most practical purposes, the leakage may be neglected at a distance of about 3Λ from the river bank. At that distance, the ratio of $\underset{s}{\gamma}$ over $\underset{s}{\gamma}(b)$ is

$$\frac{\underset{s}{\gamma}(b-3\Lambda)}{\underset{s}{\gamma}(b)} = e^{-3} \sim 0.05, \tag{5.60}$$

i.e., the leakage is reduced to about 5% of the maximum. We consider the zone $-b+3\Lambda < x < b-3\Lambda$ as the zone of equilibration; in this zone, the leakage is nearly zero, and the head in the aquifer is nearly equal (equilibrated) to the head in the river.

The Anti-symmetrical Case

We represent the potential in zone I for the anti-symmetrical case as

$$\underset{a}{\Phi} = -\underset{a}{Q_{x0}} x + \underset{a}{C}, \qquad x < -b, \tag{5.61}$$

and apply (5.33), to obtain $\underset{a}{C} = -\underset{a}{Q_{x0}} L + \Phi_1$, so that (5.61) becomes

$$\underset{a}{\Phi} = -\underset{a}{Q_{x0}}(x + L) + \Phi_1. \tag{5.62}$$

The potential $\underset{a}{\Phi}$ in zone II is anti-symmetrical with respect to $x = 0$, and thus cannot contain the cosine:

$$\underset{a}{\Phi} = B \sinh \frac{x}{\Lambda}, \qquad |x| \leq b. \tag{5.63}$$

Application of the condition of continuity of head at $x = -b$ yields

$$-\underset{a}{Q_{x0}}(L - b) + \Phi_1 = -B \sinh \frac{b}{\Lambda} \tag{5.64}$$

and application of the condition of continuity of flow at $x = -b$ gives

$$\underset{a}{Q_{x0}} = -\frac{B}{\Lambda} \cosh \frac{b}{\Lambda}, \qquad |x| \leq b. \tag{5.65}$$

We divide both sides of (5.64) by the corresponding sides of (5.65)

$$-(L - b) + \Phi_1 / \underset{a}{Q_{x0}} = \Lambda \tanh \frac{b}{\Lambda} \tag{5.66}$$

and solve for $\underset{a}{Q_{x0}}$:

$$\underset{a}{Q_{x0}} = \frac{\Phi_1}{L - b + \Lambda \tanh(b/\Lambda)} = \frac{kH(\phi_1 - \phi^*)}{L - b + \Lambda \tanh(b/\Lambda)}. \tag{5.67}$$

so that the equivalent length is $l^* = \Lambda \tanh(b/\Lambda)$. We write this equation in terms of the factor ν (compare (5.51)), which gives

$$\underset{a}{Q_{x0}} = \frac{\nu(\phi_1 - \phi^*)}{\frac{L-b}{\Lambda} + \tanh \frac{b}{\Lambda}}. \tag{5.68}$$

Note that $\underset{a}{Q_{x0}}$ does not represent the discharge into the river; some of this water will flow underneath the river. Note further that there is no net flow into the river; the discharge that enters the river on the one side leaves on the other side via leakage through the river bottom.

The flow underneath the river, $\underset{u}{Q_{x0}}$, is equal to the discharge at $x = 0$, i.e.,

$$\underset{u}{Q_{x0}} = \underset{a}{Q_x}(0) = -\frac{B}{\Lambda}\cosh\frac{0}{\Lambda} = -\frac{B}{\Lambda}. \tag{5.69}$$

We obtain an expression for B from (5.65):

$$B = -\frac{\Lambda \underset{a}{Q_{x0}}}{\cosh(b/\Lambda)} \tag{5.70}$$

so that

$$\underset{u}{Q_{x0}} = \frac{\underset{a}{Q_{x0}}}{\cosh(b/\Lambda)}. \tag{5.71}$$

This amount reduces as Λ decreases; in that case nearly all water enters and leaves the river. The flow into the river is equal to

$$\underset{a}{Q_{x0}} - \underset{u}{Q_{x0}} = \underset{a}{Q_{x0}}\left[1 - \frac{1}{\cosh(b/\Lambda)}\right]. \tag{5.72}$$

The distribution of leakage and the expression for l^* for the anti-symmetrical case become nearly identical as Λ decreases.

Equivalent Resistance

Modeling the leakage through river bottoms using computer models, as discussed in Chapter 8, takes much computational effort, especially if the leakage factor Λ is small. It is possible to model the leakage through the river bottom in an approximate manner using line-sinks, which are lines of extraction; they are similar to wells, but, rather than extracting water at a point, water is extracted along the line-sink. Line-sinks are used to model streams, and resistance-specified line-sinks are line-sinks with a leaky bottom of given resistance and a fixed head above it. Although the width of the line-sinks is not included in models in terms of geometry, the width, b^*, is included in the formulas used to compute their extraction rate. A resistance line-sink is illustrated in Figure 5.6. Our objective is to compute the resistance of the line-sink such that it withdraws the proper amount of water from the aquifer, given the head ϕ^* in the river. To do that, we compute ϕ_b below the line-sink from the given discharge Q_{x0} (see Figure 5.6). We recall the formula for the discharge Q_{x0} that flows into a river for the symmetrical case into one half of the river, i.e., into the section $-b \le x \le 0$, (5.48). We choose $L = b$, so that we obtain for Q_{x0}

$$\underset{s}{Q_{x0}} = \frac{kH(\phi_b - \phi^*)}{\Lambda\coth(b/\Lambda)}. \tag{5.73}$$

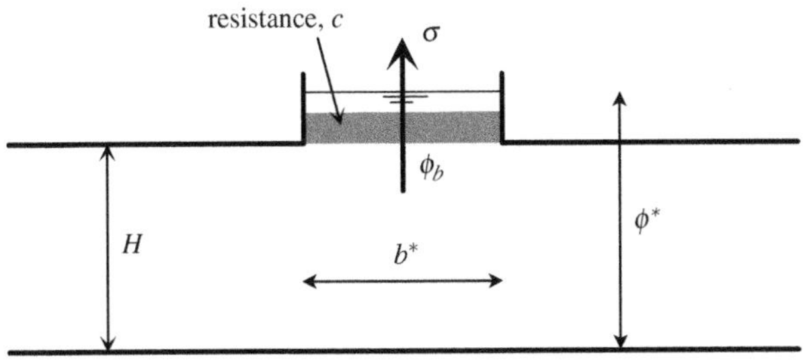

Figure 5.6 A resistance-specified line-sink; the strength of the line-sink, σ, is computed from the difference in head below the stream, ϕ_b, and in the stream, ϕ_0.

We apply the condition of continuity of flow:

$$b^*\sigma = b^*\frac{\phi_b - \phi^*}{c^*} = \frac{kH(\phi_b - \phi^*)}{kHc^*/b^*}, \tag{5.74}$$

where ϕ_b is the head along the river bank, b^* the imaginary width of the stream, and c^* the equivalent resistance of the stream. Since the stream serves to represent a river bank, rather than a stream, its width does not have any physical meaning and can be chosen arbitrarily. The two discharges in equations (5.73) and (5.74) must be the same, so that

$$\frac{c^*}{b^*} = \frac{\Lambda}{kH}\coth(b/\Lambda) = \frac{\Lambda c}{\Lambda^2}\coth(b/\Lambda) = \frac{c}{\Lambda}\coth(b/\Lambda) \tag{5.75}$$

or

$$c^* = c\frac{b^*}{\Lambda}\coth(b/\Lambda). \tag{5.76}$$

The hyperbolic cotangent of b/Λ approaches unity when b/Λ is large:

$$c^* \approx c\frac{b^*}{\Lambda}, \qquad b/\Lambda >> 1. \tag{5.77}$$

If we choose b^* to be equal to Λ, this becomes

$$c^* \approx c, \qquad b^* = \Lambda, \qquad b/\Lambda >> 1. \tag{5.78}$$

If b/Λ is small, i.e., if the leakage factor is large relative to the width of the river, the hyperbolic cotangent approaches Λ/b and (5.76) becomes

$$c^* \approx c\frac{b^*}{b}, \qquad b/\Lambda << 1. \tag{5.79}$$

If we choose b^* equal to b for this case, then

$$c^* \approx c, \qquad b^* = b, \qquad b/\Lambda << 1. \tag{5.80}$$

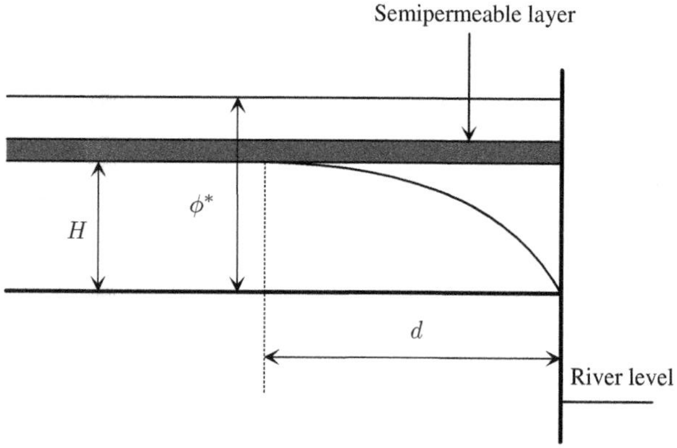

Figure 5.7 Semiconfined flow near a bluff.

5.1.5 Semiconfined Flow near a Bluff

Consider the problem of semiconfined flow near a bluff, where the level in the river is below the base of the aquifer. We choose the origin of an (x, y, z)-coordinate system at a point on the river bank, with the z-axis pointing up, the y-axis along the river bank, and the x-axis away from the aquifer. The aquifer has thickness H, hydraulic conductivity k, and the head above the aquitard is given to be ϕ^*.

We define two zones. In zone 1, the flow is semiconfined, i.e.,

$$\overset{1}{\Phi} = kH(\phi - \phi^*) = Ae^{x/\Lambda} + Be^{-x/\Lambda}, \qquad -\infty < x \leq -d. \tag{5.81}$$

The aquifer is semi-infinite, and we have

$$\overset{1}{\Phi} \to 0, \qquad x \to -\infty, \tag{5.82}$$

so that $B = 0$, and (5.81) becomes

$$\overset{1}{\Phi} = Ae^{x/\Lambda}. \tag{5.83}$$

We obtain for the discharge

$$\overset{1}{Q}_x = -\frac{A}{\Lambda}e^{x/\Lambda}. \tag{5.84}$$

The head below the semipermeable layer in zone 2 is equal to the elevation head H, because the pressure is zero there; the leakage through the upper boundary of the aquifer in the unconfined zone is:

$$N = \frac{\phi^* - \phi}{c} = \frac{\phi^* - H}{c}, \qquad -d \leq x \leq 0. \tag{5.85}$$

The expression for the potential in zone 2 is:

$$\overset{2}{\Phi} = \tfrac{1}{2}k\phi^2 = -\tfrac{1}{2}Nx^2 + ax + b, \tag{5.86}$$

and the expression for the discharge is

$$\overset{2}{Q}_x = Nx - a. \tag{5.87}$$

The boundary condition along the bluff, $x = 0$, becomes

$$\overset{2}{\Phi} = 0, \qquad x = 0, \tag{5.88}$$

and we obtain from (5.86) that $b = 0$ and

$$\overset{2}{\Phi} = -\tfrac{1}{2}Nx^2 + ax. \tag{5.89}$$

The distance between the bluff and the interzonal boundary is d, and we rewrite (5.83) as

$$\overset{1}{\Phi} = Ae^{x/\Lambda} = \alpha e^{d/\Lambda}e^{x/\Lambda} = \alpha e^{(x+d)/\Lambda}, \tag{5.90}$$

where

$$\alpha = Ae^{-d/\Lambda}. \tag{5.91}$$

The condition of continuity of flow at $x = -d$ gives, from (5.90) and (5.86),

$$\frac{d\overset{1}{\Phi}}{dx} = \frac{\alpha}{\Lambda} = \frac{d\overset{2}{\Phi}}{dx} = -Nx + a, \qquad x = -d, \tag{5.92}$$

or

$$\alpha = Nd\Lambda + a\Lambda. \tag{5.93}$$

We set the head ϕ equal to H at $x = -d$, which gives, with (5.93),

$$\overset{1}{\Phi} = \alpha = Nd\Lambda + a\Lambda = kH(H - \phi^*) \tag{5.94}$$

and

$$\overset{2}{\Phi} = \tfrac{1}{2}kH^2 = -\tfrac{1}{2}Nd^2 - ad \tag{5.95}$$

or, from (5.94)

$$a = \frac{kH(H - \phi^*)}{\Lambda} - Nd. \tag{5.96}$$

We solve (5.95) for a:

$$a = -\tfrac{1}{2}Nd - \tfrac{1}{2}\frac{kH^2}{d}. \tag{5.97}$$

We equate expressions (5.96) and (5.97) for a:

$$\frac{kH(H-\phi^*)}{\Lambda} - Nd = -\tfrac{1}{2}Nd - \tfrac{1}{2}\frac{kH^2}{d} \tag{5.98}$$

or, after multiplication by $2d$,

$$\frac{2kH(H-\phi^*)d}{\Lambda} - Nd^2 + kH^2 = 0. \tag{5.99}$$

We rearrange and divide by N,

$$d^2 + \frac{2kH(\phi^* - H)}{\Lambda N}d - \frac{kH^2}{N} = 0, \tag{5.100}$$

and use expression (5.85) for N, $N = (\phi^* - H)/c$,

$$d^2 + \frac{2kHc}{\Lambda}d - \frac{kHcH}{\phi^* - H} = 0. \tag{5.101}$$

We write kHc as Λ^2:

$$d^2 + 2\Lambda d - \Lambda^2\frac{H}{\phi^* - H} = 0, \tag{5.102}$$

or

$$\left(\frac{d}{\Lambda}\right)^2 + 2\left(\frac{d}{\Lambda}\right) - \frac{H}{\phi^* - H} = 0, \tag{5.103}$$

and solve for d/Λ:

$$\frac{d}{\Lambda} = -1 \pm \sqrt{1 + \frac{H}{\phi^* - H}}. \tag{5.104}$$

Since only positive values of d/Λ may occur, the plus sign applies, so that

$$\frac{d}{\Lambda} = \sqrt{1 + \frac{H}{\phi^* - H}} - 1 = \sqrt{\frac{\phi^*}{\phi^* - H}} - 1 = \sqrt{\frac{1}{1 - H/\phi^*}} - 1. \tag{5.105}$$

It follows that d is a linear function of Λ and that d/Λ increases as ϕ^* approaches H. For example, if H is 10 m, and ϕ^* is 12 m, then $d/\Lambda = 1.45$; when $\phi^* = 15$ m, $d/\Lambda = 0.73$.

5.1.6 Axisymmetric Flow

We derive the differential equation for axisymmetric semiconfined flow by considering continuity of flow in the volume between the cylinders of radii r and $(r+\Delta r)$ (see Figure 5.8). We define two functions $f(r)$ and $g(r)$:

$$f(r) = 2\pi r Q_r, \qquad g(r) = \pi r^2, \tag{5.106}$$

and write the condition of continuity of flow as

$$f(r+\Delta r) - f(r) = N^*[g(r+\Delta r) - g(r)], \tag{5.107}$$

where N^* is the leakage, expressed earlier by (5.10),

$$N^* = -\frac{\Phi}{\Lambda^2}. \tag{5.108}$$

We divide both sides of (5.107) by Δr, pass to the limit for $\Delta r \to 0$:

$$\frac{df}{dr} = -\frac{\Phi}{\Lambda^2}\frac{dg}{dr}, \tag{5.109}$$

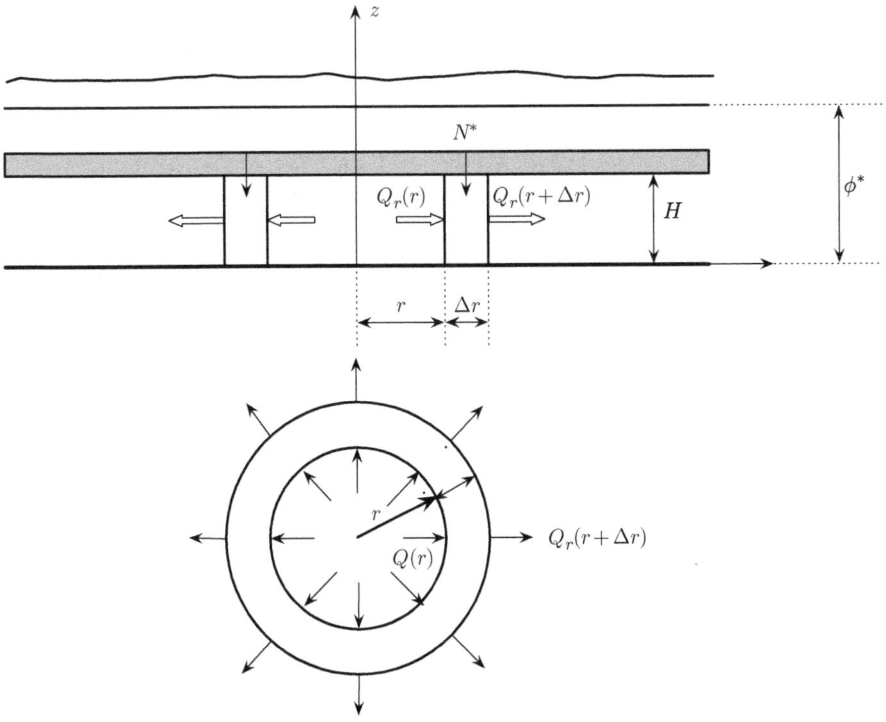

Figure 5.8 Axisymmetric semiconfined flow.

or, with (5.106),

$$2\pi \frac{d}{dr}(rQ_r) = -\frac{\Phi}{\Lambda^2}2\pi r. \tag{5.110}$$

We use Darcy's law, $Q_r = -d\Phi/dr$, and obtain the following differential equation:

$$\frac{1}{r}\frac{d}{dr}\left[r\frac{d\Phi}{dr}\right] = \frac{\Phi}{\Lambda^2}. \tag{5.111}$$

Note that (5.111) is a special case of (5.11): the Laplacian $\nabla^2\Phi$ for radial flow reduces to:

$$\nabla^2\Phi = \frac{1}{r}\frac{d}{dr}\left[r\frac{d\Phi}{dr}\right] = \frac{d^2\Phi}{dr^2} + \frac{1}{r}\frac{d\Phi}{dr}, \tag{5.112}$$

and (5.111) may be written as

$$\boxed{\frac{d^2\Phi}{dr^2} + \frac{1}{r}\frac{d\Phi}{dr} = \frac{\Phi}{\Lambda^2}} \tag{5.113}$$

We transform this equation by making the substitution

$$\xi = r/\Lambda, \tag{5.114}$$

so that, with

$$\frac{d\Phi}{dr} = \frac{d\Phi}{d\xi}\frac{d\xi}{dr} = \frac{1}{\Lambda}\frac{d\Phi}{d\xi}, \qquad \frac{d^2\Phi}{dr^2} = \frac{1}{\Lambda^2}\frac{d^2\Phi}{d\xi^2}, \tag{5.115}$$

the differential equation (5.113) becomes

$$\frac{d^2\Phi}{d\xi^2} + \frac{1}{\xi}\frac{d\Phi}{d\xi} - \Phi = 0. \tag{5.116}$$

This equation is Bessel's modified differential equation of order zero. The general solution to this equation is

$$\Phi = A\mathrm{I}_0(\xi) + B\mathrm{K}_0(\xi) = A\mathrm{I}_0(r/\Lambda) + B\mathrm{K}_0(r/\Lambda), \tag{5.117}$$

where $\mathrm{I}_0(\xi)$ and $\mathrm{K}_0(\xi)$ are the modified Bessel functions of order zero and of the first and second kind, respectively. Series expansions for these functions are available, as well as tables and subroutines for numerical evaluation. These are given in mathematical handbooks and will not be presented here. The derivatives of K_0 and I_0 are modified Bessel functions of order one:

$$\frac{d\mathrm{I}_0(\xi)}{d\xi} = \mathrm{I}_1(\xi), \qquad \frac{d\mathrm{K}_0(\xi)}{d\xi} = -\mathrm{K}_1(\xi). \tag{5.118}$$

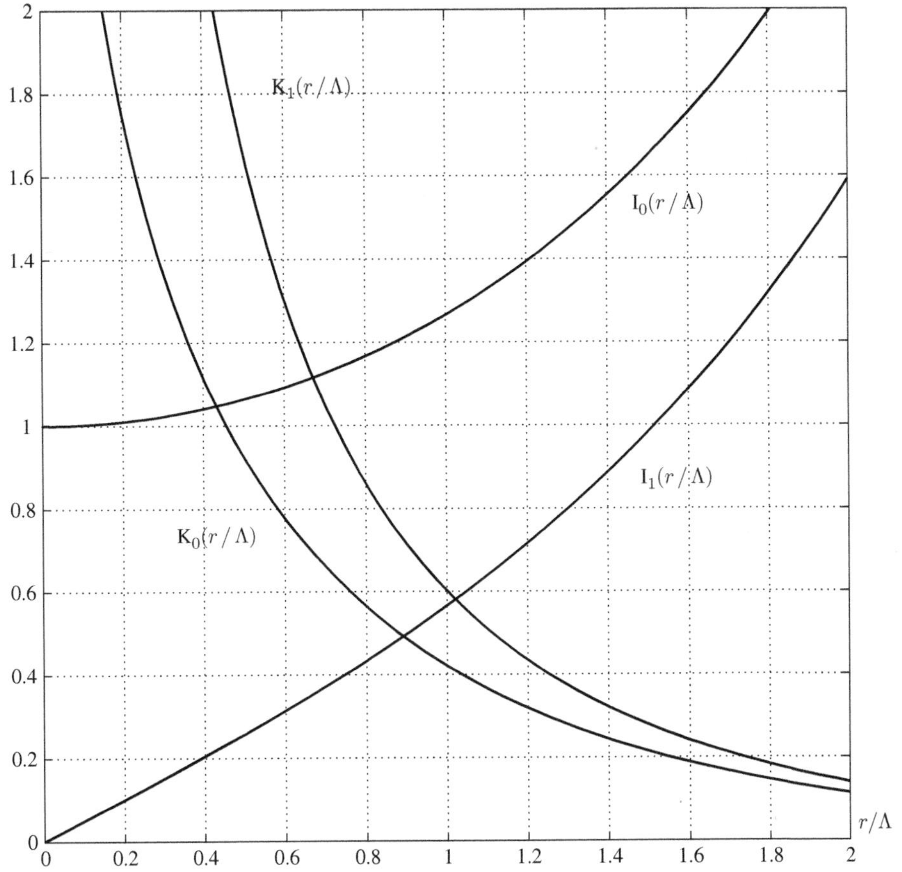

Figure 5.9 Modified Bessel functions of orders zero and one.

We obtain an expression for Q_r by differentiating (5.117) with respect to r:

$$Q_r = -\frac{A}{\Lambda}\mathrm{I}_1(r/\Lambda) + \frac{B}{\Lambda}\mathrm{K}_1(r/\Lambda). \tag{5.119}$$

Graphs of the modified Bessel functions of order one and zero are presented in Figure 5.9. The constants A and B are determined by application of the boundary conditions, as shown for an application in the following section.

5.1.7 Application: Flow toward a Well in an Infinite Leaky Aquifer

The case of radial semiconfined flow toward a well in an infinite leaky aquifer is illustrated in Figure 5.10. The discharge of the well is Q, and the origin of the radial coordinate system (r, z) is at the center of the well; the symbols for the various aquifer parameters are as introduced earlier. It is given that the head in the aquifer approaches ϕ^* at large distances from the well, i.e.,

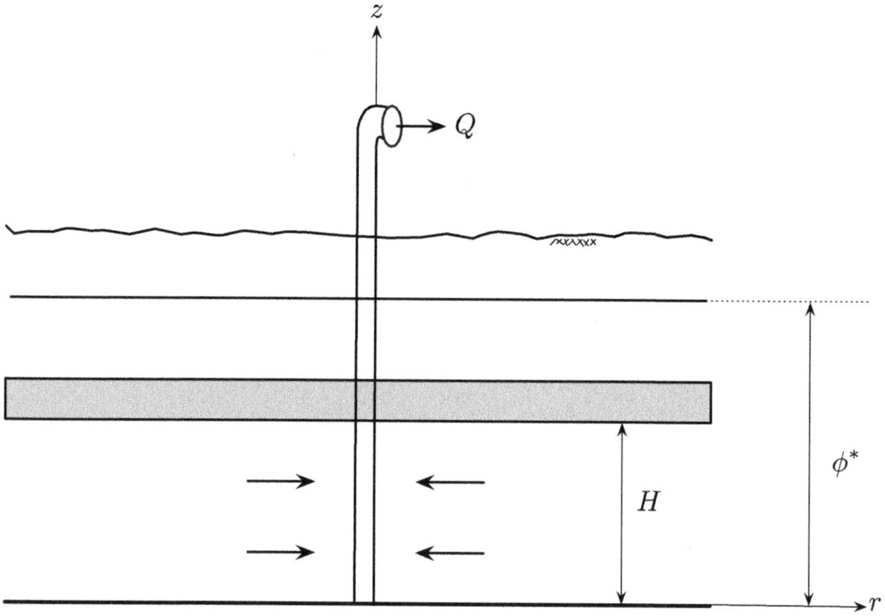

Figure 5.10 Semiconfined flow toward a well.

$$r \to \infty, \qquad \Phi \to 0. \tag{5.120}$$

The well pumps a discharge Q, and the condition at the boundary of the well, $r = r_w$, is

$$r = r_w, \qquad Q = 2\pi r_w[-Q_r(r_w)], \tag{5.121}$$

where it is noted that Q_r is negative for flow to the well.

We see from Figure 5.9 that $\mathrm{I}_0(r/\Lambda)$ approaches infinity for $r \to \infty$; the condition (5.120) will be met only if the constant A in (5.117) is zero,

$$A = 0. \tag{5.122}$$

We apply (5.121) to the expression (5.119) for Q_r:

$$Q = -2\pi r_w \frac{B}{\Lambda} \mathrm{K}_1(r_w/\Lambda) \tag{5.123}$$

or

$$B = -\frac{Q\Lambda}{2\pi r_w} \frac{1}{\mathrm{K}_1(r_w/\Lambda)}. \tag{5.124}$$

We obtain the expression for the potential by using (5.122) and (5.124) in the general solution (5.117), which yields

$$\Phi = -\frac{Q\Lambda}{2\pi r_w}\frac{K_0(r/\Lambda)}{K_1(r_w/\Lambda)}. \tag{5.125}$$

A comparison of this solution with the one obtained for confined flow is possible by using the following approximations (Abramowitz and Stegun [1965]):

$$\begin{aligned} K_0(r/\Lambda) &\approx -\ln\frac{r}{1.123\Lambda}, \qquad && r/\Lambda << 1 \\ K_1(r/\Lambda) &\approx \Lambda/r, && r/\Lambda << 1. \end{aligned} \tag{5.126}$$

These approximations are sufficiently accurate for practical purposes for $r/\Lambda < 0.2$. For many cases, Λ is larger than 100; for example, if $k/k^* = 10^3$, and $HH^* = 10\ \text{m}^2$, then $\Lambda = \sqrt{10^4}\,\text{m} = 100$ m so that the approximations (5.126) are accurate for $r < 20$ m. The potential (5.125) becomes in the neighborhood of the well:

$$\Phi \approx -\frac{Q\Lambda}{2\pi r_w}\frac{-\ln\frac{r}{1.123\Lambda}}{\Lambda/r_w} = \frac{Q}{2\pi}\ln r - \frac{Q}{2\pi}\ln(1.123\Lambda). \tag{5.127}$$

The potential behaves near the well as in a confined aquifer. Apparently, the effect of the leakage through the leaky layer inside a cylinder of radius $r/\Lambda = 0.2$ is not important, and the effect of leakage outside that cylinder is accounted for by the constant in (5.127).

Note that the solution (5.125) is valid only if the head at the well is larger than H. The potential at the well is obtained from (5.127) by setting r equal to r_w,

$$\Phi_w = kH(\phi_w - \phi^*) = \frac{Q}{2\pi}\ln\frac{r_w}{1.123\Lambda} \tag{5.128}$$

and ϕ_w must fulfill the condition

$$\phi_w = \phi^* + \frac{Q}{2\pi kH}\left[\ln\frac{r_w}{1.123\Lambda}\right] \geq H. \tag{5.129}$$

Problem

5.4 Consider the case of a semiconfined aquifer with a circular boundary of radius R, shown in Figure 5.11. The head along the boundary is ϕ_0 and the head above the semiconfining layer is ϕ^*. Determine an expression for the total leakage through the leaky layer if the hydraulic conductivity of the aquifer is k, the thickness is H, and the resistance of the leaky layer is c.

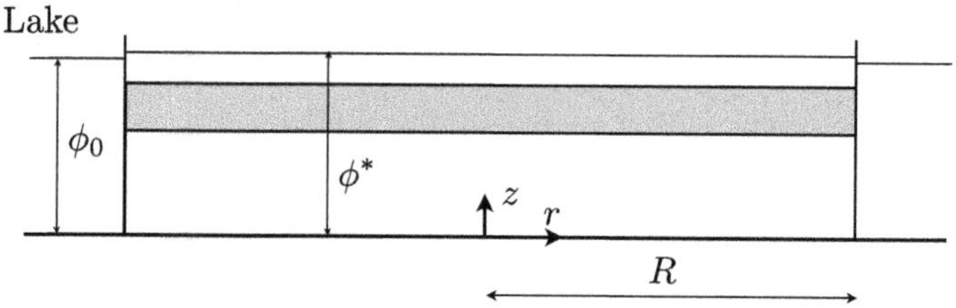

Figure 5.11 A circular semiconfined aquifer.

5.1.8 Approximate Solution for Flow toward a Well

We consider the case of a well of radius r_w, with $r_w << \Lambda$ in an infinite aquifer. The function $I_0(r/\Lambda)$ cannot be involved because the potential must vanish at infinity. Thus, the solution has the form

$$\Phi = A\mathrm{K}_0(r/\Lambda), \tag{5.130}$$

and the discharge vector component Q_r becomes

$$Q_r = -\frac{\partial \Phi}{\partial r} = \frac{A}{\Lambda}\mathrm{K}_1(r/\Lambda). \tag{5.131}$$

We use the approximate relation

$$\mathrm{K}_1(r/\Lambda) \approx \frac{\Lambda}{r}, \qquad \frac{r}{\Lambda} << 1. \tag{5.132}$$

The well has a discharge Q,

$$Q = -2\pi r_w Q_r \approx -2\pi \frac{r_w}{\Lambda} A \frac{\Lambda}{r_w} = -2\pi A, \tag{5.133}$$

so that

$$A \approx -\frac{Q}{2\pi}, \tag{5.134}$$

and the approximate (but usually very accurate) expression for the potential for a well in a semiconfined aquifer becomes

$$\Phi = -\frac{Q}{2\pi}\mathrm{K}_0(\mathrm{r}/\Lambda). \tag{5.135}$$

Problem

5.5 Answer the same question as for Problem 5.4, but now with a well of discharge Q at the center of the aquifer.

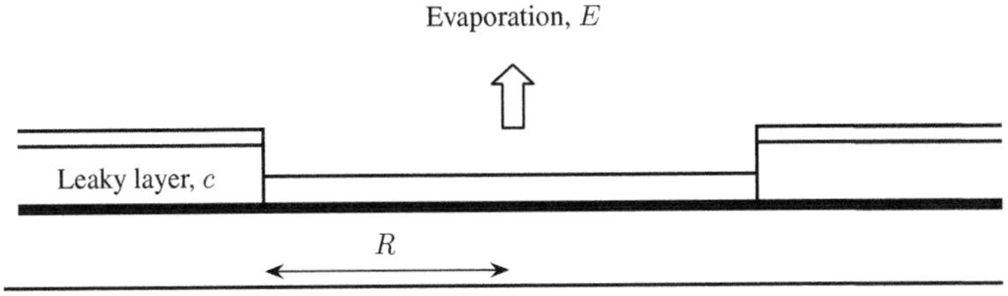

Figure 5.12 Circular lake.

Problem

5.6 It is planned to create a circular lake in the aquifer above a leaky layer of resistance c; see Figure 5.12. The leaky layer separates the upper aquifer, with constant head ϕ_0, from a lower one. Both aquifers are infinite in extent and are in equilibrium; there is no flow and the heads in the two aquifers are equal. It is planned to seal the lake off from the upper aquifer, but leakage will occur from the lower one into the upper one after the lake has been created. It is given that the evaporation from the lake is 0.2 m/y and that the radius of the lake will be R. The thickness of the aquifer is H and the hydraulic conductivity is k.

Questions:

1. Determine an expression for the head in the lake in terms of the parameters of the problem.
2. Determine an expression for the head in the aquifer.
3. Determine the numerical value of the head in the lake if it is given that $k = 10$ m/d, $H = 20$ m, $\phi_0 = 30$ m, $R = 300$ m, and $c = 100$ days. The ground surface is 35 m above the base of the aquifer. Decide whether creating the lake is feasible for the given values of the aquifer parameters.

5.1.9 Application: A Well near a River in a Semiconfined Aquifer

The differential equation (5.11) for two-dimensional semiconfined flow,

$$\nabla^2 \Phi = \frac{\Phi}{\Lambda^2}, \tag{5.136}$$

is linear so that the superposition principle holds. A number of two-dimensional problems can be solved by the method of images; as an application, consider the problem of a well in a leaky aquifer near a river, illustrated in Figure 5.13. The origin of an (x, y, z)-coordinate system is chosen at a point of the riverbank, with

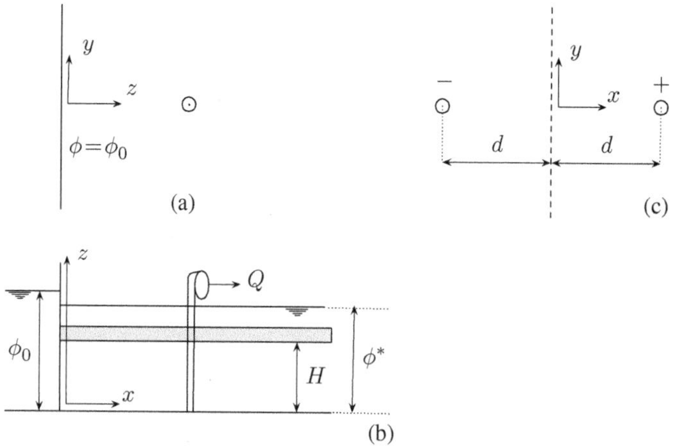

Figure 5.13 A well in a leaky aquifer near a river.

the z-axis pointing vertically upward, the y-axis along the riverbank, and the x-axis into the aquifer. The well is at $x = d,\ y = 0$; the discharge and radius of the well are Q and r_w. There are two boundary conditions. The first one is that the head is ϕ_0 along the river bank:

$$x = 0: \qquad \phi = \phi_0. \tag{5.137}$$

The corresponding value of the potential is:

$$x = 0: \qquad \Phi = \Phi_0 = kH(\phi_0 - \phi^*). \tag{5.138}$$

The second boundary condition is that the head approaches ϕ^* for $x \to \infty$:

$$x \to \infty, \qquad \phi \to \phi^*, \qquad \Phi \to 0. \tag{5.139}$$

We solve the problem in two steps: first we solve the problem illustrated in Figure 5.13(c) for flow from an image recharge well at $x = -d,\ y = 0$, toward the well at $x = d,\ y = 0$, and second we determine an additional solution to meet the boundary condition along the river bank. We obtain the potential for flow with the two wells by superposition of two solutions of the form (5.135):

$$\Phi = -\frac{Q}{2\pi}\mathrm{K}_0(r_1/\Lambda) + \frac{Q}{2\pi}\mathrm{K}_0(r_2/\Lambda), \tag{5.140}$$

where r_1 and r_2 are the radial distances from the well and the recharge well, respectively. This potential vanishes along the river bank, since r_1 equals r_2 for $x = 0$. A constant potential cannot be added to (5.140) to satisfy the boundary condition

(5.138): a constant does not fulfill the differential equation (5.136). This is in contrast to Laplace's equation.

We may, however, add a potential for uniform flow chosen such that it equals Φ_0 at $x = 0$. The boundary conditions (5.138) and (5.139) are identical to those for the problem of Figure 5.2 with the solution given by (5.25),

$$\Phi = \Phi_0 e^{-x/\Lambda}. \tag{5.141}$$

Addition of the potentials (5.140) and (5.141) yields the complete solution to the problem:

$$\Phi = \Phi_0 e^{-x/\Lambda} - \frac{Q}{2\pi}\left[\mathrm{K}_0(r_1/\Lambda) - \mathrm{K}_0(r_2/\Lambda)\right]. \tag{5.142}$$

This potential fulfills the boundary conditions (5.138) and (5.139).

Maximum Pumping Rate of a Well Near a River

We consider that the discharge of the well is limited by the condition that no river water may enter the well. The expression for the discharge potential is

$$\Phi = \frac{Q}{2\pi}\mathrm{K}_0\left(\frac{\sqrt{(x+d)^2+y^2}}{\Lambda}\right) - \frac{Q}{2\pi}\mathrm{K}_0\left(\frac{\sqrt{(x-d)^2+y^2}}{\Lambda}\right) + \Phi_0 \mathrm{e}^{-x/\Lambda}. \tag{5.143}$$

The maximum discharge of the well occurs when $Q_x(0,0) = 0$. We obtain an expression for Q_x by differentiation

$$\begin{aligned} Q_x(x,y) = {} & \frac{Q}{2\pi}\mathrm{K}_1\left(\frac{\sqrt{(x+d)^2+y^2}}{\Lambda}\right)\frac{x+d}{\Lambda\sqrt{(x+d)^2+y^2}} \\ & - \frac{Q}{2\pi}\mathrm{K}_1\left(\frac{\sqrt{(x-d)^2+y^2}}{\Lambda}\right)\frac{x-d}{\Lambda\sqrt{(x-d)^2+y^2}} + \frac{\Phi_0}{\Lambda}\mathrm{e}^{-x/\Lambda}. \end{aligned} \tag{5.144}$$

We substitute zero for x and y and obtain, replacing Q by $Q_{\max}$,

$$Q_x(0,0) = \frac{Q_{\max}}{\pi\Lambda}\mathrm{K}_1\left(\frac{d}{\Lambda}\right) + \frac{\Phi_0}{\Lambda} = 0 \tag{5.145}$$

so that

$$Q_{\max} = -\pi\frac{\Phi_0}{\mathrm{K}_1(d/\Lambda)} = \frac{\pi kH(\phi^* - \phi_0)}{\mathrm{K}_1(d/\Lambda)}. \tag{5.146}$$

We observe that pumping without capturing river water is possible only if $\phi^* > \phi_0$, i.e., if the head in the aquifer above the aquifer with the well has a head ϕ^* that is greater than that in the river. Furthermore, if Λ is large relative to d, i.e., if the semiconfining layer is relatively tight, the Bessel function can be approximated by (5.126) so that

$$Q_{\max} \approx \frac{\pi kHd(\phi^* - \phi_0)}{\Lambda}. \tag{5.147}$$

In this case the flow approaches the confined case, with $2kH(\phi^* - \phi_0)/\Lambda$ replacing the uniform flow term Q_{x0} in (2.132).

5.1.10 The Leakage Factors Λ and ν for the General Case

The distribution of leakage is governed by the space-leakage factor $\Lambda = \sqrt{kHc}$, which has the dimension of length, and the magnitudes of discharges are governed by the flux-leakage factor $\nu = kH/\Lambda = \sqrt{kH/c}$ [L/T]. We demonstrate that the properties of these factors hold for the general case of semiconfined flow. The partial differential for semiconfined flow is:

$$\frac{\partial^2 \Phi}{\partial x^2} + \frac{\partial^2 \Phi}{\partial y^2} = \frac{\Phi}{\Lambda^2}. \tag{5.148}$$

We introduce new dimensionless independent variables ξ and η, defined as

$$\xi = \frac{x}{\Lambda} \tag{5.149}$$

$$\eta = \frac{y}{\Lambda}. \tag{5.150}$$

We transform the derivatives with respect to x and y into those with respect to ξ and η as follows:

$$\frac{\partial \Phi}{\partial x} = \frac{\partial \Phi}{\partial \xi}\frac{\partial \xi}{\partial x} = \frac{\partial \Phi}{\partial \xi}\frac{1}{\Lambda}, \qquad \frac{\partial^2 \Phi}{\partial x^2} = \frac{\partial^2 \Phi}{\partial \xi^2}\frac{1}{\Lambda^2} \tag{5.151}$$

$$\frac{\partial \Phi}{\partial y} = \frac{\partial \Phi}{\partial \eta}\frac{\partial \eta}{\partial y} = \frac{\partial \Phi}{\partial \eta}\frac{1}{\Lambda}, \qquad \frac{\partial^2 \Phi}{\partial y^2} = \frac{\partial^2 \Phi}{\partial \eta^2}\frac{1}{\Lambda^2}. \tag{5.152}$$

The governing partial differential equation in terms of ξ and η becomes, after multiplication of both sides by Λ^2,

$$\frac{\partial^2 \Phi}{\partial \xi^2} + \frac{\partial^2 \Phi}{\partial \eta^2} = \Phi. \tag{5.153}$$

The solution in terms of ξ and η is independent of Λ, i.e., the leakage factor is indeed a scaling factor; for example, leakage occurring over a circle of some dimensionless radius, say ρ, in the (ξ,η)-plane corresponds to the same leakage occurring in the physical plane over a circle of radius $R = \rho\Lambda$.

The potential Φ is defined as

$$\Phi = kH(\phi - \phi^*), \tag{5.154}$$

and substitution in the differential equation (5.153) gives

$$\frac{\partial^2 kH(\phi - \phi^*)}{\partial \xi^2} + \frac{\partial^2 kH(\phi - \phi^*)}{\partial \eta^2} = kH(\phi - \phi^*). \tag{5.155}$$

Division by kH gives

$$\frac{\partial^2 (\phi - \phi^*)}{\partial \xi^2} + \frac{\partial^2 (\phi - \phi^*)}{\partial \eta^2} = \phi - \phi^*. \tag{5.156}$$

The solution to this equation consists of two parts: the constant ϕ^* plus a solution to the equation

$$\frac{\partial^2 \tilde{\phi}}{\partial \xi^2} + \frac{\partial^2 \tilde{\phi}}{\partial \eta^2} = \tilde{\phi} \tag{5.157}$$

so that

$$\phi = \tilde{\phi} + \phi^*. \tag{5.158}$$

We obtain an expression for the components of the discharge vector:

$$Q_x = -kH\frac{\partial \phi}{\partial x} = -kH\frac{\partial \phi}{\partial \xi}\frac{d\xi}{dx} = -\frac{kH}{\Lambda}\frac{\partial \phi}{\partial \xi} \tag{5.159}$$

$$Q_y = -kH\frac{\partial \phi}{\partial y} = -kH\frac{\partial \phi}{\partial \eta}\frac{d\eta}{dy} = -\frac{kH}{\Lambda}\frac{\partial \phi}{\partial \eta} \tag{5.160}$$

or, using the flux-leakage factor ν,

$$\begin{aligned} Q_x &= -\nu\frac{\partial \phi}{\partial \xi} \\ Q_y &= -\nu\frac{\partial \phi}{\partial \eta}. \end{aligned} \tag{5.161}$$

The derivatives of head with respect to the dimensionless independent variables are independent of the aquifer parameters (except, possibly, as a result of boundary values), so that indeed the discharge components appear to be linearly dependent on the flux-leakage factor ν. Equations (5.161) are similar in form to Darcy's law, but represent the discharges due to leakage; the factor ν has the same role in these equations as the hydraulic conductivity does in Darcy's law.

The flux-leakage factor ν is also a measure of the leakage that occurs over an area. The integral of the normal component of flow along a closed curve is equal to the total leakage that occurs over that area. Since the discharge vector is proportional to ν, the same holds true for the integrated leakage. For the case in which the potential is a function of x only, the integrated leakage between points x_1 and x_2 is equal to $Q_x(x_2) - Q_x(x_1)$, which is proportional to ν.

Forms of the Flux-Leakage Factor

We may represent the flux-leakage factor ν in two different ways:

$$\nu = \frac{kH}{\Lambda} = \frac{T}{\Lambda} \qquad [\mathrm{L/T}], \tag{5.162}$$

where $T = kH$ is the transmissivity of the aquifer. We use the definition of Λ to obtain an alternative form:

$$\nu = \frac{kH}{\sqrt{kHc}} = \sqrt{\frac{T^2}{cT}} = \sqrt{\frac{T}{c}} \qquad [\mathrm{L/T}]. \tag{5.163}$$

This form shows that the fluxes are controlled by the square root of the fraction T/c; the larger the transmissivity is relative to the resistance, the larger the fluxes due to leakage. The space-leakage factor, Λ, which controls the spatial distribution of leakage, has the dimensions of length, whereas the flux-leakage factor, ν, has the dimensions of specific discharge, L/T.

5.2 Flow in a System of Two Aquifers

We consider the case of leakage between two confined aquifers separated by a leaky layer of resistance c, illustrated in Figure 5.14. The aquifers are numbered from the top down[1] and have transmissivities $T_1 = k_1H_1$ and $T_2 = k_2H_2$, where k_1 and k_2 are the hydraulic conductivities and H_1 and H_2 the thicknesses of aquifers 1 and 2, respectively. We introduce discharge potentials for the two aquifers:

$$\overset{1}{\Phi} = T_1\phi, \qquad \overset{2}{\Phi} = T_2\phi, \tag{5.164}$$

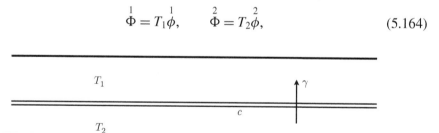

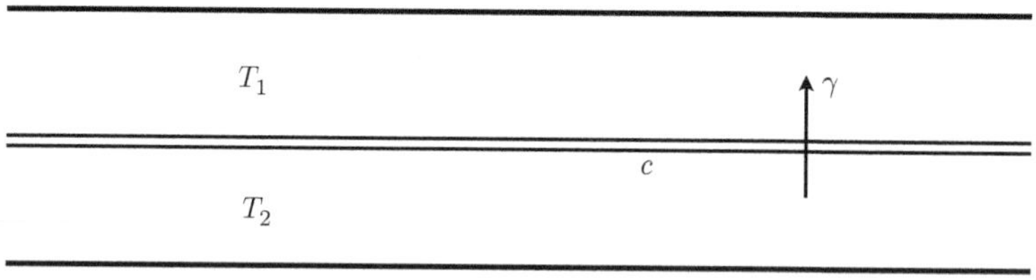

Figure 5.14 Section through a system of two aquifers separated by a leaky layer of resistance c.

[1] There is an inconsistency in numbering leaky aquifers as compared with numbering strata in stratified aquifers, where the numbering is from the bottom up. Creating regional models, we tend to include aquifers, beginning with the uppermost one where the sources of water are, and gradually include deeper aquifers as demanded by the objectives of the study.

where $\overset{1}{\phi}$ and $\overset{2}{\phi}$ are the heads of aquifers 1 and 2, respectively. We introduce a comprehensive potential Φ (Strack [1989]) as

$$\Phi = \overset{1}{\Phi} + \overset{2}{\Phi} \tag{5.165}$$

and a leakage potential G (Strack and Namazi [2014]), as

$$G = \frac{T_1 T_2}{T}(\overset{2}{\phi} - \overset{1}{\phi}) = \frac{T_1}{T}\overset{2}{\Phi} - \frac{T_2}{T}\overset{1}{\Phi}, \tag{5.166}$$

where T is the total transmissivity:

$$T = T_1 + T_2. \tag{5.167}$$

We obtain the relation between G and the leakage through the leaky layer, γ, from (5.166)

$$\gamma = \frac{\overset{2}{\phi} - \overset{1}{\phi}}{c} = \frac{T}{cT_1T_2}G = \frac{G}{\Lambda^2}, \tag{5.168}$$

where the constant Λ, the leakage factor, has the dimension of length and is defined as

$$\Lambda = \sqrt{\frac{cT_1T_2}{T}}. \tag{5.169}$$

We define the discharge vectors $(\overset{1}{Q}_x, \overset{1}{Q}_y)$ and $(\overset{2}{Q}_x, \overset{2}{Q}_y)$ for aquifers 1 and 2 and the comprehensive discharge (Q_x, Q_y) as

$$Q_x = \overset{1}{Q}_x + \overset{2}{Q}_x = -\frac{\partial \Phi}{\partial x} = -\left[\frac{\partial \overset{1}{\Phi}}{\partial x} + \frac{\partial \overset{2}{\Phi}}{\partial x}\right] \tag{5.170}$$

$$Q_y = \overset{1}{Q}_y + \overset{2}{Q}_y = -\frac{\partial \Phi}{\partial y} = -\left[\frac{\partial \overset{1}{\Phi}}{\partial y} + \frac{\partial \overset{2}{\Phi}}{\partial y}\right]. \tag{5.171}$$

We express the potentials $\overset{1}{\Phi}$ and $\overset{2}{\Phi}$ in terms of the comprehensive potential and the leakage potential using (5.165) and (5.166):

$$\overset{1}{\Phi} = \frac{T_1}{T}\Phi - G \tag{5.172}$$

$$\overset{2}{\Phi} = \frac{T_2}{T}\Phi + G. \tag{5.173}$$

Indeed, the combination $\overset{1}{\Phi}+\overset{2}{\Phi}$ yields Φ and the combination $T_1\overset{2}{\Phi}/T - T_2\overset{1}{\Phi}/T$ yields G, which is a potential; its gradient represents the contribution of leakage to the discharge vectors in the aquifers.

When the flow in the aquifers is equilibrated, i.e., when the heads in the two aquifers are equal to each other at any given point in the horizontal plane, then the function G vanishes, as follows from definition (5.168). Thus, the equilibrated state is represented by setting G equal to zero in (5.172) and (5.173), so that

$$\overset{1}{\Phi}=\frac{T_1}{T}\Phi, \qquad \overset{1}{\phi}=\overset{2}{\phi} \tag{5.174}$$

$$\overset{2}{\Phi}=\frac{T_2}{T}\Phi, \qquad \overset{1}{\phi}=\overset{2}{\phi}. \tag{5.175}$$

This implies that the discharges in the two aquifers will be distributed according to the transmissivities of the aquifers; the discharge in aquifer 1 is T_1/T times the total, and that in aquifer 2 is T_2/T times the total.

5.2.1 Governing Differential Equations

The leakage from the lower aquifer into the upper one is given by (5.168) and equals the negative of the divergence of the discharge vector in the lower aquifer, which equals the Laplacian of the discharge potential $\overset{2}{\Phi}$, so that

$$\nabla^2\overset{2}{\Phi}=\frac{G}{\Lambda^2}. \tag{5.176}$$

We use (5.173) to express $\overset{2}{\Phi}$ in terms of G:

$$\frac{T_2}{T}\nabla^2\Phi+\nabla^2 G=\frac{G}{\Lambda^2}. \tag{5.177}$$

If there is no net infiltration into the system, then $\nabla^2\Phi = 0$,

$$\nabla^2 G=\frac{G}{\Lambda^2}. \tag{5.178}$$

5.2.2 Application: Flow toward a Stream in the Upper Aquifer

We consider the case in which there is a stream in aquifer 1 at $x = 0$, with the head set at $\overset{1}{\phi}=\phi_0$, and with a total discharge $2|Q_{x0}|$; see Figure 5.15. We consider the case that the flow is symmetrical with respect to $x = 0$; the discharge from the

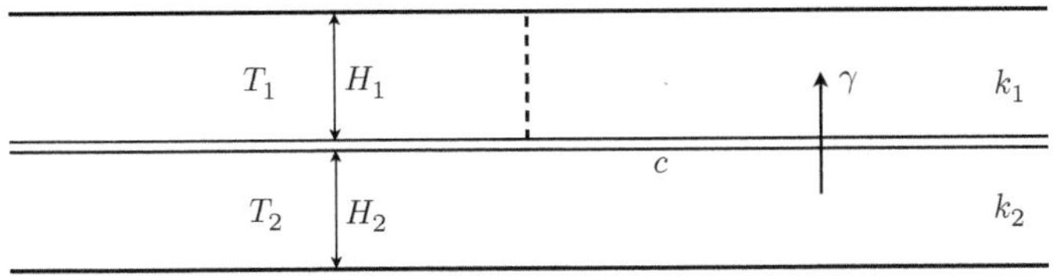

Figure 5.15 Two aquifers separated by a leaky layer with a stream in the upper aquifer.

left into the stream is thus Q_{x0}. There is no net infiltration into the system, so that (5.178) applies and the comprehensive potential is harmonic, i.e.,

$$\nabla^2 \Phi = 0. \tag{5.179}$$

The comprehensive potential for uniform flow is

$$\Phi = -Q_{x0}x + \Phi_0, \tag{5.180}$$

where Φ_0 is the value of the comprehensive potential at the origin.

The general solution to the governing equation for G for planar flow, (5.178), is

$$G = Ae^{x/\Lambda} + Be^{-x/\Lambda}. \tag{5.181}$$

We consider the following boundary conditions for $x \leq 0$:

$$x = -L,\ \overset{1}{\phi} = \phi_1, \qquad x = 0,\ \overset{1}{\phi} = \phi_0, \tag{5.182}$$

and

$$x \to -\infty,\ \gamma \to 0, \qquad x = 0,\ \overset{2}{Q}_x = 0. \tag{5.183}$$

It follows from the first condition in (5.183) that $B = 0$, so that (5.181) becomes

$$G = Ae^{x/\Lambda}. \tag{5.184}$$

It follows from (5.166) that

$$G = \frac{T_1}{T}\overset{2}{\Phi} - \frac{T_2}{T}\Phi_1 \tag{5.185}$$

so that

$$-\frac{dG}{dx} = -\frac{A}{\Lambda}e^{x/\Lambda} = -\frac{T_1}{T}\frac{d\overset{2}{\Phi}}{dx} + \frac{T_2}{T}\frac{d\overset{1}{\Phi}}{dx} = \frac{T_1}{T}\overset{2}{Q}_x - \frac{T_2}{T}\overset{1}{Q}_x. \tag{5.186}$$

We set both x and $\overset{2}{Q}_x$ equal to zero according to (5.183), and set $\overset{1}{Q}_x$ equal to the total unknown discharge Q_{x0} that flows from infinity to the stream,

$$\frac{A}{\Lambda} = \frac{T_2}{T} Q_{x0} \tag{5.187}$$

or

$$A = \frac{\Lambda T_2}{T} Q_{x0}. \tag{5.188}$$

We use (5.172) with (5.180), (5.184), and (5.188):

$$\overset{1}{\Phi} = \frac{T_1}{T}\Phi - G = \frac{T_1}{T}[-Q_{x0}x + \Phi_0] - \frac{\Lambda T_2}{T} Q_{x0} e^{x/\Lambda} \tag{5.189}$$

or

$$\overset{1}{\Phi} = -\frac{T_1}{T} Q_{x0}\left[x + \Lambda \frac{T_2}{T_1} e^{x/\Lambda}\right] + \frac{T_1}{T}\Phi_0. \tag{5.190}$$

We set the head at $x = -L$ equal to ϕ_1 in the upper aquifer, (5.182):

$$T_1\phi_1 = -\frac{T_1}{T} Q_{x0}\left[-L + \Lambda \frac{T_2}{T_1} e^{-L/\Lambda}\right] + \frac{T_1}{T}\Phi_0, \tag{5.191}$$

and set the head at the stream, $x = 0$, equal to ϕ_0:

$$T_1\phi_0 = -\frac{T_1}{T} Q_{x0}\left[\Lambda \frac{T_2}{T_1}\right] + \frac{T_1}{T}\Phi_0. \tag{5.192}$$

We subtract (5.192) from (5.191):

$$T_1(\phi_1 - \phi_0) = Q_{x0}\frac{T_1}{T}\left[L + \Lambda \frac{T_2}{T_1}\left(1 - e^{-L/\Lambda}\right)\right], \tag{5.193}$$

and solve for Q_{x0}:

$$Q_{x0} = \frac{T(\phi_1 - \phi_0)}{L + \Lambda \frac{T_2}{T_1}\left(1 - e^{-L/\Lambda}\right)}. \tag{5.194}$$

Note that the total discharge into the stream, σ, per unit length is equal to $2Q_{x0}$:

$$\sigma = 2Q_{x0}. \tag{5.195}$$

The Flux-Leakage Factor

As for the case of semiconfined flow, we introduce a flux-leakage factor ν. We divide both numerator and denominator of (5.194) by Λ:

$$Q_{x0} = \frac{\frac{T}{\Lambda}(\phi_1 - \phi_0)}{\frac{L}{\Lambda} + \frac{T_2}{T_1}\left(1 - e^{-L/\Lambda}\right)}. \tag{5.196}$$

We define the flux-leakage factor for leaky flow in a confined system of two aquifers as

$$\nu = \frac{T}{\Lambda} = T\sqrt{\frac{T}{cT_1T_2}} = \sqrt{\frac{T}{c}}\sqrt{\frac{T}{T_1}\frac{T}{T_2}}. \tag{5.197}$$

The general solution to this problem is composed of a symmetrical part and an anti-symmetrical part. The latter part does not contribute to the discharge into the stream, which may be computed for the general case, provided that ϕ_1 be redefined as

$$\phi_1 = \frac{1}{2}\left[\overset{1}{\phi}(-L) + \overset{1}{\phi}(L)\right]. \tag{5.198}$$

Special Case; $\Lambda/L >> 1$

If $\Lambda/L >> 1$, the expression for Q_{x0} simplifies; L/Λ becomes small relative to unity, and we approximate $\mathrm{e}^{-L/\Lambda}$ by the first two terms in its Taylor series expansion;

$$\mathrm{e}^{-L/\Lambda} \approx 1 - \frac{L}{\Lambda}. \tag{5.199}$$

The expression for Q_{x0}, (5.194), becomes

$$Q_{x0} \approx \frac{T(\phi_1 - \phi_0)}{L[1 + T_2/T_1]} = \frac{T(\phi_1 - \phi_0)}{LT/T_1} = \frac{T_1(\phi_1 - \phi_0)}{L}. \tag{5.200}$$

This result implies that for large values of Λ the contribution of the lower aquifer can be neglected relative to that of the upper one if the stream is in the upper aquifer.

Simplified Model; Equivalent Resistance

If the purpose of a study is to compute discharges, such as the discharge of the stream in this application, we can replace the actual two-layer system by a simplified one, based on the concept of equivalent resistance. We replace the actual aquifer system by a system of two layers, but without directly including the leaky layer of resistance c; this leads to a simplified model. We model the contribution of leakage through the leaky layer into the stream, shown in Figure 5.16(a), by

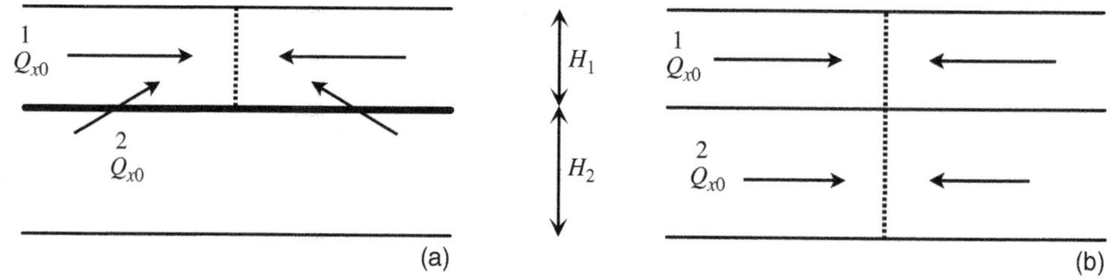

Figure 5.16 Concentrated leakage: the leakage through the leaky layer in panel (a) is replaced by concentrated leakage in panel (b).

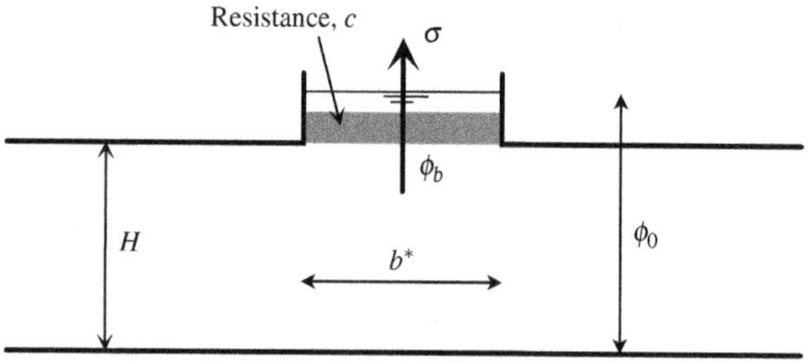

Figure 5.17 A stream modeled by a resistance-specified line-sink; the width b^* is included in the equations, but not in the geometrical representation of the stream in the model.

upward leakage concentrated below the stream. We do this by replacing the actual system by a system without a leaky layer, but with streams in both layers. The flow into the lower stream represents the upward leakage through the leaky layer, but concentrated along the line. This is illustrated in Figure 5.16(b).

We model the streams by line-sinks, which are lines in the horizontal plane along which water is extracted from the aquifer at some rate per unit length of line, σ [L^2/T]. A fixed head is associated with each line-sink, and we add resistance between the head in the line-sink and that in the aquifer directly below. We determine this resistance so that the discharges into the two streams match the actual contributions from the upper aquifer and from the lower one through leakage. The principle of a resistance line-sink is illustrated in Figure 5.17. The head in the aquifer below the line-sink is ϕ_b, which we determine from the actual model shown in Figure 5.16(a). We compute the resistance of the resistance line-sink, c^*, in combination with its width, b^*, and head, ϕ_0, such that the discharge, σb^*, matches the value obtained in the actual configuration. If the head in the aquifer of Figure 5.17 at $x = -L$ equals ϕ_1, and the head below the stream, centered along $x = 0$, is ϕ_b, then the discharge flowing from $x = -L$ to $x = 0$ is Q_{x0}:

$$Q_{x0} = \frac{T(\phi_1 - \phi_b)}{L}, \tag{5.201}$$

where T is the transmissivity of the aquifer. The upward flow into the stream through its leaky bottom is equal to Q_{x0}, by continuity of flow, and is found by application of Darcy's law to the upward flow through the leaky stream bed:

$$\frac{T(\phi_1-\phi_b)}{L}=b^*\frac{\phi_b-\phi_0}{c^*}. \tag{5.202}$$

We solve this for the unknown head below the stream, ϕ_b:

$$\phi_b=\frac{\frac{T}{L}\phi_1+\frac{b^*}{c^*}\phi_0}{\frac{T}{L}+\frac{b^*}{c^*}}. \tag{5.203}$$

Substitution of this expression for ϕ_b in (5.201) gives an equation for Q_{x0}:

$$Q_{x0}=\frac{T}{L}\left[\phi_1-\frac{\frac{T}{L}\phi_1+\frac{b^*}{c^*}\phi_0}{\frac{T}{L}+\frac{b^*}{c^*}}\right]=\frac{\frac{Tb^*}{Lc^*}(\phi_1-\phi_0)}{\frac{T}{L}+\frac{b^*}{c^*}}=\frac{T(\phi_1-\phi_0)}{L+\frac{c^*}{b^*}T}. \tag{5.204}$$

Given the discharge Q_{x0}, we compute the equivalent resistance of the stream such that this discharge is withdrawn, given ϕ_1 and ϕ_0. We use this approach to determine equivalent resistances to match the solution to the original problem. We place resistance line-sinks in each of the two aquifers and determine expressions for the discharge in each; we compute the resistances such that their application results in these discharges. The corresponding expressions for the discharges extracted from aquifers 1 and 2 through resistance line-sinks are

$$\overset{1}{Q}_{x0}=\frac{T_1(\phi_1-\phi_0)}{L+\frac{c_1^*}{b_1^*}T_1},\qquad \overset{2}{Q}_{x0}=\frac{T_2(\phi_1-\phi_0)}{L+\frac{c_2^*}{b_2^*}T_2}. \tag{5.205}$$

The discharge $\overset{2}{Q}_{x0}$ extracted by the stream from aquifer 2 represents the leakage, modeled as concentrated along $x=0$ and channeled vertically upward into the stream in aquifer 1. We computed the discharge into the stream using the system of two aquifers separated by a leaky layer, (5.194). The contributions from the semi-infinite aquifers to this discharge are $(T_1/T)Q_{x0}$ for the upper aquifer and $(T_2/T)Q_{x0}$ for the lower one. We set the contribution to the upper aquifer equal to the first equation in (5.205):

$$\overset{1}{Q}_{x0}=\frac{T_1(\phi_1-\phi_0)}{L+\Lambda\frac{T_2}{T_1}\left(1-e^{-L/\Lambda}\right)}=\frac{T_1(\phi_1-\phi_0)}{L+\frac{c_1^*}{b_1^*}T_1} \tag{5.206}$$

and determine c_1^*/b_1^* from this equality:

$$\frac{c_1^*}{b_1^*}=\frac{T_2}{T_1}\frac{\Lambda}{T_1}\left[1-e^{-L/\Lambda}\right]=\frac{T_2}{T_1}\frac{\Lambda^2}{T_1\Lambda}\left[1-e^{-L/\Lambda}\right]=\frac{T_2}{T_1}\frac{cT_1T_2}{\Lambda TT_1}\left[1-e^{-L/\Lambda}\right] \tag{5.207}$$

or

$$\frac{c_1^*}{b_1^*} = \frac{T_2}{T_1}\frac{cT_2}{\Lambda T}\left[1 - e^{-L/\Lambda}\right]. \tag{5.208}$$

We obtain a similar expression for c_2^*/b_2^*,

$$\frac{c_2^*}{b_2^*} = \frac{T_2}{T_1 T_2}\Lambda\left[1 - e^{-L/\Lambda}\right] = \frac{cT_1 T_2}{\Lambda T_1 T}\left[1 - e^{-L/\Lambda}\right] = \frac{cT_2}{\Lambda T}\left[1 - e^{-L/\Lambda}\right]. \tag{5.209}$$

We note that

$$\frac{c_1^*/b_1^*}{c_2^*/b_2^*} = \frac{T_2}{T_1}. \tag{5.210}$$

We should choose the distance L such that the leakage at $x = -L$, which is proportional to $G(-L)$, is negligible with respect to that at the origin, which is proportional to $G(0)$. This implies that $e^{-L/\Lambda} << 1$, so that (5.208) and (5.209) may be approximated as

$$\frac{c_1^*}{b_1^*} \approx \frac{T_2}{T}\frac{cT_2}{\Lambda T_1}, \qquad \frac{c_2^*}{b_2^*} \approx \frac{T_1}{T}\frac{cT_2}{\Lambda T_1}. \tag{5.211}$$

5.3 Flow to a Well of Discharge Q

We consider the case of radial flow in an infinite system of two aquifers. There is a well at $r = 0$ and the comprehensive potential is

$$\Phi = \frac{Q}{2\pi}\ln r + C, \tag{5.212}$$

where C is a constant. The leakage potential G results from a well; we write it in the following form:

$$G = \frac{Q_l}{2\pi}\mathrm{K}_0(r/\Lambda), \tag{5.213}$$

chosen in order to give a physical meaning to the constant Q_l, as we will see. For small values of r/Λ, the Bessel function K_0 can be approximated by the negative of the logarithm, so that

$$G \approx -\frac{Q_l}{2\pi}\ln r + c, \qquad r/\Lambda << 1, \tag{5.214}$$

where c is a constant.

The expressions for the potentials in the upper and lower aquifers in terms of the comprehensive and leakage potentials are (5.172) and (5.173),

$$\overset{1}{\Phi} = -G + \frac{T_1}{T}\Phi \tag{5.215}$$

and

$$\overset{2}{\Phi} = G + \frac{T_2}{T}\Phi. \tag{5.216}$$

We express the potentials in the upper and lower aquifers in terms of the general solution (5.212) and (5.213) for $r/\Lambda << 1$, which gives, with (5.214),

$$\overset{1}{\Phi} \approx \frac{Q_l}{2\pi}\ln r + \frac{T_1}{T}\frac{Q}{2\pi}\ln r + \cdots \tag{5.217}$$

$$\overset{2}{\Phi} \approx -\frac{Q_l}{2\pi}\ln r + \frac{T_2}{T}\frac{Q}{2\pi}\ln r + \frac{T_2}{T}C + \cdots. \tag{5.218}$$

We collect terms:

$$\overset{1}{\Phi} \approx \frac{Q}{2\pi}\left[\frac{Q_l}{Q} + \frac{T_1}{T}\right]\ln r + \cdots \tag{5.219}$$

$$\overset{2}{\Phi} \approx -\frac{Q}{2\pi}\left[\frac{Q_l}{Q} - \frac{T_2}{T}\right]\ln r + \cdots. \tag{5.220}$$

We now have the following three possibilities:

- Case 1: Well in the upper aquifer
- Case 2: Well in the lower aquifer
- Case 3: Well in both aquifers.

5.3.1 Case 1: Well in the Upper Aquifer

For case 1, there cannot be a well (a singularity) in the lower aquifer; the coefficient of the term $\ln r$ in the expression for $\overset{2}{\Phi}$ must be zero, i.e.,

$$Q_l = \frac{T_2}{T}Q. \tag{5.221}$$

5.3.2 Case 2: Well in the Lower Aquifer

For case 2, there cannot be a well in the upper aquifer, i.e.,

$$Q_l = -\frac{T_1}{T}Q. \tag{5.222}$$

We explain the expressions (5.221) and (5.222) using the concept of equilibrated leakage. The leakage would be equilibrated, i.e., it would be zero, if there were two wells in the system, one in aquifer 1 with discharge $Q_1 = (T_1/T)Q$ and one in aquifer 2 with discharge $Q_2 = (T_2/T)Q$. The heads would then be identical everywhere in the system. To obtain the solution for a well of discharge Q in the upper aquifer and none in the lower one, we must replace the well in the lower aquifer by leakage, i.e., the leakage component, Q_l, would be up (positive) and equal to $(T_2/T)Q$; we obtain for the reverse case a value of $Q_l = -(T_1/T)Q$, i.e., down (negative), by the same reasoning.

5.3.3 Case 3: A Well in Both Aquifers

If the well is in both aquifers, the heads in the two aquifers are equal, i.e., G must be zero at the well screen. The discharge Q represents the sum of the discharges of the wells; we compute the magnitude of Q_l from the condition:

$$G = 0, \qquad x = x_w + r_w, \qquad y = y_w, \tag{5.223}$$

where r_w is the radius of the well, and x_w and y_w the coordinates of the well.

An Abandoned Well

An abandoned well is a well that is screened in both aquifers and is no longer pumped; i.e., the total discharge is $Q = 0$. In that case the comprehensive potential due to the abandoned well is a constant (the abandoned well is not being pumped), and the leakage potential is given by

$$G = \frac{Q_l}{2\pi} \mathrm{K}_0(r/\Lambda). \tag{5.224}$$

The abandoned well removes water, say, a discharge Q_a, from one aquifer, e.g., the upper one, and injects it into the lower one. The potential in the upper aquifer is

$$\overset{1}{\Phi} = -G + \Phi_0 = -\frac{Q_l}{2\pi} \mathrm{K}_0(r/\Lambda) + \Phi_0, \tag{5.225}$$

where Φ_0 is a constant. We approximate this near the well as

$$\overset{1}{\Phi} = \frac{Q_l}{2\pi} \ln r + \cdots, \qquad r/\Lambda << 1. \tag{5.226}$$

Since the expression for a well of discharge Q is $Q/(2\pi)\ln r$, it follows that

$$Q_a = Q_l. \tag{5.227}$$

We determine the value of Q_a from the condition (5.223).

5.3.4 Far-Field Discharges

Far from the well the Bessel function K_0 is negligible, and the potential in each aquifer approaches that of a well. This is true for all three cases at large distances from the well, say, beyond $r/\Lambda = 8$. The discharge of the well in the upper aquifer is T_1/T times the total, Q, and that in the lower aquifer is T_2/T times the total.

Thus, the discharges under the equilibrium conditions, which prevail far from the well, are divided over the two aquifers according to their transmissivities. This property is universal; discharges always are distributed that way, provided that the leakage has vanished and the system is equilibrated.

5.4 Application: An Abandoned Well and a Well

Abandoned wells that connect two or more aquifers can provide a shortcut for flow between the aquifers if some other feature causes the heads in the aquifers to differ near the abandoned well. We consider the case of a pumping well in the upper aquifer, with an abandoned well somewhere else that connects the two aquifers. The pumping well is at the origin of the (x,y)-coordinate system and the abandoned well is at $x=d,\ y=0$. The discharge of the well is Q and there is no uniform flow.

The expression for the comprehensive potential contains the well only because the net discharge of the abandoned well is 0, so that

$$\Phi = \frac{Q}{4\pi}\ln[x^2+y^2]+C = \frac{Q}{2\pi}\ln r + C. \tag{5.228}$$

The leakage potential satisfies the Bessel equation; it contains two Bessel functions, one for the well, and one for the abandoned well. We assume that the radius of the well is much smaller than Λ, and represent the radial coordinate centered at the abandoned well as r_1,

$$G = \frac{Q_l}{2\pi}\mathrm{K}_0(r/\Lambda)+\frac{Q_a}{2\pi}\mathrm{K}_0(r_1/\Lambda). \tag{5.229}$$

The constant Q_l is known in terms of the discharge of the well; its expression is equal to (5.221) if the well is in the upper aquifer, and it is given by (5.222) if the well is in the lower aquifer. Q_a represents the discharge flowing through the abandoned well from the upper aquifer into the lower one. We obtain an expression for Q_a in terms of Q_l by setting $G=0$ at the abandoned well. If the distance from the well to the screen of the abandoned well is d, then the condition becomes, neglecting the radius of the abandoned well, r_a, with respect to d,

$$G = 0 = \frac{Q_l}{2\pi}\mathrm{K}_0(d/\Lambda)+\frac{Q_a}{2\pi}\mathrm{K}_0(r_a/\Lambda). \tag{5.230}$$

We solve (5.230) for the constant Q_a:

$$Q_a = -Q_l\frac{\mathrm{K}_0(d/\Lambda)}{\mathrm{K}_0(r_a/\Lambda)}. \tag{5.231}$$

This implies that if the well pumps from the upper aquifer, i.e., $Q_l=(T_2/T)Q$, then Q_a is negative; the flow is through the abandoned well upward. The expression for G becomes

$$G = \frac{Q_l}{2\pi}\left[\mathrm{K}_0(r/\Lambda) - \frac{\mathrm{K}_0(d/\Lambda)}{\mathrm{K}_0(r_a/\Lambda)}\mathrm{K}_0(r_1/\Lambda)\right]. \tag{5.232}$$

We observe that indeed G is zero if $r = d$ and $r_1 = r_a$. We obtain expressions for the potentials in the upper and lower aquifers using (5.215) and (5.216) as

$$\overset{1}{\Phi} = -\frac{Q_l}{2\pi}\left[\mathrm{K}_0(r/\Lambda) - \frac{\mathrm{K}_0(d/\Lambda)}{\mathrm{K}_0(r_a/\Lambda)}\mathrm{K}_0(r_1/\Lambda)\right] + \frac{T_1}{T}\left[\frac{Q}{2\pi}\ln r + C\right] \tag{5.233}$$

$$\overset{2}{\Phi} = \frac{Q_l}{2\pi}\left[\mathrm{K}_0(r/\Lambda) - \frac{\mathrm{K}_0(d/\Lambda)}{\mathrm{K}_0(r_a/\Lambda)}\mathrm{K}_0(r_1/\Lambda)\right] + \frac{T_2}{T}\left[\frac{Q}{2\pi}\ln r + C\right]. \tag{5.234}$$

We can obtain an expression for the constant C with some additional condition, a measured head at a point in the aquifer, for example. Remember that the constant Λ is given by either (5.221) or (5.222), depending on which aquifer contains the pumping well.

5.4.1 *Interpretation of the Results*

The magnitude of the constant Q_a is directly proportional to the discharge of the well, and further depends on the ratios r_a/Λ and d/Λ. The Bessel function $\mathrm{K}_0(d/\Lambda)$ decreases rapidly as d/Λ increases; the influence of an abandoned well is greatly affected by the distance to the cause of the leakage.

5.5 Leaky Aquifers with Infiltration

Infiltration is an important source of groundwater and often results from rainfall. Such infiltration may be modeled as being constant over large areas. Local infiltration along strips and over circular areas (center pivot systems) are common in agricultural practice, where irrigation is used to improve crop production. We begin this section with uniform infiltration, and then consider both infiltration over strips and infiltration over circular areas.

We defined potentials in the two aquifers of a leaky system in such a way that the discharge potential for each aquifer consists of two terms. The first term represents the equilibrated part of the flow, i.e., the part that does not cause leakage, and the second term represents the effect of leakage. We add a third term to each potential for the case of piecewise constant infiltration; the equilibrated part can then be separated into two terms. The first term represents equilibrated heads, and the second one gives a constant difference in head. The constant difference in head causes a constant rate of leakage, which corresponds to the equilibrated distribution of the rainfall over the system, as we will see. We write the discharge potentials in the two aquifers accordingly as

$$\overset{1}{\Phi} = \overset{1}{\underset{\text{eq}}{\Phi}} + \overset{1}{\underset{\text{ne}}{\Phi}} \tag{5.235}$$

$$\overset{2}{\Phi} = \overset{2}{\underset{\text{eq}}{\Phi}} + \overset{2}{\underset{\text{ne}}{\Phi}}, \tag{5.236}$$

where "ne" stands for nonequilibrated and "eq" for equilibrated, and where

$$\overset{1}{\underset{\text{eq}}{\Phi}} = \frac{T_1}{T}\Phi - F \tag{5.237}$$

$$\overset{2}{\underset{\text{eq}}{\Phi}} = \frac{T_2}{T}\Phi + F. \tag{5.238}$$

The first term in each of these expressions produces heads that are equal in the aquifers. The second term represents a constant difference in head that causes a constant leakage from one aquifer into the other (positive upward for positive F). The function F represents a difference in head across the leaky layer below the areas of infiltration such that this difference divided by the resistance c matches the equilibrated leakage, which we express in terms of the function F:

$$\underset{\text{eq}}{\gamma} = \frac{\overset{2}{\underset{\text{eq}}{\Phi}}}{cT_2} - \frac{\overset{1}{\underset{\text{eq}}{\Phi}}}{cT_1} = \frac{1}{c}F\left[\frac{1}{T_2} + \frac{1}{T_1}\right] = \frac{T}{cT_1T_2}F = \frac{F}{\Lambda^2}. \tag{5.239}$$

The non-equilibrated part of the potentials in the two aquifers is represented as before in terms of the function G:

$$\overset{1}{\underset{\text{ne}}{\Phi}} = -G \tag{5.240}$$

$$\overset{2}{\underset{\text{ne}}{\Phi}} = G. \tag{5.241}$$

5.5.1 Governing Equations

The divergence of the discharge vector in the lower aquifer (aquifer 2), i.e., $-\nabla^2\overset{2}{\Phi}$, is equal to $-(\underset{\text{eq}}{\gamma} + \underset{\text{ne}}{\gamma})$:

$$-\nabla^2\overset{2}{\Phi} = -\underset{\text{eq}}{\gamma} - \underset{\text{ne}}{\gamma} = -\frac{F}{\Lambda^2} - \frac{G}{\Lambda^2} \tag{5.242}$$

or, writing $\overset{2}{\Phi}$ in terms of Φ, F, and G with (5.236), (5.238), and (5.241),

$$\nabla^2\overset{2}{\Phi} = \frac{T_2}{T}\nabla^2\Phi + \nabla^2 F + \nabla^2 G = -\frac{T_2}{T}N + \nabla^2 G = \frac{F}{\Lambda^2} + \frac{G}{\Lambda^2}. \tag{5.243}$$

Since $\underset{\text{eq}}{\gamma} = -(T_2/T)N$, where N is the rate of infiltration, we have with (5.239),

$$\frac{F}{\Lambda^2} = -\frac{T_2}{T}N, \tag{5.244}$$

so that the governing equation for G becomes

$$\nabla^2 G = \frac{G}{\Lambda^2}, \tag{5.245}$$

which is the same as for the case of no infiltration.

5.5.2 *Infiltration over the Entire Aquifer*

We consider flow from infiltration over the entire upper boundary of the uppermost aquifer in the system of two aquifers, with no other features present in the system. We write the comprehensive potential as follows, taking the y-axis along the divide,

$$\Phi = -\tfrac{1}{2}N(x^2 - b^2) + \Phi_b, \tag{5.246}$$

where N is the rate of infiltration through the upper boundary, and Φ_b is the value of the comprehensive potential at $|x| = b$. We expect that flow in the two aquifers is distributed according to the transmissivities in the aquifers. This implies that the flow through the leaky layer must be T_2/T times the total infiltration; i.e., we may write

$$\underset{\text{eq}}{\gamma} = -\frac{T_2}{T}N = \frac{F}{\Lambda^2} \tag{5.247}$$

so that

$$F = -\frac{T_2\Lambda^2}{T}N. \tag{5.248}$$

The system is infinite in extent and there is constant leakage, so that the flow is equilibrated throughout. We obtain the expressions for the potentials from (5.237) and (5.238) as

$$\overset{1}{\Phi} = \underset{\text{eq}}{\overset{1}{\Phi}} = \frac{T_1}{T}\Phi + \frac{T_2\Lambda^2}{T}N \tag{5.249}$$

$$\overset{2}{\Phi} = \underset{\text{eq}}{\overset{2}{\Phi}} = \frac{T_2}{T}\Phi - \frac{T_2\Lambda^2}{T}N, \tag{5.250}$$

with Φ given by (5.246).

The solution presented above indeed satisfies all conditions: the leakage is constant, as is the difference in head between the aquifers, and the leakage equals this difference in head, divided by the resistance.

5.5.3 Application: Infiltration over the Entire Aquifer with an Abandoned Well

We can superimpose the solutions for the case of uniform infiltration and that for an abandoned well. The comprehensive potential is not affected by the well and is represented by (5.246) for the case of flow in the (x,y)-plane. The leakage potential G includes the abandoned well and equals

$$G = \frac{Q_l}{2\pi}\mathrm{K}_0(r/\Lambda) \tag{5.251}$$

so that the potentials in the two aquifers become

$$\overset{1}{\Phi} = \frac{T_1}{T}\Phi + \frac{T_2\Lambda^2}{T}N - \frac{Q_l}{2\pi}\mathrm{K}_0(r/\Lambda) \tag{5.252}$$

$$\overset{2}{\Phi} = \frac{T_2}{T}\Phi - \frac{T_2\Lambda^2}{T}N + \frac{Q_l}{2\pi}\mathrm{K}_0(r/\Lambda). \tag{5.253}$$

The heads in the two aquifers must be equal at the screen, radius r_a, of the abandoned well. This implies that the leakage at the screen of the abandoned well is zero so that the unequilibrated terms in (5.252) and (5.253) vanish:

$$-\frac{T_2\Lambda^2}{T}N + \frac{Q_l}{2\pi}\mathrm{K}_0(r_a/\Lambda) = 0 \tag{5.254}$$

or

$$Q_l = \frac{2\pi T_2\Lambda^2 N}{T\mathrm{K}_0(r_a/\Lambda)}. \tag{5.255}$$

This equation always will give a positive value for Q_l, which implies that an abandoned well acts as a discharge well in the upper aquifer; the abandoned well directs flow from the upper aquifer into the lower one. The discharge is proportional to the ratio T_2/T, i.e., the more transmissive the lower aquifer is relative to the upper one, the more water will flow through the abandoned well. The discharge is also proportional to the square of the leakage factor; the larger the resistance to flow of the leaky layer is, the more water will be carried through the abandoned well. These conclusions, drawn from the expression for Q_l, match what one might expect from the setting of the problem.

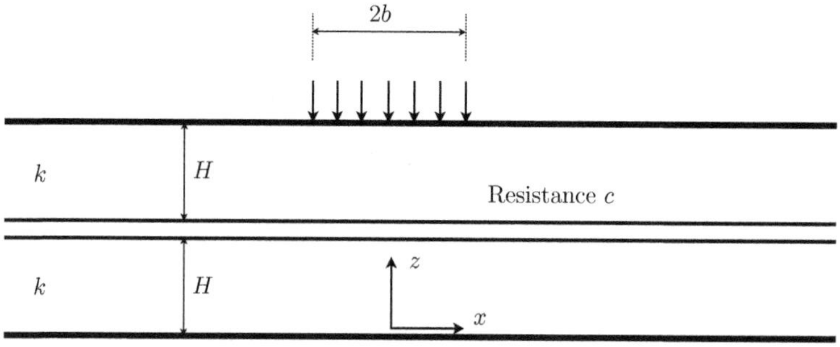

Figure 5.18 Infiltration over a strip of width $2b$ in a system of two aquifers separated by a leaky layer of resistance c.

5.5.4 *Infiltration over a Strip of Width 2b*

We consider flow from infiltration over an area of width $2b$, centered at the origin, and with the divide again coinciding with the y-axis (see Figure 5.18). The expressions for the comprehensive potential must be such that there is infiltration over the strip $-b \le x \le b$. The comprehensive potential for this case consists of three parts:

$$\Phi = -\tfrac{1}{2}N(x^2 - b^2) + \Phi_0, \quad -b \le x \le b \tag{5.256}$$

$$\Phi = -Nb(x - b) + \Phi_0, \quad b \le x < \infty \tag{5.257}$$

$$\Phi = Nb(x + b) + \Phi_0, \quad -\infty < x \le -b. \tag{5.258}$$

The potential is continuous; the value of Φ at $x = b$ and at $x = -b$ is Φ_0. The discharge vector is also continuous; the discharge flowing to each side is Nb, by continuity, and the discharge flowing toward infinity on each side is also Nb; the uniform flow is positive for $x \ge b$ and negative for $x \le b$. The problem is symmetrical; we consider only positive values of x and may obtain the solution for negative values of x simply by replacing x by $-x$ in the equations.

We represent the potentials in the two aquifers for $0 \le x \le b$ as

$$\overset{1}{\Phi} = \frac{T_1}{T}\Phi + \frac{T_2\Lambda^2}{T}N - G, \quad 0 \le x \le b \tag{5.259}$$

and

$$\overset{2}{\Phi} = \frac{T_2}{T}\Phi - \frac{T_2\Lambda^2}{T}N + G, \quad 0 \le x \le b. \tag{5.260}$$

The constant terms $(T_2\Lambda^2/T)N$ ensure that the difference in head matches the leakage in the zone below the infiltration area; these constants represent the equilibrated leakage in the zone of infiltration. We represent this term in aquifer 2 as $\underset{\text{eq}}{\Phi}$, with

$$\underset{\text{eq}}{\Phi} = -\frac{T_2\Lambda^2}{T}N, \quad |x| \leq b \tag{5.261}$$

$$\underset{\text{eq}}{\Phi} = 0, \quad |x| > b. \tag{5.262}$$

These constant terms cannot be present outside the infiltration area, i.e.,

$$\overset{1}{\Phi} = \frac{T_1}{T}\Phi - G, \quad b \leq |x| < \infty \tag{5.263}$$

and

$$\overset{2}{\Phi} = \frac{T_2}{T}\Phi + G, \quad b \leq |x| < \infty. \tag{5.264}$$

The function G must satisfy the modified Bessel equation, (5.245). The solution for $-b \leq x \leq b$ is symmetrical; we write G as

$$G = A\cosh(x/\Lambda). \tag{5.265}$$

The solution for $x \geq b$ must be finite at infinity, so that

$$G = B\mathrm{e}^{-x/\Lambda}, \quad b \leq x. \tag{5.266}$$

The boundary conditions are that both the discharge and the discharge potential are continuous at $x = b$. These conditions are met if the leakage and its derivative are continuous, i.e., if

$$\underset{\text{eq}}{\Phi^+} - \underset{\text{eq}}{\Phi^-} + G^+ - G^- = -\frac{T_2\Lambda^2}{T}N + A\cosh(b/\Lambda) - B\mathrm{e}^{-b/\Lambda} = 0, \tag{5.267}$$

where the plus sign refers to the side $x < b$ and the minus sign to the side $x > b$, and if

$$\frac{dG^+}{dx} = \frac{dG^-}{dx}, \quad x = b. \tag{5.268}$$

We apply this, with (5.265) and (5.266),

$$\frac{A}{\Lambda}\sinh(b/\Lambda) = -\frac{B}{\Lambda}\mathrm{e}^{-b/\Lambda}, \tag{5.269}$$

and use this in (5.267)

$$A\left[\cosh(b/\Lambda) + \sinh(b/\Lambda)\right] = A\mathrm{e}^{b/\Lambda} = \frac{T_2\Lambda^2}{T}N \tag{5.270}$$

or

$$A = \frac{T_2\Lambda^2}{T}Ne^{-b/\Lambda}. \tag{5.271}$$

We use (5.269) and obtain for B

$$B = -Ae^{b/\Lambda}\sinh(b/\Lambda) = -\frac{T_2\Lambda^2}{T}N\sinh(b/\Lambda). \tag{5.272}$$

We substitute the expression for A with (5.265) in expressions (5.259) and (5.260) for the potentials, which gives

$$\overset{1}{\Phi} = \frac{T_1}{T}\Phi + \frac{T_2\Lambda^2}{T}N - \frac{T_2\Lambda^2}{T}Ne^{-b/\Lambda}\cosh(x/\Lambda), \quad -b \le x \le b \tag{5.273}$$

and

$$\overset{2}{\Phi} = \frac{T_2}{T}\Phi - \frac{T_2\Lambda^2}{T}N + \frac{T_2\Lambda^2}{T}Ne^{-b/\Lambda}\cosh(x/\Lambda), \quad -b \le x \le b. \tag{5.274}$$

We use (5.263) and (5.264) to derive expressions for the potentials for the area $b \le x < \infty$, with (5.272) and (5.266)

$$\overset{1}{\Phi} = \frac{T_1}{T}\Phi + \frac{T_2\Lambda^2}{T}N\sinh(b/\Lambda)e^{-x/\Lambda}, \quad b \le x < \infty \tag{5.275}$$

and

$$\overset{2}{\Phi} = \frac{T_2}{T}\Phi - \frac{T_2\Lambda^2}{T}N\sinh(b/\Lambda)e^{-x/\Lambda}, \quad b \le x < \infty. \tag{5.276}$$

We replace x by $-x$ in (5.275) and (5.276), which gives the solution for the area $\infty < x \le -b$

$$\overset{1}{\Phi} = \frac{T_1}{T}\Phi + \frac{T_2\Lambda^2}{T}N\sinh(b/\Lambda)e^{x/\Lambda}, \quad -\infty < x \le -b \tag{5.277}$$

and

$$\overset{2}{\Phi} = \frac{T_2}{T}\Phi - \frac{T_2\Lambda^2}{T}N\sinh(b/\Lambda)e^{x/\Lambda}, \quad -\infty < x \le -b. \tag{5.278}$$

The expressions for the leakage potential become

$$G = \frac{T_2\Lambda^2}{T}Ne^{-b/\Lambda}\cosh(x/\Lambda), \quad -b \le x \le b \tag{5.279}$$

and

$$G = -\frac{T_2\Lambda^2}{T}N\sinh(b/\Lambda)e^{-x/\Lambda}, \quad b \le x < \infty. \tag{5.280}$$

We replace x by $-x$ to obtain the solution valid for $-\infty < x \le -b$

$$G = -\frac{T_2\Lambda^2}{T}N\sinh(b/\Lambda)e^{x/\Lambda}, \quad -\infty < x \le -b. \tag{5.281}$$

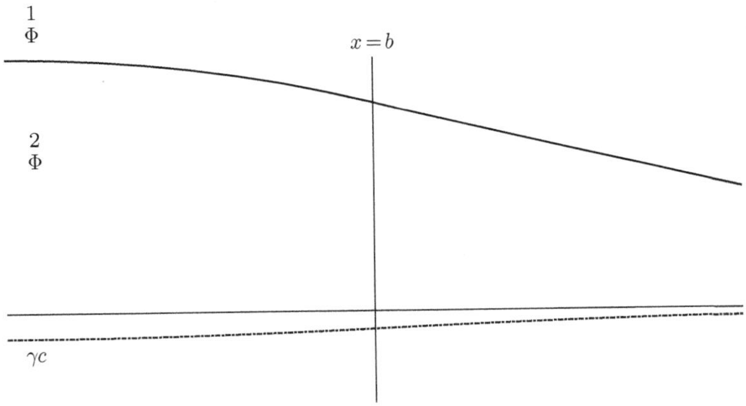

Figure 5.19 Plot of the potentials $\overset{1}{\Phi}$ and $\overset{2}{\Phi}$, and γc.

The leakage below the infiltration area is composed of the contributions of G and the constant term, each divided by Λ^2, i.e.,

$$\gamma = \frac{G}{\Lambda^2} - N\frac{T_2}{T} = \frac{T_2}{T} N e^{-b/\Lambda} \cosh(x/\Lambda) - N\frac{T_2}{T}, \quad -b \le x \le b, \tag{5.282}$$

whereas the leakage for $b \le x < \infty$ is equal to G/Λ^2, with G given by (5.281).

A plot of the potentials in the upper and lower aquifers, along with the leakage in the segment $0 \le x < 3b$ is shown in Figure 5.19.

Problem

5.7 Demonstrate that the potentials given in (5.273) through (5.276) meet the boundary conditions at $x = b$.

Problem

5.8 Consider the case of a system of two aquifers separated by a leaky layer of resistance $c = 250$ days and thickness $H^* = 2$ m. The thicknesses of both aquifers are 10 m, and the hydraulic conductivities of the upper and lower aquifers are 2 m/d and 10 m/d, respectively. The reference level (the base for measuring heads) is at the base of the aquifer system. The head at $x = -1000$ m is nearly the same in the two aquifers and is measured at $\phi_1 = 44.2$ m. There is a river at $x = 0$ with a head of 40 m; the river penetrates both aquifers.

A well of discharge Q is planned in the lower aquifer at $x = -d$, where $d = 250$ m. The radius of the well, r_w, is 0.3 m. Note that the head was measured prior to the operation of the well.

Questions:

1. Present expressions for potentials in the two aquifers for the case of flow without the well pumping.
2. Determine an expression for the discharge of the well analytically if it is given that the head at the well is 25 m.
3. Present contour plots of heads in the two aquifers (two separate plots).
4. Consider that there is an abandoned well at a distance of 50 m downstream from the well, and present an expression for the potential for this case, assuming that the discharge of the well is kept at the value computed under question 2.
5. Determine the discharge flowing through the abandoned well, if its radius is 0.5 m.
6. Present contour plots of the piezometric heads in both aquifers in two separate plots.
7. Include an area of infiltration of rate $N = 0.05$ m/d over the strip $-l-b \leq x \leq -l$; assume that the component of uniform flow remains as computed for the comprehensive potential assuming that neither are the wells present, nor the strip of infiltration. The well keeps pumping at the same rate as before, but the discharge through the abandoned well changes.
8. Present contour plots of piezometric heads in the two aquifers, choosing the width b as 100 m and the length l as 500 m.
9. Present contour plots of leakage, taking leakage as positive upward.

5.5.5 *Infiltration over a Circle*

We consider the case of uniform infiltration over a circle of radius R centered at (x_1, y_1). We make use of the local radial coordinate r_1, defined as

$$r_1 = \sqrt{(x-x_1)^2 + (y-y_1)^2}. \tag{5.283}$$

The comprehensive potential for a circular pond of infiltration rate N, centered at $r_1 = 0$ and of radius R is

$$\Phi = -\tfrac{1}{4}N(r_1^2 - R^2), \quad r_1 \leq R \tag{5.284}$$

$$\Phi = -\frac{NR^2}{2}\ln\frac{r_1}{R}, \quad r_1 > R. \tag{5.285}$$

The potentials in aquifers 1 and 2 in the area $0 \leq r_1 \leq R$ are expressed in terms of the comprehensive potential, the equilibrated leakage, and the leakage potential as for infiltration over a strip (see (5.259) and (5.260)):

$$\overset{1}{\Phi} = \frac{T_1}{T}\Phi + \frac{T_2\Lambda^2}{T}N - G_c, \quad r_1 \le R \tag{5.286}$$

$$\overset{2}{\Phi} = \frac{T_2}{T}\Phi - \frac{T_2\Lambda^2}{T}N + G_c, \quad r_1 \le R. \tag{5.287}$$

The expressions outside the area of infiltration are

$$\overset{1}{\Phi} = \frac{T_1}{T}\Phi - G_c, \quad r_1 > R \tag{5.288}$$

$$\overset{2}{\Phi} = \frac{T_2}{T}\Phi + G_c, \quad r_1 > R, \tag{5.289}$$

where the index c stands for circular infiltration area. The leakage potential G_c must have a continuous derivative at $r_1 = R$, so that

$$\left[\frac{dG_c}{dr_1}\right]^+ = \left[\frac{dG_c}{dr_1}\right]^-, \quad r_1 = R, \tag{5.290}$$

where the plus and minus signs indicate the area where the function applies; the plus sign refers to the area inside the circle, and the minus sign to the area outside it. The function G_c must jump in order to cancel the jump caused by the term $T_2\Lambda^2 N/T$, which is present inside the circle, but absent outside it. The jump in G_c must be

$$G_c^+ - G_c^- = \frac{T_2\Lambda^2}{T}N. \tag{5.291}$$

The leakage potential G_c is a solution to the modified Bessel equation. Since G_c is a radial function and since it must be finite inside the circle, we have

$$G_c^+ = A\mathrm{I}_0(r_1/\Lambda). \tag{5.292}$$

The function G_c^- must vanish at infinity; the effect of the infiltration area on the leakage reduces with distance. The function G_c^- thus must be the modified Bessel function of the second kind and order zero, i.e.,

$$G_c^- = B\mathrm{K}_0(r_1/\Lambda). \tag{5.293}$$

We apply condition (5.290) and obtain

$$A\mathrm{I}_1(R/\Lambda) = -B\mathrm{K}_1(R/\Lambda) \tag{5.294}$$

so that

$$A = -B\frac{\mathrm{K}_1(R/\Lambda)}{\mathrm{I}_1(R/\Lambda)}. \tag{5.295}$$

We apply condition (5.291),

$$A\mathrm{I}_0(R/\Lambda) - B\mathrm{K}_0(R/\Lambda) = \frac{T_2\Lambda^2}{T}N. \tag{5.296}$$

We use (5.295),

$$-B\frac{\mathrm{I}_0(R/\Lambda)\mathrm{K}_1(R/\Lambda) + \mathrm{I}_1(R/\Lambda)\mathrm{K}_0(R/\Lambda)}{\mathrm{I}_1(R/\Lambda)} = \frac{T_2\Lambda^2}{T}N \tag{5.297}$$

or

$$B = -\frac{T_2\Lambda^2 N\mathrm{I}_1(R/\Lambda)}{T[\mathrm{I}_0(R/\Lambda)\mathrm{K}_1(R/\Lambda) + \mathrm{I}_1(R/\Lambda)\mathrm{K}_0(R/\Lambda)]}. \tag{5.298}$$

We obtain the expression for A by substitution of the expression for B into (5.295) for A:

$$A = \frac{T_2\Lambda^2 N\mathrm{K}_1(R/\Lambda)}{T[\mathrm{I}_0(R/\Lambda)\mathrm{K}_1(R/\Lambda) + \mathrm{I}_1(R/\Lambda)\mathrm{K}_0(R/\Lambda)]}. \tag{5.299}$$

The expressions for the function G_c valid inside and outside the circle are

$$G_c(r_1,R,N,\Lambda,T_1,T_2) = \frac{T_2\Lambda^2 N\mathrm{K}_1(R/\Lambda)\mathrm{I}_0(r_1/\Lambda)}{T[\mathrm{I}_0(R/\Lambda)\mathrm{K}_1(R/\Lambda) + \mathrm{I}_1(R/\Lambda)\mathrm{K}_0(R/\Lambda)]}, \qquad r_1 \le R \tag{5.300}$$

$$G_c(r_1,R,N,\Lambda,T_1,T_2) = -\frac{T_2\Lambda^2 N\mathrm{I}_1(R/\Lambda)\mathrm{K}_0(r_1/\Lambda)}{T[\mathrm{I}_0(R/\Lambda)\mathrm{K}_1(R/\Lambda) + \mathrm{I}_1(R/\Lambda)\mathrm{K}_0(R/\Lambda)]}, \qquad r_1 > R. \tag{5.301}$$

The potentials presented here for a circular area of infiltration are internally consistent; all conditions along the inner boundary, the circle, are satisfied. We thus may add this solution to the other ones presented in this section, for example, to study the effect of abandoned wells on irrigation wells used for circular infiltration, which is a common practice in agriculture.

5.5.6 Application: A Circular Infiltration Area with a Well in a Leaky System

The comprehensive potential for the case of infiltration at a rate N over a circular area of radius R, centered at (x_c, y_c), with a well of discharge Q at (x_w, y_w), is

$$\Phi = \Phi_c(r_1, R) + \frac{Q}{4\pi}\ln\frac{r_2^2}{(L-x_w)^2 + y^2} + \Phi_L, \tag{5.302}$$

where

$$r_1 = \sqrt{(x-x_c)^2 + (y-y_c)^2}, \qquad r_2 = \sqrt{(x-x_w)^2 + (y-y_w)^2}. \tag{5.303}$$

The heads in the two aquifers are both approximately equal to ϕ_L at $x = L$, $y = 0$,

$$\Phi_L = T\phi_L. \tag{5.304}$$

The function Φ_c represents the discharge potential for a circular infiltration area of unit infiltration (see (5.284) and (5.285)):

$$\Phi_c = -\frac{N}{2}\left[r_1^2 - R^2\right], \quad r_1 \le R \tag{5.305}$$

$$\Phi_c = -\frac{NR^2}{2}\ln\frac{r_1}{R}, \quad r_1 > R. \tag{5.306}$$

The leakage potential consists of the contributions to the leakage of the infiltration area, given by (5.300) and (5.301) and that of the well

$$G = G_c(r_1; R, N, \Lambda, T_1, T_2) + F_c + \frac{Q_l}{2\pi}\mathrm{K}_0(r_2/\Lambda), \tag{5.307}$$

where

$$F_c = -\frac{T_2\Lambda^2}{T}N, \quad r_1 \le R \tag{5.308}$$

$$F_c = 0, \quad r_1 > R. \tag{5.309}$$

The discharge Q_l is given by T_2Q/T if the well is in the upper aquifer, and by $-T_1Q/T$ if the well is in the lower one; see (5.221) and (5.222).

If there is also an abandoned well in the system at (x_a, y_a), the comprehensive potential remains unchanged, but we must add a term to the function G, which becomes

$$G = G_c(r_1; R; N; \Lambda; T_1; T_2) + F_c + \frac{Q_l}{2\pi}\mathrm{K}_0(r_2/\Lambda) + \frac{Q_a}{2\pi}\mathrm{K}_0(r_3/\Lambda), \tag{5.310}$$

where Q_a is the discharge through the abandoned well and

$$r_3 = \sqrt{(x - x_a)^2 + (y - y_a)^2}. \tag{5.311}$$

We obtain an expression for Q_a by setting the leakage equal to zero at $x = x_a + r_a, y = y_a$.

5.5.7 The Flux-Leakage Factor

As for semiconfined flow, there exists a flux-leakage factor (see (5.197)), a proportionality factor with the dimension of specific discharge, that controls the discharge in the system due to leakage. We demonstrate that this flux-leakage factor has general application for flow in a leaky system of two aquifers. The flux-leakage factor is related to the leakage potential G. We express G in terms of the heads $\overset{1}{\phi}$, and $\overset{2}{\phi}$ in the two aquifers (see (5.185)):

$$G = \frac{T_1}{T}\overset{2}{\Phi} - \frac{T_2}{T}\overset{1}{\Phi} = \frac{T_1 T_2}{T}\left(\overset{2}{\phi} - \overset{1}{\phi}\right) = \frac{T_1 T_2}{T}\tilde{\phi}, \tag{5.312}$$

where

$$\tilde{\phi} = \overset{2}{\phi} - \overset{1}{\phi}. \tag{5.313}$$

We write the differential equation for G, the modified Helmholtz equation, in terms of the difference in head between the aquifers, $\tilde{\phi}$,

$$\nabla^2\tilde{\phi} = \frac{\tilde{\phi}}{\Lambda^2}. \tag{5.314}$$

As we did for semiconfined flow, we introduce dimensionless coordinates ξ and η:

$$\begin{aligned} \xi &= \frac{x}{\Lambda} \\ \eta &= \frac{y}{\Lambda} \end{aligned} \tag{5.315}$$

and write the modified Helmholtz equation in terms of these dimensionless independent variables:

$$\nabla^2\tilde{\phi} = \tilde{\phi}. \tag{5.316}$$

The discharges associated with leakage are:

$$\begin{aligned} \tilde{Q}_x &= -T\frac{\partial\tilde{\phi}}{\partial x} = -T\frac{\partial\tilde{\phi}}{\partial\xi}\frac{\partial\xi}{\partial x} = -\frac{T}{\Lambda}\frac{\partial\tilde{\phi}}{\partial\xi} \\ \tilde{Q}_y &= -T\frac{\partial\tilde{\phi}}{\partial y} = -T\frac{\partial\tilde{\phi}}{\partial\eta}\frac{\partial\eta}{\partial y} = -\frac{T}{\Lambda}\frac{\partial\tilde{\phi}}{\partial\eta}. \end{aligned} \tag{5.317}$$

We write this in terms of the flux-leakage factor ν that we introduced in (5.197)

$$\nu = \frac{T}{\Lambda} = \sqrt{\frac{T}{c}}\sqrt{\frac{T^2}{T_1 T_2}} \tag{5.318}$$

so that

$$\begin{aligned} \tilde{Q}_x &= \nu\frac{\partial\tilde{\phi}}{\partial\xi} \\ \tilde{Q}_y &= \nu\frac{\partial\tilde{\phi}}{\partial\eta}. \end{aligned} \tag{5.319}$$

The components $\tilde{Q}_x$ and $\tilde{Q}_y$ represent the contribution to the comprehensive discharge by leakage. These discharge components are expressed in terms of the gradients of the head difference $\tilde{\phi}$ in the dimensionless (ξ, η)-space, and are proportional to the flux-leakage factor, which is, as for the case of semiconfined flow, equal to the transmissivity divided by the leakage factor. By writing the equations in terms of the dimensionless independent variables, we achieve that the equations of the discharge vector components are independent of the distribution of leakage over space. Note that the expression for the flux-leakage factor shows that if one of the transmissivities is much smaller than the other, the fluxes due to leakage will always increase. This is true because a poorly conductive aquifer will tend to cause channeling of water to the more conductive one. This also causes the leakage factor to be small, and the leakage to be concentrated.

6

Three-Dimensional Flow

Only a few exact solutions to problems of three-dimensional flow exist. Fortunately, the concept of vertically integrated flow shows that two-dimensional solutions can be applied to the majority of practical problems. Only in cases where precise modeling of flow in three-dimensions is required do we need to resort to three-dimensional modeling. Although analytic solutions can be obtained and can be used to gain insight in three-dimensional effects, problems that need vertical modeling may require a numerical method of solutions, discussed in Chapter 10.

We derive the potential for flow in an infinite aquifer to a point-sink, and apply the method of images to solve the problem of flow toward a point-sink in a deep aquifer bounded above by a horizontal impermeable layer.

6.1 The Point-Sink

We derive the potential for flow to a point-sink in an infinite aquifer by the use of the continuity equation and Darcy's law as follows; see Figure 6.1. The flow is radial, and we denote the component of the discharge vector in the radial, r, direction as q_r, where q_r is positive in the positive r-direction, i.e., if pointing away from the point-sink. The discharge through the surface of a sphere of radius r, equal to $4\pi r^2$, around the sink must be equal to the discharge Q of the point-sink,

$$Q = -4\pi r^2 q_r. \tag{6.1}$$

We define the specific discharge potential Φ for three-dimensional flow as $k\phi$

$$\Phi = k\phi \qquad [\mathrm{L^2/T}]. \tag{6.2}$$

Darcy's law for radial flow is

$$q_r = -k\frac{d\phi}{dr} = -\frac{d\Phi}{dr}, \tag{6.3}$$

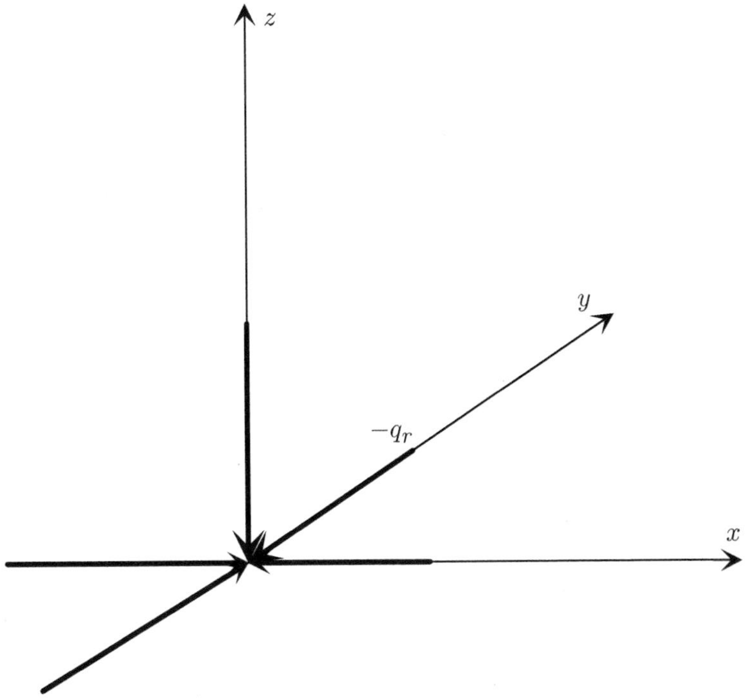

Figure 6.1 Radial flow toward a point-sink.

where Φ is a function of r alone. We combine (6.1) and (6.3),

$$\frac{d\Phi}{dr} = \frac{Q}{4\pi r^2}. \tag{6.4}$$

Integration yields

$$\Phi = -\frac{Q}{4\pi}\frac{1}{r} + \Phi_0, \tag{6.5}$$

where r is expressed as

$$r = \sqrt{(x-x_1)^2 + (y-y_1)^2 + (z-z_1)^2} \tag{6.6}$$

with x_1, y_1, and z_1 representing the coordinates of the center of the sink in a Cartesian (x, y, z)-coordinate system. As opposed to the two-dimensional case of flow to a well, the potential for the three-dimensional case approaches a constant for $r \to \infty$. It follows from (6.5) that this constant equals Φ_0,

$$r \to \infty, \qquad \Phi \to \Phi_0. \tag{6.7}$$

We explain why the potential is unbounded at infinity for two-dimensional flow toward a well. Viewed in infinite space, the two-dimensional solution for a well represents radial flow to an infinite line (called a line-sink) along the z-axis; for any value of z the radial component q_r is the same and lies in a plane parallel to the (x, y)-plane. The discharge of this infinite line-sink is infinite, as opposed to the discharge of a point-sink, which is finite. Drawing an infinite discharge appears to require that the potential increases logarithmically with the distance from the well. We always apply the two-dimensional solution to horizontal flow in an aquifer of finite thickness where the discharge is finite.

6.2 The Method of Images

The method of images can be used to solve a limited number of problems of three-dimensional flow. Consider the problem of flow toward a point-sink in a semi-infinite aquifer bounded above by a horizontal impermeable layer. We choose an (x, y, z)-coordinate system such that the plane $z = 0$ coincides with the upper boundary of the aquifer; see Figure 6.2. The point-sink has a discharge Q and is located at $x_1 = y_1 = 0$, $z_1 = -d$. We place an image sink at $x_2 = y_2 = 0$, $z_2 = d$, in order to meet the boundary condition that the plane $z = 0$ is impermeable. The potential becomes

$$\Phi = -\frac{Q}{4\pi}\left[\frac{1}{\sqrt{x^2+y^2+(z+d)^2}} + \frac{1}{\sqrt{x^2+y^2+(z-d)^2}}\right] + \Phi_0, \tag{6.8}$$

where Φ_0 represents the potential at infinity.

We include uniform flow of discharge q_{x0} in the x-direction in addition to the well by adding the potential for uniform flow, $\Phi = -q_{x0}x$, to (6.8):

$$\Phi = -q_{x0}x - \frac{Q}{4\pi}\left[\frac{1}{\sqrt{x^2+y^2+(z+d)^2}} + \frac{1}{\sqrt{x^2+y^2+(z-d)^2}}\right] + C. \tag{6.9}$$

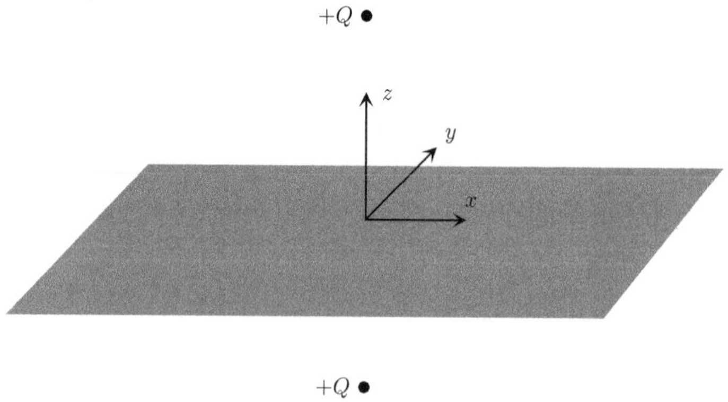

Figure 6.2 Flow toward a point-sink below an impermeable layer.

We determine the constant C by requiring that the head is ϕ_0 at some reference point $x = x_0$, $y = z = 0$,

$$\Phi_0 = k\phi_0 = -q_{x0}x_0 - \frac{Q}{2\pi}\frac{1}{\sqrt{x_0^2 + d^2}} + C. \tag{6.10}$$

We solve for C and substitute the result in (6.10):

$$\Phi = -q_{x0}(x - x_0) - \frac{Q}{4\pi}\left[\frac{1}{\sqrt{x^2 + y^2 + (z+d)^2}} + \frac{1}{\sqrt{x^2 + y^2 + (z-d)^2}} - \frac{2}{\sqrt{x_0^2 + d^2}}\right] + \Phi_0. \tag{6.11}$$

It is possible to solve the problem of flow toward a point-sink in a confined aquifer by the use of the method of images. In that case, the imaging must be continued indefinitely: an infinite number of images is necessary to meet the boundary conditions along both the upper and lower impermeable boundaries of the aquifer exactly. The procedure is somewhat involved and will not be discussed here. The reader is referred to Muskat [1937], Haitjema [1982], and Haitjema [1985] for three-dimensional modeling in confined aquifers.

Problem

6.1 Prove that the potential (6.11) meets the boundary condition that $q_z = 0$ at $z = 0$.

6.3 Partial Penetration

We consider the case of partial penetration, i.e., of features, such as wells, that penetrate the aquifer only in part. An extreme case of a partially penetrating well is a point-sink placed at the upper boundary of a confined aquifer. The capture zone of the point-sink may, or may not, cover the whole depth of the aquifer. A capture zone that penetrates only part of the aquifer is shown in Figure 6.3(a). A capture zone that penetrates the aquifer fully is shown in Figure 6.3(b). The difference between the two panels provides important information with regard to the decision whether the point-sink needs to be modeled as a three-dimensional feature, or whether a vertically integrated formulation is sufficient. The flow field approaches uniform flow at some distance from the point-sink very much in the same manner for the two cases of flow; the difference is in the shape of the capture zone, which is very different for the two cases. The capture zone may extend to a much greater depth if it is not restricted vertically, as illustrated in Figure 6.4. This illustration shows

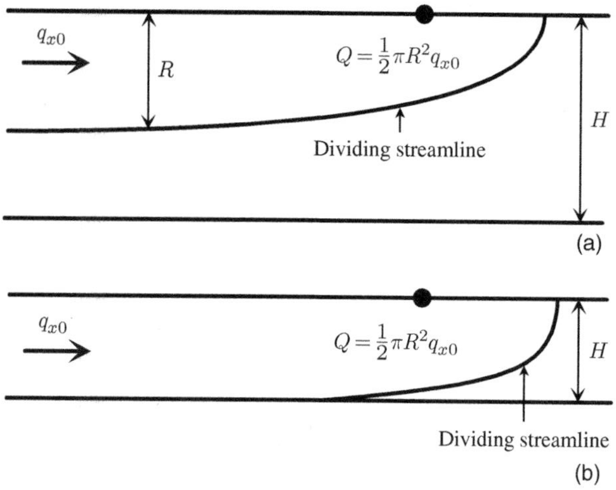

Figure 6.3 The effect of partial penetration for a partially penetrating capture zone (a) and a fully penetrating one (b).

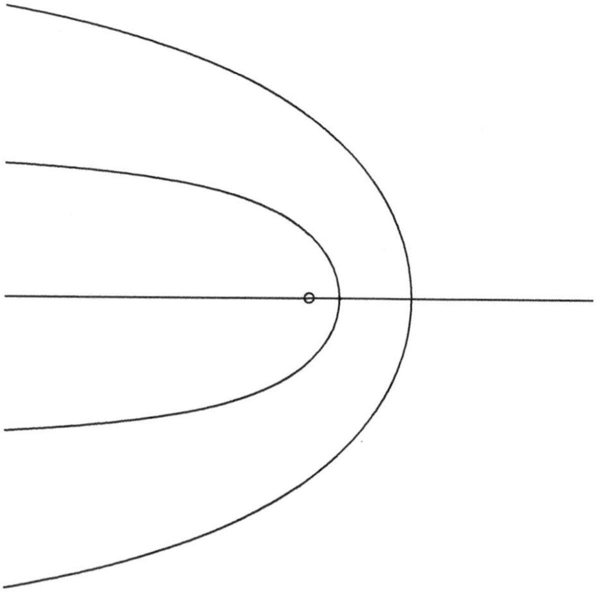

Figure 6.4 Capture zones in plan for a partially penetrating capture zone, the inner zone, and a fully penetrating one, the outer zone.

a sketch of the boundary of the capture zone at the impermeable boundary for two cases. For the one case the capture zone penetrates about halfway down the aquifer, and for the other case the lower boundary of the aquifer limits the depth of the capture zone. The inner capture zone boundary corresponds to the case of a partially penetrating capture zone, whereas the outer capture zone boundary corresponds to a fully penetrating capture zone.

6.3.1 Criterion for Three-Dimensional Modeling

If the lower boundary were not present at all, then the capture zone of the point-sink would be semicircular in cross section; the radius of this circle approaches a value of R near infinity, which we determine using continuity of flow. If the specific discharge of the uniform flow is q_{x0} and the discharge of the well is Q, then

$$Q = \tfrac{1}{2}\pi R^2 q_{x0} \rightarrow R = \sqrt{\frac{2\pi Q}{q_{x0}}}. \tag{6.12}$$

In this case the maximum width of the capture zone in horizontal projection is $2R$. By contrast, if the capture zone is limited below by the base of the aquifer, then the width b of the capture zone in the horizontal plane is, using the Dupuit-Forchheimer approximation,

$$Q = q_{x0}Hb \rightarrow b = \frac{Q}{q_{x0}H}. \tag{6.13}$$

Note that the partial penetration of the feature does not imply that the capture zone is also partially penetrating; this is controlled by the factor $Q/(q_{x0}H)$.

7

Transient Flow

The equations governing transient flow of groundwater have the same form as those governing transient flow of heat in solids. The equations that are usually applied to solve problems of transient groundwater flow in confined systems are a simplified form of the equations that govern poro-elasticity. The equations that govern transient flow in unconfined aquifers are a linearized form of the non-linear equations that apply to such flow, obtained using the Dupuit-Forchheimer approximation. A large body of solutions to the equations exist, and many are found in Carslaw and Jaeger [1959].

7.1 Transient Shallow Confined Flow

Transient effects in aquifer systems come about when boundary values or infiltration rates vary with time. A common example is a well that is switched on at some time; on starting the pump, the heads and pressures in the aquifer system change gradually until, for all practical purposes, steady-state conditions are reached. In a confined aquifer, the transient effects are caused by the compression of the grain skeleton as a result of decreasing pressures; if both the aquifer material and the fluid were incompressible, steady-state conditions would be reached instantaneously.

The problem of transient flow in a confined aquifer is a coupled one; the deformation of the grain skeleton is coupled to the groundwater flow. The problem is very complex, as the constitutive equations for soil are highly non-linear, even under dry conditions, and the coupling with the groundwater flow increases the complexity further still. Biot [1941] formulated the coupled problem mathematically, approximating the grain skeleton as a linearly elastic material. Fortunately, the pressure changes due to groundwater flow are usually small compared with the overburden stresses in the grain skeleton, which allows approximations to be made that simplify the problem considerably. We formulate the problem in terms of two equations: the equation describing storage and a simplified equation for the deformation of the grain skeleton. All strains and stresses are taken as positive for contraction and compression, respectively.

7.1.1 The Storage Equation

The storage equation expresses that water may be stored in an elementary volume V of porous material due to both an increase of V and a decrease in the volume V_w of the water. We assume that the aquifer is shallow so that all properties may be taken constant over the aquifer thickness H. A decrease in storage of an elementary volume will force water to flow out of V (see Figure 7.1) and has a similar effect on the aquifer as infiltrating water, which also forces water to leave the elementary volume; by continuity of flow the rate of outflow equals the infiltration rate N^*. By this analogy we make use of the differential equation $\nabla^2\Phi = -N^*$, interpreting N^* as the rate of decrease in storage. Let $e_0 = \Delta V/V$ and $ne_w = \Delta V_w/V$ (n is the porosity) be the volume strains of soil and water, respectively, both taken with respect to the same volume and taken as constant over the thickness of the aquifer. The rate of decrease N^* in storage equals the aquifer thickness H multiplied by the sum of the rate of increase, $\partial e_0/\partial t$, of the volume strain and the rate of decrease, $-n\partial e_w/\partial t$, of the volume strain of the water, so that

$$N^* = H\left[\frac{\partial e_0}{\partial t} - n\frac{\partial e_w}{\partial t}\right]. \tag{7.1}$$

With $\nabla^2\Phi = -N^*$, where $\Phi = kH\phi$, we obtain

$$\nabla^2\Phi = -H\left[\frac{\partial e_0}{\partial t} - n\frac{\partial e_w}{\partial t}\right]. \tag{7.2}$$

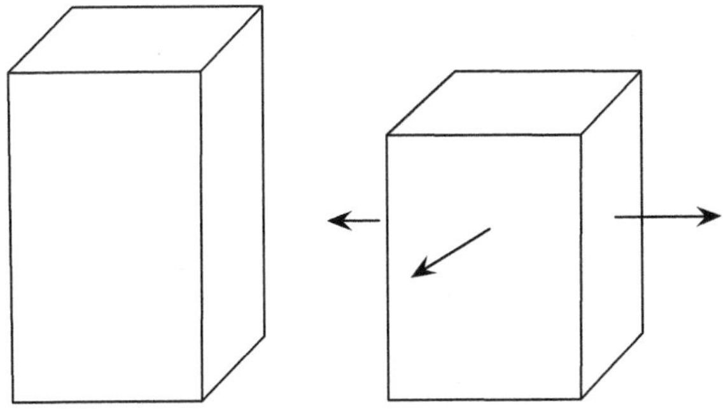

Figure 7.1 Decrease in storage of an elementary volume of porous material results in outflow.

We adopt a linear law for the compression of water,

$$\frac{\partial e_w}{\partial t} = \beta \frac{\partial p}{\partial t}, \tag{7.3}$$

so that (7.2) becomes

$$\nabla^2 \Phi = -H \left[\frac{\partial e_0}{\partial t} - n\beta \frac{\partial p}{\partial t} \right]. \tag{7.4}$$

This differential equation contains the volume strain e_0, which is dependent on the effective isotropic stress σ_0, which is generally not a function of the pressure p alone. We adopt the approximate approach introduced by Jacob [1940] and von Terzaghi [1943] in the following.

7.1.2 Simplified Stress-Strain Relation

The simplified stress-strain relation is based on the following three approximations:

1. The vertical total stress σ_{zz} does not change with time; this approximation is based on the notion that the effect of pore pressure changes is small with respect to the overburden, and implies that

$$\frac{\partial \sigma_{zz}}{\partial t} = 0. \tag{7.5}$$

2. The total stress can be written, following Terzaghi, as the sum of the effective stress σ'_{zz} and the pore pressure p:

$$\sigma_{zz} = \sigma'_{zz} + p. \tag{7.6}$$

3. The horizontal deformations in the soil are neglected.

It follows from the third approximation that

$$\frac{\partial e_0}{\partial t} = \frac{\partial e_{zz}}{\partial t} = m_v \frac{\partial \sigma'_{zz}}{\partial t}, \tag{7.7}$$

where m_v is, somewhat ambiguously, known as the coefficient of volume compressibility. It follows from (7.5) and (7.6) that

$$\frac{\partial \sigma'_{zz}}{\partial t} = -\frac{\partial p}{\partial t} \tag{7.8}$$

so that (7.7) becomes

$$\frac{\partial e_0}{\partial t} = -m_v \frac{\partial p}{\partial t}. \tag{7.9}$$

Combining (7.4) and (7.9) we may write:

$$\nabla^2\Phi = H(m_v + n\beta)\frac{\partial p}{\partial t}. \tag{7.10}$$

We neglect the variations of k, H, and ρ with time,

$$\frac{\partial \Phi}{\partial t} = kH\frac{\partial}{\partial t}\left[\frac{p}{\rho g} + Z\right] = \frac{kH}{\rho g}\frac{\partial p}{\partial t}, \tag{7.11}$$

so that (7.10) may be written in terms of Φ as

$$\nabla^2\Phi = \frac{S_s}{k}\frac{\partial \Phi}{\partial t}, \tag{7.12}$$

where S_s [1/L] is the coefficient of specific storage:

$$\boxed{S_s = \rho g(m_v + n\beta)} \tag{7.13}$$

Values of m_v range from about 10^{-7} m^2/N for clay to about 10^{-10} m^2/N for rock. The value of m_v for water, without air, is very small, and about the same as for rock. Values for sand are about 10^{-8} m^2/N. More detailed information on the properties of geological materials are found in, e.g., Freeze and Cherry [1979].

7.2 Transient Shallow Unconfined Flow

The elastic storage is several orders of magnitude less than the storage due to vertical movement of a phreatic surface, and therefore the elastic storage is usually neglected for transient shallow unconfined flow. The rate of decrease in storage, N^*, due to a rate of lowering, $-\partial\phi/\partial t$, of a phreatic surface is equal to $-s_p\partial\phi/\partial t$. The parameter s_p is the coefficient of phreatic storage; it represents the fraction of a unit volume available for storage, i.e.,

$$N^* = -s_p\frac{\partial \phi}{\partial t}. \tag{7.14}$$

The differential equation for transient unconfined flow becomes

$$\nabla^2\Phi = s_p\frac{\partial \phi}{\partial t} - N, \tag{7.15}$$

where N represents infiltration from rainfall. This differential equation, known as the Boussinesq equation, is nonlinear, and only a few exact solutions of limited practical interest exist (see Boussinesq [1904] and Aravin and Numerov [1965]).

We rewrite the right-hand side of (7.15) in terms of the potential:

$$\nabla^2\Phi = s_p\frac{d\phi}{d\Phi}\frac{\partial \Phi}{\partial t} - N, \tag{7.16}$$

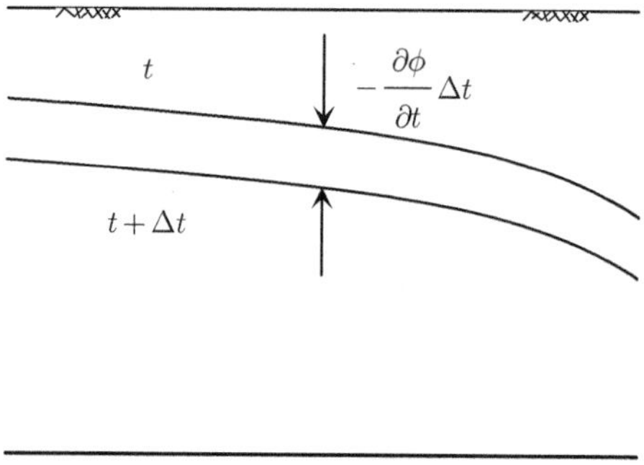

Figure 7.2 Decrease in storage due to lowering of the phreatic surface.

and we obtain for $d\Phi/d\phi$:

$$\frac{d\Phi}{d\phi} = \frac{d(\frac{1}{2}k\phi^2)}{d\phi} = k\phi. \tag{7.17}$$

We substitute this expression in (7.16),

$$\nabla^2\Phi = \frac{s_p}{k\phi}\frac{\partial \Phi}{\partial t} - N, \tag{7.18}$$

and linearize this equation by replacing the factor s_p/ϕ, which controls the rate of the release from storage with time, by a constant factor, called the specific storage S_s:

$$S_s = \frac{s_p}{\bar{h}}, \tag{7.19}$$

where $\bar{h}$ is some representative value of the saturated thickness. The specific storage in reality is variable in both time and space. We use the linearized expression for ϕ to replace s_p/ϕ in (7.18) by S_s:

$$\boxed{\nabla^2\Phi = \frac{S_s}{k}\frac{\partial \Phi}{\partial t} - N} \tag{7.20}$$

In the absence of rainfall, the linearized form of the differential equation for transient shallow unconfined flow is the same as the one for transient shallow confined flow (compare (7.12)). The solution to (7.12) given below therefore is applicable to both types of flow.

7.3 Transient Shallow Flow to a Well

The solution for flow toward a well in an aquifer of infinite extent is one of the most commonly used solutions for transient groundwater flow and is due to Theis

[1935]. The problem is axisymmetric, and we write the Laplacian $\nabla^2\Phi$ in (7.12) by the use of (5.112) in terms of the radial coordinate r:

$$\frac{\partial^2\Phi}{\partial r^2}+\frac{1}{r}\frac{\partial\Phi}{\partial r}=\frac{S_s}{k}\frac{\partial\Phi}{\partial t}. \tag{7.21}$$

The Theis solution satisfies the following boundary conditions:

$$r\to\infty, \qquad \phi=\phi_0, \qquad \Phi=\Phi_0 \tag{7.22}$$

$$\lim_{r\to 0} r\frac{\partial\Phi}{\partial r}=\frac{Q}{2\pi}, \qquad t>t_0, \tag{7.23}$$

where, for simplicity, the radius of the well is taken to be infinitely small. The solution may be obtained by application of Laplace transforms. We omit the derivation, present the solution, and demonstrate that it fulfills both the differential equation and the boundary conditions:

$$\boxed{\Phi=-\frac{Q}{4\pi}\mathrm{E}_1(u)+\Phi_0} \tag{7.24}$$

where E_1 is the exponential integral, and the dimensionless variable u is given by

$$u=\frac{S_s r^2}{4k(t-t_0)}. \tag{7.25}$$

The exponential integral (e.g., Abramowitz and Stegun [1965]) is defined as

$$\mathrm{E}_1(u)=\int_u^\infty \frac{e^{-\xi}}{\xi}d\xi. \tag{7.26}$$

In conjunction with the Theis solution, the exponential integral is often referred to as the well function W:

$$W(u)=\mathrm{E}_1(u). \tag{7.27}$$

7.3.1 Verification

To demonstrate that (7.24) is a solution to the differential equation we differentiate the exponential integral twice with respect to r and once with respect to t. We write (7.26) as

$$\mathrm{E}_1(u)=\int_u^\infty f(\xi)d\xi=F(\infty)-F(u), \tag{7.28}$$

where $f(\xi) = e^{-\xi}/\xi = dF/d\xi$, and differentiate $E_1(u)$ with respect to u:

$$\frac{dE_1(u)}{du} = -f(u) = -\frac{e^{-u}}{u}. \tag{7.29}$$

The derivatives of u with respect to r and t are

$$\frac{\partial u}{\partial r} = \frac{S_s r}{2k(t-t_0)} = \frac{2u}{r} \tag{7.30}$$

$$\frac{\partial u}{\partial t} = -\frac{S_s r^2}{4k(t-t_0)^2} = \frac{u}{t-t_0}. \tag{7.31}$$

We use this to obtain $\partial E_1/\partial r$

$$\frac{\partial E_1}{\partial r} = \frac{dE_1}{du}\frac{\partial u}{\partial r} = -\frac{e^{-u}}{u}\frac{2u}{r} = -\frac{2e^{-u}}{r}. \tag{7.32}$$

We differentiate again with respect to r:

$$\frac{\partial^2 E_1}{\partial r^2} = 2\frac{e^{-u}}{r}\frac{2u}{r} + 2\frac{e^{-u}}{r^2} = 2\frac{e^{-u}}{r^2}(2u+1) \tag{7.33}$$

and, with (7.29) and $\partial u/\partial t = -u/(t-t_0)$,

$$\frac{\partial E_1}{\partial t} = -\frac{e^{-u}}{u}\frac{-u}{t-t_0} = \frac{e^{-u}}{t-t_0}. \tag{7.34}$$

We use (7.32) through (7.34) and substitute expression (7.24) for the potential in the differential equation (7.21):

$$-\frac{Q}{2\pi}\frac{e^{-u}}{r^2}(2u+1) + \frac{Q}{2\pi}\frac{e^{-u}}{r^2} = -\frac{Q}{4\pi}\frac{S_s}{k}\frac{e^{-u}}{(t-t_0)}. \tag{7.35}$$

We multiply both sides by $(\pi/Q)r^2e^u$ and use (7.25) to obtain the identity $-u = -u$; the differential equation is fulfilled. The exponential integral vanishes for $r \to \infty$ so that (7.22) is satisfied. It follows from (7.24) and (7.32) that

$$r\frac{\partial \Phi}{\partial r} = \frac{Q}{2\pi}e^{-u}. \tag{7.36}$$

In the limit for $r \to 0$ and $t > t_0$, u equals zero and (7.36) reduces to (7.23), so that the boundary condition at the well is fulfilled.

7.3.2 Discharge Vector

We obtain the radial component of flow for a well placed at (x_1, y_1) from (7.24) as

$$Q_r = -\frac{\partial \Phi}{\partial r} = \frac{Q}{4\pi}\frac{\partial \mathrm{E}_1(u)}{\partial r}. \tag{7.37}$$

We use (7.32) for the partial derivative of E_1 with respect to r:

$$Q_r = -\frac{Q}{2\pi}\frac{\mathrm{e}^{-u}}{r}, \tag{7.38}$$

where

$$r^2 = (x-x_1)^2 + (y-y_1)^2. \tag{7.39}$$

We obtain an expression for the components (Q_x, Q_y) of the discharge vector by applying the chain rule to the discharge potential as follows:

$$Q_x = Q_r\frac{\partial r}{\partial x} = Q_r\frac{(x-x_1)}{r} = -\frac{Q}{2\pi}\frac{x-x_1}{r^2}\mathrm{e}^{-u} \tag{7.40}$$

$$Q_y = Q_r\frac{\partial r}{\partial x} = Q_r\frac{(y-y_1)}{r} = -\frac{Q}{2\pi}\frac{y-y_1}{r^2}\mathrm{e}^{-u}. \tag{7.41}$$

Note that these expressions reduce to the expressions valid for steady flow for the case in which $t \to \infty$, and $u \to 0$. Note also that we obtained the derivatives of r with respect to x and y from (7.39) as follows:

$$2r\frac{\partial r}{\partial x} = 2(x-x_1), \qquad 2r\frac{\partial r}{\partial y} = 2(y-y_1). \tag{7.42}$$

7.3.3 Series Expansion

The exponential integral may be conveniently approximated for computational purposes by various formulae listed in Abramowitz and Stegun [1965] as follows. For u in the range $0 \le u \le 1$ the formula by Allen [1954] yields results with a maximum error of $2 * 10^{-7}$:

$$\mathrm{E}_1(u) = -\ln u + \big[\{[(a_5u + a_4)u + a_3]u + a_2\}u + a_1\big]u + a_0, \tag{7.43}$$

where

$$\begin{aligned} a_0 &= -0.57721566, & a_1 &= 0.99999193, & a_2 &= -0.24991055, \\ a_3 &= 0.05519968, & a_4 &= -0.00976004, & a_5 &= 0.00107857. \end{aligned} \tag{7.44}$$

The second term in (7.43) is written such as to optimize numerical accuracy; this procedure is called Horner's rule. For u in the range $1 \le u < \infty$, Hastings [1955] presents a choice of two approximations, the more accurate one being more complex and therefore computationally slower. The first approximation is

$$ue^u\mathrm{E}_1(u) = \frac{(u+b_1)u+b_0}{(u+c_1)u+c_0} + \varepsilon(u), \qquad |\varepsilon(u)| < 5*10^{-5}, \tag{7.45}$$

where

$$\begin{aligned} b_0 &= 0.250621, & c_0 &= 1.681534 \\ b_1 &= 2.334733, & c_1 &= 3.330657. \end{aligned} \tag{7.46}$$

The second approximation is very accurate, and is given by

$$ue^u\mathrm{E}_1(u) = \frac{\{[(u+b_3)u+b_2]u+b_1\}u+b_0}{\{[(u+c_3)u+c_2]u+c_1\}u+c_0} + \varepsilon(u), \qquad |\varepsilon(u)| < 2*10^{-8}, \tag{7.47}$$

where

$$\begin{aligned} b_0 &= 0.2677737343, & c_0 &= 3.9584969228 \\ b_1 &= 8.6347608925, & c_1 &= 21.0996530827 \\ b_2 &= 18.0590169730, & c_2 &= 25.6329561486 \\ b_3 &= 8.5733287401, & c_3 &= 9.5733223454. \end{aligned} \tag{7.48}$$

The above approximations are readily implemented in a computer program as a function subroutine. The constant a_0 in (7.43) is the opposite of Euler's constant, γ,

$$a_0 = -\gamma = -0.57721566, \tag{7.49}$$

and for small values of u, $\mathrm{E}_1(u)$ may be approximated as

$$\mathrm{E}_1(u) = -\ln u - \gamma, \qquad u << 1, \tag{7.50}$$

and the potential (7.24) may be approximated as

$$\Phi = \frac{Q}{4\pi}\left[\ln\frac{S_s r^2}{4k(t-t_0)} + \gamma\right] + \Phi_0 = \frac{Q}{2\pi}\ln\frac{r}{R_i} + \Phi_0, \qquad u << 1, \tag{7.51}$$

where R_i is called the radius of influence,

$$R_i = \sqrt{\frac{4k(t-t_0)}{S_s e^\gamma}}. \tag{7.52}$$

The radius of influence is often used as an approximate measure of the distance over which the well has a noticeable effect: for $r = R_i$, Φ equals the initial value Φ_0. It is seen from (7.52) that the radius of influence increases with time. Note that the approximation is valid only for small values of u. The radius of influence is not a constant; it is an arbitrary choice. The influence of a well that pumps continuously over a long period of time will exceed the area inside the radius of influence.

7.3.4 Pumping Test

Cooper and Jacob [1946] suggested a graphical method for determining the transmissivity T,

$$T = kH, \tag{7.53}$$

and the specific storage coefficient S_s from a pumping test, using the approximation (7.51). We consider the case of a confined aquifer and rewrite (7.51) as follows (compare Lohman [1972]):

$$\phi_0 - \phi = s = \frac{Q}{4\pi T}\left[\ln e^{-\gamma} + \ln \frac{4T(t-t_0)}{Sr^2}\right], \tag{7.54}$$

where $e^{-\gamma} \approx 0.562$, s is the draw-down, and S the storativity, defined as

$$S = S_s H. \tag{7.55}$$

We write (7.54) in dimensionless form as follows (compare Verruijt [1982]):

$$\frac{\phi_0 - \phi}{\phi_1} = \frac{s}{\phi_1} = \frac{Q}{2\pi T \phi_1}\left[\ln \frac{\sqrt{4T_1(t-t_0)e^{-\gamma}/S_1}}{r} + \ln \sqrt{\frac{S_1 T}{S T_1}}\right], \tag{7.56}$$

where ϕ_1, S_1, and T_1 are arbitrary reference quantities. It follows from (7.56) that the dimensionless draw-down $(\phi_0 - \phi)/\phi_1$ when plotted versus $\ln\{[4T_1(t-t_0)/S_1 e^{\gamma})]^{1/2}/r\} = \ln \tau$, gives a straight line, as shown in Figure 7.3. By measuring the draw-down at some fixed observation point at different times and plotting these

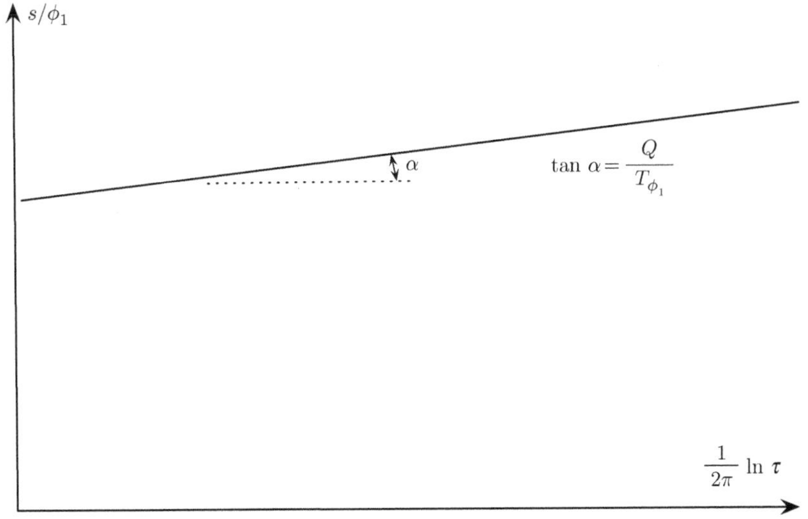

Figure 7.3 Determination of transmissivity and storativity from pumping tests.

data versus $\ln \tau$, a straight line is obtained. The slope of the line yields T, and the ordinate of the point of intersection with the draw-down axis gives the quantity $S_1T/(ST_1)$ so that S may be computed as well.

Various other graphical methods have been developed for determining S and T, both for confined and unconfined aquifers, and using the exponential integral $\mathrm{E}_1(u)$, rather than its approximation (7.50). These methods for pumping test data analysis are treated extensively in the literature, and the interested reader is referred to texts such as Freeze and Cherry [1979] and Davis and de Wiest [1966]. An alternative way of determining aquifer parameters from pumping test data is to write a simple computer program based on the Theis solution.

The Theis solution is ideally suited for pumping test analysis. It is important, in view of applications, to understand fully the implications of the boundary condition (7.22). This condition implies that the head at infinity is maintained at a constant level at all times. The well therefore draws all of its water from storage, and none from other sources. In the steady-state solution for a single well in an infinite aquifer these other sources are lumped together at infinity. As a result, the Theis solution does not converge to the steady-state solution; it will eventually draw the water levels throughout the aquifer below the aquifer base. Assuming that $u << 1$, the time at which the head at the well screen reaches the aquifer base ($\Phi_w = 0$) may be computed by the use of (7.51) as follows

$$\frac{Q}{2\pi}\ln\frac{r_w}{R_i} = -\Phi_0 \tag{7.57}$$

or

$$R_i = r_w e^{2\pi\Phi_0/Q} \tag{7.58}$$

or, with (7.52),

$$(t-t_0)_{\max} = \frac{r_w^2 S_s e^{\gamma}}{4k} e^{4\pi\Phi_0/Q}. \tag{7.59}$$

7.4 Superposition

The governing differential equation as used here is linear, and therefore solutions may be superimposed. Since any steady-state solution of $\nabla^2\Phi = 0$ satisfies the governing equation (7.12), steady-state and transient solutions may be combined. As an example, consider the case of a well, pumping under steady state a discharge Q. The well is switched off from $t = t_0$ until $t = t_1$. The potential that applies to this case may be written as

$$\Phi = \frac{Q}{2\pi}\ln\frac{r}{R} + \Phi_0 + \frac{Q}{4\pi}[\mathrm{E}_1(u_0) - \mathrm{E}_1(u_1)], \tag{7.60}$$

where

$$u_0 = \frac{S_s r^2}{4k(t-t_0)}, \qquad u_1 = \frac{S_s r^2}{4k(t-t_1)}. \tag{7.61}$$

The first term inside the brackets of (7.60) represents a recharge well that is switched on at time t_0 and cancels the discharge of the well. The second term inside the brackets represents a well switched on at time t_1; from that time onward the original discharge is restored. The two exponential integrals cancel for very large times. For sufficiently large times, u_0 and u_1 become small so that the approximation (7.51) applies:

$$\Phi = \frac{Q}{2\pi}\ln\frac{r}{R} + \Phi_0 + \frac{Q}{4\pi}\left[\ln\frac{S_s r^2}{4k(t-t_0)} - \ln\frac{S_s r^2}{4k(t-t_1)}\right] \tag{7.62}$$

or

$$\Phi = \frac{Q}{2\pi}\ln\frac{r}{R} + \Phi_0 + \frac{Q}{4\pi}\ln\frac{t-t_1}{t-t_0}. \tag{7.63}$$

The second logarithm vanishes for $t \to \infty$, so that the solution reduces to the one for steady-state flow toward a well.

The method of images may be applied to transient solutions. As an example, consider the problem of flow to a well near a river with $\Phi = \Phi_0$ along the y-axis, shown in Figure 7.4. The well is at $x = d$, $y = 0$, and starts pumping a discharge Q at time t_0. We obtain the solution by placing an image recharge well at $x = -d$, $y = 0$,

$$\Phi = -\frac{Q}{4\pi}[E_1(u_1) - E_1(u_2)] + \Phi_0, \tag{7.64}$$

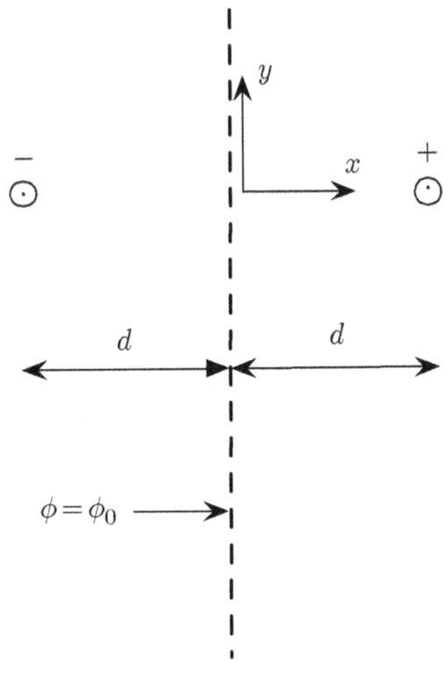

Figure 7.4 Transient flow toward a well near a river.

where

$$u_1 = \frac{S_s[(x-d)^2+y^2]}{4k(t-t_0)}, \qquad u_2 = \frac{S_s[(x+d)^2+y^2]}{4k(t-t_0)}. \tag{7.65}$$

It is left as an exercise to determine the steady-state form of (7.64) by taking the limit for $t \to \infty$.

Problem

7.1 Determine the steady-state form of (7.64) by taking the limit for $t \to \infty$.

Problem

7.2 Program the potential for a transient well in a field of uniform flow using the built-in exponential integral in the application you will use for your program. Note: The exponential integral as available in most computational programs is non-zero for negative values of u, i.e., for $t < t_0$, where t_0 is the time at which the well is turned on. You will need to program your transient well with an *if* statement that returns zero for $t < t_0$.

7.4.1 Design of a Transient City Well near a River

We consider, as an application, the case of a city well near a river in a field of uniform flow. We have seen that the city well can pump a discharge of up to $Q_{\max} = \pi d Q_{x0}$ under steady-state conditions without drawing river water into the aquifer; see (2.133). The objective of our design is to pump the well on average at a discharge considerably less than $Q_{\max}$, in order to be able to pump the well at a much higher rate than the maximum for a limited period of time. The question that we need to answer is, how long we can continue to pump at the higher rate? Let the steady-state discharge be $\underset{s}{Q}$ and the transient discharge $\underset{t}{Q}$, and let

$$\underset{s}{Q} = \alpha Q_{\max} = \alpha \pi d Q_{x0}, \qquad \alpha < 1, \tag{7.66}$$

and

$$\underset{t}{Q} + \underset{s}{Q} = \beta Q_{\max} = \beta \pi d Q_{x0}, \qquad \beta > 1, \tag{7.67}$$

so that

$$\underset{t}{Q} = (\beta - \alpha)\pi d Q_{x0}. \tag{7.68}$$

Steady-State Conditions

The steady-state conditions are those of a well of discharge $\underset{s}{Q}$ in a field of uniform flow of discharge Q_{x0} near a river of head ϕ_0. We use Cartesian coordinates with the y-axis along the river bank and the x-axis pointing into the river. The well is at $x = -d$. We obtain the steady-state solution by adding the discharge potentials for uniform flow, a well, and an image well:

$$\Phi = -Q_{x0}x + \frac{\underset{s}{Q}}{4\pi}\ln[(x+d)^2 + y^2] - \frac{\underset{s}{Q}}{4\pi}\ln[(x-d)^2 + y^2] + \Phi_0. \tag{7.69}$$

We begin our analysis by determining the location of the stagnant point, $x_s, 0$. We differentiate (7.69) with respect to x:

$$Q_x = Q_{x0} - \frac{\underset{s}{Q}}{4\pi}\frac{2(x+d)}{(x+d)^2 + y^2} + \frac{\underset{s}{Q}}{4\pi}\frac{2(x-d)}{(x-d)^2 + y^2}. \tag{7.70}$$

We set this equal to zero at $x = x_s, y = 0$:

$$0 = Q_{x0} - \frac{\underset{s}{Q}}{2\pi}\frac{1}{x_s + d} + \frac{\underset{s}{Q}}{2\pi}\frac{1}{x_s - d}. \tag{7.71}$$

We assume that the well pumps the discharge under steady-state conditions such that the stagnation point lies halfway between the river bank and the well. Thus, we compute the steady-state discharge $\underset{s}{Q}$ from the condition that the stagnation point is at $x_s = -d/2, y_s = 0$, and obtain

$$0 = Q_{x0} - \frac{\underset{s}{Q}}{2\pi}\left[\frac{1}{-d/2+d} - \frac{1}{-d/2-d}\right] \tag{7.72}$$

so that

$$\frac{\underset{s}{Q}}{\pi d}\left[1 + \frac{1}{3}\right] = Q_{x0} \rightarrow \underset{s}{Q} = \frac{3}{4}\pi d Q_{x0} \tag{7.73}$$

and

$$\alpha = 0.75. \tag{7.74}$$

Transient Conditions

We investigate at which time river water begins to enter the aquifer if we increase the discharge from $Q_s = \alpha\pi d Q_{x0}$ to $\beta\pi d Q_{x0}$ at time $t = 0$. We add both a transient well of discharge Q_t and its image to the potential, which gives

$$\Phi = -Q_{x0}x + \frac{\underset{s}{Q}}{4\pi}\ln[(x+d)^2+y^2] - \frac{\underset{s}{Q}}{4\pi}\ln[(x-d)^2+y^2] + \Phi_0$$
$$-\frac{\underset{t}{Q}}{4\pi}\mathrm{E}_1(u_1) + \frac{\underset{t}{Q}}{4\pi}\mathrm{E}_1(\underset{i}{u_1}), \tag{7.75}$$

where the subscript i stands for image, and where

$$u_1 = \frac{S_s[(x+d)^2+y^2]}{4kt}, \qquad \underset{i}{u_1} = \frac{S_s[(x-d)^2+y^2]}{4kt}. \tag{7.76}$$

We compute the time at which the discharge vector component Q_x is zero at the river bank, $x = y = 0$. We use (7.40) to obtain the expression for Q_x, with the appropriate expressions for the coordinates of the wells and the radial distance from the wells:

$$Q_x = Q_{x0} - \frac{\underset{s}{Q}}{2\pi}\frac{x+d}{(x+d)^2+y^2} + \frac{\underset{s}{Q}}{2\pi}\frac{x-d}{(x-d)^2+y^2}$$
$$-\frac{\underset{t}{Q}}{2\pi}\frac{x+d}{(x+d)^2+y^2}\mathrm{e}^{-u_1} + \frac{\underset{t}{Q}}{2\pi}\frac{x-d}{(x-d)^2+y^2}\mathrm{e}^{-\underset{i}{u_1}}. \tag{7.77}$$

We set x and y equal to zero in this expression and note that u_1 equals $\underset{i}{u_1}$ so that

$$Q_x(0,0,t) = Q_{x0} - \frac{\underset{s}{Q}}{\pi d} - \frac{\underset{t}{Q}}{\pi d}\mathrm{e}^{-u_0}, \tag{7.78}$$

where

$$u_0 = \frac{S_s d^2}{4kt}. \tag{7.79}$$

We replace $\underset{s}{Q}$ by $\alpha\pi d Q_{x0}$ and $\underset{t}{Q}$ by $(\beta-\alpha)\pi d Q_{x0}$:

$$Q_x(0,0,t) = Q_{x0} - \frac{\alpha\pi d Q_{x0}}{\pi d} - \frac{(\beta-\alpha)\pi d Q_{x0}}{\pi d}\mathrm{e}^{-u_0}, \tag{7.80}$$

and divide both sides of this equation by Q_{x0},

$$\frac{Q_x(0,0,t)}{Q_{x0}} = 1-\alpha-(\beta-\alpha)\mathrm{e}^{-u_0}. \tag{7.81}$$

We set $Q_x(0,0,t)$ equal to zero and solve the resulting equation for e^{u_0},

$$\mathrm{e}^{-u_0} = \frac{1-\alpha}{\beta-\alpha}, \tag{7.82}$$

so that

$$u_0 = \frac{S_s d^2}{4kt} = -\ln\frac{1-\alpha}{\beta-\alpha} = \ln\frac{\beta-\alpha}{1-\alpha}. \tag{7.83}$$

We find an expression for the time at which the river water begins to enter the aquifer by solving the latter equation for t, which gives

$$t = \frac{S_s d^2}{4k \ln \frac{\beta-\alpha}{1-\alpha}}. \tag{7.84}$$

If, for example, $\beta = 1.5$, i.e., after adding the transient well we pump at 1.5 times the maximum steady state value ($Q_{\max} = \pi d Q_{x0}$), and with $\alpha = 0.75$, then we obtain

$$t = \frac{S_s d^2}{4k \ln 3}. \tag{7.85}$$

Saturated Thickness

The solution presented in this application requires aquifer parameters; one of these is the saturated thickness of the aquifer. Haitjema [1995] used a numerical model to investigate how to make a good choice for the saturated thickness. Based on his analysis, he recommends taking the initial saturated thickness of the aquifer.

Problem

7.3 Consider the problem, described above, of reducing the discharge of the well back down to the steady-state value of Q_s without drawing any river water into the aquifer. This process is carried out by adding yet another set of transient wells. These wells are at the same location as the original well and its image, but with opposite signs of their discharges, and switched on at a time t_{off}. Note that the wells must be switched off *before* the time given by (7.85); it will take time for the effect of the latter transient wells to reach the river bank.

Questions:

1. Present an expression for the potential, valid after the discharge of the production well has been reduced back to Q_s.
2. Present an expression for the discharge $Q_x(0,0)$, valid after the discharge of the production well has been reduced back to Q_s.
3. Plot the discharge $Q_x(0,0)$ as a function of time for various values of t_{off}, and estimate the t_{off} such that no river water will enter the system at any time. Choose $S_s = 0.03$, $d = 200$ m, $k = 10$ m/d, and $\phi_0 = 20$ m.
4. Explain how to set up a system of two non-linear equations that will yield as a solution the time t_{off} as well as the time t at which the discharge at the origin is both a minimum and zero.

7.5 Response to a Sinusoidal Tidal Fluctuation

We consider the case of shallow unconfined transient flow induced by a fluctuating river stage. The water table at the boundary ($x = 0$) of the unconfined aquifer has a sinusoidal time variation. The boundary condition along the river bank is that a representative head in the aquifer, which we represent as $\tilde{h}$, is equal to the following expression along the river bank, $x = 0$,

$$h(0,t) = \tilde{h} + \Delta h \sin(\omega t), \tag{7.86}$$

where Δh and $\omega/(2\pi)$ are the amplitude and the frequency of the waves, respectively. The second boundary condition is that the head at infinity is not affected by the river,

$$\lim_{x\to\infty} [h(x,t)] = \tilde{h}. \tag{7.87}$$

The latter boundary condition implies that $\bar{h}$ is the average value of h. The governing differential equation for this case of flow is

$$\frac{\partial^2 \Phi}{\partial x^2} = \frac{S_p}{k\bar{h}} \frac{\partial \Phi}{\partial t}, \tag{7.88}$$

where

$$\Phi = \tfrac{1}{2}k(\phi - b)^2 = \tfrac{1}{2}kh^2; \tag{7.89}$$

h is the saturated thickness, and b the base elevation of the aquifer. We introduce $\tilde{\Phi}$ as $\frac{1}{2}k\tilde{h}^2$ and write

$$(\Phi - \tilde{\Phi}) = \tfrac{1}{2}k(h^2 - \tilde{h}^2) = \tfrac{1}{2}k(h - \tilde{h})(h + \tilde{h}) \sim k\tilde{h}(h - \tilde{h}); \tag{7.90}$$

$h + \tilde{h}$ is approximated by $2\tilde{h}$. We obtain from (7.86) and (7.90)

$$\Phi - \tilde{\Phi} = k\tilde{h}\Delta h \sin(\omega t), \qquad x = 0, \tag{7.91}$$

which is approximately true, provided that $\Delta h << \tilde{h}$.

Problems of periodic transient flow can be solved by a method known as separation of variables; this method is greatly simplified by the use of complex variables. A summary of the approach and the derivation of the solution presented here is given in Appendix A. Here, we pose the solution and demonstrate that it fulfills both the differential equation and the boundary conditions. The solution is

$$\Phi = k\tilde{h}\Delta h e^{-\mu x} \sin(\omega t - \mu x) + \tilde{\Phi}, \tag{7.92}$$

where

$$\mu = \sqrt{\frac{S_p \omega}{2k\tilde{h}}}. \tag{7.93}$$

7.5.1 Verification

We demonstrate that (7.92) satisfies the partial differential equation (7.88) as follows. We obtain for $\partial\Phi/\partial x$

$$\frac{\partial\Phi}{\partial x} = -k\tilde{h}\mu\Delta h e^{-\mu x}[\sin(\omega t-\mu x)+\cos(\omega t-\mu x)] \tag{7.94}$$

and for $\partial^2\Phi/\partial x^2$

$$\frac{\partial^2\Phi}{\partial x^2} = k\tilde{h}\mu^2\Delta h e^{-\mu x}[\sin(\omega t-\mu x)+\cos(\omega t-\mu x)] + k\tilde{h}\mu^2\Delta h e^{-\mu x}[\cos(\omega t-\mu x)-\sin(\omega t-\mu x)]. \tag{7.95}$$

We combine terms, simplify,

$$\frac{\partial^2\Phi}{\partial x^2} = 2k\tilde{h}\mu^2\Delta h e^{-\mu x}\cos(\omega t-\mu x) \tag{7.96}$$

and use (7.93):

$$\frac{\partial^2\Phi}{\partial x^2} = 2k\frac{s_p\omega}{2k\tilde{h}}\tilde{h}\Delta h e^{-\mu x}\cos(\omega t-\mu x) = s_p\omega\Delta h e^{-\mu x}\cos(\omega t-\mu x). \tag{7.97}$$

We obtain the derivative with respect to time as

$$\frac{\partial\Phi}{\partial t} = k\tilde{h}\Delta h e^{-\mu x}\omega\cos(\omega t-\mu x). \tag{7.98}$$

We see that indeed $s_p/(k\tilde{h})\partial\Phi/\partial t$ equals $\partial^2\Phi/\partial x^2$. We verify that the boundary condition at $x=0$ is satisfied by substituting zero for x in (7.92). We also observe that the potential approaches the average value for large values of x; the exponential function approaches zero.

This solution is restricted to positive values of x. If the aquifer occupies the domain $-\infty < x \leq 0$, the solution must be modified to

$$\frac{\partial\Phi}{\partial t} = k\tilde{h}\Delta h e^{\mu x}\omega\cos(\omega t+\mu x). \tag{7.99}$$

This solution is obtained simply by replacing x by $-x$. That the solution satisfies the partial differential equation follows from the general solution given in Appendix A, equation (A.14).

7.5.2 Extent of the Influence of the Periodic River Stage

The influence of the fluctuating water table reduces strongly with increasing values of x as a consequence of the factor $e^{-\mu x}$ in the solution. The factor μx is 3 for $x=3/\mu$, $e^{-\mu x}$ equals about 0.05; the fluctuations of the water table at a distance $x=3/\mu$ from the boundary are about 5% of those at $x=0$. The influence becomes less than 5% at distances x with

$$x \geq 3/\mu \tag{7.100}$$

from the boundary. Substitution of (7.93) for μ in (7.100) gives

$$x \geq 3\sqrt{\frac{2k\tilde{h}}{s_p\omega}}. \tag{7.101}$$

The influence of the fluctuating water table, or waves, decreases with the distance from the boundary and increases for increasing values of k and $\tilde{h}$ and for decreasing values of s_p and the frequency $\omega/2\pi$. The influence of waves with a higher frequency is smaller than of those with a lower frequency. For example, if $k = 10^{-5}$ m/s, $\tilde{h} = 25$ m, $s_p = 0.4$, and $\omega/2\pi$, the frequency, equals 0.1 wave/s, then (7.101) becomes

$$x \geq 0.3 \text{ m}. \tag{7.102}$$

Apparently, such short waves have only a negligible influence. However, if the period of the wave is, say, 6 hours, such as for a tide, the frequency will be $1/[6 * 3600]$ $1/\text{s}$, and (7.101) then becomes

$$x \geq 16 \text{ m}. \tag{7.103}$$

The influence of the fluctuation of the water table is still confined to a relatively small area. Plots of the potential as a function of x are shown in Figure 7.5; the plots are valid for fixed values of time t.

7.6 Falling-Head Experiment; Slug Test

The falling-head experiment and slug test are experiments that give an indication of hydraulic conductivity values in the lab or in the field. The experiment gives a

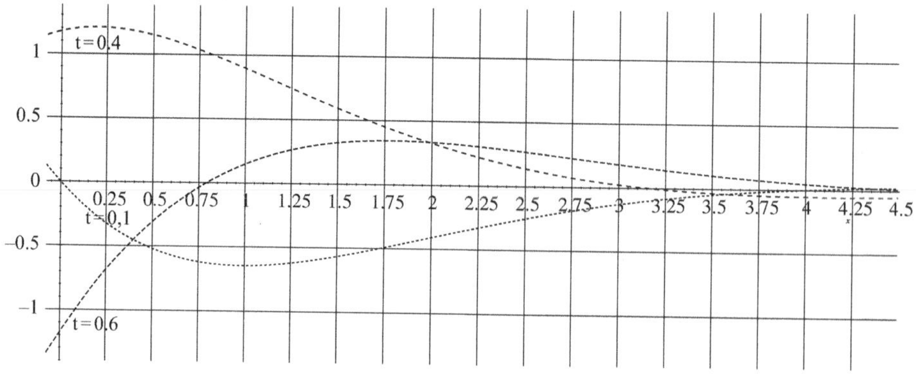

Figure 7.5 Plots of the potential as a function of x, for $s_p = 0.2$, $k = 0.1$, $\bar{h} = 10$, $\Delta h = 0.2\bar{h}$, $\omega = 2\pi$; the units are meters and days.

local value in the field; the flow induced by the test is confined to a small area. The mathematical description of the experiment, necessary for its interpretation differs from that presented so far in that the storage does not occur in the aquifer, which is considered incompressible, but in the experimental device.

7.6.1 Laboratory Experiment

The standard laboratory test for determining the hydraulic conductivity is a test, where the soil sample is connected via semipermeable dividers to two reservoirs; this test is known as the constant-head test. The heads in the reservoirs are kept constant, and the discharge through the sample is measured. With both heads and discharge known, the hydraulic conductivity can be determined. This constant-head experiment is based on the original experiment designed by Darcy.

If the soil has a low hydraulic conductivity, the discharge is small; the experiment will take a long time to perform and the results will be inaccurate. If the compressibility of the soil can be neglected, we may determine the hydraulic conductivity by an alternative experiment: the falling-head experiment shown in Figure 7.6. The water table in the standpipe is allowed to fall with time, hence the term *falling-head permeameter*. The water flows from the standpipe through the sample and into a reservoir where the water table is kept constant.

We determine the hydraulic conductivity, k, from the rate of decrease of head in the standpipe as follows. The discharge entering the sample over a period Δt is

$$-Q_z \Delta t = (h_0 - h_t) A_p, \tag{7.104}$$

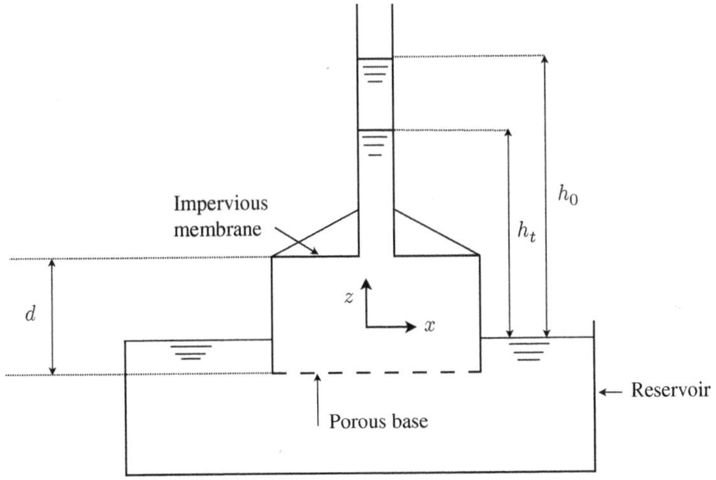

Figure 7.6 Falling-head permeameter.

where $-Q_z$ is the discharge entering the sample per unit time, Δt is the time over which the head drops from h_0 to h_t, and A_p is the area of the standpipe. We divide both sides of (7.104) by Δt and pass to the limit for $\Delta t \to 0$:

$$Q_z = \lim_{\Delta t \to 0} \frac{h_t - h_0}{\Delta t} A_p = \frac{dh}{dt} A_p. \tag{7.105}$$

We introduce an (x, y, z)-coordinate system with x- and y-axes horizontal. Darcy's law becomes, with Q_z positive for upward flow,

$$Q_z = q_z A_s, \tag{7.106}$$

where A_s is the area of the sample. The head varies linearly over the thickness d of the sample, if both the water and the soil are incompressible and if we neglect inertial effects. We rewrite (7.106), with the head at the top of the sample equal to h, and equal to zero at the bottom:

$$Q_z = -k \frac{\partial \phi}{\partial z} A_s = -k A_s \frac{h - 0}{d} = -k \frac{A_s}{d} h. \tag{7.107}$$

We combine (7.105) and (7.107):

$$\frac{dh}{dt} A_p = -k \frac{A_s}{d} h. \tag{7.108}$$

We divide both sides by h,

$$\frac{1}{h} \frac{dh}{dt} = \frac{d \ln h}{dt} = -\frac{k}{d} \frac{A_s}{A_p}, \tag{7.109}$$

integrate,

$$\int_{t_0}^{t} \frac{d \ln h}{dt} dt = \ln \frac{h_t}{h_0} = -\frac{k}{d} \frac{A_s}{A_p} (t - t_0), \tag{7.110}$$

and solve for the hydraulic conductivity, k:

$$k = \frac{A_p}{A_s} \frac{d}{t - t_0} \ln \left(\frac{h_0}{h_t} \right). \tag{7.111}$$

We compute the hydraulic conductivity k from (7.111), with the areas A_p and A_s, the sample thickness d, the initial head h_0, the head h_t, and the time $t - t_0$ known.

7.6.2 Falling-Head Experiment in the Field

An experiment similar to that described above may be performed in the field. A standpipe is connected with its bottom end to a small screen (a *well point*). This well point serves as a recharge well, i.e., a well of negative discharge.

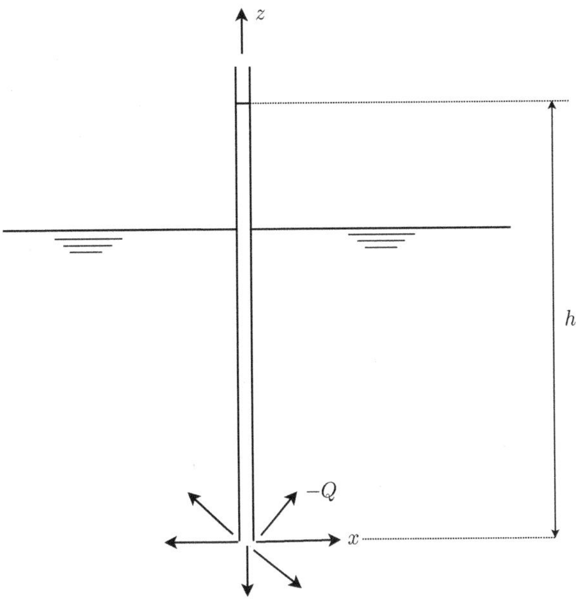

Figure 7.7 The falling-head experiment in the field (slug test).

The pipe is driven into the soil, as shown in Figure 7.7, and filled with water. The water level in the pipe will drop, supplying a discharge $(-Q)$ to the soil where, (compare (7.105)):

$$Q = \frac{dh}{dt} A_p. \tag{7.112}$$

We establish a relationship between the head, ϕ, and the flow rate, Q, in order to determine the hydraulic conductivity, k. If no flow occurs before the standpipe is placed and if the well point is sufficiently deep, the flow is nearly radial close to the well point. We assume that the soil and water are incompressible; the condition of continuity of flow implies that the discharge Q passes through any sphere of radius r around the well point. As the flow is radial, the specific discharge vector has only a radial component q_r, which is constant over the surface of the sphere. Hence, the condition of continuity of flow is

$$-Q = q_r 4\pi r^2. \tag{7.113}$$

We apply Darcy's law,

$$q_r = -k\frac{\partial \phi}{\partial r}, \tag{7.114}$$

to (7.113):

$$-Q = -k\frac{\partial \phi}{\partial r} 4\pi r^2. \tag{7.115}$$

We solve for $\partial\phi/\partial r$

$$\frac{\partial\phi}{\partial r} = \frac{Q}{4\pi k}\frac{1}{r^2} \tag{7.116}$$

and integrate

$$\phi = -\frac{Q}{4\pi k}\frac{1}{r} + C(t), \tag{7.117}$$

where C is a function of time alone. If the head far away from the well point is kept constant at a value ϕ_∞, then (7.117) becomes

$$\phi = -\frac{Q}{4\pi k}\frac{1}{r} + \phi_\infty. \tag{7.118}$$

We see from (7.118) that ϕ approaches ϕ_∞ for $r \to \infty$. The head at the well point $r = r_0$ is equal to h (r_0 is the radius of the well point):

$$h = -\frac{Q}{4\pi k}\frac{1}{r_0} + \phi_\infty. \tag{7.119}$$

We combine (7.112) and (7.119):

$$h = -\frac{1}{4\pi k r_0}A_p\frac{dh}{dt} + \phi_\infty \tag{7.120}$$

or

$$\frac{dh}{dt} + \frac{(4\pi k r_0)h}{A_p} = \frac{4\pi k r_0 \phi_\infty}{A_p}. \tag{7.121}$$

We introduce a new variable h^*:

$$h^* = h - \phi_\infty \tag{7.122}$$

and obtain, using that $d\phi_\infty/dt = 0$,

$$\frac{dh^*}{dt} + \frac{4\pi k r_0}{A_p}h^* + \frac{4\pi k r_o}{A_p}\phi_\infty = \frac{4\pi k r_0}{A_p}\phi_\infty \tag{7.123}$$

or

$$\frac{dh^*}{dt} + \frac{4\pi k r_0}{A_p}h^* = 0. \tag{7.124}$$

We divide by h^*,

$$\frac{1}{h^*}\frac{dh^*}{dt} + \frac{4\pi k r_0}{A_p} = 0, \tag{7.125}$$

and integrate:

$$\ln\frac{h_t^*}{h_0^*} = -\frac{4\pi k r_0}{A_p}(t - t_0) \tag{7.126}$$

or, solving for k,

$$k = \frac{A_p}{4\pi r_0} \frac{1}{t - t_0} \ln \frac{h_0^*}{h_t^*}. \tag{7.127}$$

We use (7.122)

$$k = \frac{A_p}{4\pi r_0} \frac{1}{t - t_0} \ln \frac{h_0 - \phi_\infty}{h_t - \phi_\infty}. \tag{7.128}$$

We must measure h_t, $t - t_0$, either estimate or measure h_0, ϕ_∞, and r_0, and know the area A_p of the standpipe in order to compute the value of the hydraulic conductivity. Note that application of the formula requires that consistent units are used. The slug test is similar to that presented above in that water is added to the standpipe, inducing flow away from the pipe. The slug test has the advantage of simplicity, but will only give a local value of the hydraulic conductivity.

8

Complex Variable Methods

8.1 Introduction

As stated in the introduction of this text, analytic solutions to practical problems, often simplified, serve to help us understand the parameters and boundary values that determine flow characteristics. Analytic solutions may be an end in themselves, but they may also be used with great advantage prior to creating a numerical model; a good understanding of the characteristics of the problem likely results in an efficient and useful computer model.

Complex variable methods have been applied extensively in the past in the area of groundwater flow. The use of complex variables has decreased over time, partly because of the popularity of powerful computer programs for groundwater flow modeling, but also because of the common misconception that complex variables are difficult to understand and that their application is limited to Laplace's equation. The word *complex* in complex variables refers to the complex nature of the complex variable: rather than representing a single independent variable, it represents two. In practice, this leads to a reduction of the number of independent variables, from x and y, to a single one, z, thus greatly simplifying the mathematical operations needed to solve problems.

Complex variables are fully supported in many computer languages, in applications in hand-held devices, in spreadsheets, and in pocket calculators. As a result, solutions formulated in terms of complex variables can be readily implemented in computer applications, without the need to separate them into real and imaginary parts.

The complex variables as introduced in this chapter are presented in terms of complex coordinates: the complex coordinate z and its complex conjugate $\bar{z}$. As a result, the reader will note that the application to Laplace's equation is a special case of complex variable applications; complex variables are useful in describing general two-dimensional problems. This view of complex variables is due to Wirtinger [1927] and is known as Wirtinger calculus.

The primary advantage of transforming the Cartesian (x, y)-coordinate system into the non-Cartesian $(\bar{z}, z)$-coordinate system, where $\bar{z} = x - \mathrm{i}y$, $z = x + \mathrm{i}y$, $(\mathrm{i}^2 = -1)$ is that it reduces the number of independent variables, x and y, to a single one, z. Since $\bar{z}$ is known once z is given, the second coordinate $\bar{z}$ does not constitute a second variable.

An additional advantage of using complex variables is that for flow problems governed by Laplace's equation, the stream function is obtained along with the discharge potential as part of the dependent complex variable.

8.2 General Complex Variables

We define a complex variable z as

$$z = x + \mathrm{i}y, \tag{8.1}$$

where i is defined as the purely imaginary number of magnitude 1, with the property that

$$\mathrm{i}^2 = -1. \tag{8.2}$$

The complex variable $\bar{z}$ is called the complex conjugate of z, and is defined as

$$\bar{z} = x - \mathrm{i}y. \tag{8.3}$$

We consider the coordinate transformation of the (x, y)-coordinate system into the $(\bar{z}, z)$-coordinate system. Application of the standard rules for coordinate transformations makes it possible to transform quantities such as vectors and tensors in terms of components in the $(\bar{z}, z)$-coordinate system. It turns out, somewhat surprisingly, that the coordinates $(\bar{z}, z)$ may be considered, for the purpose of differentiation, as independent coordinates and that the rules of partial differentiation apply to these coordinates. This is surprising, because if z is known, then both x and y must be known, so that $\bar{z}$ also is known. We use this result without further discussion, and in this way write the basic equations for groundwater flow in terms of the complex variables $\bar{z}$ and z.

One reason for using complex variables is that this reduces the number of independent variables from two to one; if one of the two independent variables is known, z for example, then the other one is known also. There are many more advantages to this formulation, which applies to any two-dimensional problem. An example is writing computer code; complex variables are implemented in many existing compilers and applications written for hand-held devices, and programming is greatly simplified and facilitated by using complex variables for two-dimensional problems.

8.2.1 Functions

Consider a function F of x and y

$$F = F(x,y) = F(x(z,\bar{z}), y(z,\bar{z})) = \tilde{F}(z,\bar{z}). \tag{8.4}$$

We obtain the partial derivative of this function with respect to z with the chain rule,

$$\frac{\partial F}{\partial z} = \frac{\partial F}{\partial x}\frac{\partial x}{\partial z} + \frac{\partial F}{\partial y}\frac{\partial y}{\partial z}. \tag{8.5}$$

We solve (8.1) and (8.3) for x and y,

$$x = \frac{1}{2}(z+\bar{z}) \tag{8.6}$$

$$y = \frac{1}{2\mathrm{i}}(z-\bar{z}) \tag{8.7}$$

so that

$$\frac{\partial x}{\partial z} = \frac{\partial x}{\partial \bar{z}} = \frac{1}{2} \tag{8.8}$$

$$\frac{\partial y}{\partial z} = -\frac{\partial y}{\partial \bar{z}} = -\frac{\mathrm{i}}{2} \tag{8.9}$$

and

$$\frac{\partial z}{\partial x} = \frac{\partial \bar{z}}{\partial x} = 1 \tag{8.10}$$

$$\frac{\partial z}{\partial y} = -\frac{\partial \bar{z}}{\partial y} = \mathrm{i}. \tag{8.11}$$

We use this with (8.5),

$$\frac{\partial F}{\partial z} = \frac{1}{2}\left[\frac{\partial F}{\partial x} + \frac{1}{\mathrm{i}}\frac{\partial F}{\partial y}\right]. \tag{8.12}$$

We rewrite $1/\mathrm{i}$ as $-\mathrm{i}$

$$\frac{1}{\mathrm{i}} = \frac{\mathrm{i}}{\mathrm{i}^2} = -\mathrm{i} \tag{8.13}$$

and simplify (8.12) to

$$2\frac{\partial F}{\partial z} = \frac{\partial F}{\partial x} - \mathrm{i}\frac{\partial F}{\partial y}. \tag{8.14}$$

We next apply this result to the case of groundwater flow by replacing the arbitrary function F by the discharge potential.

8.2.2 The Discharge Potential Φ

We replace the function F by the potential Φ in (8.14)

$$2\frac{\partial \Phi}{\partial z} = \frac{\partial \Phi}{\partial x} - \mathrm{i}\frac{\partial \Phi}{\partial y}. \tag{8.15}$$

Darcy's law in terms of the potential Φ is

$$Q_x = -\frac{\partial \Phi}{\partial x} \tag{8.16}$$

$$Q_y = -\frac{\partial \Phi}{\partial y}, \tag{8.17}$$

where (Q_x, Q_y) represents the discharge vector for some kind of flow (e.g., shallow unconfined flow) and the potential Φ is defined accordingly (e.g., $\Phi = \frac{1}{2}k\phi^2$ for unconfined flow). We apply Darcy's law to (8.15)

$$2\frac{\partial \Phi}{\partial z} = -Q_x + \mathrm{i}Q_y \tag{8.18}$$

or

$$Q_x - \mathrm{i}Q_y = -2\frac{\partial \Phi}{\partial z}. \tag{8.19}$$

This is a remarkable simplification relative to the formulation in terms of real variables: a single equation renders both components of the discharge vector, rather than just one. We obtain further simplification by introducing a new complex function, the discharge function W, defined as

$$W = Q_x - \mathrm{i}Q_y \tag{8.20}$$

so that Darcy's law, (8.16) and (8.17), becomes

$$W = -2\frac{\partial \Phi}{\partial z}. \tag{8.21}$$

The Potential for a Well at $z = z_w$

We must express the potential in terms of the complex variables z and $\bar{z}$ in order to make use of the expression for W. We may do this for any of the expressions for the potential thus far derived by the use of equations (8.6) and (8.7). We apply this to the potential for a well, which is

$$\Phi = \frac{Q}{4\pi}\ln[(x - x_w)^2 + (y - y_w)^2], \tag{8.22}$$

where x_w and y_w are the coordinates of the well, and

$$z_w = x_w + \mathrm{i}y_w. \tag{8.23}$$

We make use of the following relationship

$$(z-z_w)(\bar{z}-\bar{z}_w) = [(x-x_w)+\mathrm{i}(y-y_w)][(x-x_w)-\mathrm{i}(y-y_w)] \tag{8.24}$$

or, carrying out the multiplications

$$(z-z_w)(\bar{z}-\bar{z}_w) = (x-x_w)^2-\mathrm{i}^2(y-y_w)^2 = (x-x_w)^2+(y-y_w)^2. \tag{8.25}$$

We use this in (8.22) and obtain the following expression for the potential for a well at z_w

$$\Phi = \frac{Q}{4\pi}\ln[(z-z_w)(\bar{z}-\bar{z}_w)] = \frac{Q}{4\pi}\ln(z-z_w)+\frac{Q}{4\pi}\ln(\bar{z}-\bar{z}_w). \tag{8.26}$$

The Potential for Uniform Flow

We obtain the potential for uniform flow in a similar fashion. The potential in terms of real variables is

$$\Phi = -Q_{x0}x - Q_{y0}y \tag{8.27}$$

or, with (8.6) and (8.7) and using that $1/\mathrm{i} = -\mathrm{i}$,

$$\Phi = -Q_{x0}\frac{1}{2}(z+\bar{z}) - Q_{y0}\frac{1}{2\mathrm{i}}(z-\bar{z}) = -\tfrac{1}{2}(Q_{x0}-\mathrm{i}Q_{y0})z - \tfrac{1}{2}(Q_{x0}+\mathrm{i}Q_{y0})\bar{z}. \tag{8.28}$$

We introduce W_0 as

$$W_0 = Q_{x0} - \mathrm{i}Q_{y0}, \tag{8.29}$$

so that the expression for the potential for uniform flow reduces to

$$\Phi = -\tfrac{1}{2}W_0 z - \tfrac{1}{2}\bar{W}_0\bar{z}. \tag{8.30}$$

8.2.3 Polar Coordinates

We write a complex variable z in terms of polar coordinates (r,θ) as

$$z = x+\mathrm{i}y = r\cos\theta + \mathrm{i}r\sin\theta = r[\cos\theta+\mathrm{i}\sin\theta]. \tag{8.31}$$

We write (8.31) in a compact form using a formula known as Euler's formula, which is obtained as follows. We expand the exponential function of a purely imaginary number $\mathrm{i}\theta$ about $\theta = 0$

$$\mathrm{e}^{\mathrm{i}\theta} = 1+\mathrm{i}\theta+\frac{\mathrm{i}^2}{2!}\theta^2+\frac{\mathrm{i}^3}{3!}\theta^3+\frac{\mathrm{i}^4}{4!}\theta^4+\frac{\mathrm{i}^5}{5!}\theta^5+\cdots\frac{\mathrm{i}^n}{n!}\theta^n\cdots \tag{8.32}$$

or

$$\mathrm{e}^{\mathrm{i}\theta} = 1+\mathrm{i}\theta-\frac{1}{2!}\theta^2-\frac{\mathrm{i}}{3!}\theta^3+\frac{1}{4!}\theta^4+\frac{\mathrm{i}}{5!}\theta^5-\cdots\frac{\mathrm{i}^n}{n!}\theta^n\cdots. \tag{8.33}$$

If we separate real and imaginary parts and compare the result with the expansions of $\cos\theta$ and $\sin\theta$, we obtain Euler's formula:

$$e^{i\theta} = \cos\theta + i\sin\theta. \tag{8.34}$$

This equation is useful when expression in terms of polar coordinates is desired. For example, if the magnitude of the uniform flow is known, Q_0, and its orientation relative to the x-axis is known as well and equal to α, then we may write the complex discharge for uniform flow as:

$$W = Q_{x0} - iQ_{y0} = Q_0 e^{-i\alpha}, \tag{8.35}$$

where

$$Q_0 = \sqrt{Q_{x0}^2 + Q_{y0}^2} \tag{8.36}$$

$$\tan\alpha = \frac{Q_{y0}}{Q_{x0}}. \tag{8.37}$$

We obtain an expression for the potential by integration of $2\partial\Phi/\partial z = -W = -Q_0 e^{-i\alpha}$,

$$\Phi = -\frac{1}{2}Q_0 e^{-i\alpha} z + f(\bar{z}), \tag{8.38}$$

where $f(\bar{z})$ is an arbitrary function of $\bar{z}$. The potential must be a real function, which determines the form of $f(\bar{z})$:

$$\Phi = -\frac{1}{2}Q_0 e^{-i\alpha} z - \frac{1}{2}Q_0 e^{i\alpha}\bar{z}. \tag{8.39}$$

8.2.4 Application: A Well and a Recharge Well in a Field of Uniform Flow

We consider the case of a well and a recharge well in a field of uniform flow angled at α with respect to the x-axis. The recharge well has a recharge of Q and the discharge well has a discharge of aQ, where a is a real factor. The recharge well is at $z = -d$ and the discharge well is at $z = d$. The potential for this case is

$$\Phi = -\tfrac{1}{2}W_0 z - \frac{Q}{4\pi}\ln(z+d) + a\frac{Q}{4\pi}\ln(z-d) - \tfrac{1}{2}\bar{W}_0\bar{z} - \frac{Q}{4\pi}\ln(\bar{z}+d) + a\frac{Q}{4\pi}\ln(\bar{z}-d). \tag{8.40}$$

We wish to determine the coordinates of the stagnation point. Both components of the discharge vector are zero at the stagnation point, i.e., if the complex coordinate of the stagnation point is z_s, the condition is

$$W(z_s) = 0. \tag{8.41}$$

We obtain an expression for z_s by solving this equation for z_s. We apply Darcy's law in its complex form, (8.21), and obtain

$$W = -2\frac{\partial \Phi}{\partial z} = W_0 + \frac{Q}{2\pi}\frac{1}{z_s+d} - a\frac{Q}{2\pi}\frac{1}{z_s-d} = 0. \tag{8.42}$$

We multiply both sides of this equation by $(z_s-d)(z_s+d)$,

$$W_0(z_s^2-d^2) + \frac{Q}{2\pi}(z_s-d) - a\frac{Q}{2\pi}(z_s+d) = 0 \tag{8.43}$$

or

$$z_s^2 + \frac{Q}{2\pi W_0}(1-a)z_s - \frac{Q}{2\pi W_0}(1+a)d - d^2 = 0. \tag{8.44}$$

We solve this for z_s,

$$z_s = -\frac{Q}{4\pi W_0}(1-a) \pm \sqrt{\left[\frac{Q}{4\pi W_0}(1-a)\right]^2 + \frac{Q}{2\pi W_0}(1+a)d + d^2}. \tag{8.45}$$

Note that this expression is complex, and obtaining the real and imaginary parts by hand is not trivial. However, complex variables are implemented in many existing compilers and programs such as Excel and MATLAB.

8.2.5 Application: Transient Well

The potential for a transient well is given by (7.24):

$$\Phi = -\frac{Q}{4\pi}\mathrm{E}_1(u) + \Phi_0, \tag{8.46}$$

where

$$u \doteq \frac{S_s r^2}{4k(t-t_0)}, \tag{8.47}$$

where r is the distance from the well at z_w to the point of evaluation, i.e.,

$$r^2 = (z-z_w)(\bar{z}-\bar{z}_w) \tag{8.48}$$

so that

$$\Phi = -\frac{Q}{4\pi}\mathrm{E}_1\left(\frac{S_s(z-z_w)(\bar{z}-\bar{z}_w)}{4k(t-t_0)}\right) + \Phi_0. \tag{8.49}$$

We obtain an expression for the discharge function W by differentiation with respect to z:

$$W = -2\frac{\partial \Phi}{\partial z} = -2\frac{d\mathrm{E}_1(u)}{du}\frac{\partial u}{\partial z}. \tag{8.50}$$

The derivative of $\mathrm{E}_1(u)$ with respect to u is given by (7.29), and we obtain with the expression for u in terms of$z$

$$W=-2\frac{Q}{4\pi}\frac{e^{-u}}{u}\frac{S_s(\bar{z}-\bar{z}_w)}{4k(t-t_0)}=-\frac{Q}{2\pi}\frac{e^{-u}}{u}\frac{u}{z-z_w}=-\frac{Q}{2\pi}\frac{e^{-u}}{z-z_w}. \tag{8.51}$$

Problem

8.1 Demonstrate that expression (8.51) is identical to that obtained in real variables in Section 7.3.

8.3 Complex Variables for Harmonic Functions

There is a fundamental difference between the expressions for both the potential and the discharge function that apply to the first example given above (a well in a field of uniform flow) and the second one (a transient well). For the two functions that enter in the first example, W is a function of z only, whereas in the second example W is a function of both z and $\bar{z}$ (note that r contains both z and $\bar{z}$). For the first example, Φ is the sum of two complex conjugate functions, each of which is a function of a single complex variable, either z or $\bar{z}$. In the second example, Φ is a real function of both z and $\bar{z}$ that cannot be separated into two functions, each of either z or $\bar{z}$.

If the potential is the sum of a function of z only, and its complex conjugate, i.e., the real part of a function of z only, we can introduce a complex potential Ω:

$$\Phi=\frac{1}{2}\left[\Omega(z)+\overline{\Omega(z)}\right]. \tag{8.52}$$

We demonstrate in what follows that the real part of a complex function of a single complex variable is harmonic. This demonstration is not essential for following the remainder of the text; the reader may skip Section 8.3.1.

8.3.1 The Laplacian; Divergence and Curl

We make use of the complex conjugate of a function, defined as follows

$$\overline{\Omega(z)}=\bar{\Omega}(\bar{z}). \tag{8.53}$$

We illustrate the notation $\bar{\Omega}$ by means of the following example: if the function $\Omega(z)$ is defined as $\Omega(z)=\mathrm{i}z$, then

$$\bar{\Omega}(\bar{z})=\overline{\Omega(z)}=-\mathrm{i}\bar{z}. \tag{8.54}$$

If a function Ω implies that the argument, z, is to be multiplied by i, then the function $\bar{\Omega}$ implies that the argument be multiplied by $-$i, so that

$$\bar{\Omega}(z)=-\mathrm{i}z \tag{8.55}$$

so that indeed

$$\bar{\Omega}(\bar{z}) = -\mathrm{i}\bar{z} = \overline{\Omega(z)}. \tag{8.56}$$

We use the notation $\bar{\Omega}$ in (8.52) and obtain

$$\Phi = \frac{1}{2}\left[\Omega(z) + \bar{\Omega}(\bar{z})\right]. \tag{8.57}$$

The discharge function W is -2 times the derivative of the discharge potential with respect to z:

$$W = Q_x - \mathrm{i}Q_y = -2\frac{\partial \Phi}{\partial z} = -\frac{d\Omega(z)}{dz}. \tag{8.58}$$

This expression does not contain $\bar{z}$, so that the derivative of W with respect to $\bar{z}$ is zero:

$$\frac{\partial W}{\partial \bar{z}} = 0. \tag{8.59}$$

We express the derivative of W with respect to $\bar{z}$ in terms of derivatives with respect to x and y as

$$\frac{\partial W}{\partial \bar{z}} = \frac{\partial W}{\partial x}\frac{\partial x}{\partial \bar{z}} + \frac{\partial W}{\partial y}\frac{\partial y}{\partial \bar{z}} = \frac{\partial W}{\partial x}\frac{1}{2} + \frac{\partial W}{\partial y}\frac{\mathrm{i}}{2}, \tag{8.60}$$

where we made use of (8.8) and (8.9):

$$\frac{\partial x}{\partial \bar{z}} = \frac{1}{2}, \qquad \frac{\partial y}{\partial \bar{z}} = \frac{\mathrm{i}}{2}. \tag{8.61}$$

We multiply both sides of (8.60) by 2,

$$2\frac{\partial W}{\partial \bar{z}} = \frac{\partial W}{\partial x} + \mathrm{i}\frac{\partial W}{\partial y}, \tag{8.62}$$

and use that $W = Q_x - \mathrm{i}Q_y$,

$$2\frac{\partial W}{\partial \bar{z}} = \frac{\partial Q_x}{\partial x} - \mathrm{i}\frac{\partial Q_y}{\partial x} + \mathrm{i}\frac{\partial Q_x}{\partial y} - \mathrm{i}^2\frac{\partial Q_y}{\partial y} \tag{8.63}$$

or

$$2\frac{\partial W}{\partial \bar{z}} = \frac{\partial Q_x}{\partial x} + \frac{\partial Q_y}{\partial y} + \mathrm{i}\left[\frac{\partial Q_x}{\partial y} - \frac{\partial Q_y}{\partial x}\right]. \tag{8.64}$$

The first term to the right of the equals sign in this equation is the divergence of the discharge vector, $-\gamma$, and the second term is minus the curl or rotation of the discharge vector, $-\delta$, i.e.,

$$-2\frac{\partial W}{\partial \bar{z}} = \gamma + \mathrm{i}\delta. \tag{8.65}$$

For the special case considered here, the potential is the real part of the complex potential so that W is a function of z only, and $\partial W/\partial \bar{z} = 0$; both γ and δ are zero and the flow is both divergence-free (i.e., no infiltration, leakage, or storage) and irrotational. The potential is harmonic

$$\frac{\partial^2\Phi}{\partial x^2}+\frac{\partial^2\Phi}{\partial y^2}=0. \tag{8.66}$$

If we use Darcy's law in its real form in (8.64) we obtain, after multiplying both sides of the equation by -1

$$-2\frac{\partial W}{\partial \bar{z}}=\frac{\partial^2\Phi}{\partial x^2}+\frac{\partial^2\Phi}{\partial y^2}+\mathrm{i}\left[\frac{\partial^2\Phi}{\partial x\partial y}-\frac{\partial^2\Phi}{\partial y\partial x}\right]=\frac{\partial^2\Phi}{\partial x^2}+\frac{\partial^2\Phi}{\partial y^2}, \tag{8.67}$$

where we used the property that the order of differentiation is interchangeable for single-valued functions, such as the potential.

We substitute $-2\partial\Phi/\partial z$ for W (see (8.21)) in (8.67) and obtain

$$4\frac{\partial^2\Phi}{\partial z\partial \bar{z}}=\frac{\partial^2\Phi}{\partial x^2}+\frac{\partial^2\Phi}{\partial y^2}. \tag{8.68}$$

This equation can be integrated by parts to obtain an expression for the potential if the divergence (e.g., the infiltration) is given. We make use of this equation if there is infiltration.

8.3.2 The Stream Function

The complex potential Ω has as its real part the potential; it also has an imaginary part, which we represent as Ψ:

$$\Omega=\Phi+\mathrm{i}\Psi. \tag{8.69}$$

We have seen that W is the derivative of Ω with respect to z and

$$W=-2\frac{\partial\Phi}{\partial z}=-\frac{d\Omega}{dz}=-\frac{d[\Phi+\mathrm{i}\Psi]}{dz}=-\frac{\partial\Phi}{\partial z}-\mathrm{i}\frac{\partial\Psi}{\partial z}. \tag{8.70}$$

Note that both Φ and Ψ are functions of both z and $\bar{z}$, whereas Ω is a function of z only, so that $\partial\Omega/\partial\bar{z}=0$, or

$$\frac{\partial\Omega}{\partial\bar{z}}=0=\frac{\partial\Phi}{\partial\bar{z}}+\mathrm{i}\frac{\partial\Psi}{\partial\bar{z}}\rightarrow\frac{\partial\Phi}{\partial\bar{z}}=-\mathrm{i}\frac{\partial\Psi}{\partial\bar{z}}. \tag{8.71}$$

We take the complex conjugate of both sides, noting that Φ and Ψ are real,

$$\frac{\partial\Phi}{\partial z}=\mathrm{i}\frac{\partial\Psi}{\partial z}. \tag{8.72}$$

Equations (8.71) and (8.72) are the complex form of the Cauchy-Riemann equations; we obtain their real form by expressing the derivatives with respect to z in terms of derivatives with respect to x and y

$$\frac{\partial \Phi}{\partial x} - \mathrm{i}\frac{\partial \Phi}{\partial y} = \mathrm{i}\frac{\partial \Psi}{\partial x} - \mathrm{i}^2\frac{\partial \Psi}{\partial y}. \tag{8.73}$$

We separate the real and imaginary parts and obtain the Cauchy-Riemann equations in their standard real form:

$$\frac{\partial \Phi}{\partial x} = \frac{\partial \Psi}{\partial y} \tag{8.74}$$

$$\frac{\partial \Phi}{\partial y} = -\frac{\partial \Psi}{\partial x}. \tag{8.75}$$

Note that the complex Cauchy-Riemann conditions collapse to a single one; one is the complex conjugate of the other.

The derivatives $\partial\Phi/\partial x$ and $\partial\Phi/\partial y$ are the components of the gradient of the potential, which points opposite to the direction of flow. The derivatives $\partial\Psi/\partial x$ and $\partial\Psi/\partial y$ are, likewise, the components of the gradient of Ψ, which is the stream function. It follows from the Cauchy-Riemann conditions that

$$\frac{\partial \Phi}{\partial x}\frac{\partial \Psi}{\partial x} + \frac{\partial \Phi}{\partial y}\frac{\partial \Psi}{\partial y} = 0, \tag{8.76}$$

which is the dot product of the gradient of the potential and that of the stream function; this dot product is zero, so that these gradients are mutually orthogonal, as illustrated in Figure 8.1. The gradient of the potential points diametrically opposite to the direction of flow, and thus the gradient of the stream function is normal to the flow. The stream function is constant in the direction normal to its gradient, by definition; the stream function is constant along the streamlines, which is the reason why we call this function the stream function. The reader may verify from the relationship between the components of the gradients of Ψ and Φ that the orientations of these gradients are as shown in the figure.

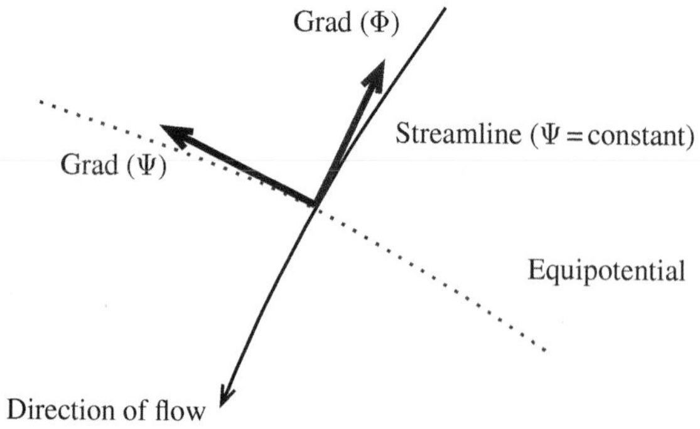

Figure 8.1 Orthogonality of the gradients of the stream function Ψ and the potential Φ.

We introduced the stream function and the Cauchy-Riemann conditions in terms of real variables in Section 2.1; see equations (2.24) and (2.25).

8.3.3 Flow Nets

A flow net is obtained by plotting contours $\Phi =$ constant and $\Psi =$ constant in a single figure; these curves are orthogonal. We obtain approximate squares, a flow net, if we choose the intervals between contours of Φ and Ψ to be equal, i.e., if we choose

$$\Delta\Phi = \Delta\Psi, \tag{8.77}$$

We see from the complex Cauchy-Riemann equations, (8.72), that if the contour intervals are equal, then so will be the line spacing; if we write the Cauchy-Riemann equations in discrete form

$$\frac{\Delta\Phi}{\Delta z_{\Phi}} = \mathrm{i}\frac{\Delta\Psi}{\Delta z_{\Psi}}, \tag{8.78}$$

then it is clear that the distance between contours of Φ is the same as the distance between contours of Ψ

$$|\Delta z_{\Phi}| \approx |\Delta z_{\Psi}|. \tag{8.79}$$

Thus, the flow net will consist of quadrilaterals that approximate squares with increasing refinement of the net.

8.3.4 Computing the Flow Passing between Two Points

We compute the discharge through some curve $\mathscr{C}$ between two points, z_1 and z_2, using complex variables for divergence-free flow as follows. This discharge is expressed as the integral of the dot product of the discharge vector and the unit vector pointing normal to the curve. If an increment along the curve is expressed as (dx, dy), as shown in Figure 8.2, then an increment pointing normal to that boundary is $(dy, -dx)$ and the integrated flow through the boundary is the dot product of the vectors (Q_x, Q_y) and $(dy, -dx)$, i.e.,

$$\int_{\mathscr{C}} Q_n ds = \int_{z_1}^{z_2} [Q_x dy - Q_y dx]. \tag{8.80}$$

We write this expression as the imaginary part of the integral

$$\int_{z_1}^{z_2} W dz = \int_{z_1}^{z_2} [(Q_x - \mathrm{i}Q_y)(dx + \mathrm{i}dy)] = \int_{z_1}^{z_2} \left[Q_x dx + Q_y dy\right] + \mathrm{i}\int_{z_1}^{z_2} \left[Q_x dy - Q_y dx\right]. \tag{8.81}$$

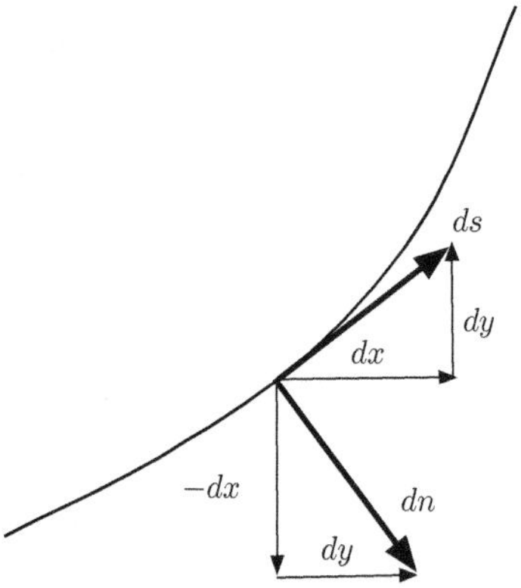

Figure 8.2 The integral $\int_{\mathscr{C}} W dz$.

We see that the imaginary part of this integral indeed equals (8.80),

$$\int_{\mathscr{C}} Q_n ds = \Im \int_{z_1}^{z_2} W dz. \tag{8.82}$$

The real part of the integrand of (8.81) is the dot product of the vectors (Q_x, Q_y) and (dx, dy) and thus equals the integrated tangential component of flow,

$$\int_{\mathscr{C}} Q_s ds = \Re \int_{z_1}^{z_2} W dz. \tag{8.83}$$

We may express W for the case of harmonic functions in terms of a complex potential

$$W = -\frac{d\Omega}{dz} \tag{8.84}$$

so that we can work out the integral (8.81) in closed form:

$$\int_{z_1}^{z_2} W dz = -\int_{z_1}^{z_2} \frac{d\Omega}{dz} dz = -[\Omega(z_2) - \Omega(z_1)] = \Phi_1 - \Phi_2 + \mathrm{i}[\Psi_1 - \Psi_2]. \tag{8.85}$$

We make two important observations from this result. First, as a consequence of (8.84), valid for complex potentials with harmonic real and imaginary parts, i.e., for divergence-free, irrotational flow, the integral is path-independent; nowhere did we need to use the geometry of the path of integration. Thus, the flow between two points is independent of the curve that connects these points. This is a direct consequence of the divergence-free character of the flow; water is neither added to the flow nor removed from it anywhere. The second observation is that the flow that passes between two points is exactly equal to the difference of the values of the stream function at these points.

Of special interest is the case where the curve $\mathscr{C}$ is closed; in that case the total flow through the curve must be zero, provided that the potential is single-valued; i.e., the complex potential has a single value at each point in the domain. We express this mathematically as

$$\oint_{\mathscr{C}} \Omega(z)dz = 0. \tag{8.86}$$

This equation is known as the Cauchy-Goursat integral theorem.

8.4 The Maximum Modulus Theorem

The maximum modulus theorem states that a function that is harmonic inside some domain $\mathscr{D}$ bounded by a boundary $\mathscr{C}$ cannot have a value inside $\mathscr{D}$ below its lowest value on $\mathscr{C}$; neither can it have a value higher than its highest value on $\mathscr{C}$. In other words, the maximum and minimum values of the potential in $\mathscr{D}$ are on the boundary $\mathscr{C}$ of $\mathscr{D}$. Knowledge of this fact is important in designing dewatering systems for cases where infiltration is negligible (this is often the case for building pits). It is sufficient to verify that conditions of maximum head are satisfied along the boundary of the pit; verification of these conditions for points inside the pit is not necessary.

We demonstrate this by the use of yet another theorem, the mean value theorem. Consider two concentric circles, $\mathscr{C}_1$ with radius R_1 and $\mathscr{C}_2$ with radius R_2, centered at the origin of an (r, θ) polar coordinate system, as shown in Figure 8.3. There are no sources of water inside the circles, and therefore the net outflow through any circle of radius r ($r \leq R_2$) is zero:

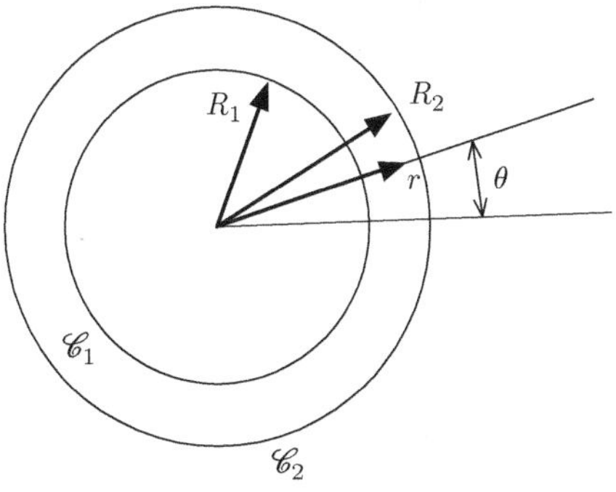

Figure 8.3 The mean value theorem.

$$\int_0^{2\pi} Q_r r d\theta = -\int_0^{2\pi} \frac{\partial \Phi}{\partial r} r d\theta = 0, \qquad r \leq R_2. \tag{8.87}$$

Division of the latter equation by r and subsequent integration from R_1 to R_2 yields

$$\int_{R_1}^{R_2}\int_0^{2\pi} \frac{\partial \Phi}{\partial r} d\theta dr = \int_0^{2\pi}\int_{R_1}^{R_2} \frac{\partial \Phi}{\partial r} dr d\theta = \int_0^{2\pi} [\Phi(R_2,\theta) - \Phi(R_1,\theta)] d\theta = 0, \tag{8.88}$$

where the change in the sequence of integration is allowed if Φ and its derivatives are continuous inside $\mathscr{C}_2$. It follows from (8.88) that

$$\int_0^{2\pi} \Phi(R_1,\theta) d\theta = \int_0^{2\pi} \Phi(R_2,\theta) d\theta. \tag{8.89}$$

We contract the circle $\mathscr{C}_1$ to a point, so that (8.89) becomes, for $R_1 \to 0$:

$$\Phi_0 = \frac{1}{2\pi R_2} \int_0^{2\pi} \Phi(R_2,\theta) R_2 d\theta, \tag{8.90}$$

where Φ_0 is the value of the potential at the center of $\mathscr{C}_2$. This result is known as the *mean value theorem* and states that the value of a harmonic function at the center of a circle equals the mean of the values along that circle. This property might explain the term *harmonic function*. The maximum modulus theorem can be deduced from the mean value theorem: Φ cannot have a maximum at any point $\mathscr{P}$ inside a domain, because it is the average of the values along a small circle of radius ε around $\mathscr{P}$: there will always be both a higher and a lower value of Φ on the circle than at its center.

8.5 Boundary Value Problems

There exist infinitely many harmonic functions. The function that applies to any given problem is determined by the boundary conditions. There are various kinds of boundaries: some boundaries are finite, some contain infinity, and in other cases the boundary consists of separate closed contours. Domains bounded by the latter types of boundaries are called multiply connected. If only one boundary exists, the domain is simply connected.

Conditions along the boundaries may be specified in many different ways. The three most common kinds of boundary value problems are the following:

1. The Dirichlet problem: all boundary conditions are given in terms of the potential Φ.
2. The Neumann problem: the derivative $d\Phi/dn$ in the direction normal to the boundary is given everywhere along the boundary.
3. The mixed boundary value problem: Φ is given along certain segments of the boundary and $d\Phi/dn$ along all the others.

In general, we may state that along each boundary segment one, and only one, condition must be specified in order to define a harmonic function fully.

Many other types of boundary conditions occur in groundwater flow problems. Closed boundaries are found mainly in two-dimensional problems in the vertical plane. Regional aquifers rarely have such boundaries and are often best modeled as infinite domains with internal boundaries consisting of rivers, creeks, or lake boundaries. If the aquifer is infinite, a condition must be specified that controls the behavior of the solution near infinity. We term such a condition the far-field condition, whereby far field refers to the flow pattern near infinity. Sometimes this far-field condition specifies a uniform far field; it always determines whether or not inflow or outflow occurs from or to infinity. We have, in Chapter 2, specified the latter condition usually by prescribing the head at some fixed point $\mathscr{P}$, which we called the reference point.

8.6 Series Expansions

Series expansions are an efficient way to express functions that represent solutions to practical problems. Many of the approaches used in the analytic element method (e.g., Strack [1989], Haitjema [1995], Strack [2003]) are based on the use of series expansions, and we make use of such expansions in this text when developing analytic elements for circular and elliptical lakes, elliptical impermeable objects, elliptical inhomogeneities, and a circular lake with a leaky bottom.

A function that is holomorphic inside a circle of radius R is a function that is expressed in terms of z only (i.e., does not contain $\bar{z}$) and, along with all of its derivatives with respect to z, has no singularities for $z\bar{z} \leq R^2$. Such a function can be expanded in a Taylor series about the center of the circle, z_0,

$$f(z) = \sum_{n=0}^{\infty} a_n (z - z_0)^n, \qquad z \in \mathscr{C}, \tag{8.91}$$

where the coefficients a_n can be expressed in terms of the derivatives of $f(z)$ computed at z_0

$$a_n = \frac{f^n(z_0)}{n!}. \tag{8.92}$$

This series converges for $(z - z_0)\overline{(z - z_0)} \leq R^2$.

Points where not all the derivatives of the function $f(z)$ exist are called singular points; the Taylor series expansion of $f(z)$ about z_0 converges to the function $f(z)$ inside a circle centered at z_0 only if it does not contain singularities of $f(z)$. The theorem works both ways: if a function can be expanded in terms of a series (8.91), then the function is holomorphic inside the circle of convergence. Note that all derivatives of $f(z)$ with respect to z exist inside the circle of convergence.

A second example of representing a complex function in terms of a power series is a function represented by its asymptotic expansion, i.e., an expansion that contains only negative powers of z,

$$f(z) = \sum_{j=0}^{\infty} \frac{a_{-j}}{(z-z_0)^j}. \tag{8.93}$$

This expansion will converge outside a circle $\mathscr{C}$ with a radius R centered at $z = z_0$ such that it does not contain any singularities of $f(z)$ outside the circle. We obtain the asymptotic expansion of the function $f(z)$, (8.93), that converges outside a circle of radius R centered at z_0 as follows. We first define a complex variable ζ as

$$\zeta = \frac{1}{z-z_0}. \tag{8.94}$$

Points z outside $\mathscr{C}$ correspond to points in a different plane, the ζ-plane, that are inside a circle centered at $\zeta = 0$ ($\zeta = 0$ corresponds to $z = \infty$) and of radius $1/R$. The function $f(\zeta)$ has no singularities inside the circle in the ζ-plane and thus can be expanded in a Taylor series in terms of ζ that converges inside the circle in the ζ-plane,

$$\tilde{f}(\zeta) = f(\zeta(z)) = \sum_{j=0}^{\infty} a_j \zeta^j. \tag{8.95}$$

We may view the two expansions (8.91) and (8.93) as the general representations of complex functions that are holomorphic in a domain bounded by a circle; one represents the general holomorphic function inside the circle, and the other the general holomorphic function outside a circle. The coefficients in the series expansions can be determined from the boundary conditions, for example, from the values of the function along the circle.

The general solution of Laplace's equation in the annulus between two circles consists of the sum of the Taylor series and the asymptotic expansion, known as the Laurent series:

$$f(z) = \sum_{n=-\infty}^{\infty} a_n (z-z_0)^n. \tag{8.96}$$

The coefficients in the Laurent series are determined from the boundary conditions along the two circles. The Laurent series converges inside the annulus, provided that $f(z)$ is holomorphic there.

8.7 Cauchy Integrals

As for the series expansions, the practical use of Cauchy integrals is not obvious, but we will see that Cauchy integrals are excellent tools to determine the constant parameters in the analytic elements that we apply in what follows. A holomorphic function can be represented as a function of a single complex variable, z. Since its derivatives with respect to $\bar{z}$ are zero, all the derivatives of a holomorphic function are themselves holomorphic functions. We thus can write a function Ω that is holomorphic inside area $\mathscr{A}$ as the derivative of another function that is holomorphic inside $\mathscr{A}$, say, F:

$$\Omega = \frac{dF}{dz}. \tag{8.97}$$

We write the integral of Ω along the boundary $\mathscr{C}$ of $\mathscr{A}$ as

$$\oint_{\mathscr{C}} \Omega dz = F(z_s) - F(z_e), \tag{8.98}$$

where z_s and z_e are the complex coordinates of the starting and end points of the integral, which are the same. Thus, the integral is equal to zero because $F(z)$ is single-valued, and the resulting formula is known as the Cauchy-Goursat integral theorem, which we introduced in (8.86),

$$\oint_{\mathscr{C}} \Omega dz = 0. \tag{8.99}$$

Another important formula applicable to holomorphic functions is the Cauchy integral, which has the form

$$\Omega(z) = \frac{1}{2\pi \mathrm{i}} \oint_{\mathscr{C}} \frac{\Omega(\delta)}{\delta - z} d\delta. \tag{8.100}$$

We obtain this identity by rewriting the integral as

$$\frac{1}{2\pi \mathrm{i}} \oint_{\mathscr{C}} \frac{\Omega(\delta)}{\delta - z} d\delta = \frac{1}{2\pi \mathrm{i}} \oint_{\mathscr{C}} \frac{\Omega(\delta) - \Omega(z)}{\delta - z} d\delta + \frac{1}{2\pi \mathrm{i}} \oint_{\mathscr{C}} \frac{\Omega(z)}{\delta - z} d\delta. \tag{8.101}$$

Since Ω is holomorphic, it can be expanded in a Taylor series as

$$\Omega(\delta) - \Omega(z) = \Omega'(z)(\delta - z) + \cdots + \Omega^{(n)}(z)(\delta - z)^n + \cdots \tag{8.102}$$

so that

$$\frac{\Omega(\delta)-\Omega(z)}{\delta-z}=\Omega'(z)+\tfrac{1}{2}\Omega''(z)(\delta-z)+\cdots, \tag{8.103}$$

which is holomorphic, so that the first integral to the right of the equals sign in (8.101) is zero; we obtain

$$\frac{1}{2\pi i}\oint_{\mathscr{C}}\frac{\Omega(\delta)}{\delta-z}d\delta=\frac{1}{2\pi i}\Omega(z)\oint_{\mathscr{C}}\frac{1}{\delta-z}d\delta=\frac{1}{2\pi i}\Omega(z)\ln(\delta-z)|_{\mathscr{C}}=\Omega(z), \qquad z\in\mathscr{A}. \tag{8.104}$$

The real part of the logarithm $\ln(\delta-z)$ is the same at the starting and end points of the contour, but the imaginary part increases by $2\pi i$ on completion of the path of integration, if point z is in area $\mathscr{A}$; otherwise, the integral is zero, i.e.,

$$\frac{1}{2\pi i}\oint_{\mathscr{C}}\frac{\Omega(\delta)}{\delta-z}d\delta=0, \qquad z\in\mathscr{A}^-, \tag{8.105}$$

where $\mathscr{A}^-$ represents the domain outside area $\mathscr{A}$.

We obtain higher order forms of the Cauchy integral (8.100) by differentiating both sides with respect to z. Repeated differentiation gives

$$\frac{d^m\Omega(z)}{dz^m}=m!\frac{1}{2\pi i}\oint_{C}\frac{\Omega(\delta)}{(\delta-z)^{m+1}}d\delta, \qquad z\in\mathscr{A}. \tag{8.106}$$

We use the Cauchy integral to determine the coefficients of the Taylor series expansion of the function $\Omega(z)$ about the origin, $z=0$; these coefficients involve the derivatives of Ω evaluated at $z=0$. We set $z=0$ in (8.106) and choose for $\mathscr{C}$ the unit circle:

$$\frac{1}{m!}\frac{d^m\Omega(0)}{dz^m}=\frac{1}{2\pi i}\oint_{C}\frac{\Omega(\delta)}{\delta^{m+1}}d\delta, \qquad \delta\in\mathscr{C}. \tag{8.107}$$

We expand the function Ω in terms of a Taylor series with coefficients $a_m, m=0,1,\ldots$, and use (8.91) and (8.92),

$$a_m=\frac{1}{2\pi i}\oint_{C}\frac{\Omega(\delta)}{\delta^{m+1}}\delta, \qquad m=0,1,\ldots, \quad \delta\in\mathscr{C}. \tag{8.108}$$

This equation expresses the coefficients of the Taylor expansion of Ω in terms of the boundary values of Ω along the circle $\mathscr{C}$. In general, these complex boundary values are not known; the function Ω is fully determined in $\mathscr{A}$ by specifying either its real or its imaginary part along $\mathscr{C}$. We address this issue by using the integral

$$\frac{1}{2\pi i}\oint_{\mathscr{C}}\Omega(\delta)\delta^n d\delta=0, \qquad \delta\in\mathscr{C}, \qquad n\geq 0, \tag{8.109}$$

which is zero because the integrand is holomorphic in $\mathscr{A}$. The complex conjugate of a complex expression that is zero is also zero,

$$\frac{1}{2\pi\mathrm{i}}\oint_{\mathscr{C}} \overline{\Omega(\delta)\delta^n d\bar{\delta}} = 0, \qquad \delta \in \mathscr{C}, \quad n \geq 0. \tag{8.110}$$

We use that $\delta\bar{\delta} = 1$ along the unit circle, replace $\bar{\delta}$ by $1/\delta$, and $d\bar{\delta}$ by $-d\delta/\delta^2$,

$$\frac{1}{2\pi\mathrm{i}}\oint_{\mathscr{C}} \overline{\Omega(\delta)}\frac{1}{\delta^{n+2}} d\delta = 0, \qquad \delta \in \mathscr{C}, \quad n \geq 0. \tag{8.111}$$

We choose $n = m - 1$ for $m > 0$ and add (8.111) to (8.108), which gives

$$a_m = \frac{1}{2\pi\mathrm{i}}\oint_{C} \frac{\Omega(\delta) + \overline{\Omega(\delta)}}{\delta^{m+1}} d\delta, \qquad m = 1, 2, \ldots, \quad \delta \in \mathscr{C}. \tag{8.112}$$

This equation gives the coefficients of the Taylor series expansion of $\Omega\ (z)$ about $z = 0$ in terms of the boundary values of the real part of the complex potential, $\Phi = \frac{1}{2}(\Omega + \bar{\Omega})$, i.e.,

$$a_m = \frac{1}{\pi\mathrm{i}}\oint_{\mathscr{C}} \frac{\Phi(\delta)}{\delta^{m+1}} d\delta, \qquad m = 1, \ldots, \quad \delta \in \mathscr{C}. \tag{8.113}$$

Alternatively, we can subtract (8.111) from (8.108), which gives

$$a_m = \frac{1}{2\pi\mathrm{i}}\oint_{\mathscr{C}} \frac{\Omega(\delta) - \overline{\Omega(\delta)}}{\delta^{m+1}} d\delta, \quad m = 2, \ldots, \quad \delta \in \mathscr{C}. \tag{8.114}$$

Since $\Omega - \bar{\Omega} = 2\mathrm{i}\Psi$, we may write this as

$$a_m = \frac{1}{\pi}\oint_{\mathscr{C}} \frac{\Psi(\delta)}{\delta^{m+1}} d\delta, \qquad m = 1, \ldots, \quad \delta \in \mathscr{C}. \tag{8.115}$$

This equation expresses the coefficients a_n in terms of the boundary values of the imaginary part of Ω. The special case in which $m = 0$ is covered by setting m in (8.108) equal to zero, which gives

$$a_0 = \frac{1}{2\pi\mathrm{i}}\oint_{\mathscr{C}} \frac{\Omega(\delta)}{\delta} d\delta, \qquad \delta \in \mathscr{C}. \tag{8.116}$$

Along the circle $\mathscr{C}$ we may write δ as

$$\delta = \mathrm{e}^{\mathrm{i}\theta}. \tag{8.117}$$

We substitute $e^{i\theta}$ for δ in (8.116),

$$a_0 = \frac{1}{2\pi i}\oint_{\mathscr{C}} \Omega(\delta)e^{-i\theta}de^{i\theta} = \frac{1}{2\pi i}\oint_{\mathscr{C}} \Omega(\delta)e^{-i\theta}ie^{i\theta}d\theta$$

$$= \frac{1}{2\pi}\int_0^{2\pi} \Omega(\delta)d\theta, \qquad \delta \in \mathscr{C}. \tag{8.118}$$

We rewrite the expressions (8.113) and (8.115) in terms of θ:

$$a_m = \frac{1}{\pi}\int_0^{2\pi} \Phi(\delta)e^{-im\theta}d\theta, \qquad m = 1,\ldots, \quad \delta \in \mathscr{C} \tag{8.119}$$

and

$$a_m = \frac{i}{\pi}\int_0^{2\pi} \Psi(\delta)e^{-im\theta}d\theta, \qquad m = 1,\ldots, \quad \delta \in \mathscr{C}. \tag{8.120}$$

These expressions for the coefficients a_m are equivalent to the expressions for the coefficient of a complex Fourier series.

8.7.1 The General Case

We obtained the latter two Cauchy integrals with real argument from the Cauchy integral of a holomorphic function Ω. For such cases the integrals of the values of Φ and Ψ along the circle produce the same holomorphic complex potential. However, in the general case the values of either one of these real functions along the circle are arbitrary, and then the resulting complex potentials are different. This is important in applications when boundary values along a circle are used to obtain the coefficients of the series expansion of Ω.

8.8 Divergence-Free and Irrotational Flow

We consider in this section a number of illustrative examples of the use of complex variables for problems of divergence-free flow, i.e., flow where no water enters the flow domain other than through its boundaries.

8.8.1 A Well

We derived the expression for the potential for a well of discharge Q at z_w (see (8.26)):

$$\Phi = \frac{Q}{4\pi}\ln(z - z_w) + \frac{Q}{4\pi}\ln(\bar{z} - \bar{z}_w). \tag{8.121}$$

This potential is the real part of the complex potential for a well:

$$\Omega = \Phi + \mathrm{i}\Psi = \frac{Q}{2\pi}\ln(z - z_w). \tag{8.122}$$

If the well is at the origin, this reduces to

$$\Omega = \Phi + \mathrm{i}\Psi = \frac{Q}{2\pi}\ln(z) = \frac{Q}{2\pi}\ln(re^{\mathrm{i}\theta}) = \frac{Q}{2\pi}[\ln(r) + \mathrm{i}\theta], \tag{8.123}$$

where we represented z in polar form

$$z = re^{\mathrm{i}\theta}. \tag{8.124}$$

We obtain the following for Φ and Ψ from (8.123):

$$\Phi = \frac{Q}{2\pi}\ln(r) \tag{8.125}$$

$$\Psi = \frac{Q}{2\pi}\theta. \tag{8.126}$$

The flow net for a single well in an infinite aquifer is shown in Figure 8.4; the values of the stream function are shown next to the corresponding streamlines and are obtained from (8.126). We see from both the figure and the expression for Ψ that the stream function is not single-valued, at least not everywhere: along the negative x-axis, we obtain $\Psi = Q/2$ if we choose $\theta = \pi$, whereas we obtain $\Psi = -Q/2$ if we choose $\theta = -\pi$. The reason for the dual values of Ψ along a line that connects the well to infinity is that the condition for Ψ to be single-valued, i.e., the condition that

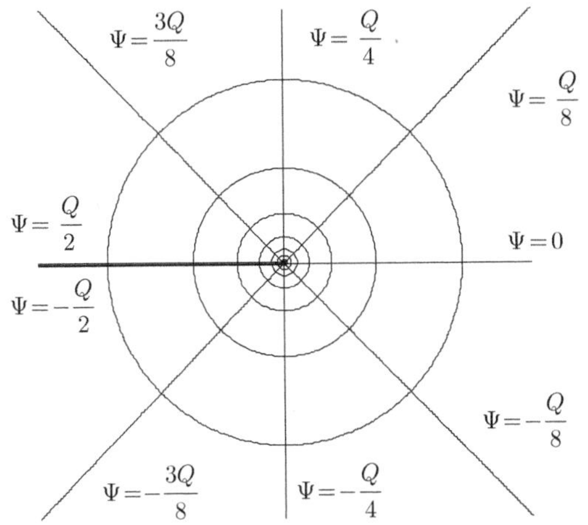

Figure 8.4 Equipotentials and streamlines for a well.

the flow is divergence-free (no infiltration or extraction), is not satisfied at the well ($z = 0$). We may view the discontinuity of Ψ as a thought experiment, as follows. We consider the negative x-axis, called the *branch cut*, as a conduit between the well and infinity. We imagine that the water that flows from infinity to the well is transferred via the branch cut back to infinity, thereby satisfying the condition that there always must be continuity of flow. The jump in the stream function is indeed compatible with flow from the well through the branch cut to infinity. Clearly, this is a mathematical observation that has nothing to do with reality.

We will, in this chapter, frequently make plots of the stream function for cases with wells; the branch cuts cannot be avoided when making contour plots (the plots shown here were made with MATLAB).

8.8.2 A Well in a Field of Uniform Flow

We obtain an expression for the complex potential for a well of discharge Q at $z = 0$ in a field of uniform flow by superposition:

$$\Omega = -Q_{x0}z + \frac{Q}{2\pi}\ln(z) + C, \tag{8.127}$$

where C is a constant. The plot in Figure 8.5 shows contours of the stream function, i.e., the imaginary part of the complex potential (8.127). The branch cut is clearly

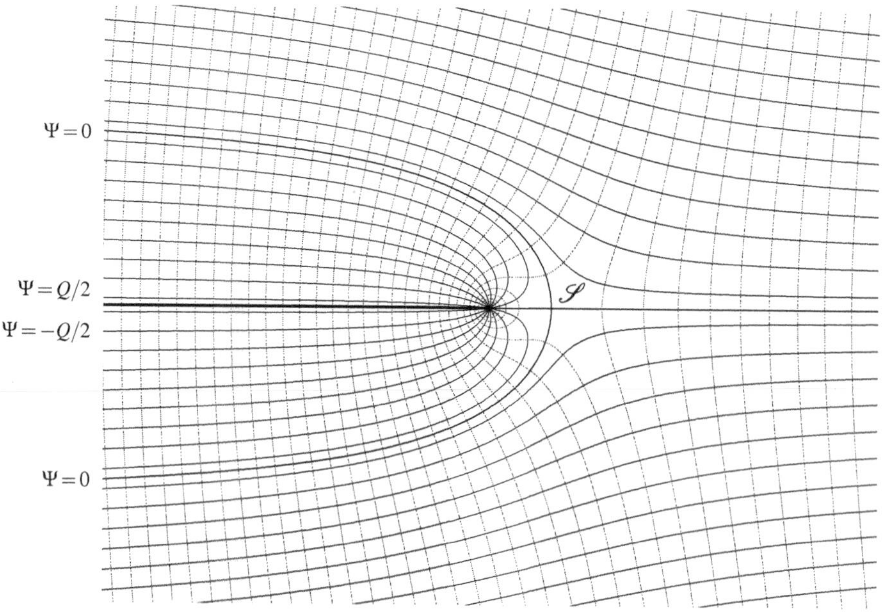

Figure 8.5 A well in a field of uniform flow.

visible in the figure. Note the special streamlines that intersect each other at right angles at the stagnation point; these streamlines are called the dividing streamlines and separate the flow drawn into the well from that bypassing the well. The stream function is zero along the positive x-axis. The dividing streamlines intersect this axis; the stream function has a single value, zero, at the point of intersection of the streamlines, which is the stagnation point. It follows that the stream function has the same value, zero, along each of the two dividing streamlines.

It seems as though this is in conflict with one of the properties of the stream function: the difference of its values at two points is equal to the discharge that flows between these points. The explanation for this apparent contradiction is the occurrence of the branch cut between the dividing streamlines; the stream function jumps there. Jumps must be taken into account when subtracting values of the stream function for computing discharges.

We obtain an expression for the complex coordinate of the stagnation point, z_s, by setting the discharge function W equal to zero at $z = z_s$,

$$W = -\frac{d\Omega}{dz} = 0 = Q_{x0} - \frac{Q}{2\pi z_s}, \tag{8.128}$$

so that

$$z_s = \frac{Q}{2\pi Q_{x0}}. \tag{8.129}$$

If we define the distance between the two dividing streamlines at infinity as b, then we may write, by continuity of flow,

$$Q = bQ_{x0}, \tag{8.130}$$

so that (8.129) becomes

$$z_s = \frac{b}{2\pi}. \tag{8.131}$$

Note that this result corresponds exactly to (2.67).

Problem

8.2 Consider the case of a well of discharge Q at $z = -d$. There is a river along the y-axis, with the head equal to ϕ_0. The flow is unconfined, the hydraulic conductivity is k, and there is uniform flow in the x-direction of magnitude Q_{x0}. The analysis must be carried out using complex variables.

Questions:

1. Determine the complex potential for this problem.

2. Determine expressions for the location(s) of the stagnation point(s) inside the aquifer, valid for all possible ratios dQ_{x0}/Q.
3. Determine an expression for the portion of the discharge of the well that comes from the river when stagnation points occur along the river bank.
4. Write a computer program using complex variables and use this to construct flow nets for the following cases:

   a. $Q = \frac{1}{2}\pi dQ_{x0}$
   b. $Q = \pi dQ_{x0}$
   c. $Q = \frac{3}{2}\pi dQ_{x0}$.

8.8.3 Application: Wells in Pump-Out Systems for Contaminant Removal

Consider the case of an aquifer with uniform flow at a rate Q_{x0}. A continuous source of contaminants exists at $x = -L, -b \leq y \leq b$. We use two wells to capture the groundwater contaminated by this source. We consider two separate methods for achieving this. For the first method, we position the wells such that their connecting line is perpendicular to the flow; for this case the spacing of the wells must be such that none of the contaminated water can escape between the wells. For the second method we space the wells such that the line connecting them is in the direction of flow.

Case 1: Wells Aligned Normal to the Flow

The complex potential for the case of uniform flow with two wells of discharge Q each and located at $z = \mathrm{i}d$ and $z = -\mathrm{i}d$ is

$$\Omega = -Q_{x0}z + \frac{Q}{2\pi}\ln(z - \mathrm{i}d) + \frac{Q}{2\pi}\ln(z + \mathrm{i}d) + C, \tag{8.132}$$

where C is a constant. The condition that no water escapes the wells by flowing between them is that at least one stagnation point exists on the x-axis. We need to determine an expression for the discharge function W in order to apply this condition,

$$W = -\frac{d\Omega}{dz} = Q_{x0} - \frac{Q}{2\pi}\frac{1}{z - \mathrm{i}d} - \frac{Q}{2\pi}\frac{1}{z + \mathrm{i}d}. \tag{8.133}$$

We find an expression for z_s, the complex coordinate of the stagnation point, by setting $W(z_s) = 0$. This gives, after multiplication by $(z_s - \mathrm{i}d)(z_s + \mathrm{i}d)$,

$$W(z_s) = 0 = Q_{x0}(z_s - \mathrm{i}d)(z_s + \mathrm{i}d) - \frac{Q}{2\pi}(z_s + \mathrm{i}d + z_s - \mathrm{i}d) \tag{8.134}$$

or, for $Q_{x0} \neq 0$,

$$z_s^2 - (\mathrm{i}d)^2 - \frac{Q}{2\pi Q_{x0}}(2z_s) = 0 \tag{8.135}$$

or

$$z_s^2 - \frac{Q}{\pi Q_{x0}} z_s + d^2 = 0. \tag{8.136}$$

This quadratic equation has two roots:

$$(z_s)_{1,2} = \frac{Q}{2\pi Q_{x0}} \pm \sqrt{\left[\frac{Q}{2\pi Q_{x0}}\right]^2 - d^2}. \tag{8.137}$$

The condition that at least one stagnation point exists along the x-axis will be satisfied if the argument of the square root is positive, i.e., if

$$\frac{Q}{2\pi d Q_{x0}} \geq 1, \qquad \frac{Q}{Q_{x0}} > 0. \tag{8.138}$$

Note that we only consider the case here that the ratio Q/Q_{x0} is positive. If $Q = 2\pi d Q_{x0}$, then there exists a single stagnation point at $z_s = Q/(2\pi d Q_{x0})$. For all other cases there exist two stagnation points:

$$(z_s)_{1,2} = (x_s)_{1,2} = \frac{Q}{2\pi Q_{x0}} \pm \sqrt{\left[\frac{Q}{2\pi Q_{x0}}\right]^2 - d^2}, \qquad \frac{Q}{2\pi d Q_{x0}} > 1. \tag{8.139}$$

and

$$(z_s)_{1,2} = x_s + \mathrm{i}(y_s)_{1,2} = \frac{Q}{2\pi Q_{x0}} \pm \mathrm{i}\sqrt{d^2 - \left[\frac{Q}{2\pi Q_{x0}}\right]^2}, \qquad \frac{Q}{2\pi d Q_{x0}} < 1. \tag{8.140}$$

Note that x_s is always positive; the stagnation point is always downstream from the wells. Two stagnation points that coincide at a single point are shown in Figure 8.6. The flow net in Figure 8.7 corresponds to the case in which $Q = 2.5\pi d Q_{x0}$; there are two stagnation points on the x-axis that are a distance D apart, where

$$D = 2\sqrt{\left[\frac{Q}{2\pi Q_{x0}}\right]^2 - d^2}. \tag{8.141}$$

A third case is illustrated in Figure 8.8: there are two stagnation points that lie symmetrically with respect to the x-axis and water escapes between the wells.

The placement of wells on a line normal to the direction of flow suffers from the drawback that the system is fragile; variation in the magnitude of flow will result in loss of the contaminant downstream. The direction of the uniform flow field is rarely known precisely, and may change over time. The design that we consider next is less sensitive to variations in the uniform flow field.

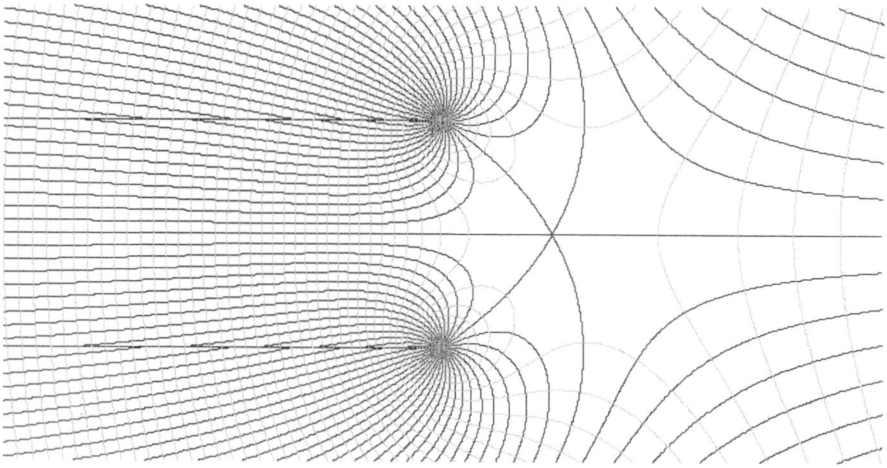

Figure 8.6 Flow with two coinciding stagnation points; $Q = 2\pi d Q_{x0}$. The streamlines are in black, the equipotentials are in gray.

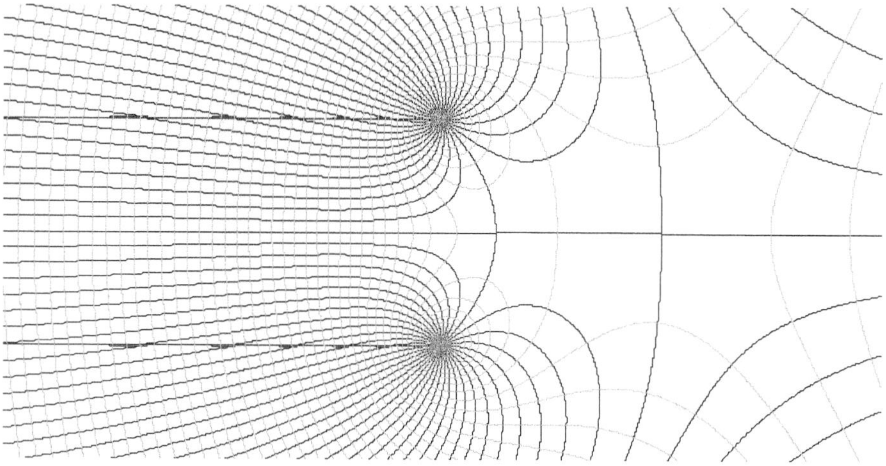

Figure 8.7 Flow with two stagnation points that lie symmetrically relative to the x-axis; $Q = 2.5\pi d Q_{x0}$. The streamlines are in black, the equipotentials are in gray.

Case 2: Wells Aligned Parallel to the Flow

The complex potential for two wells in line with the flow differs from the one for the preceding arrangement only in the arguments of the two complex logarithms and is

$$\Omega = -Q_{x0} z + \frac{Q}{2\pi} \ln(z - d) + \frac{Q}{2\pi} \ln(z - d - \Delta d) + C, \tag{8.142}$$

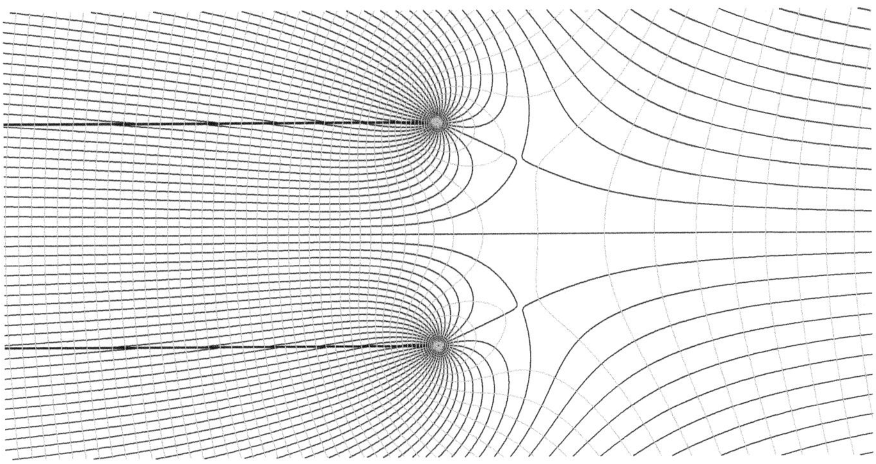

Figure 8.8 Flow with two stagnation points that lie symmetrically relative to the x-axis; $Q = 1.5\pi d Q_{x0}$. The streamlines are in black, the equipotentials are in gray.

where Δd is the distance between the wells. The expression for W is

$$W = -\frac{d\Omega}{dz} = Q_{x0} - \frac{Q}{2\pi}\frac{1}{z-d} - \frac{Q}{2\pi}\frac{1}{z-d-\Delta d}. \tag{8.143}$$

We set W equal to zero to find the locations of the stagnation points (z_s) and obtain, after multiplication by $(z_s - d)(z_s - d - \Delta d)$ and division by Q_{x0},

$$(z_s - d)(z_s - d - \Delta d) - \frac{Q}{2\pi Q_{x0}}(z_s - d - \Delta d + z_s - d) = 0 \tag{8.144}$$

or

$$z_s^2 - \left(2d + \Delta d + \frac{Q}{\pi Q_{x0}}\right) z_s + \frac{Q}{2\pi Q_{x0}}(2d + \Delta d) + d(d + \Delta d) = 0. \tag{8.145}$$

We solve this equation for z_s and obtain

$$(z_s)_{1,2} = d + \tfrac{1}{2}\Delta d + \frac{Q}{2\pi Q_{x0}} \pm \sqrt{\left(d + \tfrac{1}{2}\Delta d + \frac{Q}{2\pi Q_{x0}}\right)^2 - \frac{Q}{\pi Q_{x0}}(d + \tfrac{1}{2}\Delta d) - d(d + \Delta d)} \tag{8.146}$$

or

$$(z_s)_{1,2} = d + \tfrac{1}{2}\Delta d + \frac{Q}{2\pi Q_{x0}} \pm \sqrt{(d + \tfrac{1}{2}\Delta d)^2 + \left(\frac{Q}{2\pi Q_{x0}}\right)^2 - d(d + \Delta d)}. \tag{8.147}$$

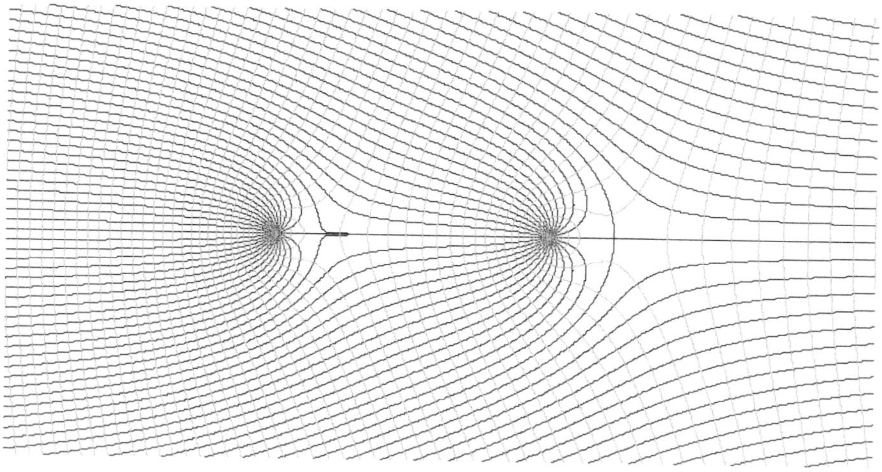

Figure 8.9 Flow with two wells aligned with the uniform flow. The discharges of the wells are both equal to $\pi d Q_{x0}$ and $d = 100$ m, $\Delta d = 50$ m.

We work out the expression under the square root sign, which simplifies, and the result is

$$(z_s)_{1,2} = d + \tfrac{1}{2}\Delta d + \frac{Q}{2\pi Q_{x0}} \pm \tfrac{1}{2}\sqrt{\left(\frac{Q}{\pi Q_{x0}}\right)^2 + (\Delta d)^2}. \tag{8.148}$$

There are always two stagnation points on the x-axis, except if $\Delta d = 0$; then the solution reduces to that of a single well of discharge $2Q$; the expression for z_s matches that obtained for the case of a single well in a field of uniform flow (see (8.129)). The flow net for $Q = \pi d Q_{x0}$ is shown in Figure 8.9.

Problem

8.3 Consider a permanent source of contamination at $x = -L$; the width of this source normal to the diretion of flow is $2b$.

Questions:

1. Consider a design with the wells on a line normal to the direction of flow, with the distance $2d$ between the wells chosen such that there exists a single stagnation point on the x-axis. Determine an expression for the discharge of the wells such that the bounding streamlines of the capture zone pass through the end points of the contamination source (i.e., $z_{1,2} = -L \pm ib$). The wells lie on the y-axis.
2. Consider a design with the wells in line with the direction of flow. For this case, space the wells a distance $2d$ apart, with d obtained from

the latter design. The upstream well is at the origin. Determine an expression for the discharge of the wells (these are equal) such that the bounding streamlines pass through the end points of the contamination zone.

3. Compare the two designs, examining the merits and disadvantages of each design.
4. Produce a flow net for each case and compare them. Comment on the differences of the flow nets.

8.8.4 A Well and a Recharge Well in a Field of Uniform Flow

Sometimes we align a well and a recharge well in a field of uniform flow in such a way that the well captures as much of the recharge well as possible. Such an application is useful, for example, if we wish to use the aquifer as a storage device for water of a temperature different from the average groundwater temperature. If, for example, groundwater is used in summer for cooling, then the water that is heated as a result may be injected and recaptured as warm water during the following winter, to be used for heating. The ideal system has no loss whatsoever, which occurs if the recharge and discharge wells are perfectly aligned with the uniform flow. The complex potential for this system is

$$\Omega = -Q_{x0}z - \frac{Q}{2\pi}\ln(z+d) + \frac{Q}{2\pi}\ln(z-d) + C, \tag{8.149}$$

where there is uniform flow of discharge Q_{x0} in the x-direction, there is a well of discharge Q at $z = d$, and a recharge well of recharge Q at $z = -d$. We determine the locations of the stagnation points by setting W equal to zero, i.e.,

$$W = Q_{x0} + \frac{Q}{2\pi}\frac{1}{z+d} - \frac{Q}{2\pi}\frac{1}{z-d}. \tag{8.150}$$

We set this expression to zero at $z = z_s$ and obtain, after multiplication by $(z_s - d)(z_s + d)$ and division by Q_{x0},

$$z_s^2 - d^2 + \frac{Q}{2\pi Q_{x0}}(z_s - d - z_s - d) = 0. \tag{8.151}$$

We solve this for z_s,

$$(z_s)_{1,2} = \pm\sqrt{d^2 + \frac{Qd}{\pi Q_{x0}}}. \tag{8.152}$$

There are always two stagnation points, which lie symmetrically with respect to the point midway between the wells. The flow net is shown in Figure 8.10 for $Q = \pi d Q_{x0}$.

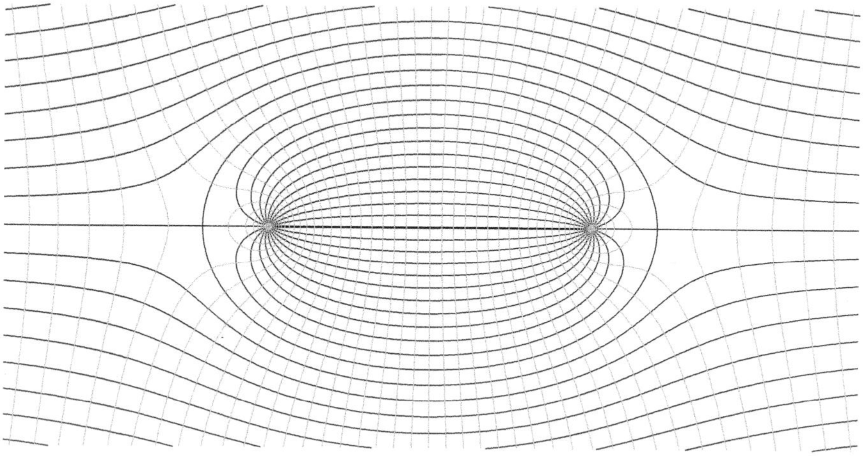

Figure 8.10 A recharge well and a well in a field of uniform flow for $Q = \pi d Q_{x0}$.

Problem

8.4 Consider the case of a recharge well of recharge Q at $z = -d$ and a discharge well at $z = d$ of discharge aQ, where $a > 1$. The system is in a field of uniform flow of discharge Q_0 oriented at an angle α to the x-axis.

Questions:

1. Determine the complex potential for this problem.
2. Determine expressions for the location(s) of the stagnation point(s) in terms of the parameters of the problem and write a computer program to implement this.
3. Determine the factor a such that the discharge well captures all of the water recharged when $\alpha = 30$ degrees. Choose the parameters such that $Q = 2\pi d Q_{x0}$ where $2d$ is the distance between the well and the recharge well. Solve the problem by ensuring that the two stagnation points lie on the same streamline, i.e., that the stream function has the same value at the two stagnation points.
4. Produce a flow net for the case considered to verify the accuracy of your solution.

8.9 Far-Field Conditions

The flow in aquifers is controlled by many features; lakes, rivers, zones of infiltration, and wells are examples. We often study local effects, such as the effect of a lake on the flow in an aquifer or the effect of pumping wells near a river. We usually leave out many of the features that control the flow, if they are far away and their effect does not need to be included in detail; we call the combined effect of these features the *far field*. If the problem concerns an area that is part of a very large system that we model as an infinite aquifer, then we should include the effect of the far field; we attempt to do this in such a way that the far field is not affected by the local elements that we will manipulate, such as wells and lakes.

8.9.1 Far-Field Components

One way to represent the far field is to estimate the potential, Φ_∞, along a large circle, roughly centered at the area of interest and with a radius, R_∞, that is much larger than the dimensions of the area of interest. We estimate the potential along this circle from given features such as lakes, river sections, and other known boundaries. We may use estimated values along the sections of this circle where known, and interpolated values elsewhere. We apply the Cauchy integral theorem to compute the two first coefficients in the Taylor series expansion of the holomorphic function that has Φ_∞ as its boundary values

$$a_0 = \tilde{\Phi}_\infty = \frac{1}{2\pi} \oint\limits_{\mathscr{C}} \Phi_\infty(R_\infty, \theta) d\theta. \qquad (8.153)$$

We obtain an estimate of the next coefficient in a similar manner

$$a_1 = \frac{1}{\pi} \oint\limits_{\mathscr{C}} \Phi_\infty(R_\infty, \theta) \mathrm{e}^{-\mathrm{i}\theta} d\theta. \qquad (8.154)$$

Since the first term in the Taylor series expansion is the constant and the second one the uniform flow term, we obtain

$$\Omega_\infty = -Q_0 \mathrm{e}^{-\mathrm{i}\alpha_\infty} z + \Phi_\infty, \qquad (8.155)$$

where

$$Q_0 \mathrm{e}^{-\mathrm{i}\alpha} = a_1. \qquad (8.156)$$

8.10 A Straight Infinite Inhomogeneity

Inhomogeneities in the hydraulic conductivity are a common occurrence in groundwater flow practice. We consider a few elementary cases to gain an understanding of the effect of inhomogeneities and to outline a general approach for solving problems involving inhomogeneities.

8.10.1 Uniform Flow in the Lower Half Plane

We consider an inhomogeneity that occupies the entire half plane $y \geq 0$. We label the lower half plane, $y < 0$, as region 1, with hydraulic conductivity k_1, and the upper half plane, $y \geq 0$, as region 2, with hydraulic conductivity k_2. There is uniform flow at a rate W_0 in the lower half plane, region 1,

$$\underset{1}{\Omega} = -W_0 z + \Phi_0, \qquad y \leq 0, \tag{8.157}$$

where

$$W_0 = Q_{x0} - \mathrm{i}Q_{y0}, \qquad \Phi_0 = k_1 f(\phi_0), \qquad y \geq 0, \tag{8.158}$$

and $f(\phi_0)$ represents the function that relates head to potential; this function is $\frac{1}{2}\phi^2$ for unconfined flow. We write the potential in the upper half plane, region 2, as

$$\underset{2}{\Omega} = -W_0 z + Az + B + \Phi_0, \qquad \Im A = \Im B = 0. \tag{8.159}$$

Note that the solution satisfies the condition of continuity of flow; the function $Az + B$ does not generate a component of flow in the direction normal to the boundary of the inhomogeneity. We determine the constants A and B from the condition that the head is continuous, which we express in terms of the potential on the positive and negative sides of the boundary of the inhomogeneity:

$$\Phi^+ - \Phi^- = \frac{k_2 - k_1}{k_1}\Phi^-. \tag{8.160}$$

We determine the real parts of expressions (8.157) and (8.159) and use these in (8.160):

$$Ax + B = \frac{k_2 - k_1}{k_1}\left[-Q_{x0}x + Q_{y0}y + \Phi_0\right], \qquad y = 0. \tag{8.161}$$

We solve this for A and B:

$$\begin{aligned} A &= -\frac{k_2 - k_1}{k_1} Q_{x0} \\ B &= \frac{k_2 - k_1}{k_1}\Phi_0. \end{aligned} \tag{8.162}$$

Streamlines and equipotentials are shown in Figure 8.11 for flow with the hydraulic conductivity in the upper half plane ten times that in the lower one. Note the effect of refraction of the streamlines, which have a discontinuous slope at the interface. The uniform flow makes an angle of $\pi/3$ with the x-axis. The change in direction of flow at the interface is such that the flow rate in the highly conductive zone is larger than that in the zone with the lower hydraulic conductivity.

We determined the solution to this problem by separating the complex potential into two parts. One part is continuous throughout the flow domain; the other one has a real part that jumps across the boundary, but satisfies the condition of continuity of flow. The discontinuous part of the solution is $Ax + B$, and was chosen so that it met the condition of continuity of flow; it does not have a component of flow normal to the boundary. We multiplied the potential in the upper half plane by a

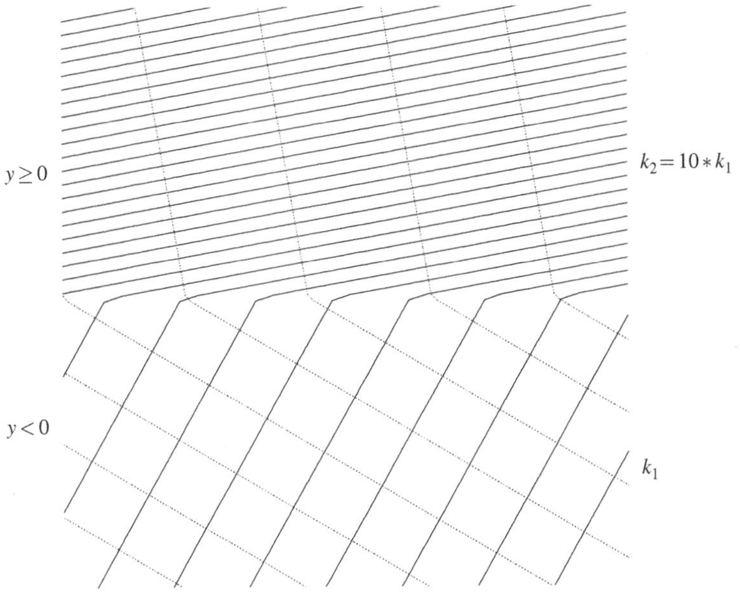

Figure 8.11 Streamlines and equipotentials for the case of uniform flow in the lower half plane.

factor k_1/k_2 for plotting purposes in order to eliminate the discontinuity across the boundary; the dotted contours in the upper half plane represent a scaled potential. We follow this practice for inhomogeneities in what follows.

8.10.2 A Well in the Upper Half Plane

We consider the case of a well in the upper half plane at $z = z_w$ and of discharge Q as a second example. We use the same notation for the aquifer parameters as in the preceding subsection. We again divide the solution into two parts, one part that is continuous throughout the domain, and a second part that has a discontinuous real part, but a continuous imaginary one, across the real axis, so that it satisfies the condition of continuity of flow. The complex potentials are Ω, valid for all values of z (identified by the symbol $\forall z$),

$$\Omega = \frac{\alpha Q}{2\pi} \ln(z - z_w), \qquad \forall z, \tag{8.163}$$

where α is a real constant, and $\underset{2}{\Omega}$, valid in $y \geq 0$, is

$$\underset{2}{\Omega} = \frac{(1-\alpha)Q}{2\pi}[\ln(z - z_w) + \ln(z - \bar{z}_w)], \qquad y \geq 0. \tag{8.164}$$

The latter function has a constant imaginary part along the real axis, and thus meets the condition of continuity of flow. The jump in potential, $\Re\Omega_2$, must be equal to $(k_2-k_1)/k_1\Phi^-$ along $y=0$:

$$\Delta\Phi = \frac{(1-\alpha)Q}{\pi}\ln|x-z_w| = \frac{k_2-k_1}{k_1}\frac{\alpha Q}{2\pi}\ln|x-z_w|. \tag{8.165}$$

The well is not on the real axis, so that $x \neq z_w$; we divide by $\ln|x-z_w|$, and obtain

$$(1-\alpha) = \frac{k_2-k_1}{2k_1}\alpha \rightarrow \left(\frac{k_2}{2k_1}-\frac{1}{2}+1\right)\alpha = 1 \rightarrow \alpha = \frac{2k_1}{k_2+k_1}. \tag{8.166}$$

We may add both a constant and uniform flow to this solution by combining this complex potential with the one for uniform flow.

An example is shown in Figure 8.12 for a hydraulic conductivity in the domain with the well that is ten times the one in the lower half plane. Another example is shown in Figure 8.13 for a hydraulic conductivity in the domain with the well that is one-tenth of the one in the lower half plane.

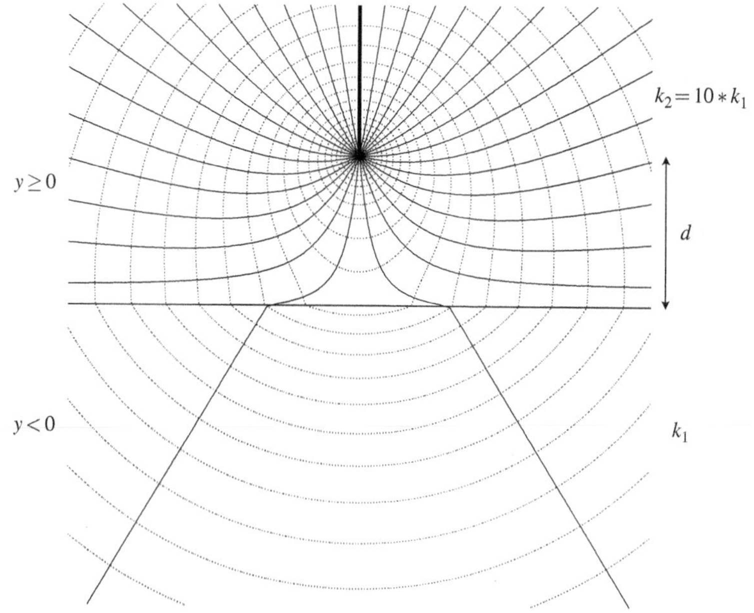

Figure 8.12 A well in the upper half plane; the hydraulic conductivity in the upper half plane is ten times that in the lower one.

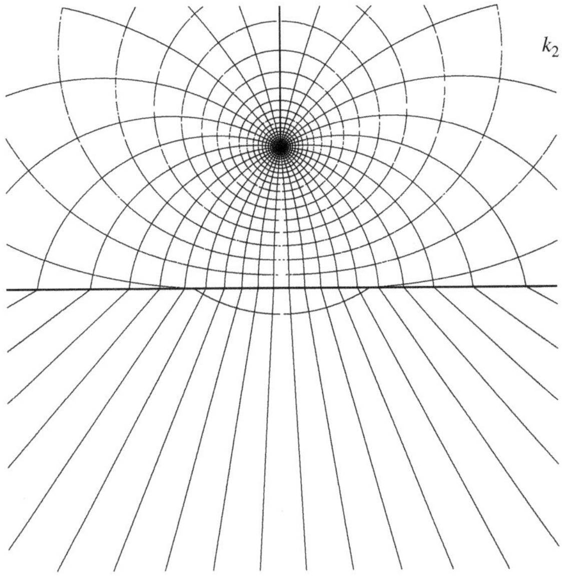

Figure 8.13 A well in the upper half plane; the hydraulic conductivity in the upper half plane is one-tenth that in the lower one.

8.11 A Circular Lake

We consider the effect of a circular lake in direct contact with an aquifer, which may be either confined or unconfined. The direct contact implies in terms of boundary conditions that the head along the lake boundary is equal to that in the lake, so that the discharge potential along the lake boundary is Φ_0. We choose the origin of the coordinate system at the center of the lake; the radius of the lake is R. We first consider radial flow, so that the lake acts as a well of radius R and the complex potential is

$$\Omega = \frac{Q_0}{2\pi} \ln \frac{z}{R} + \Phi_0. \tag{8.167}$$

We next consider the effect of a well somewhere in the aquifer, which draws some water from the lake.

8.11.1 A Circular Lake with a Well

We consider flow with a well of discharge Q at $z = z_w$; the well is outside the lake boundary, so that:

$$z_w = de^{i\alpha}, \qquad d > R. \tag{8.168}$$

We demonstrate that it is possible to choose the location of an image well of recharge Q inside the lake, i.e., outside the flow domain, in such a way that the boundary condition,

$$\Phi = \Phi_0, \qquad z\bar{z} = R^2, \tag{8.169}$$

is met. We choose the location, z_{wi}, of the image recharge well so that it lies on the line through the center of the lake and the well, but inside the lake and at a distance p from the origin,

$$z_{wi} = p\mathrm{e}^{\mathrm{i}\alpha}, \qquad p < R. \tag{8.170}$$

The expression for the complex potential is composed of that for the lake by itself, (8.167), that for the well, and that for the image well,

$$\Omega = \frac{Q_0}{2\pi}\ln\frac{z}{R} + \frac{Q}{2\pi}\ln\frac{z - z_w}{z - p\mathrm{e}^{\mathrm{i}\alpha}} + C. \tag{8.171}$$

We determine an expression for the real part, Φ, and require that Φ be constant along the lake boundary. The first term in (8.171) satisfies this condition, and the second part is

$$\Phi = \frac{Q}{4\pi}\ln\frac{(z - d\mathrm{e}^{\mathrm{i}\alpha})(\bar{z} - d\mathrm{e}^{-\mathrm{i}\alpha})}{(z - p\mathrm{e}^{\mathrm{i}\alpha})(\bar{z} - p\mathrm{e}^{-\mathrm{i}\alpha})} + C = \frac{Q}{4\pi}\ln\frac{z\bar{z} - d(\mathrm{e}^{\mathrm{i}\alpha}\bar{z} + \mathrm{e}^{-\mathrm{i}\alpha}z) + d^2}{z\bar{z} - p(\mathrm{e}^{\mathrm{i}\alpha}\bar{z} + \mathrm{e}^{-\mathrm{i}\alpha}z) + p^2} + C. \tag{8.172}$$

We choose an arbitrary point on the boundary by setting z equal to $R\mathrm{e}^{\mathrm{i}\theta}$:

$$\Phi = \frac{Q}{4\pi}\ln\frac{R^2 - dR\left[\mathrm{e}^{\mathrm{i}(\theta-\alpha)} + \mathrm{e}^{-\mathrm{i}(\theta-\alpha)}\right] + d^2}{R^2 - pR\left[\mathrm{e}^{\mathrm{i}(\theta-\alpha)} + \mathrm{e}^{-\mathrm{i}(\theta-\alpha)}\right] + p^2} + C. \tag{8.173}$$

We write $\mathrm{e}^{\mathrm{i}(\theta-\alpha)} + \mathrm{e}^{-\mathrm{i}(\theta-\alpha)}$ as $2\cos(\theta - \alpha)$:

$$\Phi = \frac{Q}{4\pi}\ln\frac{R^2 - 2dR\cos(\theta-\alpha) + d^2}{R^2 - 2pR\cos(\theta-\alpha) + p^2} + C. \tag{8.174}$$

We choose $p = R^2/d$ and demonstrate that this choice renders a constant value, Φ_0, for the potential along the boundary:

$$\begin{aligned}\Phi_0 &= \frac{Q}{4\pi}\ln\frac{R^2 - 2dR\cos(\theta-\alpha) + d^2}{R^2 - 2\frac{R^2}{d}R\cos(\theta-\alpha) + \frac{R^4}{d^2}} + C \\ &= \frac{Q}{4\pi}\ln\frac{R^2 - 2dR\cos(\theta-\alpha) + d^2}{\left[d^2 - 2dR\cos(\theta-\alpha) + R^2\right]\frac{R^2}{d^2}} + C = \frac{Q}{4\pi}\ln\frac{d^2}{R^2} + C.\end{aligned} \tag{8.175}$$

We solve for the constant C:

$$C = \Phi_0 + \frac{Q}{4\pi}\ln\frac{R^2}{d^2} = \Phi_0 + \frac{Q}{2\pi}\ln\frac{R}{d}. \tag{8.176}$$

We substitute this expression for C in equation (8.171) for the complex potential,

$$\Omega = \frac{Q_0}{2\pi}\ln\frac{z}{R} + \frac{Q}{2\pi}\ln\left[\frac{z-z_w}{z-R^2/\bar{z}_w}\frac{R}{d}\right] + \Phi_0, \tag{8.177}$$

where

$$z_{wi} = pe^{i\alpha} = \frac{R^2}{d}e^{i\alpha} = \frac{R^2}{de^{-i\alpha}} = \frac{R^2}{\bar{z}_w}, \tag{8.178}$$

and the complex potential for the case of a well near a circular boundary of zero potential is

$$\Omega = \frac{Q}{2\pi}\ln\left[\frac{z-z_w}{z-R^2/\bar{z}_w}\frac{R}{|z_w|}\right]. \tag{8.179}$$

8.11.2 A Circular Lake in a Field of Uniform Flow with Zero Net Discharge

We make use of the solution for a circular lake with a well to construct the solution for a lake in a field of uniform flow. We consider a lake that does not remove any water from the aquifer. We choose a recharge well at $z = -d$, and a discharge well at $z = d$, both with their images inside the lake, i.e.,

$$\Omega = -\frac{Q}{2\pi}\ln\frac{z+d}{z-d} + \frac{Q}{2\pi}\ln\frac{z+R^2/d}{z-R^2/d} + C. \tag{8.180}$$

We attempt to simulate uniform flow by letting the distance d as well as the discharge Q approach infinity such that their ratio remains finite, i.e.,

$$\lim_{\substack{Q\to\infty \\ d\to\infty}} \frac{Q}{d} = A \tag{8.181}$$

We write (8.180) in a form suitable for taking the limit,

$$\Omega = -\frac{Q}{2\pi}\ln\left[-\frac{1+\frac{z}{d}}{1-\frac{z}{d}}\right] + \frac{Q}{2\pi}\ln\frac{1+\frac{R^2}{dz}}{1-\frac{R^2}{dz}} + C, \tag{8.182}$$

expand $\ln(1+\varepsilon)$ for small ε about $\varepsilon = 0$, ignore all terms of order higher than ε, i.e., $\ln(1+\varepsilon) \approx \varepsilon$, and apply this to the expression for Ω,

$$\Omega \approx -\frac{Q}{2\pi}\ln(-1) - \frac{Q}{\pi d}z + \frac{Q}{\pi d}\frac{R^2}{z} + C, \qquad d\to\infty. \tag{8.183}$$

The first term is a purely imaginary constant and affects only the stream function; this constant term is immaterial and we leave it out, so that

$$\Omega \approx -\frac{Q}{\pi d}\left[z - \frac{R^2}{z}\right] + C, \qquad d\to\infty. \tag{8.184}$$

The flow approaches uniform flow far from the lake,

$$\lim_{\substack{Q\to\infty \\ d\to\infty}} \frac{Q}{\pi d} = Q_{x0}, \tag{8.185}$$

so that, in the limit,

$$\Omega = -Q_{x0}\left[z - \frac{R^2}{z}\right] + C. \tag{8.186}$$

We determine the value of the constant from the boundary condition, which we do by setting $z\bar{z} = R^2$, or $R^2/z = \bar{z}$,

$$\Omega = \Phi_0 + \mathrm{i}\Psi = -Q_{x0}(z - \bar{z}) + C = -2\mathrm{i}yQ_{x0} + C. \tag{8.187}$$

The first term is purely imaginary, and it follows that the boundary condition is met, with

$$C = \Phi_0. \tag{8.188}$$

The final expression for the complex potential for a circular lake in a field of uniform flow is

$$\Omega = -Q_{x0}\left[z - \frac{R^2}{z}\right] + \Phi_0. \tag{8.189}$$

A flow net for the case governed by (8.189) is shown in Figure 8.14. This is a special case, where the head at the lake boundary happens to be identical to the head that would occur in the aquifer if the lake were not present; we refer to this head as

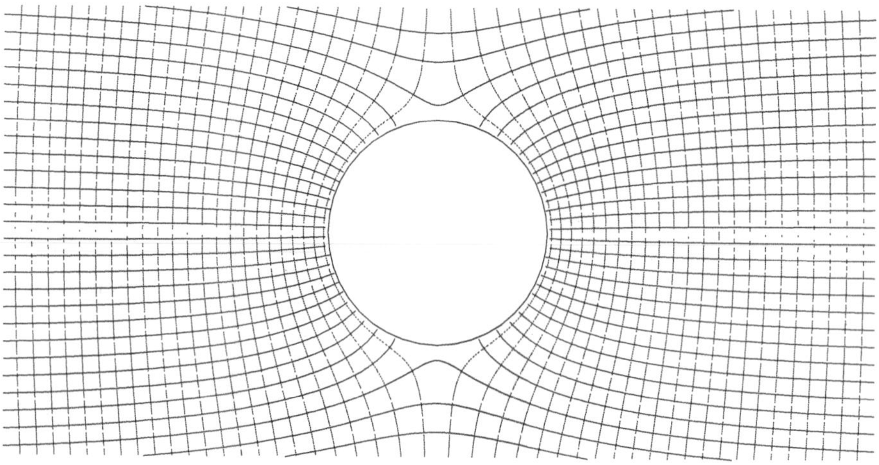

Figure 8.14 Flow net for the case governed by (8.189).

the equilibrium value. We observe from the flow net that the water captured by the lake along the upstream half of its boundary leaves the lake along the downstream half of the boundary. We compute the discharge passing through the lake as the difference in value of the stream function Ψ between points $z = \mathrm{i}R$ and $z = -\mathrm{i}R$. We use (8.187),

$$-Q_{\text{lake}} = \Psi(\mathrm{i}R) - \Psi(-\mathrm{i}R) = -2Q_{x0}(R - (-R)) = -4Q_{x0}R, \tag{8.190}$$

so that the discharge flowing through the lake is

$$Q_{\text{lake}} = 4Q_{x0}R. \tag{8.191}$$

The lake captures a discharge equal to the uniform flow multiplied by twice the diameter of the lake.

8.11.3 A Circular Lake in a Field of Uniform Flow with Net Discharge

Lakes may add or remove water from an aquifer; this happens if the head in the lake differs from the head that would naturally occur at the center of the lake if it were absent. We must incorporate the far field in this case; the values of Φ_∞ and R_∞ will affect the value of the potential along the boundary of the lake. We cover this case by adding a term of the form (8.167)

$$\Omega = -Q_{x0}\left(z - \frac{R^2}{z}\right) + \frac{Q_0}{2\pi}\ln\frac{z}{R_\infty} + \tilde{\Phi}_\infty. \tag{8.192}$$

The real part of the first term does not contribute to the average value of the potential along the far-field circle, and the real part of the second term vanishes so that the far-field condition is met. If there were no lake present, the head at the center of the lake would be Φ_∞ as a result of the mean value theorem applied to the far-field circle. Thus, if $Q_0 = 0$, i.e., if the lake neither adds nor subtracts water from the aquifer, then the potential at the lake will be identical to the average far-field potential. We will call such a lake a *neutral lake*. If water is removed or added to the lake, then the potential along the lake boundary is

$$\Phi_0 = -\frac{Q_0}{2\pi}\ln\frac{R_\infty}{R} + \Phi_\infty. \tag{8.193}$$

Thus, if $-Q_0 > 0$, i.e., if water is added to the lake, then the potential along the lake boundary will be larger than the far-field potential; otherwise it will be less. In some cases the discharge Q_0 is known, for example, from a water balance of the lake, but it may also be computed by considering the deviation of the lake level from the equilibrium value. In that case we solve (8.193) for Q_0:

$$Q_0 = 2\pi \frac{\Phi_\infty - \Phi_0}{\ln(R_\infty/R)}. \tag{8.194}$$

The flow net shown in Figure 8.15 corresponds to $Q_0 = 0.1RQ_{x0}$; the lake captures more water than it passes on to the flow. A lake contributing more to the flow than it captures is illustrated in Figure 8.16. The flow net shown in Figure 8.17 illustrates a case where a well is placed downstream from the lake. For this case $d = 2R$, $Q_0 = 0.1RQ_{x0}$, and $Q = 0.1RQ_{x0}$.

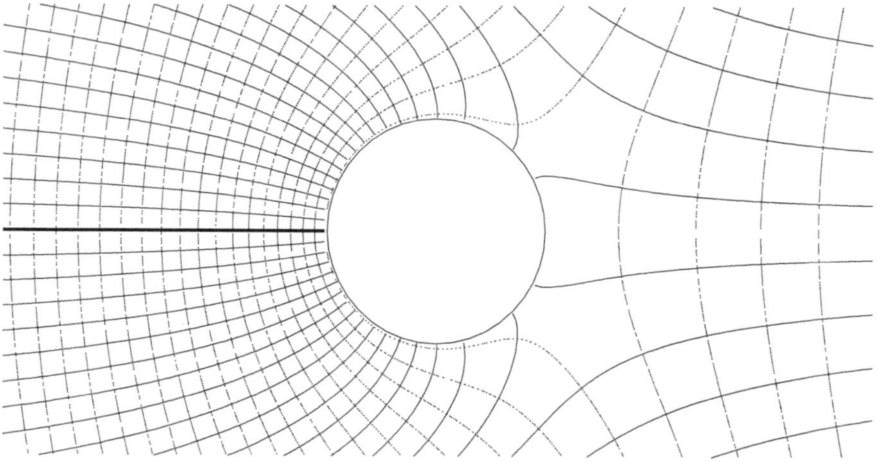

Figure 8.15 Flow net for $Q_0 = 10RQ_{x0}$.

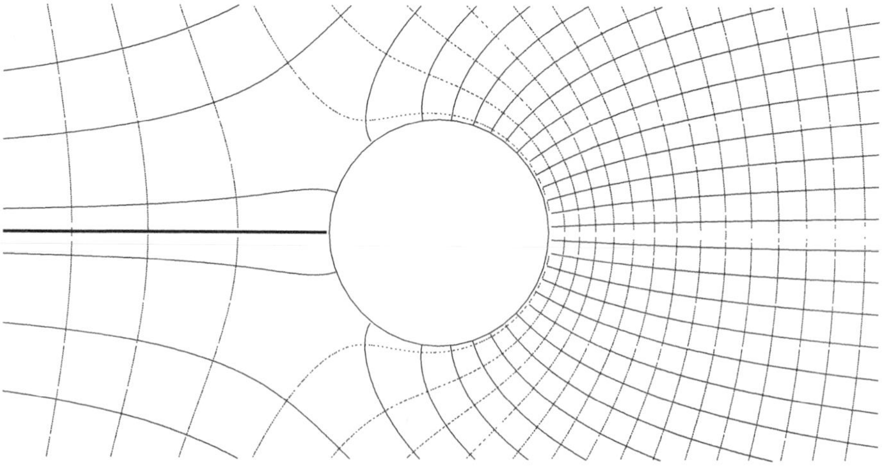

Figure 8.16 Flow net for $Q_0 = -10RQ_{x0}$.

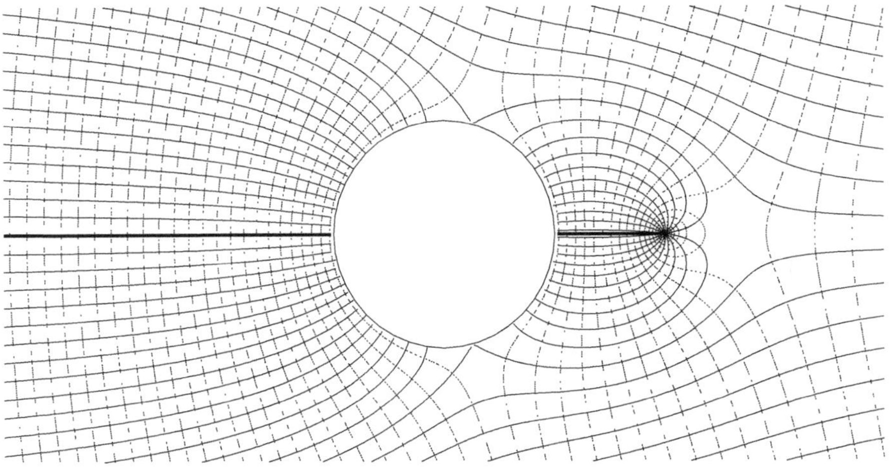

Figure 8.17 Flow from a circular lake to a well; $d = 2R$, $Q_0 = 0.1RQ_{x0}$, and $Q = 0.1RQ_{x0}$.

8.11.4 Stagnation Points

We obtain the coordinates of the stagnation points by differentiating the complex potential (8.192) with respect to z:

$$W = Q_{x0}\left(1 + \frac{R^2}{z^2}\right) - \frac{Q_0}{2\pi}\frac{1}{z}. \tag{8.195}$$

We set this equal to zero at the stagnation point, $z = z_s$, and obtain, after multiplying both sides of the equation by $z_s^2/(R^2 Q_{x0})$,

$$\zeta^2 - q_0\zeta + 1 = 0, \tag{8.196}$$

where we introduced the dimensionless quantities ζ and q_0,

$$\zeta = \frac{z_s}{R}, \qquad q_0 = \frac{Q_0}{2\pi Q_{x0} R}. \tag{8.197}$$

We solve (8.196) for ζ and obtain

$$\zeta_{1,2} = \frac{q_0}{2} \pm \sqrt{\left(\frac{q_0}{2}\right)^2 - 1}. \tag{8.198}$$

We observe that when $q_0 = 0$, i.e., when the lake level is at the equilibrium level in the aquifer, then there are two stagnation points, at $z_s = \pm iR$; this case corresponds to Figure 8.14. If $|q_0/2|$ is less than one, then there are two stagnation points, located symmetrically with respect to the x-axis. The value of q_0 for the flow nets shown in Figures 8.15 and 8.16 are plus and minus $5/\pi$, respectively. For the case of Figure 8.15, $\zeta_{1,2} = 0.79 \pm 0.61$, and for the case of Figure 8.16, $\zeta_{1,2} = -0.79 \pm 0.61$.

8.11.5 A Circular Lake with a Well

We next consider the possibility of creating a lake at a location where the natural groundwater level is below the desired level of the lake. We must add water to the lake for this case. We could do this by obtaining the necessary discharge, $-|Q_0|$, from some source and inject it into the lake. Alternatively, we could draw water from the aquifer, e.g., on either the downstream or upstream side of the lake, and inject it into the lake. We cover this case by adding a well at a point $z = z_w$. The image recharge well inside the lake, necessary to maintain the lake level constant in the expression for the complex potential, simulates the injection of water into the lake to maintain its level as desired.

We retain the term with Q_0 in the equation; it represents the net amount drawn from the lake, for example, by evaporation, and does not include the injection of water pumped from the aquifer. The corresponding expression for the potential is (compare (8.179))

$$\Omega = -Q_{x0}\left(z - \frac{R^2}{z}\right) + \frac{Q_0}{2\pi}\ln\frac{z}{R_\infty} + \frac{Q}{2\pi}\ln\left[\frac{z - z_w}{z - R^2/\bar{z}_w}\frac{R}{|z_w|}\right] + C, \tag{8.199}$$

where

$$z_w = d\mathrm{e}^{\mathrm{i}\beta}. \tag{8.200}$$

The far-field condition implies that the average potential along the far-field circle is Φ_∞. The first two terms in (8.199) do not contribute, and the third term approximates $Q/(2\pi)\ln(R/d)$, since $(z - z_w)/(z - R^2/\bar{z}_w)$ approaches 1 for large values of $|z|$. We apply the far-field condition, with good approximation, as

$$\Phi_\infty = \frac{Q}{2\pi}\ln\frac{R}{d} + C \tag{8.201}$$

and solve for the constant C:

$$C = \Phi_\infty - \frac{Q}{2\pi}\ln\frac{R}{d}. \tag{8.202}$$

The potential along the lake boundary is now

$$\Phi_0 = \Phi_\infty + \frac{Q}{2\pi}\ln\frac{d}{R} - \frac{Q_0}{2\pi}\ln\frac{R_\infty}{R}. \tag{8.203}$$

We can apply this equation to compute the discharge of the well, necessary to obtain the desired value of Φ_∞, given the net discharge Q_0 from the lake, possibly due to evaporation.

Flow from the Far Field

The well draws all of its water from the lake for the case in which $Q_0 = 0$; the complex potential then reduces to uniform flow plus a constant and terms that decrease with distance. For the general case, i.e., $Q_0 \neq 0$, the complex potential approaches $Q_0/(2\pi)\ln(z)+c$, where c is a constant for $z \to \infty$; i.e., the total flow from infinity is always Q_0. The total flow from the aquifer across the lake boundary and into the lake is $Q_0 - Q$.

Problem

8.5 Consider the case of a field of uniform flow in an unconfined aquifer. A circular lake of radius R is being planned. We choose the origin of the z-plane as its center. The heads in the aquifer are measured at $z = -L$ as ϕ_1 and at $z = L$ as ϕ_2. The hydraulic conductivity is k. The head in the lake is planned to be above the level that would correspond to the natural level at $z = 0$, which is obtained from the average value of the potential at $z = 0$, Φ_{av}:

$$\Phi_{av} = \tfrac{1}{2}(\Phi_1 + \Phi_2), \tag{8.204}$$

so that

$$\phi_{av} = \sqrt{\frac{2\Phi_{av}}{k}}. \tag{8.205}$$

It is given that $L = 1000$ m, $\phi_1 = 24$ m, $\phi_2 = 18$ m, $R = 100$ m, $k = 10$ m/day, and $\phi_0 = \phi_{av} + 1.5$ m. The well should be placed within 200 m from the lake boundary. It is given that the head at $x = -L$ may change, but not the head at $x = L$.

Questions:

1. Determine the amount Q_0 that would have to be added to the lake per unit time in order to raise the level to the desired amount; this water would not be supplied by the aquifer.
2. Examine the possibility of using a well in the aquifer to pump water from the aquifer and then via a pipe into the lake to raise the level to the desired amount. If such a solution is indeed feasible, examine different locations of the well and the corresponding discharges that will keep the lake at the desired level, while keeping the head at $x = L$ constant.
3. Compare your solution, if it exists, with that obtained for question 1 by comparing flow nets.

8.11.6 Uniform Flow at an Angle to the x-Axis

We oriented the x-axis in the direction of the uniform flow when determining the complex potential for a circular lake in a field of uniform flow, for reasons of convenience. This is not possible if the direction of uniform flow is not known a priori, for example. For such cases, the angle of the uniform flow must be included in the complex potential. We achieve this by introducing a complex variable $\zeta = \xi + i\eta$ such that the ξ-axis is in the direction α of the uniform flow. The relation between the complex variables z and ζ is such that the ξ-axis makes an angle α with the x-axis. Thus, the argument of z must be α when the argument of ζ is zero:

$$z = \zeta e^{i\alpha}, \qquad \zeta = z e^{-i\alpha}. \tag{8.206}$$

We obtain the expression for the circular lake in a field of uniform flow in terms of ζ from (8.189) by replacing z by ζ

$$\Omega = -Q_u\left[\zeta - \frac{R^2}{\zeta}\right] + \Phi_0, \tag{8.207}$$

where we replaced Q_{x0}, which corresponds to flow in the x-direction, by the general discharge rate Q_u. We obtain the expression in terms of z by replacing ζ by $ze^{-i\alpha}$

$$\Omega = -Q_u\left[ze^{-i\alpha} - \frac{R^2 e^{i\alpha}}{z}\right] + \Phi_0. \tag{8.208}$$

This expression indeed satisfies the condition that $\Omega \to -Q_u z e^{-i\alpha}$ for $z \to \infty$ and that the potential is constant along the boundary; we verify the latter statement by setting z equal to $Re^{i\theta}$, which yields

$$\Omega = -Q_u R\left[e^{i(\theta-\alpha)} - e^{-i(\theta-\alpha)}\right] + \Phi_0 = -2iQ_u R\sin(\theta-\alpha) + \Phi_0. \tag{8.209}$$

The first term is purely imaginary; the potential is indeed equal to Φ_0 along the boundary of the lake.

8.12 A Cylindrical Impermeable Object

We consider a cylindrical impermeable object in a uniform flow field with wells. We first study the case of an infinite aquifer with a well and a cylindrical impermeable object, then present the complex potential for an impermeable cylindrical object in a field of uniform flow. Note that problems with cylindrical objects can be solved using the circle theorem; see Milne-Thomson [1958].

8.12.1 A Cylindrical Impermeable Object with a Well

We apply the method of images to flow with a well outside the impermeable object; the image of the well has the same sign as the well: both well and image well have a positive discharge. Since the boundary of the inclusion is impermeable, the discharge well inside the circle must be balanced by a recharge well. This recharge well is at the center, and could be considered as the image of the source at infinity that supplies the water for the well. We demonstrate that the complex potential obtained in this way satisfies the boundary conditions; the expression for Ω is

$$\Omega = \frac{Q}{2\pi}\ln(z - z_w) + \frac{Q}{2\pi}\ln\left(z - \frac{R^2}{\bar{z}_w}\right) - \frac{Q}{2\pi}\ln(z) + C. \tag{8.210}$$

We combine the three logarithms into a single one:

$$\Omega = \frac{Q}{2\pi}\ln\left[\frac{(z - z_w)(z - R^2/\bar{z}_w)}{z}\right] + C = \frac{Q}{2\pi}\ln\left[(z - z_w)(R^2/z - \bar{z}_w)(-1/\bar{z}_w)\right] + C \tag{8.211}$$

or

$$\Omega = \frac{Q}{2\pi}\ln\left[(z - z_w)(R^2/z - \bar{z}_w\right] + \frac{Q}{2\pi}\ln\frac{-1}{\bar{z}_w} + C. \tag{8.212}$$

We choose the constant C as

$$C = -\frac{Q}{2\pi}\ln\frac{-1}{\bar{z}_w} + C_1, \qquad \Im C_1 = 0, \tag{8.213}$$

so that the expression for Ω becomes

$$\Omega = \frac{Q}{2\pi}\ln\left[(z - z_w)\left(\frac{R^2}{z} - \bar{z}_w\right)\right] + C_1, \qquad \Im C_1 = 0. \tag{8.214}$$

This complex potential indeed satisfies the condition that the boundary of the object, where $z\bar{z} = R^2$, is a streamline, i.e., that Ψ is constant:

$$\Omega = \frac{Q}{2\pi}\ln[(z - z_w)(\bar{z} - \bar{z}_w)] + C_1 = \frac{Q}{2\pi}\ln|z - z_w|^2 + C_1. \tag{8.215}$$

8.12.2 A Cylindrical Impermeable Object in a Field of Uniform Flow

We construct the complex potential for an impermeable cylindrical object in a field of uniform flow as for the case of the circular lake. The result is

$$\Omega = -Q_{x0}\left(z + \frac{R^2}{z}\right) + C. \tag{8.216}$$

We verify that this solution indeed satisfies the boundary condition by setting $z\bar{z} = R^2$, or $R^2/z = \bar{z}$,

$$\Omega = -Q_{x0}(z + \bar{z}) + C = -2Q_{x0}x + C. \tag{8.217}$$

This expression is indeed real, provided that C is real; the stream function is constant along the boundary; the boundary is a streamline, as asserted.

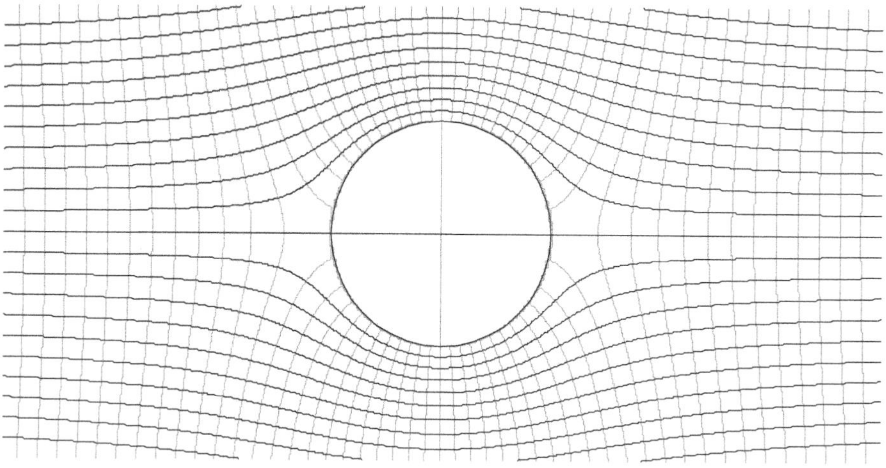

Figure 8.18 An impermeable cylinder in a field of uniform flow.

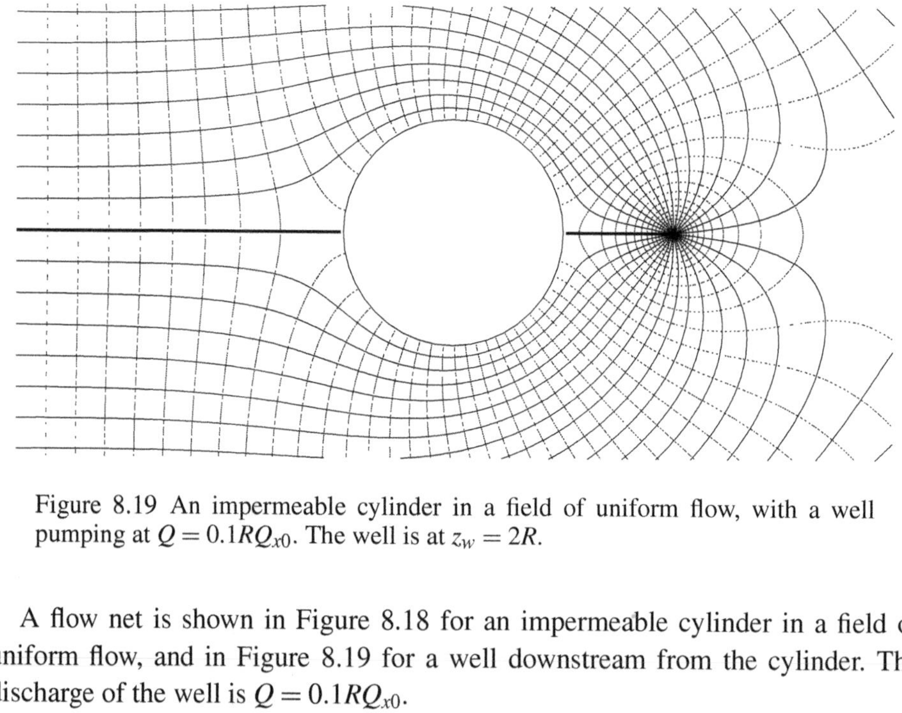

Figure 8.19 An impermeable cylinder in a field of uniform flow, with a well pumping at $Q = 0.1RQ_{x0}$. The well is at $z_w = 2R$.

A flow net is shown in Figure 8.18 for an impermeable cylinder in a field of uniform flow, and in Figure 8.19 for a well downstream from the cylinder. The discharge of the well is $Q = 0.1RQ_{x0}$.

Problem

8.6 Present the complex potential for an impermeable cylinder of radius R centered at z_0 in a field of uniform flow that makes an angle α with the x-axis. Demonstrate that the stream function is constant along the boundary of the impermeable cylinder.

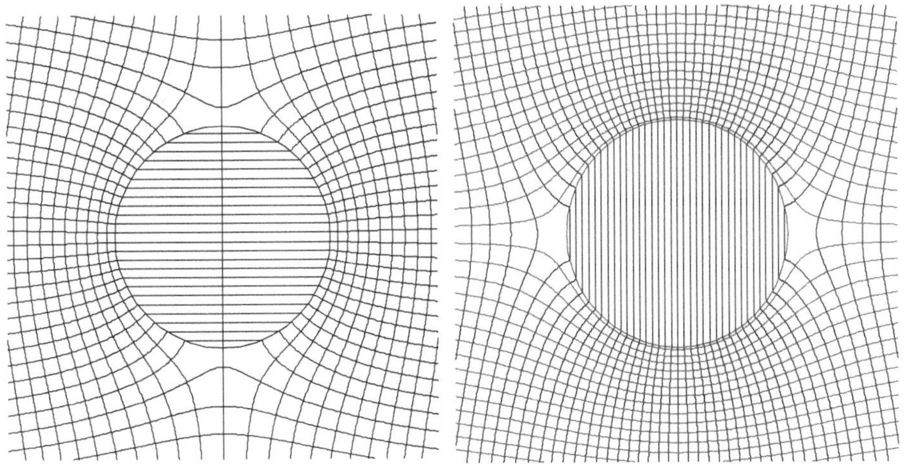

Figure 8.20 A cylindrical inhomogeneity in a field of uniform flow. The hydraulic conductivity inside the cylinder is 100 times that on the outside for case (a), and 1/100 times that on the outside for case (b).

8.13 A Cylindrical Inhomogeneity

We consider the case of a cylindrical inhomogeneity in a field of uniform flow as a final example of cylindrical objects. Two cases of a cylindrical inhomogeneity in a confined aquifer in a field of uniform flow are illustrated in Figure 8.20. The hydraulic conductivity inside the inhomogeneity is 100 times that on the outside for the case of Figure 8.20(a); it is 1/100 times that on the outside for Figure 8.20(b). We refer to the hydraulic conductivities inside and outside the inhomogeneity as k_1 and k, respectively. The origin is at the center of the inhomogeneity. The uniform flow is in the x-direction at a rate Q_{x0}. The cases of flow we considered up to this point – unconfined flow, confined flow, and interface flow – all have discharge potentials associated with them that can be written in the form

$$\Phi = kf(\phi), \tag{8.218}$$

where $f(\phi)$ is some function of the head; for unconfined flow this is $\frac{1}{2}\phi^2$, and for confined flow it is $H\phi - \frac{1}{2}H^2$. The head must be continuous across the boundary of the inhomogeneity and therefore the condition is

$$\frac{\Phi^+}{\Phi^-} = \frac{k^+ f(\phi)}{k^- f(\phi)} = \frac{k^+}{k^-} = \frac{k_1}{k}, \qquad z\bar{z} = R^2, \tag{8.219}$$

where the plus sign indicates the inside of the inhomogeneity and the minus sign the outside. There is a jump in potential across the boundary, which we represent as λ:

$$\lambda = \Phi^+ - \Phi^- = \frac{k^+ - k^-}{k^-}\Phi^- = \frac{k_1 - k}{k}\Phi^-, \qquad z\bar{z} = R^2. \tag{8.220}$$

The second condition along the boundary is that there is continuity of flow; i.e., there must not be a jump in the stream function,

$$\Psi^+ - \Psi^- = 0. \tag{8.221}$$

We obtain the solution to this problems in two steps, as we did for the case of a straight infinite inhomogeneity; see Section 8.10. First, we meet the condition of continuity of flow by creating a solution with a jump in potential and a constant value of the stream function along the boundary of the inhomogeneity. We derive this solution from the complex potential for flow around an impermeable cylinder, (8.216):

$$\underset{1}{\Omega} = -aQ_{x0}\left(z + \frac{R^2}{z}\right), \qquad z\bar{z} \geq R^2$$
$$\underset{1}{\Omega} = C_1, \qquad z\bar{z} < R^2, \qquad \Im C_1 = 0. \tag{8.222}$$

We obtain an expression for the jump in potential created by this function by setting $R^2/z = \bar{z}$:

$$\lambda = \Phi^+ - \Phi^- = C_1 + 2aQ_{x0}x, \qquad z = Re^{i\theta}. \tag{8.223}$$

We add a second complex potential $\underset{2}{\Omega}$ throughout the flow domain. The potential far away must be uniform; there cannot be a singularity inside the circle (R^2/z is singular (infinite) at $z = 0$); and the sum of the two complex potentials at the outside must approach that for uniform flow. The function that together with $\underset{1}{\Omega}$ fulfills these conditions is

$$\underset{2}{\Omega} = (a-1)Q_{x0}z + \Phi_0. \tag{8.224}$$

The complex potential on the outside is $\underset{1}{\Omega} + \underset{2}{\Omega}$

$$\Omega = \underset{1}{\Omega} + \underset{2}{\Omega} = -aQ_{x0}\left(z + \frac{R^2}{z}\right) + (a-1)Q_{x0}z + \Phi_0$$
$$= -Q_{x0}\left(z + a\frac{R^2}{z}\right) + \Phi_0, \qquad z\bar{z} \geq R^2. \tag{8.225}$$

The condition of continuity of flow across the boundary of the cylinder is met, because both $\underset{1}{\Omega}$ and $\underset{2}{\Omega}$ satisfy this condition. We apply the condition of continuity of head, (8.220), and obtain with (8.223)

$$\lambda = C_1 + 2aQ_{x0}x = -\frac{k_1-k}{k}Q_{x0}\Re(z+a\bar{z}) + \frac{k_1-k}{k}\Phi_0$$
$$= -\frac{k_1-k}{k}Q_{x0}(1+a)x + \frac{k_1-k}{k}\Phi_0, \qquad z\bar{z} = R^2. \tag{8.226}$$

We choose a and C_1 such that this equation is satisfied for all x:

$$2a = -\frac{k_1-k}{k}(1+a) \tag{8.227}$$

$$C_1 = \frac{k_1-k}{k}\Phi_0. \tag{8.228}$$

We solve (8.227) for a:

$$\frac{2k+k_1-k}{k}a = -\frac{k_1-k}{k} \tag{8.229}$$

or

$$a = -\frac{k_1-k}{k_1+k}. \tag{8.230}$$

The expression for the complex potential on the outside is given by (8.225), which becomes, with the latter expression for a,

$$\Omega = -Q_{x0}\left(z - \frac{k_1-k}{k_1+k}\frac{R^2}{z}\right) + \Phi_0, \qquad z\bar{z} \geq R^2. \tag{8.231}$$

The complex potential on the inside consists of $\underset{2}{\Omega}$ plus C_1 so that, with $a-1 = (-k_1+k-k_1-k)/(k_1+k) = -2k_1/(k_1+k)$,

$$\Omega = -\frac{2k_1}{k_1+k}Q_{x0}z + C_1 + \Phi_0 = -\frac{2k_1}{k_1+k}Q_{x0}z + \frac{k_1}{k}\Phi_0, \qquad z\bar{z} < R^2. \tag{8.232}$$

Note that Φ_0 is the potential along the line $x = 0$ for $|y| \geq R$. We see from (8.232) that the complex potential inside the inhomogeneity is that for uniform flow, with a discharge equal to

$$\overset{\text{in}}{Q}_{x0} = \frac{2k_1}{k_1+k}Q_{x0}. \tag{8.233}$$

8.13.1 Limiting Cases of Hydraulic Conductivity Contrast

The latter equation provides insight in the effect of inhomogeneities on flow fields. We consider the two extreme cases of contrast of hydraulic conductivity.

The Case $k_1/k \to \infty$

We consider that $k_1 \gg k$; the expression for the uniform flow inside the inhomogeneity then approaches the following:

$$\overset{\text{in}}{Q}_{x0} = \frac{2}{1+k/k_1} Q_{x0} \to 2Q_{x0}, \qquad k_1/k \to \infty. \tag{8.234}$$

We see that in the limit as $k_1/k \to \infty$, the discharge vector inside the inhomogeneity approaches two times the discharge vector of the uniform far field.

The Case $k_1/k \to 0$

The effect of the inhomogeneity is quite different if the ratio k_1/k approaches zero, i.e., if the hydraulic conductivity inside the inhomogeneity is much smaller than that outside. In this case, the discharge vector inside approaches

$$\overset{\text{in}}{Q}_{x0} = \frac{2k_1/k}{1+k_1/k} Q_{x0} \to 2\frac{k_1}{k} Q_{x0}, \qquad k_1/k \to 0. \tag{8.235}$$

The discharge inside the inhomogeneity continues to decrease with decreasing ratio k_1/k, which is quite different from the behavior observed for the limit of $k_1/k \to \infty$, where an upper bound of $2Q_{x0}$ exists. This observation helps us to understand why in contaminant transport the break-through curve is relatively steep, but has a long tail, due to the long travel times experienced inside zones of low hydraulic conductivity and near stagnation points.

8.13.2 A Cylindrical Inhomogeneity with a Well Inside the Inhomogeneity

We determine the complex potential for the case of a cylindrical inhomogeneity of radius R centered at $z = 0$, but now with a well rather than a uniform flow field. The well is inside the inhomogeneity at z_w and has a discharge Q. We again solve the problem by first constructing a function $\underset{1}{\Omega}$ with the property that $\underset{1}{\Psi} = \Im \underset{1}{\Omega}$ is equal to a constant for $z\bar{z} = R^2$. $\underset{1}{\Omega}$ is defined as zero for $|z| < R$. We define $\underset{1}{\Omega}$ outside the inhomogeneity as the complex potential for a well of discharge Q_1 inside an impermeable cylindrical boundary of radius R (compare (8.214)) and outside the inhomogeneity as zero:

$$\underset{1}{\Omega} = \frac{Q_1}{2\pi} \ln\left[(z - z_w)(R^2/z - \bar{z}_w)\right] + \Re C_1, \qquad z\bar{z} \le R^2 \tag{8.236}$$

$$\underset{1}{\Omega} = 0, \qquad z\bar{z} > R^2, \tag{8.237}$$

where

$$z_w = |z_w| e^{i\alpha}, \qquad |z_w| < R. \tag{8.238}$$

The function $\underset{1}{\Omega}$ is purely real along the boundary of the inhomogeneity, as we verify by setting $R^2/z = \bar{z}$. We write $\underset{1}{\Omega}$ inside the inhomogeneity in a different form as

$$\underset{1}{\Omega} = \frac{Q_1}{2\pi}\left\{\ln(z-z_w) + \ln\left[(R^2/\bar{z}_w - z)\bar{z}_w\right] - \ln(z)\right\} + \Re C_1, \qquad z\bar{z} \le R^2. \quad (8.239)$$

We add to $\underset{1}{\Omega}$ a function $\underset{2}{\Omega}$, defined throughout the domain, such that there is a well of discharge Q at z_w inside the inhomogeneity and there is no well at the center of the inhomogeneity,

$$\underset{2}{\Omega} = \frac{Q-Q_1}{2\pi}\ln(z-z_w) + \frac{Q_1}{2\pi}\ln\frac{z}{R} + C, \quad (8.240)$$

where C is a real constant:

$$\Im C = 0. \quad (8.241)$$

8.13.3 The Complex Potential Inside the Inhomogeneity

The complex potential inside the inhomogeneity is the sum of (8.239) and $\underset{2}{\Omega}$, which gives

$$\Omega = \frac{Q}{2\pi}\ln(z-z_w) + \frac{Q_1}{2\pi}\ln\left[\left(\frac{R^2}{\bar{z}_w} - z\right)\frac{\bar{z}_w}{R}\right] + \Re C_1 + C, \qquad z\bar{z} \le R^2. \quad (8.242)$$

We write this in a slightly different form,

$$\Omega = \frac{Q}{2\pi}\ln(z-z_w) + \frac{Q_1}{2\pi}\ln\left[\left(z - \frac{R^2}{\bar{z}_w}\right)\frac{\bar{z}_w}{-R}\right] + \Re C_1 + C, \quad z\bar{z} \le R^2. \quad (8.243)$$

The real part of the second term in this equation is equal to the potential for the image of a well of discharge Q_1 at z_w to match the boundary condition of zero potential for a circular lake of radius R. The complex potential for a well at z_w of discharge Q_1 and a circular lake at the origin with radius R is

$$\Omega_{\text{lake}} = \frac{Q_1}{2\pi}\ln\left[\frac{z-z_w}{z-R^2/\bar{z}_w|}\frac{R}{|z_w|}\right]. \quad (8.244)$$

The real part of this complex potential is zero at the boundary of the lake, i.e., for $z\bar{z} = R^2$, so that

$$\ln\left[\left|z - \frac{R^2}{\bar{z}_w}\right|\frac{|z_w|}{R}\right] = \ln|z-z_w|, \qquad z\bar{z} = R^2. \quad (8.245)$$

We use this relation to write the real part of (8.243) at $z\bar{z} = R^2$, i.e., for Φ^+ as

$$\Phi^+ = \frac{Q+Q_1}{2\pi}\ln|z-z_w| + \Re C_1 + C. \quad (8.246)$$

The complex potential on the outside of the inhomogeneity is due solely to $\underset{2}{\Omega}$, given by (8.240); the real part of this equation is, for $z\bar{z} = R^2$,

$$\Phi^- = \frac{Q - Q_1}{2\pi} \ln|z - z_w| + C. \tag{8.247}$$

The ratio Φ^+/Φ^- must be equal to k_1/k,

$$\frac{\frac{Q+Q_1}{2\pi} \ln|z - z_w| + \Re C_1 + C}{\frac{Q-Q_1}{2\pi} \ln|z - z_w| + C} = \frac{k_1}{k} \tag{8.248}$$

so that

$$k\frac{Q+Q_1}{2\pi} \ln|z - z_w| + k[\Re C_1 + C] = k_1 \frac{Q - Q_1}{2\pi} \ln|z - z_w| + k_1 C. \tag{8.249}$$

The coefficient of $\ln|z - z_w|$ must be the same on both sides of the equal sign,

$$k(Q + Q_1) = k_1(Q - Q_1) \tag{8.250}$$

or

$$Q_1(k + k_1) = Q(k_1 - k) \tag{8.251}$$

and

$$Q_1 = \frac{k_1 - k}{k_1 + k} Q. \tag{8.252}$$

We choose the real part of C_1 such that the constants on both sides of (8.249) match

$$\Re C_1 = \frac{k_1 - k}{k} C. \tag{8.253}$$

The potential $\underset{2}{\Omega}$ contains a term $Q - Q_1$, which we can now write as

$$Q - Q_1 = \frac{k_1 + k - k_1 + k}{k_1 + k} Q = \frac{2k}{k_1 + k} Q. \tag{8.254}$$

We obtain the expressions for the complex potential inside the inhomogeneity from (8.243) with expression (8.252) for Q_1,

$$\Omega = \frac{Q}{2\pi} \ln(z - z_w) + \frac{k_1 - k}{k_1 + k} \frac{Q}{2\pi} \ln\left[\left(z - \frac{R^2}{\bar{z}_w}\right) \frac{\bar{z}_w}{-R}\right] + \frac{k_1}{k} C, \qquad z\bar{z} \le R^2. \tag{8.255}$$

The complex potential outside the inhomogeneity is given by (8.240), with the expressions for Q and $Q - Q_1$, (8.252) and (8.254),

$$\Omega = \frac{2k}{k_1 + k} \frac{Q}{2\pi} \ln(z - z_w) + \frac{k_1 - k}{k_1 + k} \frac{Q}{2\pi} \ln\frac{z}{R} + C, \qquad z\bar{z} > R^2. \tag{8.256}$$

Note that this solution reduces to that for a single well in an infinite aquifer for $k_1 = k$.

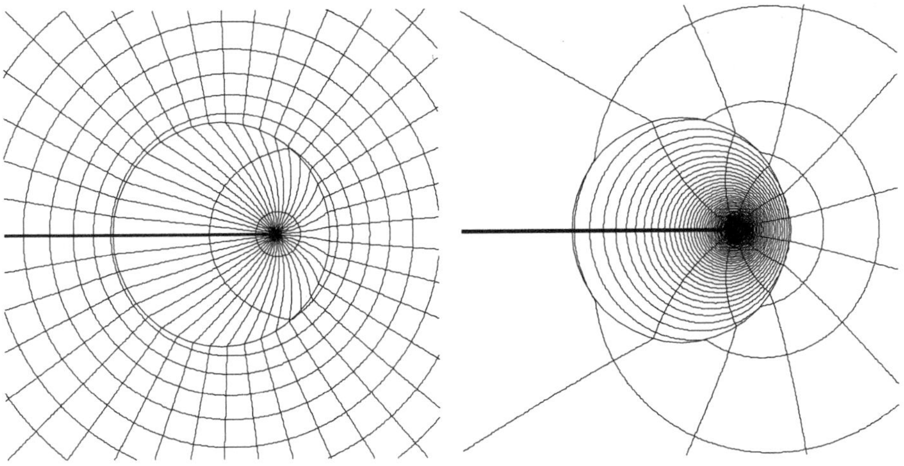

Figure 8.21 Flow net for the case of a cylindrical inhomogeneity with a well inside.

Flow nets for two ratios of k_1/k are shown in Figure 8.21; the ratios are $k_1/k = 10$ for case (a) and $k_1/k = 0.1$ for case (b). The presence of the inhomogeneity has a profound effect on the efficiency of the well, i.e., the head required at the well screen in order to maintain a certain discharge. We observe from Figure 8.21 that for case in which $(k_1/k = 10)$ there is minimal draw-down inside the inhomogeneity, whereas the opposite is true for the case in which $k_1/k = 0.1$. Since inhomogeneities with much higher contrasts than a factor of 10 occur in nature, we expect the efficiency of wells to vary greatly with position of the well for inhomogeneous aquifers. This is indeed the case; well drillers often report vast differences in productivity of wells that are placed relatively close to one another, but in different zones of conductivity.

As an additional illustration, we combine the effect of uniform flow with that of the well. We obtain the complex potential for that case simply by adding the complex potential for an inhomogeneity in a field of uniform flow (8.231) and (8.232) to that of a well inside the inhomogeneity. The result is

$$\Omega = -Q_{x0}\left(z - \frac{k_1 - k}{k_1 + k}\frac{R^2}{z}\right) + \frac{2k}{k_1 + k}\frac{Q}{2\pi}\ln(z - z_w) + \frac{k_1 - k}{k_1 + k}\frac{Q}{2\pi}\ln\frac{z}{R} + \Re C, \qquad |z| > R \tag{8.257}$$

$$\Omega = -\frac{2k_1}{k_1 + k}Q_{x0}z + \frac{Q}{2\pi}\ln(z - z_w) + \frac{k_1 - k}{k_1 + k}\frac{Q}{2\pi}\ln\left(R - z\frac{\bar{z}_w}{R}\right) + \frac{k_1}{k}\Re C, \qquad |z| \le R. \tag{8.258}$$

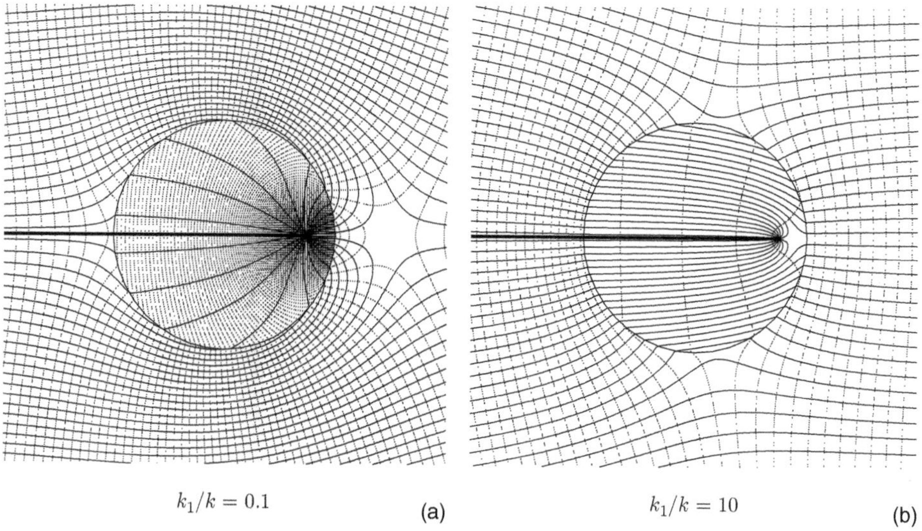

Figure 8.22 Flow net for the case of a cylindrical inhomogeneity with a well inside in a field of uniform flow of $Q_x = Q_{x0}$; the discharge of the well is $Q = 2RQ_{x0}$ where R is the radius of the inhomogeneity.

An example of an inhomogeneity with a well inside in a field of uniform flow is shown in Figure 8.22, which applies to the case in which $Q = 2RQ_{x0}$ for two different ratios of k_1/k. Note that the capture zone, bounded by the streamlines that limit the flow into the well, is entirely outside the inhomogeneity for the case of the lower hydraulic conductivity and is inside the cylinder for the case of the higher hydraulic conductivity. The bounding streamlines, however, have the same asymptotes because the discharges of the wells for the two cases are the same.

8.13.4 A Cylindrical Inhomogeneity with a Well Outside the Inhomogeneity

We next consider a well outside the inhomogeneity. We again define the potential $\underset{1}{\Omega}$ such that its imaginary part is constant along the boundary; we use the complex potential for a cylindrical impermeable object, (8.214):

$$\underset{1}{\Omega} = \frac{Q_1}{2\pi} \ln\left[(z - z_w)\left(\frac{R^2}{z} - \bar{z}_w\right)\right] + \Re C_1, \qquad |z| \geq R \tag{8.259}$$

$$\underset{1}{\Omega} = 0, \qquad |z| < R. \tag{8.260}$$

The well is outside the inhomogeneity, so that

$$z_w = |z_w| e^{i\alpha}, \qquad |z_w| > R. \tag{8.261}$$

The function $\underset{2}{\Omega}$ is defined everywhere and must combine with $\underset{1}{\Omega}$ to represent a well at z_w of discharge Q, i.e.,

$$\underset{2}{\Omega} = \frac{Q - Q_1}{2\pi} \ln(z - z_w) + C. \tag{8.262}$$

We obtain an expression for $\Omega^+ - \Omega^-$ along the boundary $z\bar{z} = R^2$ from (8.259),

$$\Omega^+ - \Omega^- = \Phi^+ - \Phi^- = -\frac{Q_1}{2\pi} \ln[(z - z_w)(\bar{z} - \bar{z}_w)] - \Re C_1 = -\frac{Q_1}{\pi} \ln|z - z_w| - \Re C_1. \tag{8.263}$$

The potential inside the inhomogeneity is the real part of $\underset{2}{\Omega}$ so that

$$\Phi^+ = \underset{2}{\Phi} = \Re \underset{2}{\Omega} = \frac{Q - Q_1}{2\pi} \ln|z - z_w| + \Re C, \qquad |z| = R. \tag{8.264}$$

We now express the condition of continuity of head as

$$\lambda = \Phi^+ - \Phi^- = \frac{k^+ - k^-}{k^+} \Phi^+ = \frac{k_1 - k}{k_1} \underset{2}{\Phi} = \Re \underset{2}{\Omega}, \qquad |z| = R. \tag{8.265}$$

Note that we choose to express the jump in terms of the potential just inside the inhomogeneity for reasons of convenience. We use (8.263) and (8.264),

$$-\frac{Q_1}{\pi} \ln|z - z_w| - \Re C_1 = \frac{k_1 - k}{k_1} \frac{Q - Q_1}{2\pi} \ln|z - z_w| + \frac{k_1 - k}{k_1} \Re C, \qquad |z| = R. \tag{8.266}$$

The terms that contain the independent variable z must cancel so that

$$-2Q_1 = \frac{k_1 - k}{k_1} Q - \frac{k_1 - k}{k_1} Q_1. \tag{8.267}$$

We solve this for Q_1,

$$Q_1 \left(-2 + 1 - \frac{k}{k_1} \right) = \frac{k_1 - k}{k_1} Q \rightarrow Q_1 = -\frac{k_1 - k}{k_1 + k} Q. \tag{8.268}$$

which is the negative of the expression we obtained for the case of the well inside the inhomogeneity. We obtain the following relations for the constants from (8.266)

$$\Re C_1 = -\frac{k_1 - k}{k_1} \Re C \tag{8.269}$$

and the term $Q - Q_1$ is

$$Q - Q_1 = \frac{k_1 + k + k_1 - k}{k_1 + k} Q = \frac{2k_1}{k + k_1} Q. \tag{8.270}$$

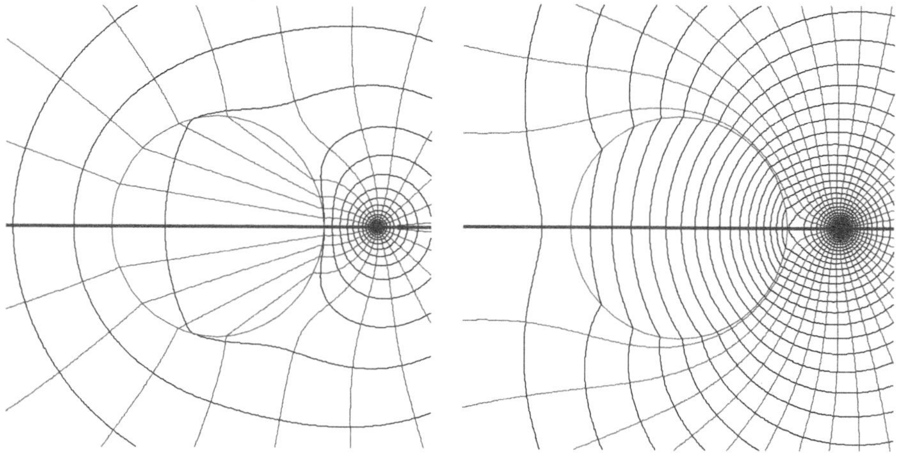

$k_1/k = 10$ (a) $k_1/k = 0.1$ (b)

Figure 8.23 Flow net for the case of a cylindrical inhomogeneity with a well outside.

We obtain expressions for the potential inside and outside the inhomogeneity by adding Ω_1 and Ω_2, which gives

$$\Omega = \frac{Q}{2\pi}\ln(z - z_w) - \frac{k_1 - k}{k_1 + k}\frac{Q}{2\pi}\ln(R^2/z - \bar{z}_w) + \frac{k}{k_1}\Re C, \qquad |z| > R \qquad (8.271)$$

$$\Omega = \frac{2k_1}{k_1 + k}\frac{Q}{2\pi}\ln(z - z_w) + \Re C, \qquad z \le R. \qquad (8.272)$$

Flow nets for two ratios of k_1/k are shown in Figure 8.23; the ratios are $k_1/k = 0.1$ for case (a) and $k_1/k = 10$ for case (b). Note that the well pumps the same discharge for the two cases; the well captures the same width of the uniform flow field for both cases. As expected, the effect of the inhomogeneity on the efficiency of the well is far less than if the well is inside the inhomogeneity.

8.14 Hydraulic Barrier for Waste Isolation

The purpose of using a barrier to isolate waste is to reduce the flow inside the barrier to some acceptable level. We may achieve this reduction either by creating a fully penetrating cylindrical barrier of a hydraulic conductivity that is much lower than the ambient hydraulic conductivity or, surprisingly, by creating a cylindrical fully penetrating barrier of a hydraulic conductivity that is much higher than the ambient one. We examine this by constructing complex potentials in the zone outside the barrier, the zone inside the barrier itself, i.e., in the annulus between the

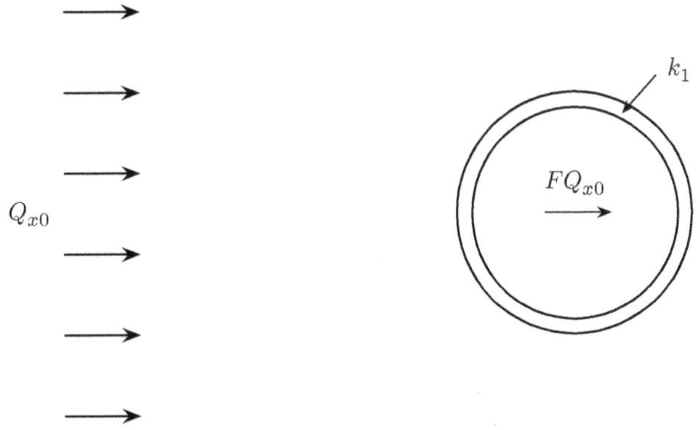

Figure 8.24 A cylindrical barrier in a field of uniform flow.

bounding cylinders, and in the domain inside the barrier. We represent the hydraulic conductivity inside the barrier as k_1 and the ambient one as k. The radius of the outer cylinder is R and that of the inner one is R_1. There is a uniform flow of discharge Q_{x0} and the head along the y-axis is ϕ_0. The cylinders are centered at $z = 0$. We construct the solution in a form inspired by the solutions we obtained for a cylindrical inhomogeneity as follows:

$$\Omega = -Q_{x0}\left(z + a\frac{R^2}{z}\right) + \Phi_0, \qquad R^2 \le z\bar{z} \tag{8.273}$$

$$\Omega = -Q_{x0}\left(bz + c\frac{R_1^2}{z}\right) + \frac{k_1}{k}\Phi_0, \qquad R_1^2 \le z\bar{z} \le R^2 \tag{8.274}$$

$$\Omega = -Q_{x0}dz + \Phi_0, \qquad z\bar{z} \le R_1^2. \tag{8.275}$$

The boundary conditions are that the stream function and the head are both continuous at the two boundaries We denote the inside of the cylinders as the positive side and the outside as the negative side and write the boundary conditions as

$$\Psi^+ - \Psi^- = 0, \qquad \Phi^+ = \frac{k_1}{k}\Phi^-, \qquad z\bar{z} = R^2 \tag{8.276}$$

$$\Psi^+ - \Psi^- = 0, \qquad \Phi^+ = \frac{k}{k_1}\Phi^-, \qquad z\bar{z} = R_1^2. \tag{8.277}$$

We obtain expressions for the complex potentials on the outside and on the inside of the inner cylinder by replacing R_1^2/z by $\bar{z}$, as $z\bar{z} = R_1^2$,

$$\Omega^- = -Q_{x0}(bz + c\bar{z}) + \frac{k_1}{k}\Phi_0, \qquad z\bar{z} = R_1^2 \tag{8.278}$$

$$\Omega^+ = -Q_{x0}dz + \Phi_0, \qquad z\bar{z} = R_1^2. \tag{8.279}$$

We subtract the imaginary parts of the latter two equations and set the result equal to zero

$$\Psi^{+} - \Psi^{-} = -Q_{x0}y\,(d - b + c) = 0 \tag{8.280}$$

so that

$$d = b - c. \tag{8.281}$$

We apply the second condition in (8.277) to the real parts of (8.278) and (8.279)

$$-Q_{x0}xd + \Phi_0 = \frac{k}{k_1}\left[-Q_{x0}x(b+c) + \frac{k_1}{k}\Phi_0\right] \tag{8.282}$$

so that

$$d = \frac{k}{k_1}(b+c). \tag{8.283}$$

We subtract the latter equation from (8.281),

$$0 = b\left(1 - \frac{k}{k_1}\right) - c\left(1 + \frac{k}{k_1}\right) \tag{8.284}$$

so that

$$b = \frac{k_1 + k}{k_1 - k}c. \tag{8.285}$$

The complex potentials Ω^{-} and Ω^{+} along $z\bar{z} = R^2$ are, with $R^2/z = \bar{z}$

$$\Omega^{-} = -Q_{x0}\,(z + a\bar{z}) + \Phi_0 \tag{8.286}$$

$$\Omega^{+} = -Q_{x0}\left(bz + c\frac{R_1^2}{R^2}\bar{z}\right) + \frac{k_1}{k}\Phi_0. \tag{8.287}$$

We set the jump in the stream function equal to zero,

$$\Psi^{+} - \Psi^{-} = -Q_{x0}y\left(b - c\frac{R_1^2}{R^2}\right) + Q_{x0}y(1-a) = 0 \tag{8.288}$$

or

$$1 - a = b - c\frac{R_1^2}{R^2} \rightarrow a = 1 - b + c\frac{R_1^2}{R^2}. \tag{8.289}$$

The final boundary condition is that the head is continuous across $z\bar{z} = R^2$, which gives, with the real parts of (8.286) and (8.287),

$$-Q_{x0}x\left(b + c\frac{R_1^2}{R^2}\right) + \frac{k_1}{k}\Phi_0 = \frac{k_1}{k}[-Q_{x0}x(1+a) + \Phi_0] \tag{8.290}$$

so that

$$b + c\frac{R_1^2}{R^2} = \frac{k_1}{k}(1+a) \rightarrow a = \frac{k}{k_1}\left(b + c\frac{R_1^2}{R^2}\right) - 1. \tag{8.291}$$

We subtract the latter equation from (8.289)

$$2-b\left(1+\frac{k}{k_1}\right)+c\frac{R_1^2}{R^2}\left(1-\frac{k}{k_1}\right)=0. \tag{8.292}$$

We solve this for b

$$b=2\frac{k_1}{k_1+k}+c\frac{R_1^2}{R^2}\frac{k_1-k}{k_1+k}=c\frac{k_1+k}{k_1-k}, \tag{8.293}$$

where we used (8.285) to express b in terms of c. We solve the latter equation for c and obtain

$$c\left[\frac{R_1^2}{R^2}\frac{k_1-k}{k_1+k}-\frac{k_1+k}{k_1-k}\right]=-2\frac{k_1}{k_1+k} \tag{8.294}$$

so that

$$c=\frac{-2\frac{k_1}{k_1+k}}{\frac{R_1^2}{R^2}\frac{k_1-k}{k_1+k}-\frac{k_1+k}{k_1-k}}=\frac{2\frac{k_1}{k_1-k}}{\left(\frac{k_1+k}{k_1-k}\right)^2-\frac{R_1^2}{R^2}}. \tag{8.295}$$

We obtain two expressions for b from the two expressions for c,

$$b=\frac{-2\frac{k_1}{k_1-k}}{\frac{R_1^2}{R^2}\frac{k_1-k}{k_1+k}-\frac{k_1+k}{k_1-k}}=\frac{2\frac{k_1}{k_1-k}\frac{k_1+k}{k_1-k}}{\left(\frac{k_1+k}{k_1-k}\right)^2-\frac{R_1^2}{R^2}}. \tag{8.296}$$

The constant d is given by (8.281), $d=b-c$, and we obtain with the second terms of (8.295) and (8.296),

$$d=\frac{\frac{2k_1}{k_1-k}\left(\frac{k_1+k}{k_1-k}-1\right)}{\left(\frac{k_1+k}{k_1-k}\right)^2-\frac{R_1^2}{R^2}}=\frac{\frac{2k_1}{k_1-k}\left(\frac{k_1+k-k_1+k}{k_1-k}\right)}{\left(\frac{k_1+k}{k_1-k}\right)^2-\frac{R_1^2}{R^2}}=\frac{\frac{4k_1k}{(k_1-k)^2}}{\left(\frac{k_1+k}{k_1-k}\right)^2-\frac{R_1^2}{R^2}}. \tag{8.297}$$

We write this expression in a different form by using the relationship

$$4k_1k=(k_1+k)^2-(k_1-k)^2 \tag{8.298}$$

so that the expression for d becomes

$$d=\frac{\left(\frac{k_1+k}{k_1-k}\right)^2-1}{\left(\frac{k_1+k}{k_1-k}\right)^2-\frac{R_1^2}{R^2}}. \tag{8.299}$$

The final constant to be determined is a; we use (8.289),

$$a=1-b+c\frac{R_1^2}{R^2}, \tag{8.300}$$

where b and c are given by (8.296) and (8.295), and obtain

$$a = \frac{\left(\frac{k_1+k}{k_1-k}\right)^2 - \frac{R_1^2}{R^2} - \frac{2k_1}{k_1-k}\left(\frac{k_1+k}{k_1-k} - \frac{R_1^2}{R^2}\right)}{\left(\frac{k_1+k}{k_1-k}\right)^2 - \frac{R_1^2}{R^2}} = \frac{\frac{k_1+k}{k_1-k}\left(\frac{k_1+k}{k_1-k} - \frac{2k_1}{k_1-k}\right) + \frac{R_1^2}{R^2}\left(\frac{2k_1}{k_1-k} - 1\right)}{\left(\frac{k_1+k}{k_1-k}\right)^2 - \frac{R_1^2}{R^2}}. \tag{8.301}$$

The first term in parentheses in the numerator equals -1 and the second one equals $(k_1+k)/(k_1-k)$ so that

$$a = -\frac{\frac{k_1+k}{k_1-k}\left(1 - \frac{R_1^2}{R^2}\right)}{\left(\frac{k_1+k}{k_1-k}\right)^2 - \frac{R_1^2}{R^2}}. \tag{8.302}$$

The complex potential inside the inhomogeneity is given by (8.275), which becomes, with expression (8.299) for d,

$$\Omega = -Q_{x0}\frac{\left(\frac{k_1+k}{k_1-k}\right)^2 - 1}{\left(\frac{k_1+k}{k_1-k}\right)^2 - \frac{R_1^2}{R^2}} z + \Phi_0. \tag{8.303}$$

It follows that the flow inside the barrier is uniform and that its magnitude is equal to a factor less than one times the uniform flow far-field; the ratio R_1/R is always less than one, so that the denominator is always larger in magnitude than the numerator.

8.14.1 Flow Reduction

The barrier will always reduce the flow; the amount of reduction depends on the ratio of the hydraulic conductivities and the ratio of the radii of the inner and outer boundaries. The factor of reduction, D, is given by

$$D(k_1,k) = \frac{\left(\frac{k_1+k}{k_1-k}\right)^2 - 1}{\left(\frac{k_1+k}{k_1-k}\right)^2 - \frac{R_1^2}{R^2}} = \frac{\left(\frac{k_1/k+1}{k_1/k-1}\right)^2 - 1}{\left(\frac{k_1/k+1}{k_1/k-1}\right)^2 - \frac{R_1^2}{R^2}} = \frac{\left(\frac{k/k_1+1}{k/k_1-1}\right)^2 - 1}{\left(\frac{k/k_1+1}{k/k_1-1}\right)^2 - \frac{R_1^2}{R^2}}. \tag{8.304}$$

The factor D is a function of both k and k_1 and has the property

$$D(k_1,k) = D(k,k_1), \tag{8.305}$$

which implies that we obtain the same factor of reduction of flow by increasing the hydraulic conductivity inside the barrier by a factor D relative to that outside as we obtain by decreasing it by the same factor. Thus, if the ambient hydraulic conductivity is low, it will be more efficient to create a barrier of high hydraulic conductivity, whereas if the ambient hydraulic conductivity is high, a barrier of

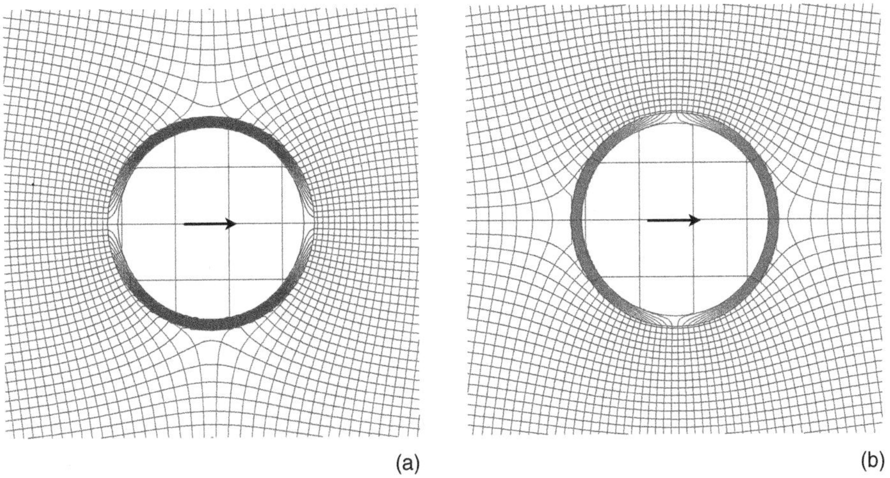

Figure 8.25 Flow in an aquifer with a barrier that is 100 times more permeable (a) and hundred times less permeable (b) than the ambient hydraulic conductivity.

low hydraulic conductivity, e.g., a slurry wall, will be more efficient. Two cases are shown in Figure 8.25(a) and (b). The hydraulic conductivity inside the barrier is one-hundredth of the ambient hydraulic conductivity for case (a), and for case (b) the ratio is 100, rather than 0.01. The contour level intervals are identical and it is clear from the figures that the magnitude of flow inside the barrier is the same for the two cases. Note also that in the former case the flow is diverted around the barrier, whereas in the second case it is channeled through the barrier.

An important observation is that a barrier of high hydraulic conductivity is much less insensitive to flaws than a slurry wall is; if there is an obstruction in a barrier of high hydraulic conductivity, it will barely affect the ability (especially since the flow is three-dimensional) of the barrier to transmit the water around the interior, whereas flaws in a slurry wall are much more prone to let water pass through the interior, especially if flaws occur on opposite sides of the barrier.

8.15 Infiltration over an Area with a Cylindrical Inhomogeneity

We noted that the effect of inhomogeneities in hydraulic conductivity extends only several times the largest dimension of the inhomogeneity, except when water is removed from the aquifer by a well. The same holds true in the case of infiltration. We may add global infiltration of rate N to an infinite aquifer with a circular inhomogeneity by adding the radial solution to the Poisson equation in such a way that its contribution to the potential along the boundary of the inhomogeneity vanishes; it will not contribute to the potential there and thus does not affect the condition of continuity of head. We first consider uniform infiltration and radial flow. The corresponding expression for the potential for the case of an inhomogeneity centered at z_0 is

$$\Phi = \underset{N}{\Phi} + \underset{C}{\Phi} = -\frac{N}{4}\left[(z-z_0)(\bar{z}-\bar{z}_0) - R^2\right] + C, \qquad (8.306)$$

where $\underset{N}{\Phi}$ is the contribution of the infiltration and $\underset{C}{\Phi}$ is the potential at the boundary of the inhomogeneity, which must meet the condition of continuity of head. This implies that the ratio of the potentials inside and outside the inhomogeneity at the boundary is k_1/k, so that the constant must jump across the boundary,

$$\underset{c}{\Phi} = \frac{k_1}{k}\Phi_0, \qquad (z-z_0)(\bar{z}-\bar{z}_0) \le R^2 \qquad (8.307)$$

$$\underset{c}{\Phi} = \Phi_0, \qquad (z-z_0)(\bar{z}-\bar{z}_0) > R^2, \qquad (8.308)$$

where Φ_0 is the constant value of the potential just outside the boundary of the inhomogeneity. Equipotentials are shown in Figure 8.26 for $k = 1$ m/day, $k_1 = 0.02k$, $N = 0.05k_1$, $R = 100$ m; the head inside the inhomogeneity increases relative to that outside; the increase is nearly 5 m. The opposite effect is apparent when the hydraulic conductivity inside the inhomogeneity is higher than that outside. This is illustrated in Figure 8.27 for $k_1/k = 50$; the other data are the same.

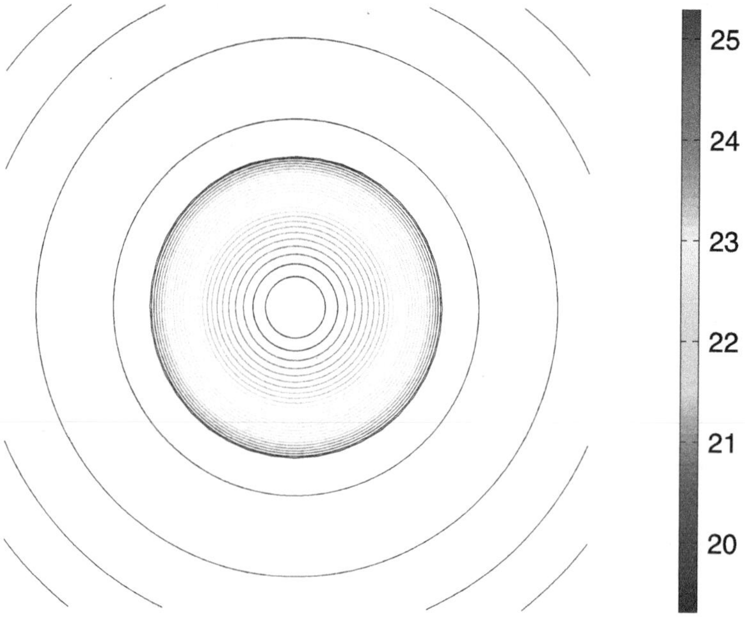

Figure 8.26 recharge due to rainfall on a cylindrical inhomogeneity; k_1 = 0.02 m/d, $k = 1$ m/d, $N = 10^{-3}$ m/d, $R = 100$ m.

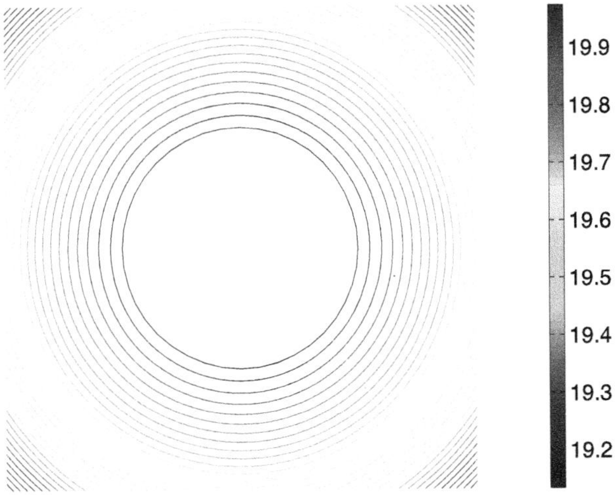

Figure 8.27 Recharge due to rainfall on a cylindrical inhomogeneity; $k_1 = 50$ m/d, $N = 10^{-3}$ m/d, $R = 100$ m.

The head inside the inhomogeneity is now nearly constant. Note that the total amount infiltrated is the same for both cases.

The effect of uniform flow on the system can be accounted for by adding the complex potential for uniform flow with an inhomogeneity, i.e., a function of the form (8.231) and (8.232). Note that the solution includes the constant contribution to the potential. We obtain an expression for the potential for the circular inhomogeneity with rainfall and uniform flow by adding to (8.306) the real part of the following complex potential

$$\Omega = -Q_{x0}\left(z - \frac{k_1 - k}{k_1 + k}\frac{R^2}{z}\right), \qquad (z - z_0)(\bar{z} - \bar{z}_0) \geq R^2 \tag{8.309}$$

$$\Omega = -\frac{2k_1}{k_1 + k}Q_{x0}z, \qquad (z - z_0)(\bar{z} - \bar{z}_0) < R^2. \tag{8.310}$$

The solution is illustrated in Figure 8.28 with a uniform flow at a rate of $Q_{x0} = 0.4$ m^2/day.

Differences in hydraulic conductivity may influence the amount of infiltration in practice. In that case the infiltration rate on the inside of the inhomogeneity differs from that on the outside; a new expression for the potential must be used. Let the infiltration rate inside the inhomogeneity be N_1 and outside N. The expression for the potential is

$$\Phi = -\frac{N_1}{4}\left[(z - z_0)(\bar{z} - \bar{z}_0) - R^2\right], \qquad (z - z_0)(\bar{z} - \bar{z}_0) \leq R^2 \tag{8.311}$$

Figure 8.28 Rainfall on a cylindrical inhomogeneity with uniform flow of 0.4 m^2/d; $k_1 = 0.02$ m/d, $k = 1$ m/d, $N = 10^{-3}$ m/day, $R = 100$ m.

$$\Phi = -\frac{N}{4}\left[(z-z_0)(\bar{z}-\bar{z}_0) - R^2\right] + \frac{Q}{4\pi}\ln\frac{(z-z_0)(\bar{z}-\bar{z}_0)}{R^2}, \quad (z-z_0)(\bar{z}-\bar{z}_0) > R^2. \tag{8.312}$$

The logarithmic term in the expression for the potential on the outside is added to meet the condition of continuity of flow; note that we set the potential to zero along the boundary to meet the condition of continuity of head. The expressions for the discharge function inside and outside the inhomogeneity are, with $W = -2\partial\Phi/\partial z$,

$$W = \frac{N_1}{2}(\bar{z}-\bar{z}_0), \qquad (z-z_0)(\bar{z}-\bar{z}_0) \le R^2 \tag{8.313}$$

$$W = \frac{N}{2}(\bar{z}-\bar{z}_0) - \frac{Q}{2\pi}\frac{1}{z-z_0}, \qquad (z-z_0)(\bar{z}-\bar{z}_0) > R^2. \tag{8.314}$$

A point on the boundary is represented by $\bar{z}-\bar{z}_0 = R\mathrm{e}^{-\mathrm{i}\theta}$ and we write the complex discharge as $W = |W|\mathrm{e}^{-\mathrm{i}\alpha}$, so that on the boundary

$$|W|\mathrm{e}^{-\mathrm{i}\alpha} = \frac{N_1}{2}R\mathrm{e}^{-i\theta}, \qquad (z-z_0)(\bar{z}-\bar{z}_0) \le R^2 \tag{8.315}$$

$$|W|\mathrm{e}^{-\mathrm{i}\alpha} = \frac{N}{2}R\mathrm{e}^{-i\theta} - \frac{Q}{2\pi}\frac{1}{R}\mathrm{e}^{-\mathrm{i}\theta}, \qquad (z-z_0)(\bar{z}-\bar{z}_0) > R^2. \tag{8.316}$$

It follows that α equals θ, i.e., the discharge is normal to the boundary, as expected; the flow is radial. Continuity of flow requires that

$$\frac{N_1}{2}R = \frac{N}{2}R - \frac{Q}{2\pi}\frac{1}{R} \tag{8.317}$$

so that

$$Q = \pi R^2 (N - N_1). \tag{8.318}$$

We substitute this expression for Q in (8.311) and (8.312)

$$\Phi = -\frac{N_1}{4}\left[(z-z_0)(\bar{z}-\bar{z}_0) - R^2\right], \qquad (z-z_0)(\bar{z}-\bar{z}_0) \le R^2 \tag{8.319}$$

$$\Phi = -\frac{N}{4}\left[(z-z_0)(\bar{z}-\bar{z}_0) - R^2\right] + \frac{(N-N_1)R^2}{4}\ln\frac{(z-z_0)(\bar{z}-\bar{z}_0)}{R^2}, \quad (z-z_0)(\bar{z}-\bar{z}_0) > R^2. \tag{8.320}$$

8.16 Cylindrical Analytic Elements

The solutions discussed thus far in this chapter are valid for cylindrical objects in fields of uniform flow, with wells. It is possible to modify these solutions in such a way that they can be superimposed, i.e., that we can combine more than a single cylindrical object in a field of flow. We call such modified functions analytic elements (see e.g., Strack [1989]; Haitjema [1995]; Strack [2003]), and we obtain them by adding degrees of freedom for each cylindrical element. We discuss such cylindrical analytic elements first for the case of cylindrical lakes, and then for cylindrical inhomogeneities.

8.16.1 An Analytic Element for a Circular Lake

We consider the case of flow with an arbitrary number of circular equipotentials and other elements with given properties, such as uniform flow and wells with given discharge. The head at some reference point z_0 is given; the corresponding value of the potential is Φ_0.

We apply the analytic element method, the superposition of analytic elements to satisfy the boundary conditions. We first construct an analytic element for a cylindrical lake centered at $\underset{m}{z}$ and of radius $\underset{m}{R}$. The complex potential for this element is holomorphic outside the circle and thus can be expanded in an asymptotic expansion in terms of the local dimensionless complex variable $\underset{m}{Z}$, defined as

$$\underset{m}{Z} = \frac{z - \underset{m}{z}}{\underset{m}{R}} \tag{8.321}$$

plus a logarithmic term that represents the net flow into the lake, $\underset{m}{Q}$. The complex potential for the mth element is

$$\underset{m}{\Omega} = \sum_{n=1}^{\infty} \underset{m}{\alpha_n} \underset{m}{Z}^{-n} + \frac{\underset{m}{Q}}{2\pi} \ln \frac{z - \underset{m}{z}}{\left| z_0 - \underset{m}{z} \right|}. \tag{8.322}$$

Note that we have written the term representing the net flow into the lake such that its contribution to the potential at z_0 is zero. The reason for this choice is that the term that represents the net discharge into the lake dominates; its effect grows with distance, whereas the effect of the series decreases with distance. We achieve in this way that the effect of the lakes on the potential at the reference point is reduced, which enhances the process of solving for the unknowns.

We construct the solution for N of these elements by adding one function, $\underset{g}{\Omega}$, that consists entirely of given elements (e.g., wells, ponds, and uniform flow), a constant C, and N complex potentials of the form (8.322):

$$\Omega = \underset{g}{\Omega} + \sum_{m=1}^{N} \underset{m}{\Omega}(\underset{m}{Z}) + C. \tag{8.323}$$

We choose the function $\underset{g}{\Omega}$ for now as the complex potential for uniform flow, i.e.,

$$\underset{g}{\Omega} = -\underset{0}{W} z, \tag{8.324}$$

where $\underset{0}{W}$ is a complex discharge vector, defined as

$$\underset{0}{W} = \underset{0}{Q} e^{-i\alpha}. \tag{8.325}$$

$\underset{0}{Q}$ is the magnitude of the discharge vector and α the angle between the discharge vector and the x-axis. We require that the circles do not intersect each other and determine the constants $\underset{m}{\alpha_n}$ in such a way that the cylinder walls are equipotentials, i.e., that $\Re\Omega = \Phi$ is constant along each of the cylinder walls.

8.16.2 Method of Solution

We solve the problem iteratively, i.e., we solve for the constants in the complex potential for one analytic element at a time, treating the constants in the complex potentials for all the other elements as knowns. The complex potential for any of the analytic elements, say, element j, is holomorphic outside its boundary; it is holomorphic inside the circle associated with the analytic element, say, element m, whose unknowns we solve for. The complex potential for element j thus can be expanded in a Taylor series about the center of circle m. We show that this Taylor series allows us to solve for the unknown coefficients in the expansion (8.322) that represents element m on the circle as well as everywhere outside it.

We write the complex potential as the sum of the complex potential for element m, $\underset{m}{\Omega}$, plus a function $\underset{\text{other}}{\Omega}$, which is the sum of the complex potentials for all other elements, plus the uniform flow term

$$\Omega = \underset{m}{\Omega} + \underset{\text{other}}{\Omega}, \tag{8.326}$$

where $\underset{m}{\Omega}$ is given by (8.322), and where

$$\underset{\text{other}}{\Omega}(z) = \sum_{\substack{j=1 \\ j\neq m}}^{\infty} \underset{j}{\Omega}(z) + \underset{g}{\Omega} + C. \tag{8.327}$$

We expand $\underset{\text{other}}{\Omega}$ about the center of the cylinder, $\underset{m}{z}$, in a Taylor series, written in terms of the local independent variable $\underset{m}{Z}$,

$$\underset{\text{other}}{\Omega} = \sum_{n=0}^{\infty} \underset{m}{a_n} \underset{m}{Z}^n \tag{8.328}$$

and obtain expressions for the complex constants $\underset{m}{a_n}$ by the use of (8.119),

$$\underset{m}{a_n} = \frac{1}{\pi} \int_0^{2\pi} \underset{\text{other}}{\Phi}(\delta) e^{-in\theta} d\theta \qquad n = 1, \ldots, \quad \delta \in \mathscr{C}. \tag{8.329}$$

where

$$\underset{\text{other}}{\Phi} = \frac{1}{2}\left(\underset{\text{other}}{\Omega} + \underset{\text{other}}{\overline{\Omega}} \right) \tag{8.330}$$

and

$$a_0 = \frac{1}{2\pi} \int_0^{2\pi} \underset{\text{other}}{\Omega}(z(\delta)) d\theta. \tag{8.331}$$

Evaluation of this integral for different values of n produces the values of the unknown coefficients $\underset{m}{a_n}$ in the expansion of $\underset{\text{other}}{\Omega}$, valid along the boundary of the circular equipotential. Although it is possible to perform the integration analytically, it is usually preferred to carry out the integration numerically; a numerical evaluation of the integral is presented in Appendix B.

Solving for the Constants of Element j

The complex potential for element m is

$$\underset{m}{\Omega} = \sum_{n=1}^{\infty} \underset{m}{\alpha_n} \underset{m}{Z}^{-n} + \frac{\underset{m}{Q}}{2\pi} \ln \frac{z - \underset{m}{z}}{\left| z_0 - \underset{m}{z} \right|}. \tag{8.332}$$

The boundary condition along the cylinder wall is that the potential is constant. The complex potential along the cylinder wall is the sum of the complex potential (8.328) and the latter equation, i.e.,

$$\Omega = \sum_{n=0}^{\infty} \underset{m}{a_n} \underset{m}{Z}^{n} + \sum_{n=1}^{\infty} \underset{m}{\alpha_n} \underset{m}{Z}^{-n} + \frac{\underset{m}{Q}}{2\pi} \ln \frac{z - \underset{m}{z}}{\left| z_0 - \underset{m}{z} \right|}, \qquad \underset{m}{Z}\underset{m}{\bar{Z}} = 1. \tag{8.333}$$

We use the condition that $\underset{m}{Z}$ represents a point on the boundary of the cylinder, i.e,

$$\underset{m}{Z}^{-n} = \underset{m}{\bar{Z}}^{n} \tag{8.334}$$

so that (8.333) becomes

$$\Omega = \sum_{n=0}^{\infty} \underset{m}{a_n} \underset{m}{Z}^{n} + \sum_{n=1}^{\infty} \underset{m}{\alpha_n} \underset{m}{\bar{Z}}^{n} + \frac{\underset{m}{Q}}{2\pi} \ln \frac{z - \underset{m}{z}}{\left| z_0 - \underset{m}{z} \right|}, \qquad \underset{m}{Z}\underset{m}{\bar{Z}} = 1. \tag{8.335}$$

The boundary condition is that the potential Φ is constant along the cylinder wall; this condition is met if the sum of the first two terms in (8.335) is purely imaginary, i.e., if the constants $\underset{m}{\alpha_n}$ satisfy the following conditions

$$\underset{m}{\alpha_n} = -\underset{m}{\bar{a}_n}, \qquad m = 1, 2, \ldots; \; n = 1, 2, \ldots . \tag{8.336}$$

We obtain an expression for the potential along the cylinder wall by taking the real part of (8.335), and setting it equal to the given value of the potential along the cylinder wall, $\underset{m}{\Phi}$:

$$\Phi = \underset{m}{\Phi} = \Re \underset{m}{a_0} + \frac{\underset{m}{Q}}{2\pi} \ln \frac{\underset{m}{R}}{\left| z_0 - \underset{m}{z} \right|}. \tag{8.337}$$

We solve for the unknown net flow into the lake, $\underset{m}{Q}$, and obtain

$$\underset{m}{Q} = 2\pi \frac{\underset{m}{\Phi} - \Re \underset{m}{a_0}}{\ln \left[\underset{m}{R} \Big/ \left| z_0 - \underset{m}{z} \right| \right]}. \tag{8.338}$$

We determine the coefficients $\underset{m}{a_n}$ for each of the cylinders in turn, and use the values of these coefficients to determine the values of $\underset{m}{\alpha_n}$. The updated coefficients are then used in the complex potential $\underset{\text{other}}{\Omega}$ when applying the procedure to the next cylinder. The process is repeated until the coefficients $\underset{m}{\alpha_n}$ do not change from the previous iteration beyond a certain chosen value. Note that the constant C is also an unknown quantity; we may solve for this constant after looping over all the cylinders. Sometimes the evaluation of the constant is a process that does not converge, especially if the reference point is close to the cylinders (this condition should be avoided); in such cases convergence may be obtained by solving for the constant each time after the coefficients of a cylinder have been determined.

8.16.3 Hybrid Iterative Approach

Convergence is slow, because the effect of the discharges of the lakes on each other grows with distance; small errors in the discharges have a large effect. We improve this process by adopting a hybrid iterative scheme, suggested by Bandilla et al. [2007]. Part of the problem is solved iteratively, while a direct solution is applied before each iteration cycle to determine the discharges involved in the system. All terms that involve discharge are solved for together; we keep the coefficients $\underset{m}{\alpha_n}$ constant before each new iteration cycle, while determining the discharges $\underset{m}{Q}$ such that all of the logarithmic terms together produce the correct *average* value of the potential along the boundary of each element. The corresponding system of equations is

$$\Phi_m = \frac{1}{2\pi}\int_0^{2\pi} \Re\Omega^*\left(\underset{m}{z}(Z)\right) d\theta = \sum_{n=1}^{M} A_{mn}\underset{n}{Q}, \qquad m = 1,2,\ldots, M. \tag{8.339}$$

The integral in this equation represents the contribution to the average value of the current, and complete, potential along the boundary of element m, with the discharges $\underset{m}{Q}$ all set to zero (Ω^*). The matrix A_{mn} is defined as

$$A_{mn} = \frac{1}{2\pi}\int_0^{2\pi} \ln\left|\frac{\underset{m}{(z(Z))}}{z_\sigma - \underset{n}{z}}\right| d\theta, \qquad \underset{m}{Z} = e^{i\theta}, \quad m = 1,2,\ldots, M; \quad n = 1,2,\ldots, M. \tag{8.340}$$

A_{nm} represents the integral of the function that $\underset{n}{Q}$ is multiplied by, integrated along the boundary of element m. This integral is divided by the circumference of the unit circle, 2π, to obtain the average value. The integral is the contribution of $\underset{n}{Q}$ to the average potential along the boundary of element m.

8.17 An Elliptical Lake

The solutions presented above for a variety of cases of flow for circular boundaries offer interesting results, but including a second degree of freedom to the boundary shape would make it possible to examine the effect of orientation of the object relative to the direction of flow. Cylindrical objects bounded by an ellipse make this possible; we examine the effect of such cylindrical objects on the flow field in what follows. Dealing with elliptical objects is greatly facilitated by expressing the solution in terms of a parameter χ that we choose such that the ellipse corresponds to a circle in terms of χ. The transformations of the ellipse into the various domains are shown in Figure 8.29. We define the variable $Z = X + \mathrm{i}Y$ as

$$Z = X + \mathrm{i}Y = \frac{z - \frac{1}{2}(z_1 + z_2)}{\frac{1}{2}(z_2 - z_1)}, \tag{8.341}$$

where z_1 and z_2 are the complex coordinates of the foci of the ellipse. We observe from (8.341), on substitution, that the points $z = z_1$ and $z = z_2$ correspond to $Z = -1$ and $Z = 1$, respectively. The distance L between the foci in the z-plane is transformed to a distance 2 in the Z-plane, where

$$z_2 - z_1 = L e^{\mathrm{i}\alpha}. \tag{8.342}$$

The variable χ is defined in terms of Z by

$$\frac{\chi}{\nu} = Z + \sqrt{Z^2 - 1}, \tag{8.343}$$

where ν is a parameter that defines the shape of the ellipse. We see from (8.343) that

$$\begin{aligned}\frac{\nu}{\chi} &= \frac{1}{Z + \sqrt{Z^2 - 1}} = \frac{Z - \sqrt{Z^2 - 1}}{(Z - \sqrt{Z^2 - 1})(Z + \sqrt{Z^2 - 1})} \\ &= \frac{Z - \sqrt{Z^2 - 1}}{Z^2 - Z^2 + 1} = Z - \sqrt{Z^2 - 1}.\end{aligned} \tag{8.344}$$

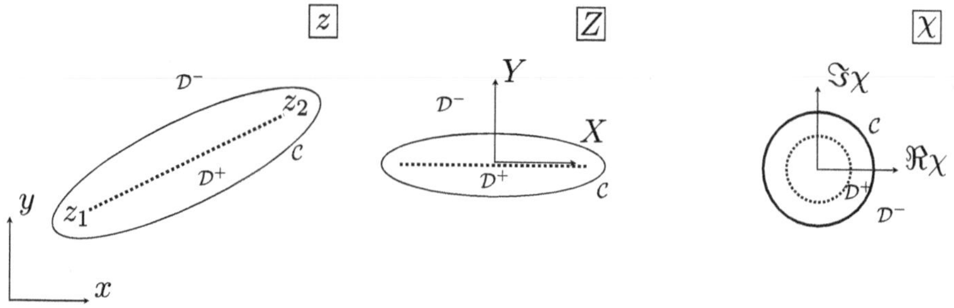

Figure 8.29 The ellipse in the z-, Z-, and χ-planes.

We eliminate the square root from (8.343) and (8.344) by addition, which gives an expression for Z in terms of χ,

$$Z = \frac{1}{2}\left(\frac{\chi}{\nu} + \frac{\nu}{\chi}\right). \tag{8.345}$$

We verify that indeed the unit circle in the χ-plane corresponds to an ellipse in the Z-plane by setting χ equal to $\mathrm{e}^{\mathrm{i}\theta}$ in (8.345), which gives

$$Z = \frac{1}{2}\left(\frac{1}{\nu}\mathrm{e}^{\mathrm{i}\theta} + \nu\mathrm{e}^{-\mathrm{i}\theta}\right) = \frac{1}{2}\left(\frac{1}{\nu} + \nu\right)\cos\theta + \frac{\mathrm{i}}{2}\left(\frac{1}{\nu} - \nu\right)\sin\theta \tag{8.346}$$

so that

$$X = \frac{1}{2}\left(\frac{1}{\nu} + \nu\right)\cos\theta \tag{8.347}$$

$$Y = \frac{1}{2}\left(\frac{1}{\nu} - \nu\right)\sin\theta. \tag{8.348}$$

We use the relationship $\cos^2\theta + \sin^2\theta = 1$ and obtain an equation for an ellipse:

$$\left[\frac{X}{\frac{1}{2}(\frac{1}{\nu} + \nu)}\right]^2 + \left[\frac{Y}{\frac{1}{2}(\frac{1}{\nu} - \nu)}\right]^2 = 1. \tag{8.349}$$

The major and minor principal axes of the ellipse in the Z-plane are A and B with

$$A = \frac{1}{2}\left(\frac{1}{\nu} + \nu\right), \qquad B = \frac{1}{2}\left(\frac{1}{\nu} - \nu\right) \tag{8.350}$$

so that

$$A - B = \nu, \qquad A + B = \frac{1}{\nu}, \qquad A^2 - B^2 = 1. \tag{8.351}$$

Note that ν is defined by choosing only the size of the major axis in the Z-plane; since $B^2 = A^2 - 1$ we have

$$\nu = A - B = A - \sqrt{A^2 - 1}. \tag{8.352}$$

The family of ellipses defined by the transformation all share the foci at $Z = 1$ and $Z = -1$; the shape of the ellipse is thus determined by choosing either A or B. Note that the ellipse will approach a circle for $\nu \to 0$, but can never reach it: the third equation in (8.351) does not allow the solution $A = B$.

We obtain the relation between $2A$ and the size of the major axis $2a$ in the z-plane by noticing that the distance L between the foci in the z-plane becomes 2 in the Z-plane, so that the scaling from Z to z is by a factor $L/2$. Hence

$$A = \frac{2a}{L}. \tag{8.353}$$

We solve the problem of flow from infinity to the elliptical lake in the χ-plane, where the flow is radial:

$$\Omega = \frac{Q_0}{2\pi} \ln \chi + \Phi_0, \tag{8.354}$$

where Q_0 is the total flow into the lake [L^3/T]. The lake corresponds to $\chi = e^{i\theta}$,

$$\Omega = \Phi + i\Psi = \frac{Q_0}{2\pi} i\theta + \Phi_0, \qquad \chi\bar{\chi} = 1, \tag{8.355}$$

so that the potential is equal to Φ_0 along the lake boundary.

8.17.1 The Inverse Transformation; Computing χ from z

We write χ in (8.354) in terms of Z by (8.343), which gives

$$\Omega = \frac{Q_0}{2\pi} \ln\left[\nu\left(Z + \sqrt{Z-1}\sqrt{Z+1}\right)\right] + \Phi_0. \tag{8.356}$$

Note that we wrote the term $\sqrt{Z^2-1}$ in a special manner,

$$\sqrt{Z^2-1} = \sqrt{Z-1}\sqrt{Z+1}. \tag{8.357}$$

The form on the right of the equal sign *must* be used, both when interpreting the mapping function and when implementing it in a computer program. The reason for this is as follows. When $Z = X,\ X < -1,\ Y = 0^+$, the term $\sqrt{Z-1}$ is equal to $\sqrt{|Z-1|e^{i\pi}} = i\sqrt{|Z-1|}$, and $\sqrt{Z+1} = \sqrt{|Z+1|e^{i\pi}} = i\sqrt{|Z+1|}$, so that $\sqrt{Z-1}\sqrt{Z+1} = i^2\sqrt{|Z^2-1|} = -\sqrt{|Z^2-1|}$. However, when we evaluate $\sqrt{Z^2-1}$ for $Z = X < -1$ the term Z^2-1 is positive, and so is $\sqrt{Z^2-1}$. The reader may verify that by splitting the square root term in two parts, as suggested, the mapping function indeed meets the conditions that the exterior of the ellipse is mapped onto the exterior of the unit circle.

8.17.2 The Complex Potential for an Elliptical Lake with a Well

One of the advantages of using complex variables to determine solutions to the Laplace equation is that any function of z only, i.e., any function that does not depend on $\bar{z}$, is a solution. Since the function $\chi(z)$ does not contain $\bar{z}$, any function in terms of χ only is also a function of z only. Thus, we may solve the problem in the χ-plane where the boundary condition is that the circle $\chi\bar{\chi} = 1$ is an equipotential. We may use the solution for a well outside a circular lake (see (8.179)) in the χ-plane to obtain the one for the case of a well outside an elliptical lake. This gives, if we denote the location of the well in the χ-plane as χ_w,

$$\Omega = \frac{Q}{2\pi}\ln\left[\frac{\chi - \chi_w}{\chi - 1/\bar{\chi}_w}\frac{1}{|\chi_w|}\right] + \Phi_0. \tag{8.358}$$

This function satisfies the boundary condition that $\Re\Omega$ is zero along the boundary of the lake. The well draws all of its water from the lake; the complex potential reduces to Φ_∞ for $\chi \to \infty$. If that is not the case and the lake draws an amount Q_0 from infinity, we must add a term $Q_0/(2\pi)\ln\chi$,

$$\Omega = \frac{Q_0}{2\pi}\ln\chi + \frac{Q}{2\pi}\ln\left[\frac{\chi - \chi_w}{\chi - 1/\bar{\chi}_w}\frac{1}{|\chi_w|}\right] + \Phi_0, \tag{8.359}$$

where Q_0 is the flow from infinity, and Φ_0 the potential along the lake boundary. We can compute the discharge Q_0 from a given value of the potential at some given point, or from the average value of the potential some distance away from the lake boundary.

Problems with wells inside an elliptical inhomogeneity can be solved, but the solution is far more complex than for the case of the well outside. The reason is the presence of singularities of the mapping function inside the inhomogeneity, at points $z = z_1$ and $z = z_2$. In order to place wells inside the ellipse, the mapping of the interior of a circle onto the interior of the ellipse is required, which involves elliptic integrals; this solution is beyond the scope of this text.

8.17.3 An Elliptical Lake in Uniform Flow

In order to solve the problem of an elliptical lake in a field of uniform flow in the χ-plane, we must establish how the uniform flow transforms in the χ-plane; we consider the far field, i.e., the complex potential for uniform flow far away from the lake. The complex potential in the z-plane will approach that for uniform flow far away from the lake, i.e.,

$$\Omega \to -Q_u e^{-i\beta} z, \qquad z \to \infty, \tag{8.360}$$

where Q_u is the discharge of the uniform flow and β the angle it makes with the x-axis. We use that $z \to \frac{1}{2}(z_2 - z_1)Z = \frac{1}{2}Le^{i\alpha}Z$ for $z \to \infty$ and that $Z \to \frac{1}{2}\chi/\nu$ and obtain

$$\Omega \to -\frac{Q_u L}{2}e^{i(\alpha-\beta)}Z \to -\frac{Q_u}{4\nu}Le^{i(\alpha-\beta)}\chi, \qquad z \to \infty. \tag{8.361}$$

We solve the problem of flow in the χ-plane by placing an image for the uniform flow to maintain constant head along the unit circle; the result is

$$\Omega = -\frac{Q_u L}{4\nu}\left[e^{i(\alpha-\beta)}\chi - \frac{e^{-i(\alpha-\beta)}}{\chi}\right] + \frac{Q_0}{2\pi}\ln\chi + \Phi_0. \tag{8.362}$$

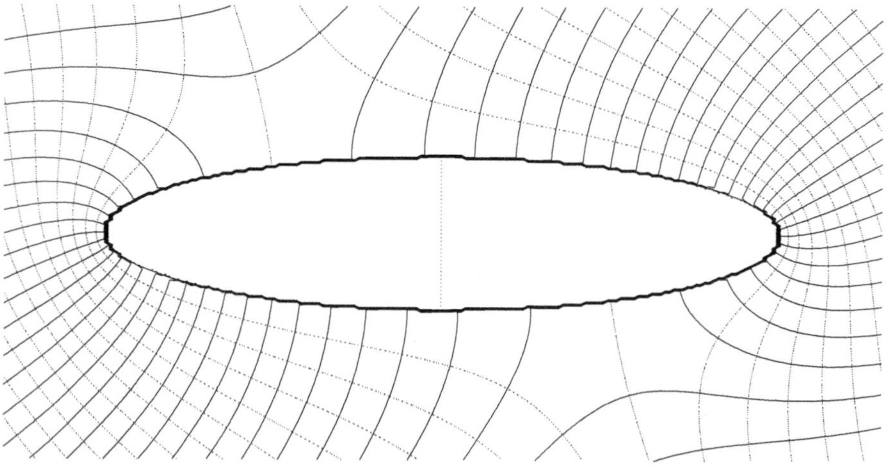

Figure 8.30 An elliptical lake in a field of uniform flow that makes an angle of 30 degrees with the x-axis; there is no net inflow into the lake.

This function approaches the complex potential for uniform flow near infinity and includes a net inflow into the lake at a rate Q_0. We verify that the boundary conditions are indeed met by setting χ equal to a point on the boundary, i.e., $\chi = e^{i\theta}$, which gives

$$\Omega = -\frac{Q_u L}{4\nu}\left(e^{i(\alpha-\beta)} - e^{-i(\alpha-\beta)}\right) + \Phi_0 = -\frac{Q_u L}{2\nu} i \sin(\alpha-\beta) + \Phi_0, \quad \chi\bar{\chi} = 1. \tag{8.363}$$

The real part of this expression is indeed constant and equal to Φ_0. Figure 8.30 is an illustration of a lake in a field of uniform flow that makes an angle of 30 degrees with the major axis of the ellipse. The case of the elliptical lake of Figure 8.30, but now with a well present, is shown in Figure 8.31. The discharge of the well is $Q = 1.3Q_{x0}a$. Note the branch cut that connects the well to the lake boundary; it represents a jump in the stream function and shows that in this case the well is drawing all of its water from the lake. The cut is the image in the z-plane of the branch cut that connects the well to its image inside the unit circle in the χ-plane.

8.18 An Impermeable Ellipse in Uniform Flow

We obtain the complex potential for an impermeable elliptical object in a similar manner as for the case of an elliptical lake. The complex potential for this case is

$$\Omega = -\frac{Q_u L}{4\nu}\left[e^{i(\alpha-\beta)}\chi + \frac{e^{-i(\alpha-\beta)}}{\chi}\right] + C, \tag{8.364}$$

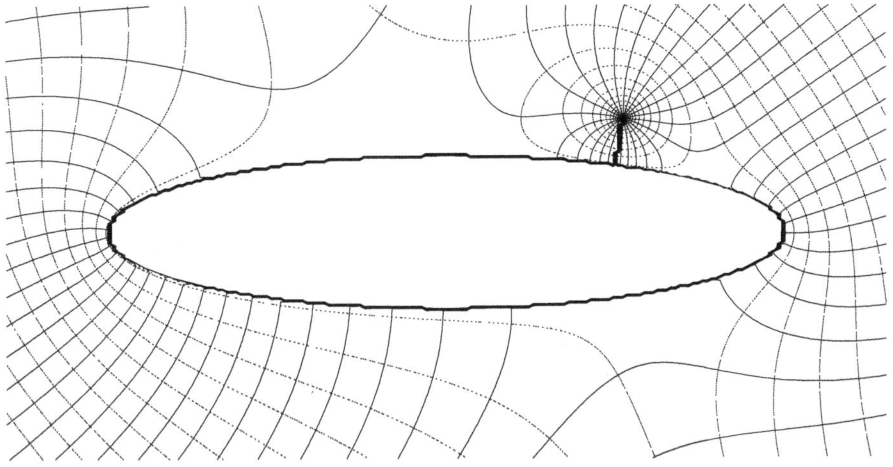

Figure 8.31 An elliptical lake in a field of uniform flow with a well.

where C is a constant. The complex potential along the boundary is

$$\Omega = -\frac{Q_u L}{4\nu}(e^{i(\theta+\alpha-\beta)} + e^{-i(\theta+\alpha-\beta)} + \Phi_0 = \Omega = -\frac{Q_u L}{2\nu}\cos(\theta+\alpha-\beta) + \Phi_0. \tag{8.365}$$

This expression is purely real, so that indeed the stream function is equal to a constant.

8.18.1 The Complex Potential for an Elliptical Impermeable Object with a Well

We use the method of images in the χ-plane (see (8.211)), to obtain the solution for flow with an impermeable elliptical object with a well and obtain

$$\Omega = \frac{Q}{2\pi}\ln\left[\frac{(\chi-\chi_w)(\chi-1/\bar{\chi}_w)}{\chi}\right] + C, \tag{8.366}$$

which satisfies the boundary condition that the stream function is constant along the ellipse.

The case of an impermeable elliptical object in a field of uniform flow is shown in Figure 8.32; the data are the same as for the case shown in Figure 8.31. Note the branch cut running from the well to infinity; in this case the well draws all of its water from infinity.

8.19 Linear Objects

We consider the case in which the minor principal axis of the ellipse is zero so that the ellipse reduces to a slot. We consider both the case of a finite canal and an impermeable linear object.

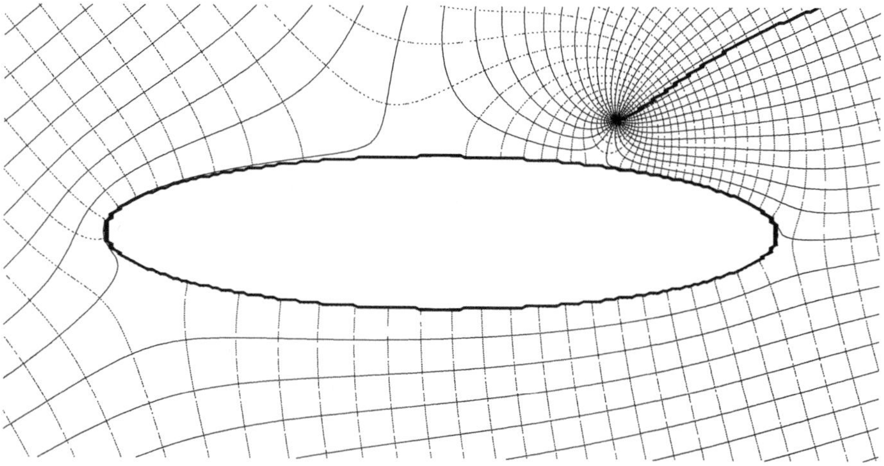

Figure 8.32 An elliptical impermeable object in a field of uniform flow that makes and angle of 30 degress with the x-axis.

8.19.1 A Finite Canal

We use the solution for an elliptical lake in a field of uniform flow with a well to solve the problem of a finite ditch or canal of constant head with a well. The ellipse reduces to a slot when the minor principal axis reduces to zero, i.e., $B = 0$. That case occurs when $\nu = 1$ (see (8.350)) and the transformations (8.343) and (8.345) reduce to

$$Z = \tfrac{1}{2}\left(\chi + \frac{1}{\chi}\right) \tag{8.367}$$

and

$$\chi = Z + \sqrt{Z^2 - 1}. \tag{8.368}$$

The solution for a canal is obtained from the equations presented above simply by using the latter transformations in the expressions for the complex potential. The flow net shown in Figure 8.33 is obtained by setting $\nu = 1$ in the expression for the complex potential used to produce Figure 8.31.

8.19.2 A Thin Impermeable Wall

We can use the solutions presented above for an impermeable elliptical object for a thin impermeable wall by using the transformations (8.367) and (8.368). A flow net is shown in Figures 8.33 and 8.34 for the case of a thin wall in a field of uniform flow with a well. The flow net was obtained by setting $\nu = 1$ in the complex potentials used to produce Figures 8.31 and 8.32.

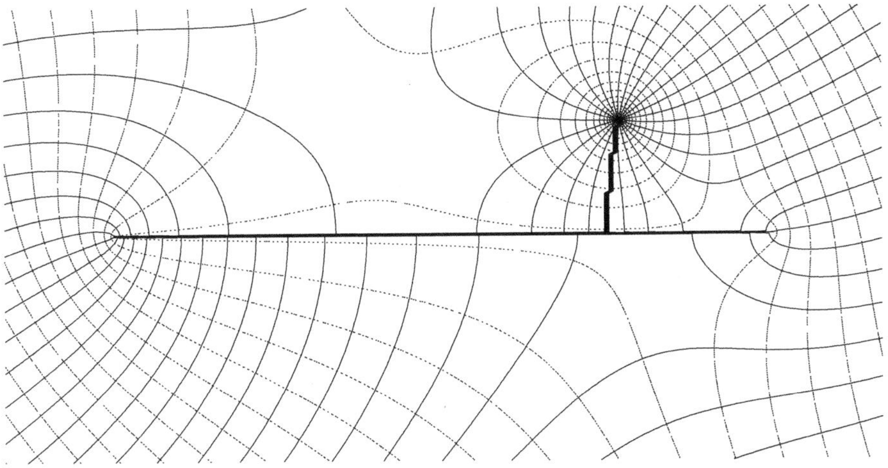

Figure 8.33 Flow net for the case of a canal with a well in a field of uniform flow; it was obtained by setting $\nu = 1$ in the complex potential used to produce Figure 8.31.

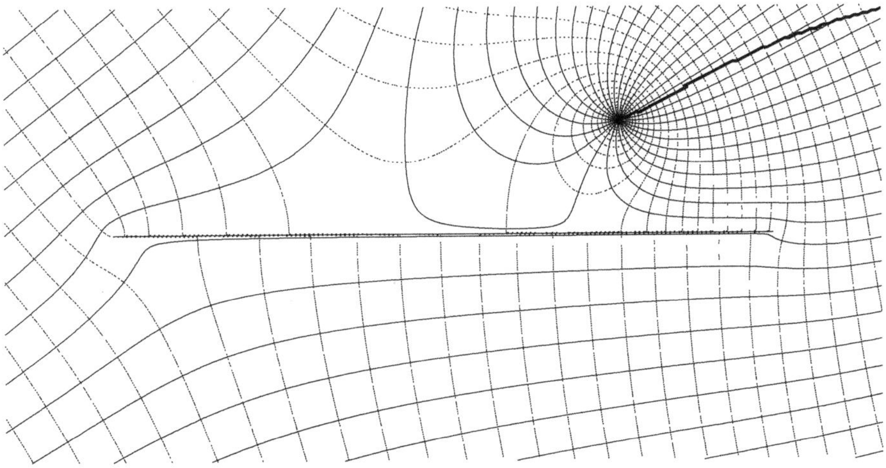

Figure 8.34 Flow net for the case of a wall with a well in a field of uniform flow; it was obtained by setting $\nu = 1$ in the complex potential used to produce Figure 8.32.

8.20 An Elliptical Inhomogeneity

We solve the problem for flow with an elliptical inhomogeneity in a field of uniform flow in the χ-plane, using the complex potential for an elliptical impermeable object in a field of uniform flow, (8.364), as the basis for our analysis and define Ω_1 as

$$\underset{1}{\Omega} = \tau\chi + \frac{\bar{\tau}}{\chi}, \qquad \chi\bar{\chi} \geq 1 \tag{8.369}$$

$$\underset{1}{\Omega} = C_1, \quad \Im C_1 = 0, \quad \chi\bar{\chi} < 1. \tag{8.370}$$

We determine the complex constant τ later and express the complex potential $\underset{1}{\Omega}$ along the boundary $\chi\bar{\chi} = 1$ as

$$\underset{1}{\Omega} = \underset{1}{\Phi} + \mathrm{i}\underset{1}{\Psi} = (\tau\chi + \bar{\tau}\bar{\chi}), \qquad \chi\bar{\chi} = 1 \tag{8.371}$$

so that

$$\underset{1}{\Phi} = (\tau\chi + \bar{\tau}\bar{\chi}), \qquad \underset{1}{\Psi} = 0, \quad \chi\bar{\chi} = 1. \tag{8.372}$$

We add a complex potential $\underset{2}{\Omega}$, defined throughout the z-plane, such that addition of $\underset{2}{\Omega}$ to $\underset{1}{\Omega}$ produces the uniform far field with a discharge Q_u angled at β to the real axis (see (8.361))

$$\begin{aligned}\underset{2}{\Omega} &= -\left(\frac{Q_u L}{2}\mathrm{e}^{\mathrm{i}(\alpha-\beta)} + 2\tau\nu\right)Z + C \\ &= -\frac{Q_u L}{2}\mathrm{e}^{\mathrm{i}(\alpha-\beta)}Z - \tau\nu\left(\frac{\chi}{\nu} + \frac{\nu}{\chi}\right) + C, \qquad \Im C = 0, \qquad \forall z.\end{aligned} \tag{8.373}$$

Note that we write this function in terms of $Z = \frac{1}{2}(\chi/\nu + \nu/\chi)$, rather than in terms of χ, which is singular at $Z = 1$ and $Z = -1$ (see (8.343)); otherwise, we are likely to introduce undesirable singularities into our solution. The discontinuity in potential across the boundary, λ, is equal to

$$\lambda = \Phi^+ - \Phi^- = C_1 - (\tau\chi + \bar{\tau}\bar{\chi}), \qquad \chi\bar{\chi} = 1. \tag{8.374}$$

We express the potential on the inside in terms of χ

$$\begin{aligned}\Phi^+ = C_1 + \frac{1}{2}(\underset{2}{\Omega} + \underset{2}{\bar{\Omega}}) &= -\frac{1}{2}\left[\frac{Q_u L}{4}\mathrm{e}^{\mathrm{i}(\alpha-\beta)} + \tau\nu\right]\left[\frac{\chi}{\nu} + \nu\bar{\chi}\right] \\ &\quad - \frac{1}{2}\left[\frac{Q_u L}{4}\mathrm{e}^{-\mathrm{i}(\alpha-\beta)} + \bar{\tau}\nu\right]\left[\frac{\bar{\chi}}{\nu} + \nu\chi\right] + C + C_1, \quad \chi\bar{\chi} = 1.\end{aligned} \tag{8.375}$$

We combine terms and write this as

$$\Phi^+ = \chi P + \bar{\chi}\bar{P} + C + C_1, \tag{8.376}$$

where

$$P = -\frac{Q_u L}{8}\left(\frac{\mathrm{e}^{\mathrm{i}(\alpha-\beta)}}{\nu} + \nu\mathrm{e}^{-\mathrm{i}(\alpha-\beta)}\right) - \tfrac{1}{2}\tau - \tfrac{1}{2}\bar{\tau}\nu^2. \tag{8.377}$$

The condition for continuity of head is

$$\lambda = \frac{k^+ - k^-}{k^+}\Phi^+ = \frac{k_1 - k}{k_1}\Phi^+ \tag{8.378}$$

and we obtain with expression (8.374) for λ and (8.376) for Φ^+

$$-(\tau\chi + \bar{\tau}\bar{\chi}) + C_1 = \frac{k_1 - k}{k_1}\left\{P\chi + \bar{P}\bar{\chi} + C + C_1\right\}, \qquad \chi\bar{\chi} = 1. \tag{8.379}$$

This equation must be satisfied for all values of χ with $\chi\bar{\chi} = 1$ so that

$$(\tau + fP)\chi + (\bar{\tau} + f\bar{P})\bar{\chi} = 0, \qquad f = \frac{k_1 - k}{k_1}, \qquad \chi\bar{\chi} = 1. \tag{8.380}$$

We must choose the constant τ as$-fP$ to satisfy the latter equation, so that, with (8.377)

$$\tau = f\frac{Q_u L}{8}\left(\frac{e^{i(\alpha-\beta)}}{\nu} + \nu e^{-i(\alpha-\beta)}\right) + \frac{1}{2}f\tau + \frac{1}{2}f\bar{\tau}\nu^2 \tag{8.381}$$

or, after rearrangement and multiplication by 2

$$\tau(2-f) - f\bar{\tau}\nu^2 = f\frac{Q_u L}{4}\left(\frac{e^{i(\alpha-\beta)}}{\nu} + \nu e^{-i(\alpha-\beta)}\right) = fD, \tag{8.382}$$

where we introduced the constant D as

$$D = \frac{Q_u L}{4}\left(\frac{e^{i(\alpha-\beta)}}{\nu} + \nu e^{-i(\alpha-\beta)}\right). \tag{8.383}$$

We divide both sides of (8.382) by f,

$$\tau\mu - \bar{\tau}\nu^2 = \frac{Q_u L}{4}\left(\frac{e^{i(\alpha-\beta)}}{\nu} + \nu e^{-i(\alpha-\beta)}\right) = D, \tag{8.384}$$

where

$$\mu = \frac{2-f}{f} = \frac{k_1 + k}{k_1}\frac{k_1}{k_1 - k} = \frac{k_1 + k}{k_1 - k}. \tag{8.385}$$

We take the complex conjugate of (8.384) and eliminate $\bar{\tau}$ from that and (8.384) to solve for τ:

$$\tau = \frac{D\mu + \bar{D}\nu^2}{\mu^2 - \nu^4} = \Re D\frac{\mu + \nu^2}{\mu^2 - \nu^4} + i\Im D\frac{\mu - \nu^2}{\mu^2 - \nu^4} = \frac{\Re D}{\mu - \nu^2} + i\frac{\Im D}{\mu + \nu^2}, \tag{8.386}$$

or, with expression (8.385) for μ

$$\tau = \Re D\frac{k_1 - k}{(k_1 + k) - \nu^2(k_1 - k)} + i\Im D\frac{k_1 - k}{(k_1 + k) + \nu^2(k_1 - k)}. \tag{8.387}$$

We use expression (8.383) for D and obtain

$$\tau=\frac{Q_uL}{4}\left\{\left(\frac{1}{\nu}+\nu\right)\cos(\alpha-\beta)\frac{(k_1-k)}{k_1+k-\nu^2(k_1-k)}\right.$$
$$\left.+\mathrm{i}\left(\frac{1}{\nu}-\nu\right)\sin(\alpha-\beta)\frac{(k_1-k)}{k_1+k+\nu^2(k_1-k)}\right\} \qquad (8.388)$$

or, with $1/\nu+\nu=2A$ and $1/\nu-\nu=2B$,

$$\tau=\frac{Q_uL}{2}\left\{A\cos(\alpha-\beta)\frac{(k_1-k)}{k_1+k-\nu^2(k_1-k)}\right.$$
$$\left.+\mathrm{i}B\sin(\alpha-\beta)\frac{(k_1-k)}{k_1+k+\nu^2(k_1-k)}\right\}. \qquad (8.389)$$

We obtain the expression for the complex potential valid outside the ellipse by adding $\underset{1}{\Omega}$, (8.369), to $\underset{2}{\Omega}$, (8.373),

$$\Omega=-\frac{Q_uL}{2}\mathrm{e}^{\mathrm{i}(\alpha-\beta)}Z+\frac{\bar{\tau}-\tau\nu^2}{\chi}+C, \qquad \chi\bar{\chi}\geq 1. \qquad (8.390)$$

The complex potential on the inside is equal to C_1 plus $\underset{2}{\Omega}$:

$$\Omega=-\frac{Q_uL}{2}\mathrm{e}^{\mathrm{i}(\alpha-\beta)}Z-\tau\nu\left(\frac{\chi}{\nu}+\frac{\nu}{\chi}\right)+C_1+C, \qquad \chi\bar{\chi}<1. \qquad (8.391)$$

We express C_1 in terms of C by setting the constants equal to each other in (8.379)

$$C_1=\frac{k_1-k}{k_1}(C+C_1)\rightarrow\frac{k}{k_1}C_1=\frac{k_1-k}{k_1}C\rightarrow C_1=\frac{k_1-k}{k}C\rightarrow C_1+C=\frac{k_1}{k}C. \qquad (8.392)$$

We use this in (8.391), replace $\chi/\nu+\nu/\chi$ by $2Z$, and use expression (8.389) for τ

$$\Omega=-\frac{Q_uL}{2}\left\{\mathrm{e}^{\mathrm{i}(\alpha-\beta)}\right.$$
$$+2\left[A\cos(\alpha-\beta)\frac{(k_1-k)\nu}{(k_1+k)-(k_1-k)\nu^2}\right.$$
$$\left.\left.+\mathrm{i}B\sin(\alpha-\beta)\frac{(k_1-k)\nu}{(k_1+k)+(k_1-k)\nu^2}\right]\right\}Z+\frac{k_1}{k}C, \qquad \chi\bar{\chi}<1. \qquad (8.393)$$

We divide both numerator and denominator in the fractions by ν,

$$\Omega = -\frac{Q_u L}{2}\left\{ e^{i(\alpha-\beta)} + 2\left[A\cos(\alpha-\beta)\frac{k_1 - k}{(k_1+k)/\nu - (k_1-k)\nu} \right.\right.$$
$$\left.\left. + iB\sin(\alpha-\beta)\frac{k_1 - k\nu}{(k_1+k)/\nu + (k_1-k)\nu} \right]\right\} Z + \frac{k_1}{k}C, \quad \chi\bar{\chi} < 1. \tag{8.394}$$

We write $\nu = A - B$ and $1/\nu = A + B$ and collect terms,

$$\Omega = -\frac{Q_u L}{2}\left\{ e^{i(\alpha-\beta)} + \left[A\cos(\alpha-\beta)\frac{k_1-k}{Ak+Bk_1} + iB\sin(\alpha-\beta)\frac{k_1-k}{Ak_1+Bk} \right]\right\} Z$$
$$+ \frac{k_1}{k}C, \qquad \chi\bar{\chi} < 1. \tag{8.395}$$

We combine the terms in the braces,

$$\Omega = -\frac{Q_u L}{2}\left[\cos(\alpha-\beta)\frac{Ak+Bk_1+Ak_1-Ak}{Ak+Bk_1} \right.$$
$$\left. + i\sin(\alpha-\beta)\frac{Ak_1+Bk+Bk_1-Bk}{Ak_1+Bk} \right] Z + \frac{k_1}{k}C, \qquad \chi\bar{\chi} < 1. \tag{8.396}$$

and simplify

$$\Omega = -\frac{Q_u L}{2}\left[\cos(\alpha-\beta)\frac{(A+B)k_1}{Ak+Bk_1} + i\sin(\alpha-\beta)\frac{(A+B)k_1}{Ak_1+Bk} \right] Z + \frac{k_1}{k}C, \qquad \chi\bar{\chi} < 1. \tag{8.397}$$

Flow nets are shown in Figures 8.35 and 8.36 for two cases of an ellipse in a field of uniform flow that makes an angle of 30 degress with the x-axis; for the former case the hydraulic conductivity inside the inhomogeneity is 10 times that outside, whereas it is 0.1 times that outside for the latter case. As for the circular inhomogeneity, the discharge inside the inhomogeneity is constant; the flow is uniform. We differentiate the complex potential for the inside with respect to z in order to determine the flow inside:

$$W = -\frac{d\Omega}{dz} = -\frac{d\Omega}{dZ}\frac{dZ}{dz} = -\frac{2}{L}\frac{d\Omega}{dZ}, \qquad \chi\bar{\chi} < 1. \tag{8.398}$$

We apply this to (8.397)

$$W = Q_u\left[\cos(\alpha-\beta)\frac{(A+B)k_1}{Ak+Bk_1} + i\sin(\alpha-\beta)\frac{(A+B)k_1}{Ak_1+Bk} \right], \qquad \chi\bar{\chi} < 1. \tag{8.399}$$

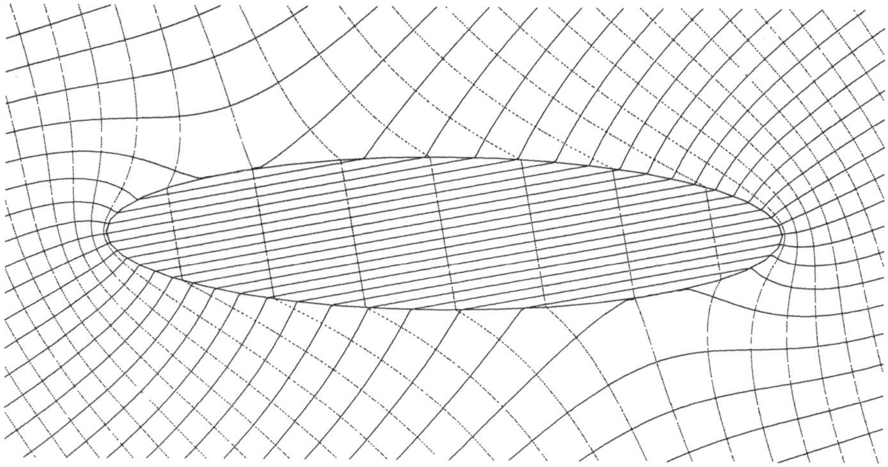

Figure 8.35 An elliptical inhomogeneity in a field of uniform flow that makes an angle of 30 degrees with the x-axis; $k_1 = 10k, \nu = 0.8$.

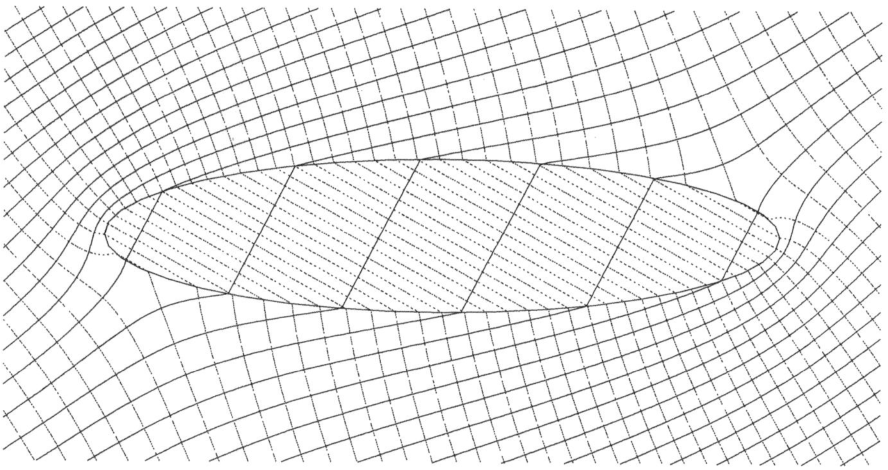

Figure 8.36 An elliptical inhomogeneity in a field of uniform flow that makes an angle of 30 degrees with the x-axis; $k_1 = 0.1k, \nu = 0.8$.

8.20.1 Limiting Case: $\nu \to 0$

The solution for the elliptical inhomogeneity should reduce to that for a circular inhomogeneity; to verify this we set $A = B$ in the expression for W and obtain

$$W = Q_u[\cos(\alpha-\beta)+\mathrm{i}\sin(\alpha-\beta)]\frac{2k_1}{k+k_1} = Q_u \mathrm{e}^{\mathrm{i}(\alpha-\beta)}\frac{2k_1}{k+k_1}, \qquad \chi\bar{\chi} \leq 1, \quad (8.400)$$

which matches the solution we obtained for the case of a circular inhomogeneity.

8.20.2 Limiting Case $k_1 \to \infty$

In the limit, for $k_1 \to \infty$, the expression for W becomes

$$W \to Q_u(A+B)\left[\cos(\alpha-\beta)\frac{1}{B} + \mathrm{i}\sin(\alpha-\beta)\frac{1}{A}\right], \qquad k_1 \to \infty, \qquad \chi\bar{\chi} \le 1. \tag{8.401}$$

Thus, the discharge approaches a maximum for $k_1 \to \infty$; this maximum depends on the orientation of the uniform flow relative to the orientation of the ellipse and the eccentricity of the ellipse. If the flow is in the direction of the major principal axis, i.e., if $\alpha - \beta = 0$, then we find an approximate expression for the total flow captured by the ellipse for large values of k, from (8.401) as

$$Wb = Q_x b \approx Q_u \frac{(A+B)}{B} b = Q_u \frac{a+b}{b} b = Q_u(a+b), \qquad \chi\bar{\chi} \le 1. \tag{8.402}$$

If the flow is in the direction of the minor principal axis, i.e., if $\alpha = 0$, $\beta = \pi/2$, then we obtain for Wa, the magnitude of which equals the flow through the ellipse,

$$Wa = -\mathrm{i}Q_y a \approx iQ_u \frac{A+B}{A} a = \mathrm{i}Q_u \frac{a+b}{a} a = \mathrm{i}Q_u(a+b), \qquad \chi\bar{\chi} \le 1. \tag{8.403}$$

We see that the flow captured by the elliptical inhomogeneity for $k_1 \to \infty$ is the same for the case in which the flow is parallel to the major axis and the case in which the flow is parallel to the minor axis. For the case where $a = b = R$, the result obtained for a circular inhomogeneity is retrieved.

8.20.3 Limiting Case $k_1 \to 0$

If the hydraulic conductivity inside the ellipse approaches zero, the expression for W becomes

$$W \to Q_u(A+B)\frac{k_1}{k}\left[\frac{\cos(\alpha-\beta)}{A} + \mathrm{i}\frac{\sin(\alpha-\beta)}{B}\right], \qquad \chi\bar{\chi} \le 1. \tag{8.404}$$

As for the case of a circular inhomogeneity, the flow inside the inhomogeneity approaches a maximum for $k_1/k \to \infty$, but will decrease as $k_1/k \to 0$ in proportion to this ratio.

8.21 Elliptical Analytic Elements

We presented circular cylindrical analytic elements in Section 8.16, where we applied a hybrid iterative approach to solve for the parameters in the expressions for individual circular lakes. Extension to elliptical lakes is not difficult, and the result is more easily applied to actual lakes that interact with both one another and other analytic elements. The main difference between lakes bounded by circles and lakes bounded by ellipses is that the parameter χ replaces Z in formulating the solution to the problem.

We use the dimensionless variables $\underset{m}{Z}$ and $\underset{m}{\chi}$,

$$\underset{m}{Z} = Z(z, \underset{m}{z_1}, \underset{m}{z_2}), \qquad \underset{m}{\chi} = \chi(\underset{m}{Z}). \tag{8.405}$$

The complex potential for an elliptical lake is holomorphic outside the bounding ellipse, except for the logarithmic singularity at infinity, which reflects the discharge of the lake. The complex potential for element m is thus holomorphic outside the unit circle in the $\underset{m}{\chi}$-plane. We represent the discharge captured by lake m as $\underset{m}{Q}$,

$$\underset{m}{\Omega}(\underset{m}{\chi}) = \sum_{j=1}^{\infty} \alpha_j \underset{m}{\chi}^{-j} + \frac{\underset{m}{Q}}{2\pi} \ln \frac{\underset{m}{\chi}}{\underset{m}{\chi_0}}, \tag{8.406}$$

where $\underset{m}{\chi_0}$ represents the location in the $\underset{m}{\chi}$-plane that corresponds to the reference point, i.e., the point in the physical plane where the head is fixed. Both the head at and the location of the reference point control the amount of flow from infinity. If the location of the reference point is z_0, then

$$\underset{m}{\chi_0} = \chi\left[\underset{m}{Z}\left(z_0, \underset{m}{z_1}, \underset{m}{z_2}\right)\right]. \tag{8.407}$$

We chose the denominator in the logarithm of (8.406) such that the influence of the discharge of lake m is zero at the reference point.

We write the complete complex potential as the sum of M lakes plus a constant and a function Ω_g, which represents the influence of other analytic elements that we consider given:

$$\Omega = \sum_{m=1}^{M} \sum_{j=1}^{\infty} \underset{m}{\alpha_j} \underset{m}{\chi}^{-j} + \sum_{m=1}^{M} \underset{m}{Q} \frac{1}{2\pi} \ln \frac{\underset{m}{\chi}}{\underset{m}{\chi_0}} + C + \Omega_g. \tag{8.408}$$

We truncate the asymptotic expansion in (8.408) to a finite number, N, of terms, and consider the problem of determining the M times N complex constants $\underset{m}{\alpha_j}$ in (8.408) and the M constants $\underset{m}{Q}$. We prefer to solve problems of this kind iteratively, as the number of unknowns in the system can be large. The advantage of an iterative approach is that it circumvents the need to solve a large system of equations, but its disadvantage is its slow convergence: the influence of the logarithm in (8.406) grows with distance from the element and is sensitive to the location of the reference point. We deal with this issue by separating the problem of solving for the discharges from that of solving for the coefficients.

We expand the complex potential, composed of all elements except element m, $\underset{\neq m}{\Omega}$, about the center of the circle in the $\underset{m}{\chi}$-plane:

$$\underset{\neq m}{\Omega} = \sum_{j=1}^{\infty} \underset{m}{a_j} \underset{m}{\chi}^j. \tag{8.409}$$

We used a Cauchy integral of the complex potential when solving for the constants for each element when modeling circular cylindrical lakes; see (8.329). We apply the integral to the present case:

$$\underset{m}{a_j} = \frac{1}{\pi} \int_0^{2\pi} \mathrm{e}^{-\mathrm{i}j\theta} \Re \underset{\neq m}{\Omega}\left(z(\underset{m}{\chi})\right) d\theta, \qquad j = 1, 2, \ldots, \infty, \tag{8.410}$$

and

$$\underset{m}{a_0} = \frac{1}{2\pi} \int_0^{2\pi} \Re\left(\underset{\neq m}{\Omega}\left(z(\underset{m}{\chi})\right)\right) d\theta. \tag{8.411}$$

The complete complex potential on the boundary of ellipse m is the sum of $\underset{\neq m}{\Omega}$ and $\underset{m}{\Omega}$,

$$\Omega = \underset{\neq m}{\Omega} + \underset{m}{\Omega} = \underset{m}{a_0} + \sum_{j=1}^{\infty} \underset{m}{a_j} \underset{m}{\chi}^j + \sum_{j=1}^{\infty} \underset{m}{\alpha_j} \underset{m}{\chi}^{-j} + \frac{\underset{m}{Q}}{2\pi} \ln \frac{\underset{m}{\chi}}{\underset{m}{\chi_0}}. \tag{8.412}$$

The boundary of ellipse m corresponds to $\underset{m}{\chi}\underset{m}{\bar{\chi}} = 1$ so that

$$\Omega = \sum_{j=1}^{\infty} \left[\underset{m}{a_j} \underset{m}{\chi}^j + \underset{m}{\alpha_j} \underset{m}{\bar{\chi}}^j \right] + \frac{\underset{m}{Q}}{2\pi} \ln \frac{\underset{m}{\chi}}{\underset{m}{\chi_0}}. \tag{8.413}$$

The real part of this expression is constant if

$$\underset{m}{\alpha_j} = -\underset{m}{\bar{a}_j}, \qquad j = 1, 2, \ldots, \tag{8.414}$$

so that the potential is constant along the boundary of ellipse m. The boundary condition is

$$\Re\Omega = \Phi_m, \qquad \underset{m}{\chi}\underset{m}{\bar{\chi}} = 1, \tag{8.415}$$

where Φ_m is the specified value of the potential. We choose the discharge $\underset{m}{Q}$ such that the latter equation is satisfied. We cycle through all elements to complete an iteration loop, and continue with the loops until the constants do not vary more than some specified amount.

8.21.1 Hybrid Iterative Approach

As for the cylindrical analytic elements, we apply a hybrid iterative approach that is a minor modification of that proposed by Bandilla et al. [2007]. The difference between that approach and the fully iterative one is that we solve directly for the discharges of all elements prior to each full iteration loop. We do this in such a way that the average value of the potential along each circle matches the specified value. Note that this implies that the value along the corresponding ellipse is a weighted average, with emphasis placed on the areas near the foci.

If there are no head-specified elements other than the elliptical lakes, the system of equations for the direct solve has the form

$$\Phi_m = \frac{1}{2\pi}\int_0^{2\pi} \Re\Omega^*\left(z(\underset{m}{\chi})\right) d\theta = \sum_{n=1}^{M} A_{mn}\underset{n}{Q}, \qquad m = 1,2,\ldots, M. \tag{8.416}$$

The integral in this equation represents the contribution to the average value of the current potential along the boundary of element m, but with the discharges $\underset{m}{Q}$ all set to zero (represented as Ω^* in (8.416)). We define the matrix A_{mn} as

$$A_{mn} = \frac{1}{2\pi}\int_0^{2\pi} \ln\left|\frac{\underset{n}{\chi}(z(\underset{m}{\chi}))}{\underset{n}{\chi_0}}\right| d\theta, \qquad \underset{m}{\chi} = \mathrm{e}^{\mathrm{i}\theta}, \quad m = 1,\ 2,\ldots,\ M, \quad n = 1,\ 2,\ldots,\ M. \tag{8.417}$$

A_{nm} represents the coefficient function of $\underset{n}{Q}$,[1] integrated along the boundary of element m. This integral is divided by the circumference of the unit circle, 2π, and is the contribution of $\underset{n}{Q}$ to the average potential along the boundary of element m.

The complex potential obtained in this way is illustrated in Figure 8.37, where a set of ten lakes is shown; the values of the discharge potentials divided by kH^2 are indicated inside the ellipses. The potential at the reference point, $\Phi_0/(kH^2)$, is 20; the discharge $Q_{x0}/(kH)$ is 0.04; the number of terms in the expansions is 20; and the number of intervals to evaluate the integrals is 100. The coordinates of the window shown in the plot are $-100 \le x/H \le 100$ and $-100 \le y/H \le 100$. Note the branch cuts in the stream function, which appear as heavy black lines in the figure.

8.21.2 Analytic Elements for Impermeable Elliptical Objects in a Field of Uniform Flow

We apply the analytic element presented for lakes with minor modification to solve problems with impermeable cylindrical objects. The functions have the same form as for the lakes, but without the logarithmic term,

[1] We use the term *coefficient function* of some unknown coefficient, Q_m in this case, to indicate the function that the coefficient is multiplied by.

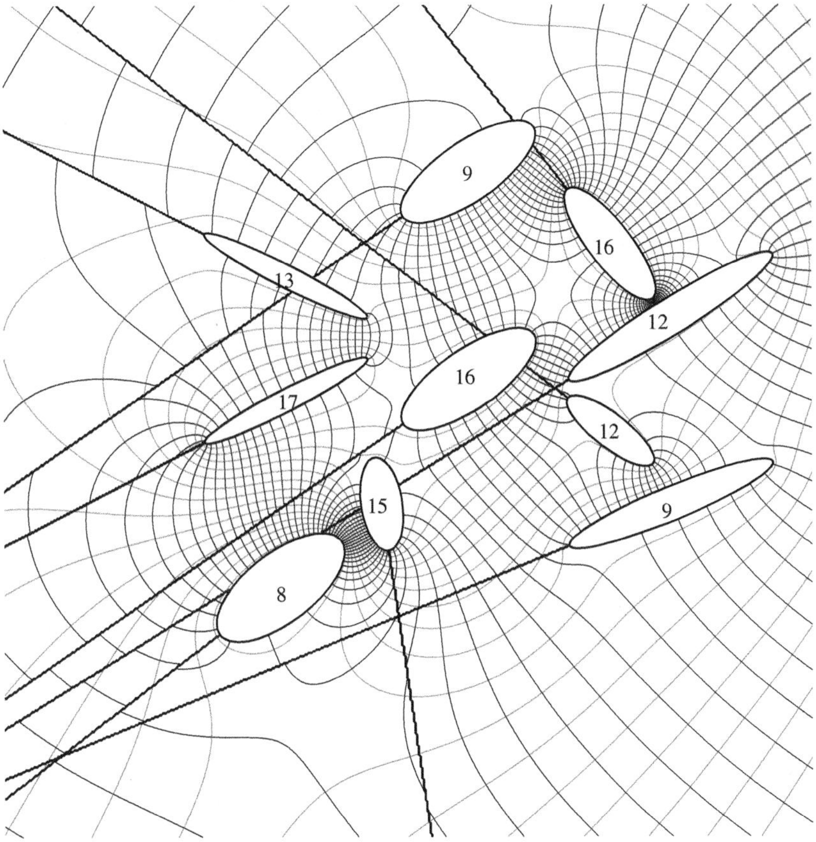

Figure 8.37 Flow net for ten elliptical lakes; the data are given in the text.

$$\underset{m}{\Omega}(\underset{m}{\chi}) = \sum_{j=1}^{\infty} \alpha_j \underset{m}{\chi}^{-j}. \tag{8.418}$$

The total complex potential is

$$\Omega = \sum_{m=1}^{M}\sum_{j=1}^{\infty} \underset{m}{\alpha_j} \underset{m}{\chi}^{-j} + C + \Omega_g. \tag{8.419}$$

We obtain equations for the complex constants $\underset{m}{a_j}$ in the expansion of $\underset{\neq m}{\Omega}$ from the Cauchy integral in terms of the stream function along the boundary of each ellipse:

$$\underset{m}{a_j} = \frac{\mathrm{i}}{\pi}\int_0^{2\pi} \mathrm{e}^{-\mathrm{i}j\theta}\Im \underset{\neq m}{\Omega}\left(z(\underset{m}{\chi})\right) d\theta, \qquad j = 1,2,\ldots,\infty. \tag{8.420}$$

The condition that the stream function is constant along the boundary of ellipse m requires that

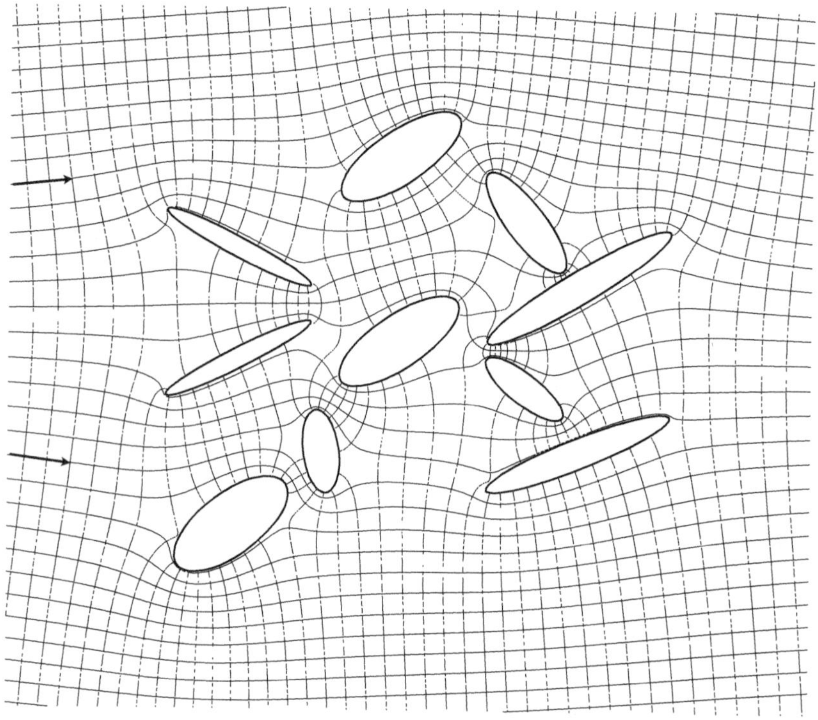

Figure 8.38 A field of uniform flow in the x-direction with ten impermeable elliptical objects.

$$\underset{m}{\alpha_j} = -\underset{m}{\bar{a}_j}, \qquad j = 1, 2, \ldots. \tag{8.421}$$

A case of ten impermeable ellipses in a field of uniform flow is illustrated in Figure 8.38. The data are the same as for the elliptical lakes, Figure 8.37. Note that this approach needs to be modified if elements are present that extract or supply water; the branch cuts must then be taken into account.

8.21.3 An Analytic Element for an Inhomogeneity

The elliptic analytic element can be applied to inhomogeneities in the hydraulic conductivity. Strack [1989] introduced these elements; later implementations are presented in Suribhadla et al. [2004] and Strack [2005]. The forms of the functions are the same as for the elliptical lakes, except that we include the inside of the ellipse in the complex potential.

The circle of radius ν in the χ-plane represents a slot in the z-plane; representing the complex potential inside the circle by a Taylor series will cause an unrealistic jump across this slot. This jump can be avoided by writing the series inside the ellipse as a Laurent series (see (8.96)), as the general representation of a function that is holomorphic in the annulus between two circles:

$$\Omega = \sum_{m=-\infty}^{\infty} \alpha_m \chi^m = \alpha_0 + \sum_{m=1}^{\infty} \left[\alpha_m \chi^m + \alpha_{-m} \chi^{-m}\right]. \tag{8.422}$$

The coefficients in this series are determined from the boundary conditions along the circles $\chi\bar{\chi} = \nu^2$ and $\chi\bar{\chi} = 1$. The boundary condition along the inner circle is that there be no discontinuity across the slot that corresponds to the circle. Since $-1 \leq X \leq 1$, $Y = 0^+$, corresponds to the semi-circle $\Im\chi \geq 0$, and $-1 \leq X \leq 1$, $Y = 0^-$, corresponds to $\Im\chi \leq 0$, the condition is:

$$\Omega(\chi) = \Omega(\bar{\chi}), \qquad \chi\bar{\chi} = \nu^2. \tag{8.423}$$

We apply this condition to (8.422):

$$\alpha_0 + \sum_{m=1}^{\infty} \left[\alpha_m \chi^m + \alpha_{-m} \chi^{-m}\right] = \alpha_0 + \sum_{m=1}^{\infty} \left[\alpha_m \nu^{2m} \chi^{-m} + \frac{\alpha_{-m}}{\nu^{2m}} \chi^m\right]. \tag{8.424}$$

This condition is fulfilled if

$$\alpha_{-m} = \nu^{2m} \alpha_m \tag{8.425}$$

so that

$$\Omega = \sum_{m=0}^{\infty} \alpha_m \left[\chi^m + \frac{\nu^{2m}}{\chi^m}\right], \qquad \chi\bar{\chi} \leq 1. \tag{8.426}$$

Note that we redefined α_0. This complex potential is continuous across the slot in the z-plane, since replacing χ by $\bar{\chi} = \nu^2/\chi$ does not change the expression for Ω.

We represent the complex potential outside the ellipse as an asymptotic expansion:

$$\Omega = \sum_{m=1}^{\infty} \beta_m \chi^{-m}, \qquad \chi\bar{\chi} > 1. \tag{8.427}$$

There must be continuity of flow across the ellipse; i.e., the stream function must be continuous across $\chi\bar{\chi} = 1$. The jump in complex potential across the ellipse is

$$\Omega^+ - \Omega^- = \sum_{m=0}^{\infty} \alpha_m \left[\chi^m + \frac{\nu^{2m}}{\chi^m}\right] - \sum_{m=1}^{\infty} \beta_m \chi^{-m}, \qquad \chi\bar{\chi} = 1, \tag{8.428}$$

or, with $1/\chi = \bar{\chi}$,

$$\Omega^+ - \Omega^- = 2\alpha_0 + \sum_{m=1}^{\infty} \left[\alpha_m \chi^m + \left(\alpha_m \nu^{2m} - \beta_m\right) \bar{\chi}^m\right], \qquad \chi\bar{\chi} = 1. \tag{8.429}$$

Since the stream function is continuous, the jump in complex potential must be real:

$$\alpha_m \nu^{2m} - \beta_m = \bar{\alpha}_m \to \beta_m = \alpha_m \nu^{2m} - \bar{\alpha}_m, \qquad m = 1, 2, \ldots, \infty, \qquad \Im\alpha_0 = 0, \tag{8.430}$$

so that the complex potential outside the inhomogeneity is

$$\Omega = \sum_{m=1}^{\infty} (\alpha_m \nu^{2m} - \bar{\alpha}_m)\chi^{-m}, \qquad \chi\bar{\chi} > 1. \tag{8.431}$$

The boundary condition is that the jump in potential equals $(k_{\text{in}} - k_{\text{out}})/k_{\text{out}})\Phi^-$,

$$\Phi^+ - \Phi^- = \sum_{m=0}^{\infty} (\alpha_m \chi^m + \bar{\alpha}_m \bar{\chi}^m) = \frac{k_{\text{in}} - k_{\text{out}}}{k_{\text{out}}} \Phi^-, \qquad \chi\bar{\chi} = 1. \tag{8.432}$$

The potential on the outside is the sum of the contribution of the element considered and the contributions of all other elements, which are represented by a Taylor expansion about the center of the circle, with complex coefficients a_m:

$$\Phi^- = \sum_{m=1}^{\infty} \{\tfrac{1}{2}[(\alpha_m \nu^{2m} - \bar{\alpha}_m)\,\bar{\chi}^m\,(\bar{\alpha}_m \nu^{2m} - \alpha_m)\,\chi^m]\} + \sum_{m=0}^{\infty} \tfrac{1}{2}(a_m \chi^m + \bar{a}_m \bar{\chi}^m), \qquad \chi\bar{\chi} = 1. \tag{8.433}$$

We define the factor μ as

$$\mu = \frac{k_{\text{in}} - k_{\text{out}}}{k_{\text{out}}}. \tag{8.434}$$

The condition (8.432) must be satisfied for each power $m \geq 1$:

$$\begin{aligned} \alpha_m - \tfrac{1}{2}\mu\left(\bar{\alpha}_m \nu^{2m} - \alpha_m\right) &= \tfrac{1}{2}\mu a_m, \qquad m = 1, 2, \ldots \\ \bar{\alpha}_m - \tfrac{1}{2}\mu\left(\alpha_m \nu^{2m} - \bar{\alpha}_m\right) &= \tfrac{1}{2}\mu \bar{a}_m, \qquad m = 1, 2, \ldots. \end{aligned} \tag{8.435}$$

We solve this for α_m:

$$\alpha_m = \tfrac{1}{2}\mu \frac{\left(1 + \tfrac{1}{2}\mu\right) a_m + \tfrac{1}{2}\mu \nu^{2m} \bar{a}_m}{(1 + \tfrac{1}{2}\mu)^2 - \left(\tfrac{1}{2}\mu \nu^{2m}\right)^2}, \qquad m = 1, 2, \ldots. \tag{8.436}$$

We again use a Cauchy integral to compute the coefficients a_m. We compute the coefficient α_0 separately from (8.428) with (8.433) for $m = 0$. This gives

$$\alpha_0 = \tfrac{1}{4}\mu(a_0 + \bar{a}_0). \tag{8.437}$$

We implemented this in a MATLAB progam; a flow net is shown for a case of ten ellipsoidal inhomogeneities in Figure 8.39.

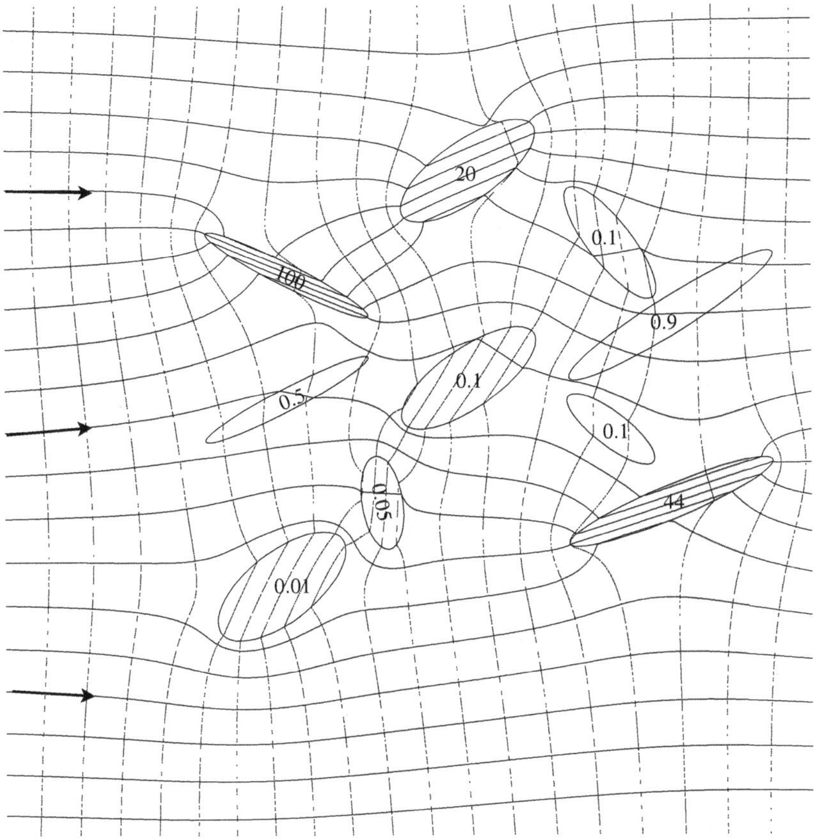

Figure 8.39 Flow in the x-direction through a domain with ten elliptical inhomogeneities. The numbers inside the ellipses represent the ratio $\rho = k_{in}/k_{out}$. The potentials inside the ellipses are multiplied by a factor $1/\rho$ in order to obtain a continuous function.

8.22 An Elliptical Pond

We consider an elliptical area of infiltration, an elliptical pond, as the final application of elements bounded by an ellipse. Elements with given infiltration differ fundamentally from elements with boundary conditions that are affected by other elements, e.g., boundaries of given head. Elements of given infiltration, such as circular ponds, maintain their boundary conditions if they are combined with other elements; these boundary conditions involve continuity of flow and head, and each element satisfies these conditions by itself. We call these elements elliptical ponds; they are more useful than circular ponds because their boundaries can be made to match more closely the actual ponds they represent. The solution for an elliptical pond was presented by Strack [2009a]. We refer to the latter reference for a detailed description of the approach used to obtain the solution; a presentation of the solution along with the demonstration that the solution meets the boundary conditions suffices here. The conditions along the boundary of the ellipse are that there be continuity of flow and continuity of head along the boundary of the ellipse, which corresponds to the unit circle in the χ-plane:

$$W^+ = W^-, \qquad \chi\bar{\chi} = 1. \tag{8.438}$$

Since the derivative $d\chi/dz$ is continuous across the boundary of the ellipse, we may replace (8.438) by

$$\left(\frac{\partial \Phi}{\partial \chi}\right)^+ = \left(\frac{\partial \Phi}{\partial \chi}\right)^- . \tag{8.439}$$

There is infiltration inside the ellipse; the potential must satisfy the Poisson equation there:

$$\nabla^2 \Phi = 4\frac{\partial^2 \Phi}{\partial z \partial \bar{z}} = 4\frac{\partial^2 \Phi}{\partial Z \partial \bar{Z}}\frac{dZ}{dz}\frac{d\bar{Z}}{d\bar{z}} = \frac{16}{L^2}\frac{\partial^2 \Phi}{\partial Z \partial \bar{Z}} = -N \tag{8.440}$$

so that

$$\frac{\partial^2 \Phi}{\partial Z \partial \bar{Z}} = -\frac{L^2}{16}N, \qquad \chi\bar{\chi} \le 1. \tag{8.441}$$

The potential satisfies Laplace's equation outside the ellipse and we demonstrate that it can be represented as the real part of the following complex potential,

$$\Phi = \Re\Omega, \qquad \Omega = -\frac{Q}{2\pi}\left[\ln\chi + \tfrac{1}{2}\left(\frac{\nu}{\chi}\right)^2\right], \qquad \chi\bar{\chi} \ge 1. \tag{8.442}$$

The potential inside the ellipse is given by

$$\begin{aligned}\Phi = -\frac{L^2}{64}N\Bigg\{&\left(\frac{\chi}{\nu}+\frac{\nu}{\chi}\right)\left(\frac{\bar{\chi}}{\nu}+\frac{\nu}{\bar{\chi}}\right) - \frac{\nu^2}{2}\left[\left(\frac{\chi}{\nu}\right)^2+\left(\frac{\nu}{\chi}\right)^2+\left(\frac{\bar{\chi}}{\nu}\right)^2+\left(\frac{\nu}{\bar{\chi}}\right)^2\right]\\ &-\left(\frac{1}{\nu^2}+\nu^2\right)\Bigg\}, \qquad \chi\bar{\chi} \le 1.\end{aligned} \tag{8.443}$$

We use the relationship between χ and Z,

$$Z = \tfrac{1}{2}\left(\frac{\chi}{\nu}+\frac{\nu}{\chi}\right), \qquad Z^2 = \tfrac{1}{4}\left\{\left(\frac{\chi}{\nu}\right)^2 + 2 + \left(\frac{\nu}{\chi}\right)^2\right\} \tag{8.444}$$

to rewrite the expression for Φ inside the ellipse in terms of Z:

$$\Phi = -\frac{NL^2}{16}\left\{Z\bar{Z} - \frac{\nu^2}{2}[Z^2+\bar{Z}^2-4] - \frac{1}{\nu^2} - \nu^2\right\}. \tag{8.445}$$

We obtain on differentiation with respect to Z and then $\bar{Z}$:

$$\frac{\partial^2 \Phi}{\partial Z \partial \bar{Z}} = -\frac{L^2}{16} N, \tag{8.446}$$

which matches (8.441); the differential equation is satisfied.

We verify that the discharge function is continuous by first differentiating Ω, valid for $\chi \bar{\chi} > 1$, with respect to χ:

$$\frac{\partial \Phi}{\partial \chi} = \frac{1}{2} \frac{d\Omega}{d\chi} = -\frac{Q}{4\pi} \left[\frac{1}{\chi} - \nu^2 \frac{1}{\chi^3} \right], \qquad \chi \bar{\chi} \geq 1. \tag{8.447}$$

We next differentiate expression (8.443), valid for $\chi \bar{\chi} \leq 1$, with respect to χ,

$$\frac{\partial \Phi}{\partial \chi} = -\frac{L^2}{64} N \left\{ \left(\frac{1}{\nu} - \frac{\nu}{\chi^2} \right) \left(\frac{\bar{\chi}}{\nu} + \frac{\nu}{\bar{\chi}} \right) - \nu^2 \left[\frac{\chi}{\nu^2} - \frac{\nu^2}{\chi^3} \right] \right\}, \qquad \chi \bar{\chi} \leq 1. \tag{8.448}$$

The latter two equations must match for $\chi \bar{\chi} = 1$; we replace $\bar{\chi}$ by $1/\chi$:

$$\frac{\partial \Phi}{\partial \chi} = -\frac{L^2}{64} N \left\{ \left(\frac{1}{\nu} - \frac{\nu}{\chi^2} \right) \left(\frac{1}{\chi \nu} + \nu \chi \right) - \nu^2 \left[\frac{\chi}{\nu^2} - \frac{\nu^2}{\chi^3} \right] \right\}, \qquad \chi \bar{\chi} \leq 1. \tag{8.449}$$

We simplify this

$$\frac{\partial \Phi}{\partial \chi} = -\frac{L^2}{64} N \left[\frac{1}{\nu^2 \chi} + \chi - \frac{1}{\chi^3} - \frac{\nu^2}{\chi} - \chi + \frac{\nu^4}{\chi^3} \right] \tag{8.450}$$

or

$$\frac{\partial \Phi}{\partial \chi} = -\frac{L^2}{64} N \left\{ \frac{1}{\chi} \left(\frac{1}{\nu^2} - \nu^2 \right) - \frac{1}{\chi^3} (1 - \nu^4) \right\}. \tag{8.451}$$

We take a term $(1 - \nu^4)/\nu^2$ outside the braces:

$$\frac{\partial \Phi}{\partial \chi} = -\frac{L^2 N}{64} \frac{1 - \nu^4}{\nu^2} \left\{ \frac{1}{\chi} - \frac{\nu^2}{\chi^3} \right\}. \tag{8.452}$$

We use (8.350) through (8.353), which relate ν and L to the axes A and B in the Z-plane:

$$A = \frac{1 + \nu^2}{2\nu}, \quad B = \frac{1 - \nu^2}{2\nu}, \quad a = \tfrac{1}{2} L A, \quad b = \tfrac{1}{2} L B, \quad \tfrac{1}{4} L^2 A B = ab \tag{8.453}$$

and

$$\frac{1 - \nu^4}{\nu^2} = 4AB = \frac{16ab}{L^2}. \tag{8.454}$$

We note that $Q = \pi a b N$ and write (8.350) as

$$\frac{\partial \Phi}{\partial \chi} = -\frac{Q}{4\pi} \left[\frac{1}{\chi} - \nu^2 \frac{1}{\chi^3} \right]. \tag{8.455}$$

This equation matches (8.447), so that the complex discharge is indeed continuous across the boundary of the ellipse. Because both the tangential and normal components of flow are continuous, the potential can only jump by a constant amount. We verify that this constant jump is zero, by setting $\chi = \mathrm{e}^{\mathrm{i}\theta}$ in both equations for Φ; the real part of (8.442) is

$$\Phi = -\frac{Q}{4\pi}\nu^2\cos(2\theta). \tag{8.456}$$

We work out (8.443)

$$\Phi = -\frac{L^2N}{64}\left\{\frac{\chi\bar{\chi}}{\nu^2} + \frac{\chi}{\bar{\chi}} + \frac{\bar{\chi}}{\chi} + \frac{\nu^2}{\chi\bar{\chi}} - \frac{\nu^2}{2}(\chi^2+\bar{\chi}^2)\left(\frac{1}{\nu^2}+\nu^2\right) - \frac{1}{\nu^2} - \nu^2\right\}, \tag{8.457}$$

and set $\chi\bar{\chi} = 1$ and $\chi = \mathrm{e}^{\mathrm{i}\theta}$:

$$\Phi = -\frac{L^2N}{64}\left\{\nu^2 + \frac{1}{\nu^2} + 2\cos(2\theta) - (1+\nu^4)\cos(2\theta) - \frac{1}{\nu^2} - \nu^2\right\} \tag{8.458}$$

or

$$\Phi = -\frac{L^2N}{64}(1-\nu^4)\cos(2\theta) = -\frac{abN}{4}\nu^2\cos(2\theta) = -\frac{Q}{4\pi}\nu^2\cos(2\theta). \tag{8.459}$$

This matches the expression for Φ obtained from the potential on the outside.

8.22.1 Application: An Elliptical Pond in a Field of Uniform Flow

We consider as an application the case of an elliptical pond in a field of uniform flow; the major axis of the pond makes an angle α with the x-axis, and the uniform flow field is given by W_0. We obtain the solution by adding the complex potential for uniform flow to the solution for the elliptical pond. The complex potential outside the pond is:

$$\Omega = -W_0 z - \frac{Q}{2\pi}\left[\ln\chi + \tfrac{1}{2}\left(\frac{\nu}{\chi}\right)^2\right], \qquad \chi\bar{\chi} \geq 1. \tag{8.460}$$

We write the potential inside the pond in terms of z and Z (see (8.445)),

$$\Phi = -\tfrac{1}{2}(W_0 z + \bar{W}_0\bar{z}) - \frac{NL^2}{16}\left\{Z\bar{Z} - \frac{\nu^2}{2}\left[Z^2+\bar{Z}^2-4\right] - \frac{1}{\nu^2} - \nu^2\right\}, \qquad \chi\bar{\chi} < 1. \tag{8.461}$$

A flow net outside the pond and equipotentials inside the pond are shown in Figure 8.40; the data are $W_0/(kH) = Q_{x0}/(kH) = 0.4$, $N/k = 0.01$, $a/H = 105$, $b/H = 30$, $z_1/H = -(100+10\mathrm{i})$, and $z_2/H = -z_1/H$.

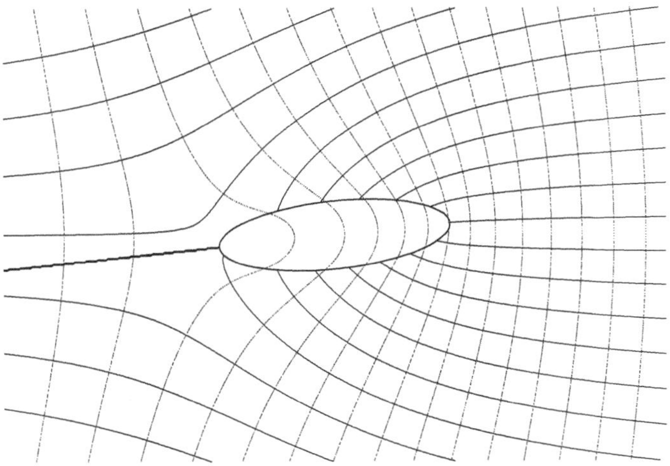

Figure 8.40 Elliptical pond in a field of uniform flow.

Stagnation Points

We determine whether stagnation points exist, and, if so, where they are. We assume that the stagnation point is outside the pond and we apply the chain rule to differentiate the complex potential (8.460) with respect to z:

$$W = -\frac{d\Omega}{dz} = W_0 - \frac{d\Omega}{d\chi}\frac{d\chi}{dZ}\frac{dZ}{dz} = W_0 + \frac{Q}{2\pi}\left[\frac{1}{\chi} - \frac{\nu^2}{\chi^3}\right]\frac{d\chi}{dZ}\frac{2}{L}\mathrm{e}^{-\mathrm{i}\alpha}, \quad \chi\bar{\chi} \geq 1, \tag{8.462}$$

where α is the angle between the major principal axis of the ellipse and the x-axis. We differentiate the first equation in (8.444) with respect to Z,

$$\frac{dZ}{d\chi} = \frac{1}{2}\left[\frac{1}{\nu} - \frac{\nu}{\chi^2}\right] = \frac{1}{2}\frac{\chi^2 - \nu^2}{\nu\chi^2}. \tag{8.463}$$

We use this in (8.462), with $d\chi/dZ = 1/(dZ/d\chi)$,

$$W = W_0 + \frac{Q}{2\pi}\left[\frac{1}{\chi} - \frac{\nu^2}{\chi^3}\right]\frac{2\nu\chi^2}{\chi^2 - \nu^2}\frac{2}{L}\mathrm{e}^{-\mathrm{i}\alpha} = W_0 + \frac{2Q}{\pi L}\frac{\chi^2 - \nu^2}{\chi^3}\frac{\nu\chi^2}{\chi^2 - \nu^2}\mathrm{e}^{-\mathrm{i}\alpha}, \quad \chi\bar{\chi} \geq 1. \tag{8.464}$$

and simplify:

$$W = W_0 + \frac{2Q}{\pi L}\frac{\nu}{\chi}\mathrm{e}^{-\mathrm{i}\alpha}, \quad \chi\bar{\chi} \geq 1. \tag{8.465}$$

The stagnation point occurs where $W = 0$,

$$\chi_s = -\frac{Q\nu}{\pi W_0}\frac{2}{L}\mathrm{e}^{-\mathrm{i}\alpha}, \qquad \chi_s\bar{\chi}_s \geq 1. \tag{8.466}$$

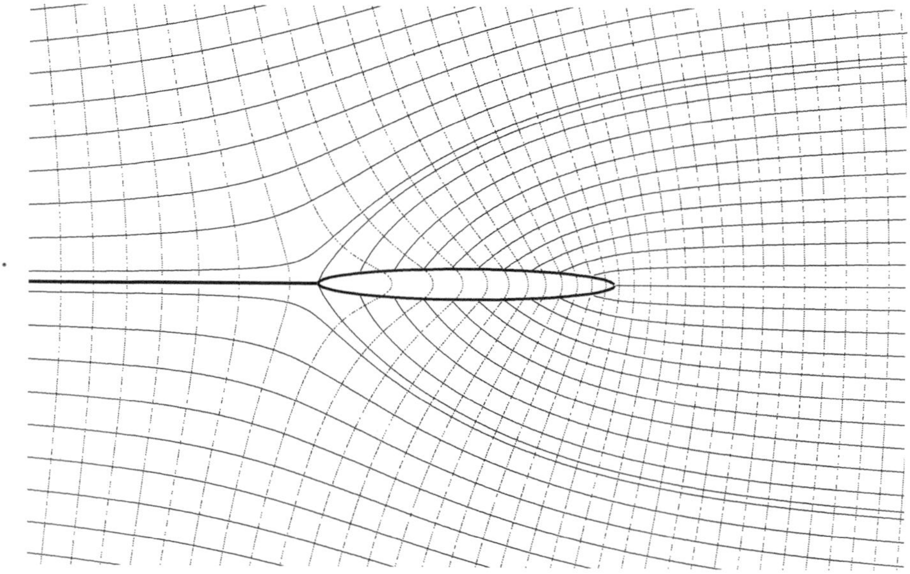

Figure 8.41 Flow net for a pond with its major axis aligned with the flow with the stagnation point on the boundary.

A stagnation point exists outside the pond if

$$\frac{Q\nu}{\pi|W_0|}\frac{2}{L} \geq 1. \tag{8.467}$$

The stagnation point reaches the boundary of the pond when the equal sign applies.

The stagnation point on the inside will reach the boundary under the same conditions as for the one on the outside, because W is continuous at the boundary. Therefore, either there will be no stagnation points or there will be both one on the outside and one on the inside of the ellipse. The condition (8.467) is independent of the angle that the uniform flow makes with the major axis of the ellipse. We illustrate two cases where the stagnation point is on the boundary. For the first case, shown in Figure 8.41, the ellipse is aligned with the flow. The data for this case are $W_0 = 0.1$ m^2/d, $L = 100$ m, $a = 50.2$ m, $b = 5.02$ m, $N = 0.0198$ m/d. The flow net in Figure 8.42 is constructed with the same data as Figure 8.41, but the pond is oriented normal to the flow. Again, the stagnation point is on the boundary. For the limit, where $a \to b = R$, i.e., when the ellipse becomes a circle, we have

$$\lim_{\substack{\nu\to 0\\ L\to 0}} a = \lim_{\substack{\nu\to 0\\ L\to 0}} \frac{AL}{2} = \lim_{\substack{\nu\to 0\\ L\to 0}} \tfrac{1}{4}\left(\nu + \frac{1}{\nu}\right)L = \lim_{\substack{\nu\to 0\\ L\to 0}} \frac{L}{4\nu} = R, \tag{8.468}$$

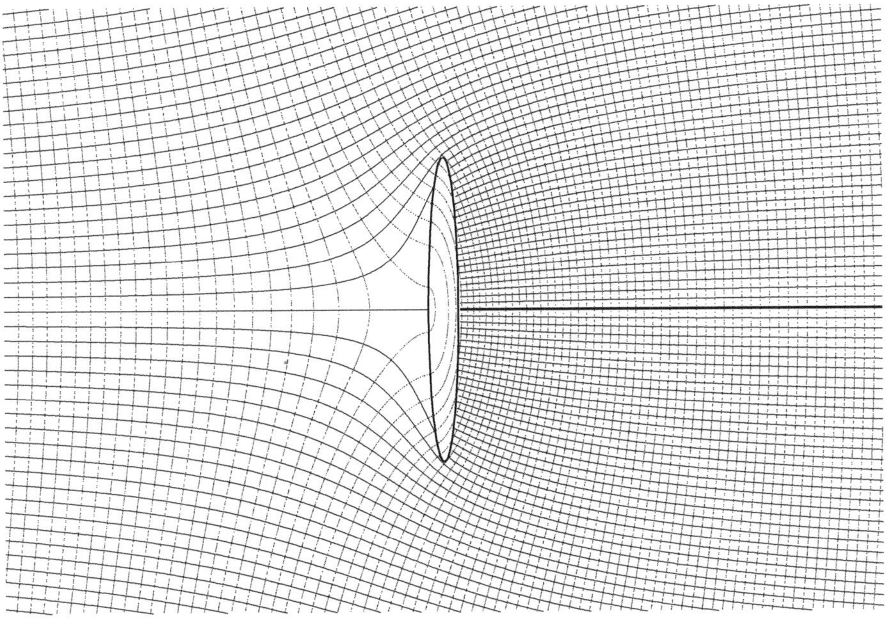

Figure 8.42 Flow net for a pond with its major axis normal to the flow with the stagnation point on the boundary.

so that the limiting condition is

$$\lim_{\substack{\nu\to 0\\ L\to 0}} \frac{L}{\nu} = 4R. \tag{8.469}$$

We apply this limit to (8.467),

$$\frac{NR}{2|W_0|} \geq 1. \tag{8.470}$$

This result matches the one obtained for a circular pond, (2.268).

8.23 The Line-Sink

A line-sink captures groundwater flow along a line. The rate of inflow into the line may vary in reality but we consider here only line-sinks of constant extraction rate, called line-sinks of constant strength. Probably the first to apply the line-sink to a groundwater flow problem was Irmay [1960]. Applications of line-sinks to boundary value problems were given by van der Veer [1978] in terms of a complex integral and by Liggett [1977] in terms of a real one. Strack and Haitjema [1981] used line-sinks for modeling creeks and rivers in a regional flow model.

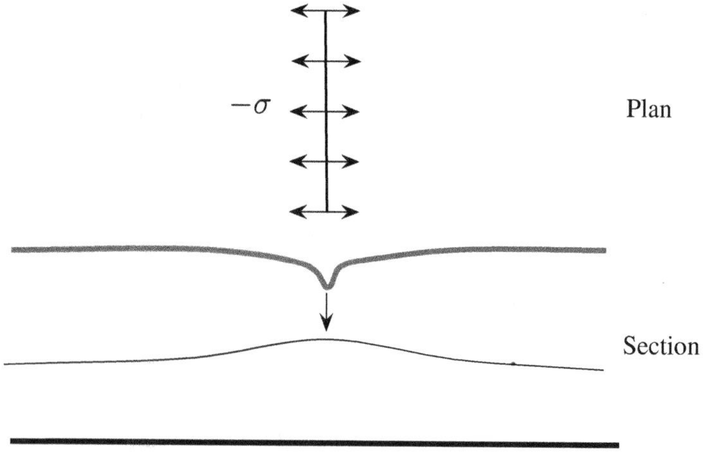

Figure 8.43 A narrow stream above the water table.

We may view line-sinks as the limiting case of a series of wells that are distributed along the line-sink; as opposed to a well, the inflow takes place uniformly along the line rather than being concentrated at a point. As for wells, the sign of the discharge governs whether inflow or outflow occurs; a line-source is a line-sink along which water flows into the aquifer. An example of an application of a line-sink is a narrow creek with its bottom above an unconfined aquifer. The infiltration rate into the aquifer is constant in space if both the resistance of the creek bed and the head in the creek are constant; the infiltration rate equals the difference in head across the bottom of the creek bed. The head below the creek bed is equal to the elevation of the underside of the creek bed assuming atmospheric pressure. If the creek is horizontal and the elevation of the bottom of the creek bed above the base is H_0, then the infiltration rate through the creek bed is

$$\sigma = -\frac{\phi_0 - H_0}{c} b \qquad [L^2/\mathrm{T}], \tag{8.471}$$

where c is the resistance of the creek bed [L], and b its width [L]. A line-sink used to model such a stream is illustrated in Figure 8.43. The rate of infiltration per unit length of line-sink is $-\sigma$. Note that σ is positive for extraction.

8.23.1 The Complex Potential

Consider a uniformly spaced row of n wells along a straight line as shown in Figure 8.44(a). The line is inclined at an angle α with respect to the x-axis, its length is L, and the end points are $\overset{1}{z}$ and $\overset{2}{z}$; see Figure 8.44(b). The wells all have a discharge $\sigma \Delta\xi$, where $\Delta\xi$ is the distance between the wells (see Figure 8.44(a)), and the mth well is at $z = \delta_m$. The corresponding complex potential is

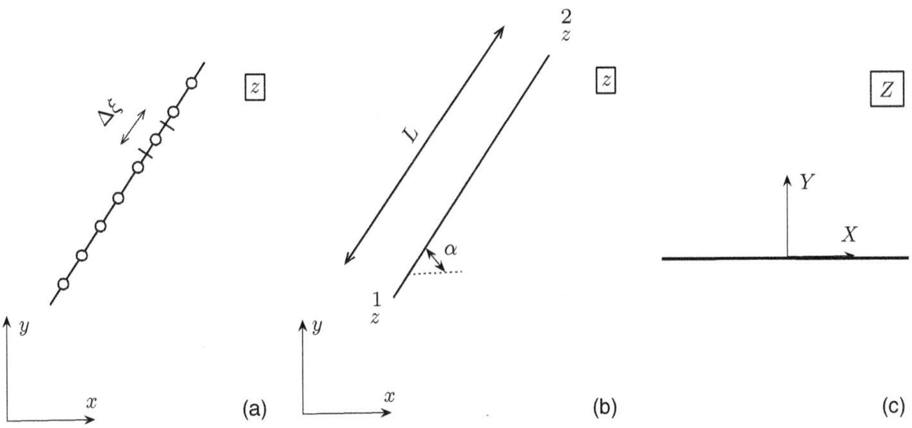

Figure 8.44 A line-sink of constant strength.

$$\Omega = \sum_{m=1}^{n} \frac{\sigma}{2\pi} \ln \frac{z - \delta_m}{\frac{1}{2}(\overset{2}{z} - \overset{1}{z})} \Delta\xi. \tag{8.472}$$

Note that we made the argument of the logarithm dimensionless by dividing the numerator by $\frac{1}{2}(\overset{2}{z} - \overset{1}{z})$; we can do this because it amounts to adding a constant to the complex potential, which does not affect the discharge vector. We made this choice in order to obtain a compact result, as will become clear in what follows. The sum becomes an integral in the limit, for $\Delta\xi \to 0$ and $n \to \infty$,

$$\Omega = \int_{-\frac{1}{2}L}^{\frac{1}{2}L} \frac{\sigma}{2\pi} \ln \frac{z - \delta}{\frac{1}{2}(\overset{2}{z} - \overset{1}{z})} d\xi, \tag{8.473}$$

where ξ is $-\frac{1}{2}L$ at $\overset{1}{z}$ and $\frac{1}{2}L$ at $\overset{2}{z}$. The complex number δ denotes a point of the line, which we express in terms of ξ as follows:

$$\delta = \xi e^{i\alpha} + \tfrac{1}{2}\left(\overset{1}{z} + \overset{2}{z}\right). \tag{8.474}$$

We introduce the following dimensionless variables:

$$Z = X + iY = \frac{z - \frac{1}{2}\left(\overset{1}{z} + \overset{2}{z}\right)}{\frac{1}{2}\left(\overset{2}{z} - \overset{1}{z}\right)}, \qquad \Delta = \Delta_1 + i\Delta_2 = \frac{\delta - \frac{1}{2}\left(\overset{1}{z} + \overset{2}{z}\right)}{\frac{1}{2}\left(\overset{2}{z} - \overset{1}{z}\right)}, \tag{8.475}$$

where

$$\overset{2}{z} - \overset{1}{z} = Le^{i\alpha}. \tag{8.476}$$

The complex variables Z and Δ are real along the element and vary between -1 at $\overset{1}{z}$ to $+1$ and $\overset{2}{z}$; see Figure 8.44(c). Note that the complex number $\frac{1}{2}\left(\overset{1}{z}+\overset{2}{z}\right)$ represents the center of the element and that $\overset{2}{z}-\overset{1}{z}$ represents a vector pointing from $\overset{1}{z}$ to $\overset{2}{z}$, which has length L and inclination α. We enter these variables in (8.473), with $d\xi = e^{-\mathrm{i}\alpha} d\delta = \frac{1}{2} L d\Delta$,

$$\Omega = \int_{-1}^{+1} \frac{\sigma}{2\pi} \ln(Z-\Delta) \tfrac{1}{2} L d\Delta. \tag{8.477}$$

The extraction rate σ is constant, and Z is kept constant during integration, so that we may write:

$$\Omega = -\frac{\sigma L}{4\pi} \int_{Z+1}^{Z-1} \ln(Z-\Delta) d(Z-\Delta), \tag{8.478}$$

where we replaced $d\Delta$ by $-d(Z-\Delta)$ in the second integral. We work out the integral in (8.478):

$$\boxed{\Omega = \frac{\sigma L}{4\pi}[(Z+1)\ln(Z+1) - (Z-1)\ln(Z-1) - 2]} \tag{8.479}$$

which represents the complex potential for a line-sink of extraction rate σ and length L, written in terms of the local complex variable Z. We refer to the extraction rate σ as the strength of the line-sink.

8.23.2 The Far Field

We examine the behavior of the complex potential (8.479) for large values of Z by first writing it in a suitable form as

$$\Omega = \frac{\sigma L}{4\pi} \left\{ Z \ln \frac{1+1/Z}{1-1/Z} + \ln \left[Z^2 \left(1 - \frac{1}{Z^2} \right) \right] - 2 \right\}, \tag{8.480}$$

and then expanding the function about $1/Z = 0$:

$$\Omega = \frac{\sigma L}{4\pi} \left\{ 2\ln Z + Z \left[\frac{2}{Z} + O\left(\frac{1}{Z^3}\right) \right] - 2 + O\left(\frac{1}{Z^2}\right) \right\}, \qquad Z \to \infty,$$

where $O(1/Z^3)$ means *order of* $1/Z^3$ *or higher*. Note that a higher order means a smaller absolute value, for $Z \to \infty$. We disregard all terms of order higher than $1/Z^2$,

$$\Omega = \frac{\sigma L}{2\pi}\ln Z + O\left(\frac{1}{Z^2}\right), \qquad Z \to \infty. \tag{8.481}$$

The complex potential approaches that of a well of discharge $Q = \sigma L$, located at the center of the line-sink for $Z \to \infty$. This is to be expected; σL represents the total discharge of the line-sink.

8.23.3 The Discharge Function

We introduce the local coordinates ξ and η and determine the discharge vector components Q_ξ and Q_η, which are parallel and normal, respectively, to the line-sink as shown in Figure 8.45. The local coordinates ξ and η differ from X and Y only in scale; the latter coordinates are dimensionless; X equals 1 at $\overset{1}{z}$, where ξ equals $L/2$,

$$\zeta = \xi + i\eta = \frac{L}{2}Z. \tag{8.482}$$

We define the discharge function W^* as:

$$W^* = Q_\xi - iQ_\eta = -\frac{d\Omega}{d\zeta} = -\frac{d\Omega}{dZ}\frac{dZ}{d\zeta} = -\frac{2}{L}\frac{d\Omega}{dZ}. \tag{8.483}$$

We use the expression for Ω in terms of Z, (8.479):

$$Q_\xi - iQ_\eta = \frac{\sigma}{2\pi}\ln\frac{Z-1}{Z+1} = \frac{\sigma}{2\pi}\left[\ln\left|\frac{Z-1}{Z+1}\right| + i\theta\right]. \tag{8.484}$$

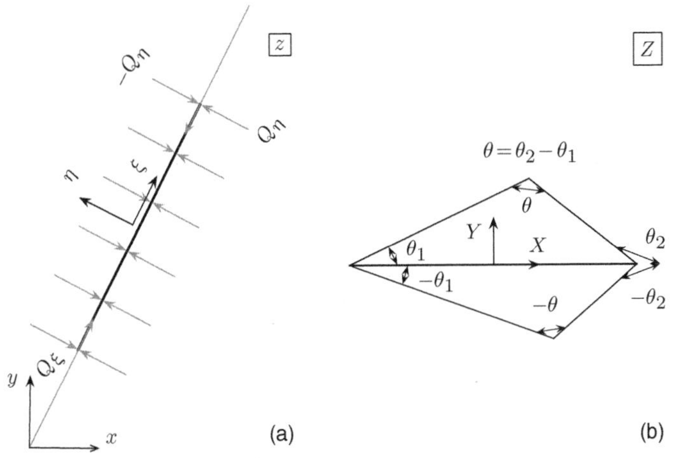

Figure 8.45 Tangential and normal components of flow along the element.

where

$$\theta = \arg\left[\frac{Z-1}{Z+1}\right] = \theta_2 - \theta_1 = \arg(Z-1) - \arg(Z+1). \tag{8.485}$$

We choose the branch cut for the angle θ as shown in Figure 8.45(b):

$$\begin{aligned} Y &\geq 0^+, & 0 &\leq \theta \leq \pi \\ Y &\leq 0^-, & -\pi &\leq \theta \leq 0, \end{aligned} \tag{8.486}$$

where the superscripts + and − refer to the side of the element where Y is positive or negative, respectively. We denote Q_η along the positive and negative sides of the element as Q_η^+ and Q_η^-, and obtain from (8.484) and (8.486):

$$Q_\eta^+ = -\frac{\sigma}{2}, \qquad Q_\eta^- = \frac{\sigma}{2}, \qquad Q_\eta^+ - Q_\eta^- = -\sigma. \tag{8.487}$$

The normal component, Q_η, is discontinuous across the element; there is an inflow $-(Q_\eta^+ - Q_\eta^-)$ equal to σ. The tangential component, Q_ξ, is continuous,

$$Q_\xi = \frac{\sigma}{2\pi} \ln\left|\frac{Z-1}{Z+1}\right|, \tag{8.488}$$

but singular at the end points of the element, which are singular points. The tangential component is zero at the center of the element, and increases in magnitude toward the tips.

8.23.4 The Stream Function

The stream function Ψ is the imaginary part of the complex potential, (8.479). We examine the behavior of this function along the X-axis and obtain:

$$\begin{aligned} \Psi &= \frac{\sigma L}{4\pi}[(X+1)\theta_1 - (X-1)\theta_2] \\ &= \frac{\sigma L}{4\pi}[-X(\theta_2 - \theta_1) + \theta_1 + \theta_2], \qquad Y = 0. \end{aligned} \tag{8.489}$$

We use that $\theta_2 - \theta_1 = \theta$ (see Figure 8.45(b)) and express Ψ as

$$\Psi = \frac{\sigma L}{4\pi}(-X\theta + \theta_1 + \theta_2), \qquad Y = 0. \tag{8.490}$$

We choose the branch cuts for θ_1 and θ_2 such that

$$Y \geq 0^+, \quad \left\{\begin{matrix} 0 \leq \theta_1 \leq \pi \\ 0 \leq \theta_2 \leq \pi \end{matrix}\right\}, \qquad Y \leq 0^-, \quad \left\{\begin{matrix} -\pi \leq \theta_1 \leq 0 \\ -\pi \leq \theta_2 \leq 0 \end{matrix}\right\}. \tag{8.491}$$

We write Ψ as Ψ^+ along $Y=0^+$, and as Ψ^- along $Y=0^-$, and obtain with $\theta=0$ for $|X|\geq 1$, $Y=0$:

$$-\infty < X < -1, \qquad Y=0^+, \qquad \Psi^+ = \frac{\sigma L}{4\pi}(+2\pi) \tag{8.492}$$

$$-\infty < X < -1, \qquad Y=0^-, \qquad \Psi^- = \frac{\sigma L}{4\pi}(-2\pi) \tag{8.493}$$

$$-1 \leq X \leq 1, \qquad Y=0^+, \qquad \Psi^+ = \frac{\sigma L}{4\pi}(1-X)\pi \tag{8.494}$$

$$-1 \leq X \leq 1, \qquad Y=0^-, \qquad \Psi^- = \frac{\sigma L}{4\pi}(X-1)\pi \tag{8.495}$$

$$1 < X < \infty, \qquad \begin{Bmatrix} Y=0^+ \\ Y=0^- \end{Bmatrix}, \qquad \begin{Bmatrix} \Psi^+ \\ \Psi^- \end{Bmatrix} = 0. \tag{8.496}$$

The stream function jumps both across the negative X-axis and across the element:

$$-\infty < X < -1, \qquad Y=0, \qquad \Psi^+ - \Psi^- = \sigma L \tag{8.497}$$

$$-1 \leq X \leq 1, \qquad Y=0, \qquad \Psi^+ - \Psi^- = \tfrac{1}{2}\sigma L(1-X). \tag{8.498}$$

The jump σL along the negative X-axis outside the element occurs because the element extracts an amount $Q=\sigma L$. The jump $\frac{1}{2}\sigma L(1-X)$ along the element varies linearly with X and corresponds to the constant rate of extraction. Indeed, this jump corresponds to the inflow:

$$Q_\eta^+ - Q_\eta^- = \frac{\partial}{\partial \xi}(\Psi^+ - \Psi^-) = \frac{2}{L}\frac{\partial}{\partial X}(\Psi^+ - \Psi^-) = -\sigma, \tag{8.499}$$

where we applied the Cauchy-Riemann equation $\partial\Psi/\partial\xi = -\partial\Phi/\partial\eta = Q_\eta$.

The stream function is finite and double-valued at $z = \overset{1}{z}$, i.e., at $Z=-1$. The value of this function at $X=-1$, $Y=0$, (8.489) is

$$\Psi = \frac{\sigma L}{4\pi}(2\theta_2) = \frac{\sigma L}{2\pi}\theta_2, \qquad X=-1,\ Y=0. \tag{8.500}$$

The angle θ_2 is π if $Y=0^+$, and $-\pi$ if $Y=0^-$. The stream function jumps along the semi-infinite segment $-\infty < X \leq 1$, $Y=0$: the line-sink has a branch cut extending from the end point at $\overset{2}{z}$ via the end point at $\overset{1}{z}$ to infinity. The branch cut extends to infinity because the line-sink extracts a net amount of water from the aquifer; see Figure 8.46.

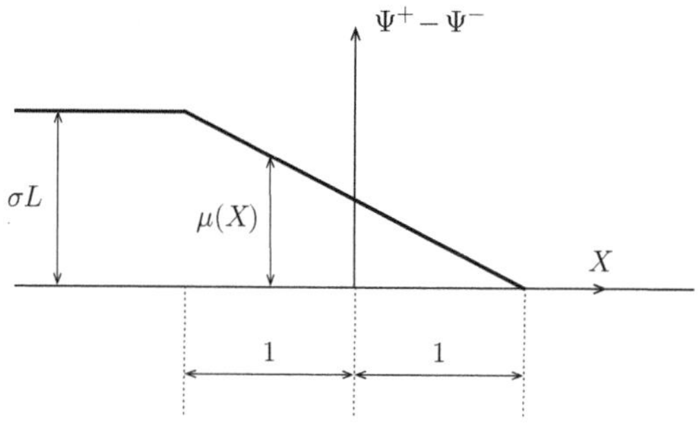

Figure 8.46 Jump in the stream function along the line-sink.

8.23.5 The Potential

The potential is single-valued throughout the domain. We verify this by first rearranging (8.479):

$$\Omega = \frac{\sigma L}{4\pi}\left\{-Z\ln\frac{Z-1}{Z+1} + \ln\left[(Z-1)(Z+1)\right] - 2\right\} \tag{8.501}$$

and then taking the real part:

$$\Phi = \frac{\sigma L}{4\pi}\left[-X\ln\left|\frac{Z-1}{Z+1}\right| + Y\theta + \ln|Z^2-1| - 2\right]. \tag{8.502}$$

Indeed, the only function that is discontinuous is θ; this function jumps only when Y passes through zero, and the factor Y in (8.502) ensures that this jump does not cause the potential to jump.

The complex potential is finite at the tips of the element; the terms $(Z-1)\ln(Z-1)$ and $(Z+1)\ln(Z+1)$ in (8.479) vanish at $Z=1$ and $Z=-1$, respectively; the complex potential is finite at the end points, but not holomorphic; none of the derivatives of Ω exist there. A plot of equipotentials and streamlines is reproduced in Figure 8.47.

8.23.6 Some Applications of Line-Sinks

We apply line-sinks to model a variety of boundary conditions. The superposition of this kind of analytic function to solve boundary value problems is an elementary form of the analytic element method; see, for example, Strack [1989], Haitjema [1995], and Strack [2003]. We consider three different cases: the first is the case of a narrow stream above the groundwater table, the second concerns a boundary with the head given, and the third a boundary where there is a resistance between the head in the feature, a stream bed, for example, and that in the aquifer.

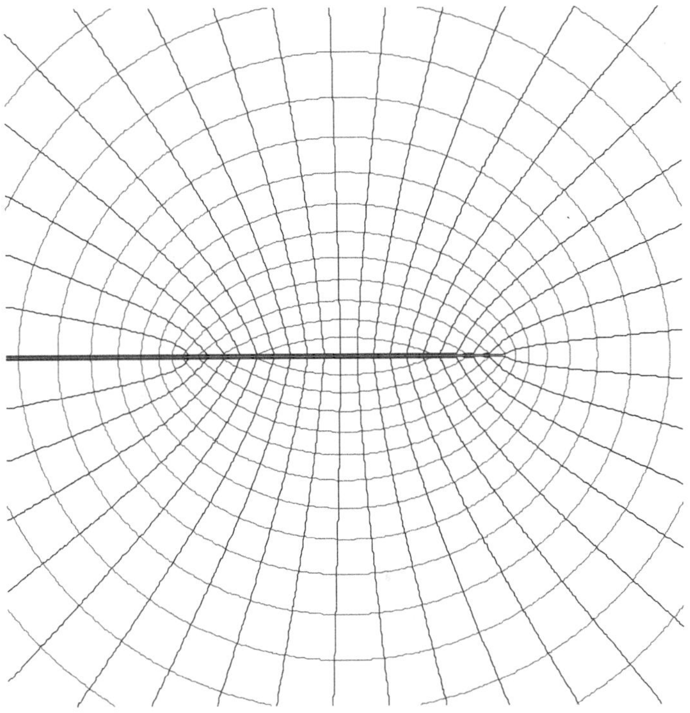

Figure 8.47 Equipotentials and streamlines for a line-sink.

8.23.7 Application 1: A Stream with Given Infiltration or Extraction Rate

We use line-sinks to model the infiltration through the bottoms of creeks that are not in direct contact with the groundwater, as illustrated in Figure 8.43. In such cases the strengths of the line-sinks are known. If there are n line-sinks between points $\overset{1}{z}_j$ and $\overset{2}{z}_j$ $(j = 1, 2, \ldots, n)$ of length L_j, and with strength σ_j, and if there is a uniform flow in the aquifer of discharge Q_0 in a direction α, then the complex potential becomes

$$\Omega = -Q_0 z e^{-i\alpha} + \sum_{j=1}^{n} \frac{\sigma_j L_j}{4\pi} \Big\{ (Z_j + 1)\ln(Z_j + 1) - (Z_j - 1)\ln(Z_j - 1) - 2 \Big\} + C, \tag{8.503}$$

where L_j is the length of the jth line-sink, where C is a real constant, and Z_j is defined by (8.475), with $\overset{1}{z}_j$ and $\overset{2}{z}_j$ replacing $\overset{1}{z}$ and $\overset{2}{z}$

$$Z_j = \frac{z - \frac{1}{2}\left(\overset{2}{z}_j + \overset{1}{z}_j\right)}{\frac{1}{2}\left(\overset{2}{z}_j - \overset{1}{z}_j\right)}. \tag{8.504}$$

8.23.8 Application 2: Modeling Boundaries with Given Head Using Line-Sinks

Creeks or rivers in direct contact with the groundwater control the head along their bottoms; we make the approximation that the head in the aquifer in the vertical section through the creek or river bank is constant and use line-sinks to model the creek or river in an approximate fashion.

We present the details of such an approximate model in what follows. We introduce the real function $\Lambda_{ls}\left(z, \overset{1}{z}_j, \overset{2}{z}_j\right)$ for the sake of compactness as

$$\Lambda_{ls}\left(z, \overset{1}{z}_j, \overset{2}{z}_j\right) = \frac{L_j}{4\pi}\Re\left\{(Z_j+1)\ln(Z_j+1) - (Z_j-1)\ln(Z_j-1) - 2\right\}. \tag{8.505}$$

We write the potential for a uniform far field and n line-sinks as

$$\Phi = \sum_{j=1}^{n} \sigma_j \Lambda_{ls}\left(z, \overset{1}{z}_j, \overset{2}{z}_j\right) + C + \Phi_g(z), \tag{8.506}$$

where $\Phi_g(z)$ is a function that represents the uniform flow:

$$\Phi_g(z) = \Re(-Q_0 z e^{-i\alpha}). \tag{8.507}$$

This given potential may also account for other functions with given parameters, wells, for example. The strengths σ_j are unknown, as opposed to the strengths in (8.503), and are determined by requiring that the potential Φ be equal to the values Φ_m, $m = 1, 2, \ldots, n$, at the centers $\overset{c}{z}_m$ of the n line-sinks. Note that the values Φ_m are obtained from the given values of the head along the creeks; the result is n conditions:

$$\Phi_m = \sum_{j=1}^{n} \sigma_j \Lambda_{ls}\left(\overset{c}{z}_m, \overset{1}{z}_j, \overset{2}{z}_j\right) + C + \Phi_g\left(\overset{c}{z}_m\right), \quad m = 1, 2, \ldots, n. \tag{8.508}$$

An additional constraint may be that the potential is equal to a fixed value Φ_0 at a reference point z_0,

$$\Phi_0 = \sum_{j=1}^{n} \sigma_j \Lambda_{ls}\left(z_0, \overset{1}{z}_j, \overset{2}{z}_j\right) + C + \Phi_g(z_0). \tag{8.509}$$

We write the system of equations in matrix form as follows. We introduce $\overset{c}{z}_{n+1}$:

$$\overset{c}{z}_{n+1} = z_0 \tag{8.510}$$

and the column vector b_m, $m = 1, \ldots, n+1$:

$$b_m = \Phi_m - \Phi_g(\overset{c}{z}_m), \qquad m = 1, \ldots, n+1. \tag{8.511}$$

We further introduce the matrix $A_{m,j}$:

$$A_{m,j} = \Lambda_{ls}(\overset{c}{z}_m, \overset{1}{z}_j, \overset{2}{z}_j), \qquad m = 1,\ldots,n+1, \qquad j = 1,\ldots,n, \tag{8.512}$$
$$A_{m,n+1} = 1$$

and an $(n+1)$th unknown strength σ_{n+1}:

$$\sigma_{n+1} = C. \tag{8.513}$$

We write the system of equations in terms of the notation we introduced,

$$b_m = \sum_{j=1}^{n+1} \sigma_j A_{mj}, \qquad m = 1,\ldots,n+1, \tag{8.514}$$

or, in symbolic form, replacing σ_j by **s**, A_{mj} by **A**, and b_m by **b**

$$\mathbf{As} = \mathbf{b} \tag{8.515}$$

with the solution

$$\mathbf{s} = \mathbf{A}^{-1}\mathbf{b}. \tag{8.516}$$

Equations (8.515) are usually solved numerically, for example, by Gaussian elimination.

In some cases equation (8.509) is replaced by a continuity equation. For example, if there is no net inflow from infinity, the sum of the strengths of the line-sinks is zero, and (8.509) would be replaced by

$$\sum_{j=1}^{n} \sigma_j L_j = 0. \tag{8.517}$$

Although (8.509) and (8.517) are different in form, they both represent a far-field condition and regulate the amount of flow from infinity.

Problem

8.7 A well of given discharge $Q = 800$ m^3/d is located at a point $z_w = d + id$ ($\Im d = 0$), and draws all of its water from a finite canal that extends from $z_s = -2d$ to $z_e = 2d$. The potential along the canal is Φ_0. Model the canal by two line-sinks of equal length and determine their strengths in terms of the aquifer parameters such that the head at the center of each line-sink is ϕ_0=25 m. The head at a reference point at $z_0 = 1000$ m is $\phi_1 = 28$ m. There is uniform flow at a rate $Q_{x0} = 0.4$ m/d. The flow is unconfined; the hydraulic conductivity is k=10 m/d. Choose d =100 m.

Questions:

1. Solve for the strengths of the two line-sinks.

2. Program the line-sinks, produce a flow net, and check your solution by verifying that the heads at the centers of the line-sinks and at the reference point are as required.
3. Determine what portion of the discharge of the well is drawn from the line-sinks, and what portion from infinity.

8.23.9 Application 3: Modeling a Stream in Indirect Contact with the Aquifer

We consider a stream that is separated from the aquifer by some leaky substance. We denote the resistance of the stream bed as c and represent the head in the stream as ϕ_0. If the head in the aquifer below the stream bed is ϕ_a, then the flow from the aquifer into the stream bed, q_z^*, is

$$q_z^* = \frac{\phi_a - \phi_0}{c}. \tag{8.518}$$

We express the strength σ of the line-sink in terms of the width b of the stream bed, the resistance c, and the heads ϕ_0 and ϕ_a:

$$\sigma = bq_z^* = \frac{b(\phi_a - \phi_0)}{c}. \tag{8.519}$$

If the flow is unconfined, we write the difference in value of the potential in the creek and below the creek as

$$\Phi_a - \Phi_0 = \tfrac{1}{2}k(\phi_a^2 - \phi_0^2) = \tfrac{1}{2}k(\phi_a + \phi_0)(\phi_a - \phi_0) \approx k\phi_0(\phi_a - \phi_0) \tag{8.520}$$

so that the linearized expression for σ becomes

$$\sigma = \frac{b(\Phi_a - \Phi_0)}{kc\phi_0}. \tag{8.521}$$

We obtain the system of equations as follows. We require that (8.521) be met at the center $\overset{c}{z}_m$ of the mth line-sink, which gives

$$\sigma_m = \left(\frac{b}{kc\phi_0}\right)_m \left\{\sum_{j=1}^{n} \sigma_j \Lambda_{ls}(\overset{c}{z}_m, \overset{1}{z}_j, \overset{2}{z}_j) + C + \Phi_g(\overset{c}{z}_m) - (\Phi_0)_m\right\}, \tag{8.522}$$

where the subscript m refers to the mth line-sink.

We demonstrate in what follows that the resistance-specified line-sinks may be used to model a partially penetrating stream.

8.23.10 Application 4: Modeling a Partially Penetrating Stream

The resistance caused by partial penetration of a stream into an aquifer may be included as expressed by (8.522). In that case the resistance may be computed from the equivalent length; see Verruijt [1970]. The concept of equivalent length is based on comparing the discharge flowing into an equipotential of depth d in an aquifer of thickness H with that flowing into an equipotential of depth H in the same aquifer, i.e., a fully penetrating one. The two discharges are made equal by adding the equivalent length, L_{eq}, to the aquifer with the fully penetrating stream. If the discharge into the stream is Q_0, the head in the stream is ϕ_0, and the thickness of the aquifer is H, then

$$Q_0 = \frac{kH(\phi_a - \phi_0)}{L_{\text{eq}}} = \frac{b(\phi_a - \phi_0)}{c}, \tag{8.523}$$

where the expression to the right represents the flow into a stream of width b and resistance c. We solve the latter equation for c:

$$c = \frac{bL_{\text{eq}}}{kH}. \tag{8.524}$$

The expression for L_{eq} is given in Strack [1989, 388–389]:

$$L_{\text{eq}} = \frac{2H}{\pi} \ln \sin \frac{\pi d}{2H} \tag{8.525}$$

and substitution of this expression for L_{eq} in (8.524) gives

$$c = -\frac{2b}{\pi k} \ln \sin \left(\frac{\pi d}{2H} \right), \tag{8.526}$$

where d is the depth of penetration, H is the thickness of the aquifer, and b the width of the stream.

8.24 The Line-Dipole

The complex potential for a line-sink jumps along the extension of its line to infinity, which may either simply lead to unattractive graphical representations or cause difficulties that must be corrected, as in the case of the area-sink, presented later in this text. This problem can be avoided by using line-dipoles, which are dipoles integrated along lines. A dipole is an abstract mathematical concept that appears to be useful in applications, and is the limiting case of a well and a recharge well that are a distance d apart approaching one another such that the limit of Qd for $d \to 0$ and $Q \to \infty$ is finite. The dipole has an orientation, the orientation of the line that connects the recharge well with the well. The dipole is aligned along the line of integration for the case of a line-dipole.

8.24.1 A Dipole

A dipole is a combination of a well and a recharge well in the limit as their centers move together indefinitely, while their discharges approach infinity; see

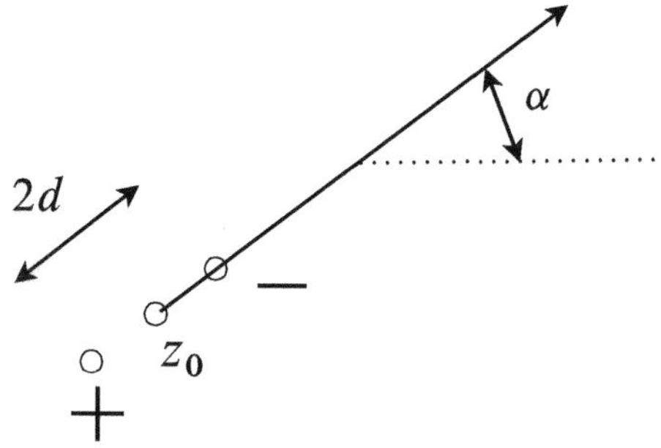

Figure 8.48 A dipole as the limit for a recharge well and a discharge well moving together.

Figure 8.48. The complex potential for a recharge well at $z_0 + de^{i\alpha}$ and a discharge well at $z_0 - de^{i\alpha}$ is

$$\Omega = \frac{Q}{2\pi} \ln \frac{z - z_0 + de^{i\alpha}}{z - z_0 - de^{i\alpha}}. \qquad (8.527)$$

We let Q increase and d decrease in such a way that their product remains constant,

$$\lim_{\substack{Q\to 0\\ d\to 0}} Qd = \frac{s}{2}, \qquad (8.528)$$

where s is a constant. We determine Ω in the limit, noting that $\ln(1+\varepsilon)$ approaches ε in the limit for $\varepsilon \to 0$

$$\lim_{\substack{Q\to 0\\ d\to 0}} \Omega = \frac{Q}{2\pi} \lim_{\substack{Q\to 0\\ d\to 0}} \ln \frac{1 + de^{i\alpha}/(z - z_0)}{1 - de^{i\alpha}/(z - z_0)} = \lim_{\substack{Q\to 0\\ d\to 0}} \frac{Q}{2\pi} \frac{2de^{i\alpha}}{z - z_0} = \frac{s}{2\pi} \frac{e^{i\alpha}}{z - z_0}. \qquad (8.529)$$

The complex potential for a dipole at z_0 and of strength s is

$$\Omega = \frac{s}{2\pi} \frac{e^{i\alpha}}{z - z_0}. \qquad (8.530)$$

Equipotentials and streamlines for a dipole are shown in Figure 8.49, where the dipole is placed at a point on a line. The dipole is oriented along the line and points from the starting point to the end point.[2]

8.24.2 The Complex Potential for a Line-Dipole

We obtain the complex potential for a line-dipole by integrating the potential for a dipole along a line. The starting and end points of the line are again $\overset{1}{z}$ and $\overset{2}{z}$, and the length and orientation of the element are given by L and α:

[2] The second term inside the brackets in the expression for the complex potential for a circular lake in a field of uniform flow, (8.189), represents a dipole also; this dipole cancels precisely the effect of uniform flow along the boundary of the circular lake, thus creating an equipotential.

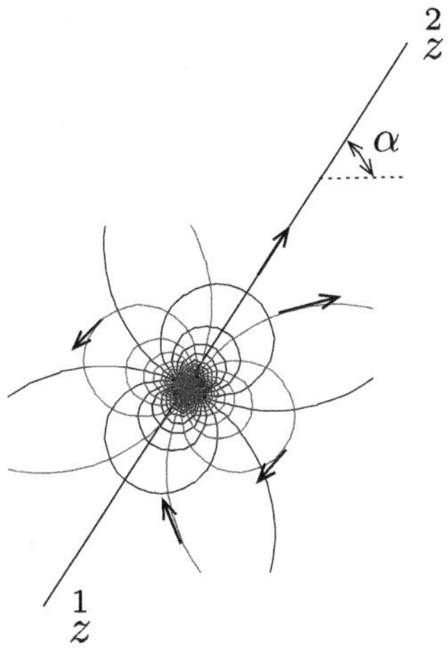

Figure 8.49 A line-dipole.

$$\overset{2}{z} - \overset{1}{z} = Le^{i\alpha}. \tag{8.531}$$

We integrate (8.530), replacing z_0 by δ and s by $-\mu$,

$$\Omega = -\frac{1}{2\pi} \int\limits_{-\frac{1}{2}L}^{\frac{1}{2}L} \frac{\mu e^{i\alpha}}{z-\delta} d\xi, \tag{8.532}$$

where ξ is defined as for the line-sink; we use (8.474) ($e^{i\alpha} d\xi = d\delta$) and obtain

$$\Omega = -\frac{1}{2\pi} \int\limits_{\overset{1}{z}}^{\overset{2}{z}} \frac{\mu}{z-\delta} d\delta. \tag{8.533}$$

We write this in dimensionless form in terms of Z and Δ, defined in (8.475):

$$\Omega = -\frac{1}{2\pi} \int\limits_{-1}^{+1} \frac{\mu(\Delta)}{Z-\Delta} d\Delta. \tag{8.534}$$

We limit the analysis to a linear density distribution μ:

$$\mu = \mu(Z) = \tfrac{1}{2}(\mu_2 - \mu_1)Z + \tfrac{1}{2}(\mu_1 + \mu_2) = aZ + b, \qquad \Im a = \Im b = \Im \mu_1 = \Im \mu_2 = 0, \tag{8.535}$$

where μ_1 is the value of μ at $Z = -1$ and μ_2 is the value of μ at $Z = 1$. The complex potential for the line dipole, (8.533), becomes with this density distribution:

$$\Omega = -\frac{1}{2\pi}\int_{-1}^{1}\frac{a\Delta+b}{Z-\Delta}d\Delta = -\frac{1}{2\pi}\int_{-1}^{1}\frac{a(\Delta-Z)+aZ+b}{Z-\Delta}d\Delta \tag{8.536}$$

or

$$\Omega = \frac{a}{2\pi}\int_{1}^{1} d\Delta + \frac{1}{2\pi}(aZ+b)\int_{Z+1}^{Z-1}\frac{d(Z-\Delta)}{Z-\Delta}. \tag{8.537}$$

We integrate and obtain

$$\boxed{\Omega = \frac{\mu(Z)}{2\pi}\ln\frac{Z-1}{Z+1} + \frac{\mu_2-\mu_1}{2\pi}} \tag{8.538}$$

where $a = \frac{1}{2}(\mu_2-\mu_1)$; see (8.535).

8.24.3 The Behavior of $\Omega(Z)$ along the Element

The strength μ, which we refer to as the density distribution, is real along the element. We use the angle θ (compare (8.485)):

$$\theta = \arg\frac{Z-1}{Z+1}, \qquad -\pi \le \theta \le \pi, \tag{8.539}$$

with a branch cut along the element as shown in Figure 8.45(b) (page 319), we may represent Ω along the positive and negative sides of the element as

$$-1 \le X \le 1, \qquad Y = 0^+, \qquad \Omega^+ = \frac{\mu(X)}{2\pi}\left[\ln\left|\frac{X-1}{X+1}\right| + i\pi\right] + \frac{\mu_2-\mu_1}{2\pi} \tag{8.540}$$

$$-1 \le X \le 1, \qquad Y = 0^-, \qquad \Omega^- = \frac{\mu(X)}{2\pi}\left[\ln\left|\frac{X-1}{X+1}\right| - i\pi\right] + \frac{\mu_2-\mu_1}{2\pi}. \tag{8.541}$$

The potential Φ is continuous across the element, but the stream function jumps:

$$-1 \le X \le 1, \qquad Y = 0, \qquad \left\{\begin{array}{l}\Phi^+ - \Phi^- = 0\\ \Psi^+ - \Psi^- = \mu(X)\end{array}\right\}. \tag{8.542}$$

The angle θ is zero along the X-axis outside the element, and therefore the stream function does not jump there:

$$\left\{\begin{array}{r}-\infty \le X < -1\\ 1 < X < \infty\end{array}\right\}, \qquad Y = 0, \qquad \left\{\begin{array}{l}\Phi^+ - \Phi^- = 0\\ \Psi^+ - \Psi^- = 0\end{array}\right\}. \tag{8.543}$$

The behavior of the stream function along the element is illustrated in Figure 8.50, where the jump $\Psi^+ - \Psi^-$ is plotted along the X-axis. There is no net inflow or

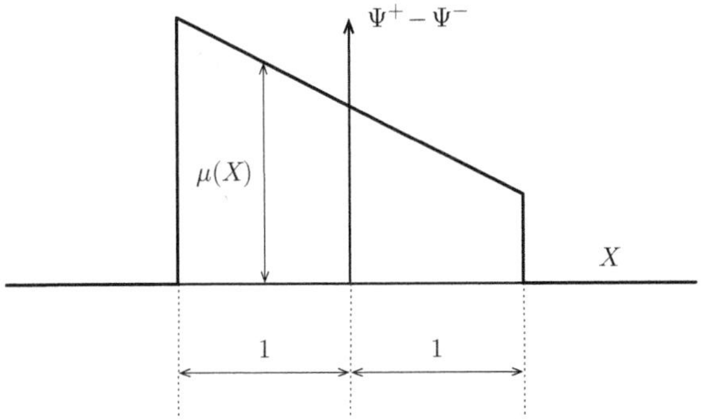

Figure 8.50 The jump in the stream function along the X-axis.

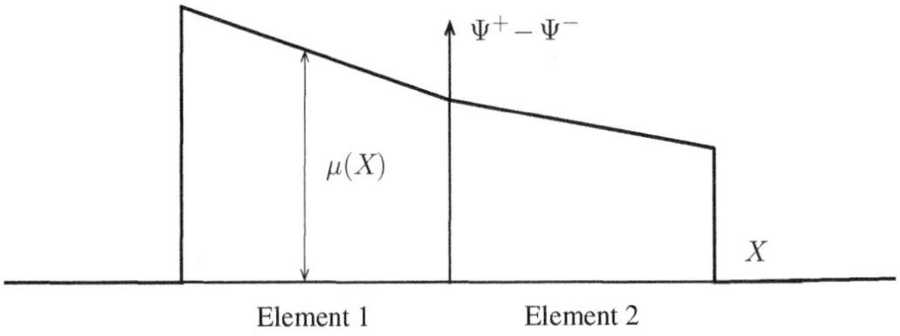

Figure 8.51 Two adjoining line-dipoles.

outflow associated with a dipole, and the same is true for a line-dipole. This is the reason why the branch cut does not extend to infinity; the stream function only jumps along the element. The difference between a line-dipole and a line-sink is that the line-dipole does not extract a net amount of fluid, whereas the line-sink does. It appears from (8.538) that the complex potential for a line-dipole has logarithmic singularities at the tips of the element. These singularities correspond to a well at $Z = 1$ of discharge $\mu(1)$ and a recharge well at $Z = -1$ of discharge $\mu(-1)$. If line-dipoles are strung together, and the density distribution μ is continuous at the nodal points where the elements meet, then the singularities cancel; the recharge well at the end point of the one line-dipole will cancel with the well at the adjoining starting point of the next one, as illustrated in Figure 8.51, where the jump $\Psi^+ - \Psi^-$ is plotted along two adjoining elements.

8.25 The Relation between the Line-Dipole and the Line-Sink

We examine the relationship between the line-dipole and the line-sink. For reasons of convenience, we use (8.533), rather than the dimensionless form (8.534). We integrate (8.533) by parts, and obtain

$$\Omega = -\frac{1}{2\pi}\int_{\overset{1}{z}}^{\overset{2}{z}} \frac{\mu}{z-\delta}d\delta = \frac{1}{2\pi}\int_{\overset{1}{z}}^{\overset{2}{z}} \mu d\ln(z-\delta)$$

$$= \frac{1}{2\pi}\left[\mu\left(\overset{2}{z}\right)\ln\left(z-\overset{2}{z}\right) - \mu\left(\overset{1}{z}\right)\ln\left(z-\overset{1}{z}\right)\right] - \frac{1}{2\pi}\int_{\overset{1}{z}}^{\overset{2}{z}} \frac{d\mu}{d\delta}\ln(z-\delta)d\delta. \qquad (8.544)$$

We use (8.474) and write

$$d\delta = e^{i\alpha}d\xi, \qquad \frac{d\mu}{d\delta} = \frac{d\mu}{d\xi}e^{-i\alpha}, \qquad (8.545)$$

so that (8.544) becomes

$$\Omega = \frac{1}{2\pi}\left[\mu\left(\overset{2}{z}\right)\ln\left(z-\overset{2}{z}\right) - \mu\left(\overset{1}{z}\right)\ln\left(z-\overset{1}{z}\right)\right] - \frac{1}{2\pi}\int_{\overset{1}{z}}^{\overset{2}{z}} \frac{d\mu}{d\xi}\ln(z-\delta)d\xi. \qquad (8.546)$$

We note that $\mu = \Psi^+ - \Psi^-$ along the element and that $\partial(\Psi^+ - \Psi^-)/\partial\xi = -\sigma$ (see (8.499)), and obtain

$$\Omega = \frac{1}{2\pi}\left[\mu_2\ln\left(z-\overset{2}{z}\right) - \mu_1\ln\left(z-\overset{1}{z}\right)\right] + \frac{1}{2\pi}\int_{\overset{1}{z}}^{\overset{2}{z}} \sigma\ln(z-\delta)d\xi, \qquad (8.547)$$

where

$$\sigma = -\frac{d\mu}{d\xi}, \qquad \mu_1 = \mu\left(\overset{1}{z}\right), \qquad \mu_2 = \mu\left(\overset{2}{z}\right). \qquad (8.548)$$

The integral in (8.547) is the complex potential for a line-sink, (8.473). The first two terms represent a well of discharge μ_2 and a recharge well of discharge $-\mu_1$. Thus, we obtain the complex potential for a line-sink from that for a line-dipole by adding the appropriate terms for wells at the end points:

$$\frac{1}{2\pi}\int_{\overset{1}{z}}^{\overset{2}{z}} \sigma\ln(z-\delta)d\xi = -\frac{1}{2\pi}\int_{\overset{1}{z}}^{\overset{2}{z}} \frac{\mu}{z-\delta}d\delta - \frac{1}{2\pi}\left[\mu_2\ln\left(z-\overset{2}{z}\right) - \mu_1\ln\left(z-\overset{1}{z}\right)\right].$$

(8.549)

For example, we obtain the complex potential for a line-sink of constant strength from (8.549) by letting μ be a linear function along the element; $\mu = -\sigma\xi + \mu_0$, where μ_0 is a constant.

The main difference between line-dipoles and line-sinks lies in the way the branch cut is handled. Consider, for example, a string of three straight line-sinks

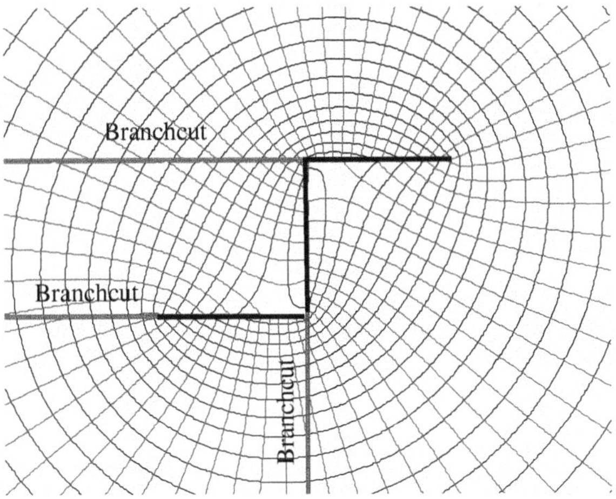

Figure 8.52 Branch cuts for a string of three line-sinks of constant strength.

with nodal points z_1, z_2, z_3, z_4 as shown in Figure 8.52. The branch cut of each line-sink falls along the element and extends to infinity, as shown in the figure by the heavy line that extends to infinity for each line-sink.

Alternatively, the string may be modeled by a series of line-dipoles of linearly varying strength with $\mu_1 = 0$, chosen such that the extraction rate per unit length is the same as for the string of line-sinks, plus a well at point z_4 of discharge μ_4, where μ_4 equals the total amount of inflow into the string. In this case, however, the branch cut follows the string, and there exists only one branch cut in the field that extends to infinity, namely, the branch cut emanating from the well at point z_4, as shown in Figure 8.53.

Some features in aquifers can only be modeled with line-dipoles. Thin zones of very high hydraulic conductivity, such as cracks, are examples of this. Both inflow and outflow occur along a crack so that the stream-function jumps across the crack, but there is neither net inflow nor outflow.

Problem

8.8 Solve Problem 8.7 again, but now using line-dipoles of linearly varying strength and one well to model the canal.

8.26 Infiltration or Evaporation Inside Polygons

We consider the case of uniform infiltration or extraction over areas bounded by polygons and construct the potential that can be used to model such cases. The corresponding analytic elements are called area-sinks and can be used to model

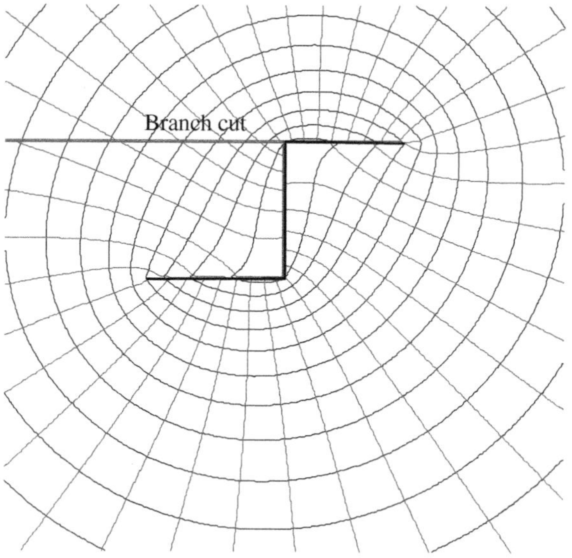

Figure 8.53 A string of line-dipoles.

the infiltration through the bottoms of lakes that are above the groundwater table, and the evaporation through swamps. Area-sinks are especially useful to model recharge from rainfall, which depends not only on average rainfall, but also on land use.

We construct the potential from the following expression for the constant extraction rate γ over an area bounded by a polygon with n sides with vertices $z_j, j = 1, \ldots, n$. We define z_{n+1} as

$$z_{n+1} = z_1. \tag{8.550}$$

We number the vertices sequentially in such a way that the inside of the polygon is to the left of the direction from z_j to z_{j+1}; the positive side of the line elements is on the inside. The extraction rate γ is constant inside the polygon, labeled as D^+, and zero outside, labeled as D^-

$$\begin{aligned} \gamma &= \gamma_0, \qquad z \in D^+ \\ \gamma &= 0, \qquad z \in D^-. \end{aligned} \tag{8.551}$$

We demonstrate that we may represent this function as

$$\gamma = \frac{\gamma_0}{4\pi \mathrm{i}} \sum_{j=1}^{n} \left[\ln \frac{z - z_{j+1}}{z - z_j} - \ln \overline{\frac{z - z_{j+1}}{z - z_j}} \right]. \tag{8.552}$$

We choose the branch cuts for each of the logarithms in the sum to fall along the side of the polygon it represents. The function inside the brackets is purely imaginary

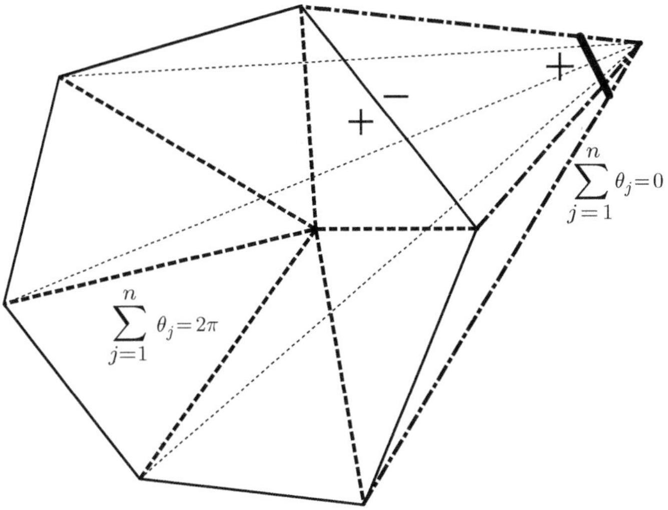

Figure 8.54 The angles θ_j.

and equal to two times the sum of the angles θ_j that are formed between the lines drawn from any point z to the end points of the respective sides, i.e.,

$$\gamma = \frac{\gamma_0}{2\pi} \sum_{j=1}^{n} \theta_j. \tag{8.553}$$

These angles are shown in Figure 8.54 and are positive for points to the left of each line and negative if the point is to the right. We observe from the figure that the sum of the angles is zero for all points outside the polygon, and 2π for points inside the polygon. The function γ is thus equal to γ_0 for points inside the polygon and zero for points outside the polygon.

The function γ is equal to minus the divergence of the discharge vector, and equals the Laplacian of the discharge potential Φ inside the polygon:

$$\nabla^2 \Phi = 4 \frac{\partial^2 \Phi}{\partial z \partial \bar{z}} = \gamma, \qquad z \in D^+. \tag{8.554}$$

We obtain the function Φ by integration of the following equation, obtained from (8.552) with (8.554)

$$4 \frac{\partial^2 \Phi}{\partial z \partial \bar{z}} = \frac{\gamma_0}{4\pi \mathrm{i}} \sum_{j=1}^{n} \left[\ln \frac{z - z_{j+1}}{z - z_j} - \ln \overline{\frac{z - z_{j+1}}{z - z_j}} \right]. \tag{8.555}$$

We introduce the local complex variable $Z_j, j = 1, \ldots, n$

$$Z_j = \frac{2z - z_j - z_{j+1}}{z_{j+1} - z_j} \tag{8.556}$$

$$z = \tfrac{1}{2}(z_{j+1} - z_j) Z_j + \tfrac{1}{2}(z_j + z_{j+1}). \tag{8.557}$$

We introduce the complex number ν_j and the coordinate of the center of the line element, $\underset{c}{z_j}$ as

$$\nu_j = (z_{j+1} - z_j) = L_j e^{i\alpha_j}, \qquad \underset{c}{z_j} = \tfrac{1}{2}(z_{j+1} + z_j), \tag{8.558}$$

so that we may simplify (8.556) and (8.557):

$$Z_j = \frac{z - \underset{c}{z_j}}{\nu_j} \tag{8.559}$$

$$z = 2\nu_j Z_j + \underset{c}{z_j}. \tag{8.560}$$

We see that

$$Z_j - 1 = \frac{2z - z_j - z_{j+1} - z_{j+1} + z_j}{z_{j+1} - z_j} = 2\frac{z - z_{j+1}}{z_{j+1} - z_j} \tag{8.561}$$

$$Z_j + 1 = \frac{2z - z_j - z_{j+1} - z_{j+1} + z_j}{z_{j+1} - z_j} = 2\frac{z - z_j}{z_{j+1} - z_j} \tag{8.562}$$

and use these two equations to write (8.555) in terms of the local complex variables as

$$\frac{\partial^2 \Phi}{\partial z \partial \bar{z}} = \frac{\gamma_0}{16\pi i} \sum_{j=1}^{n} \left[\ln\frac{Z_j - 1}{Z_j + 1} - \ln\frac{\bar{Z}_j - 1}{\bar{Z}_j + 1} \right]. \tag{8.563}$$

We introduce the functions $f(Z_j)$ and $F(Z_j)$:

$$f(Z_j) = \ln\frac{Z_j - 1}{Z_j + 1} \tag{8.564}$$

and

$$F(Z_j) = \int f(Z_j) dZ_j. \tag{8.565}$$

We integrate (8.563) first with respect to $\bar{z}$ and use (8.560) to write $d\bar{z}$ as $\frac{1}{2}\bar{\nu}_j d\bar{Z}_j$:

$$\frac{\partial \Phi}{\partial z} = \frac{\gamma_0 \bar{\nu}_j}{32\pi i} \sum_{j=1}^{n} \left[\bar{Z}_j f(Z_j) - F(\bar{Z}_j) + h_0(Z_j) \right], \tag{8.566}$$

where $h_0(Z_j)$ is an arbitrary function of Z_j and where

$$\begin{aligned} F(Z_j) &= \int f(Z_j) dZ_j = \int \ln\frac{Z_j - 1}{Z_j + 1} dZ_j \\ &= (Z_j - 1)\ln(Z_j - 1) - (Z_j + 1)\ln(Z_j + 1) + 2 + C_j, \end{aligned} \tag{8.567}$$

where C_j is a constant of integration that we will choose in such a way that the far field of the function reduces to the complex potential for a well at the center of the line plus a function that is holomorphic at infinity, i.e., a function that can be represented in terms of a Laurent expansion with only negative powers of Z_j. We obtain the expression for C_j by rewriting (8.567) as

$$F(Z_j) = Z_j \ln \frac{Z_j - 1}{Z_j + 1} - \ln(Z_j^2 - 1) + 2 + C_j \tag{8.568}$$

or

$$F(Z_j) = Z_j \ln \frac{Z_j - 1}{Z_j + 1} - \ln \frac{Z_j^2 - 1}{Z_j^2} - 2\ln Z_j + 2 + C_j, \qquad z \to \infty. \tag{8.569}$$

The first term in the expansion of the first logarithm about $1/Z_j = 0$ gives $-2/Z_j$ and the lowest-order term in the second logarithm, which equals $\ln(1 - Z_j^{-2})$, is $-1/Z_j^2$, so that the largest term in (8.569) is

$$F(Z_j) \to Z_j \frac{-2}{Z_j} - 2\ln Z_j + 2 + C_j = C_j - 2\ln \frac{z - \frac{1}{2}(z_{j+1} + z_{j-1})}{\frac{1}{2}(z_{j+1} - z_j)}, \qquad z \to \infty. \tag{8.570}$$

We choose the constant C_j as

$$C_j = -2\ln[\tfrac{1}{2}(z_{j+1} - z_j)], \tag{8.571}$$

so that the function F for $z \to \infty$ reduces to a well of discharge 2π at the center of the line element. The expression for the potential F now becomes

$$F(Z_j) = (Z_j - 1)\ln(Z_j - 1) - (Z_j + 1)\ln(Z_j + 1) + 2 - 2\ln\left[\tfrac{1}{2}\left(z_{j+1} - z_j\right)\right]. \tag{8.572}$$

This function has the same form as the complex potential for a line-sink, which has the property that the real part is continuous across the element and the imaginary part jumps. The discharge function

$$W = -2\frac{\partial \Phi}{\partial z} \tag{8.573}$$

must be continuous across the boundary of the polygon; we must choose the function $h_0(Z_j)$ in (8.566) such that this condition is satisfied. We fulfill this condition by choosing $h_0(Z_j)$ as

$$h_0(Z_j) = -Z_j f(Z_j) - F(Z_j), \tag{8.574}$$

and substitute this expression in (8.566):

$$\frac{\partial \Phi}{\partial z} = \frac{\gamma_0}{32\pi \mathrm{i}} \sum_{j=1}^{n} \bar{\nu}_j \left[(\bar{Z}_j - Z_j) f(Z_j) - F(\bar{Z}_j) - F(Z_j)\right] \tag{8.575}$$

and obtain an expression for the discharge function by the use of (8.573):

$$W = -2\frac{\partial\Phi}{\partial z} = -\frac{\gamma_0}{16\pi\mathrm{i}}\sum_{j=1}^{n}\bar{\nu}_j\left[(\bar{Z}_j - Z_j)f(Z_j) - F(\bar{Z}_j) - F(Z_j)\right]. \tag{8.576}$$

This function is indeed continuous across the sides of the polygon; the factor $Z_j - \bar{Z}_j$ is zero along the side because Z_j is real there, and the real part of the function $F(Z_j)$, i.e, $\frac{1}{2}[F(Z_j) + F(\bar{Z}_j)]$, is continuous across the side.

We obtain an expression for the function Φ by integration of (8.575) with respect to z, which gives, with $dz = \frac{1}{2}\nu_j dZ_j$,

$$\Phi = \frac{\gamma_0}{64\pi\mathrm{i}}\sum_{j=1}^{n}\nu_j\bar{\nu}_j\left[\bar{Z}_jF(Z_j) - Z_jF(\bar{Z}_j) - \int Z_jf(Z_j)dZ_j - \int F(Z_j)dZ_j + g(\bar{Z}_j)\right], \tag{8.577}$$

where $g(\bar{Z}_j)$ is an arbitrary function of $\bar{Z}_j$. We work out the first integral by parts:

$$\int Z_jf(Z_j)dZ_j = \int Z_j dF(Z_j) = Z_jF(Z_j) - \int F(Z_j)dZ_j. \tag{8.578}$$

Substitution of this expression in (8.577) results in cancellation of the integrals of $F(Z_j)$, and we obtain

$$\Phi = \frac{\gamma_0}{64\pi\mathrm{i}}\sum_{j=1}^{n}\nu_j\bar{\nu}_j\left[\bar{Z}_jF(Z_j) - Z_jF(\bar{Z}_j) - Z_jF(Z_j) + g(\bar{Z}_j)\right]. \tag{8.579}$$

We choose the function $g(\bar{Z}_j)$ such that the potential is continuous, which we achieve by setting $g(Z_j) = \bar{Z}_jF(\bar{Z}_j)$, and the final result is

$$\Phi = \frac{\gamma_0}{64\pi\mathrm{i}}\sum_{j=1}^{n}L_j^2(\bar{Z}_j - Z_j)\left[F(Z_j) + F(\bar{Z}_j)\right], \tag{8.580}$$

where we used (8.558) to write $\nu_j\bar{\nu}_j$ as L_j^2. We simplify this expression somewhat by replacing $\bar{Z}_j - Z_j$ by $-2\mathrm{i}Y_j$, which gives

$$\Phi = -\frac{\gamma_0}{32\pi}\sum_{j=1}^{n}Y_jL_j^2\left[F(Z_j) + F(\bar{Z}_j)\right]. \tag{8.581}$$

8.26.1 Implementation

The potential for the area-sink, given by (8.581), is useful in cases where infiltration is piecewise constant. In that case, the potential consists of a sum of functions of the form (8.581). We tesselate the domain with polygons composed of line-segments, but we include the line-segments that are common to two polygons only once, rather than for each of the polygons. We must introduce the end points of each of the line-segments separately; i.e, if $\underset{1}{z_j}$ represents the starting point of line-segment j and $\underset{2}{z_j}$ the end point, then we redefine Z_j as

$$Z_j = \frac{2z - \underset{1}{z_j} - \underset{2}{z_j}}{\underset{2}{z_j} - \underset{1}{z_j}}. \tag{8.582}$$

The coefficient γ_0 in (8.581) must be replaced for each line segment by the extraction rate to the left with respect to the line from $\underset{1}{z_j}$ to $\underset{2}{z_j}$ minus that to the right; i.e, the expression for the potential becomes

$$\Phi = -\frac{1}{32\pi}\sum_{j=1}^{n}\left(\gamma_j^+ - \gamma_j^-\right) Y_j L_j^2 \left[F(Z_j) + F(\bar{Z}_j)\right]. \tag{8.583}$$

Application of this approach must be done with some care; if the jumps of the extraction rates are not carefully determined, the potential will be in error and will not be harmonic outside the area of extraction or infiltration. To ensure that the discontinuities are properly entered, the extraction rate should be contoured, and it should be verified that it is indeed piecewise constant. This can be done by contouring the function γ defined as

$$\gamma = \frac{1}{2\pi \mathrm{i}}\sum_{j=1}^{n}\left(\gamma_j^+ - \gamma_j^-\right)\ln\frac{Z_j - 1}{Z_j + 1}. \tag{8.584}$$

Contours of the discharge potential are shown as an example in Figure 8.55; the values of the extraction rate γ are shown in Figure 8.56, where the extraction rate is plotted using the expression (8.584). Note the jagged lines along the line segments; these are the result of the contouring of a function that jumps.

8.26.2 Far-Field Behavior

Although the discharge potential given by (8.583) is ready for implementation, it is not at all clear that the function indeed represents flow due to piecewise constant infiltration. The remaining part of this section is devoted to demonstrating that the conditions are indeed all met and to presenting a complex potential that is holomorphic outside the area of infiltration. For this purpose, it is necessary to consider the polygon, rather than limiting the discussion to a single side.

The discharge potential for the area-sink should approach the discharge potential for a well at large distances. We demonstrate that the discharge potential (8.581) indeed has this behavior by writing the function $F(Z_j)$ in a different form:

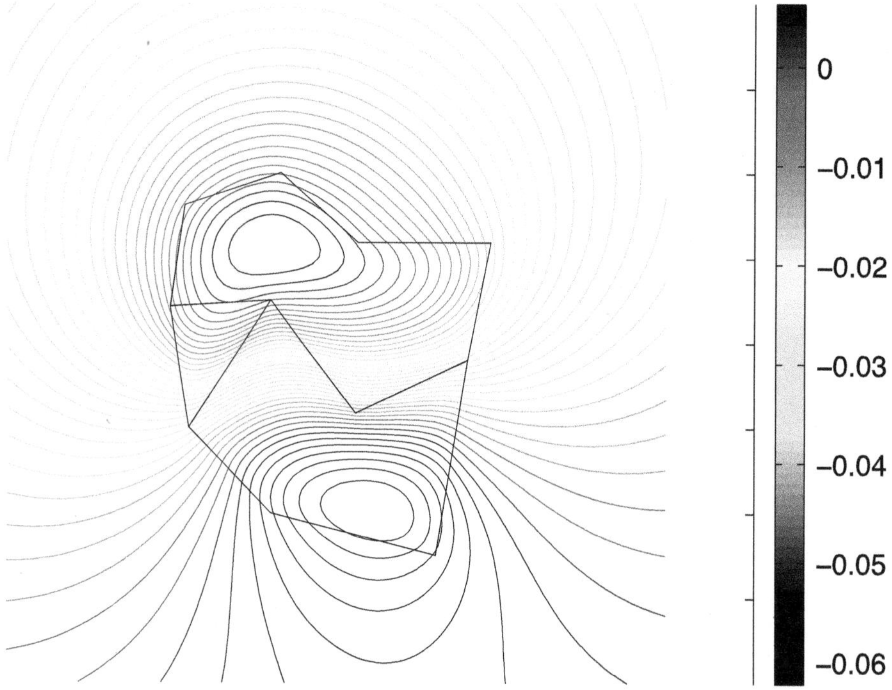

Figure 8.55 Contours of the discharge potential for a piecewise constant distribution of the extraction rate; the extraction rates are shown in Figure 8.56.

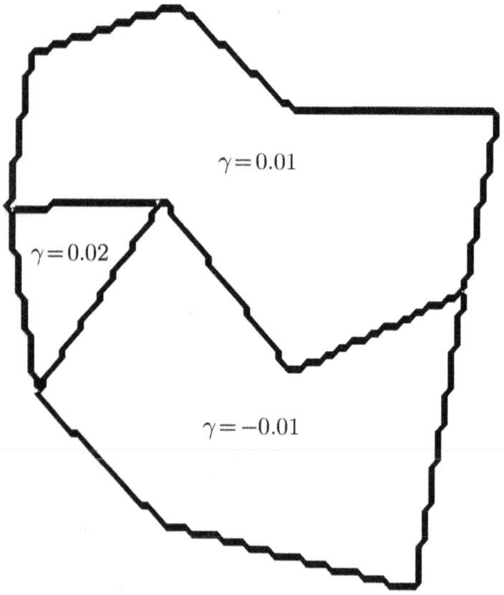

Figure 8.56 Piecewise constant extraction rates.

$$F(Z_j) = Z_j \ln \frac{Z_j - 1}{Z_j + 1} - \ln(Z_j - 1) - \ln(Z_j + 1) + 2 + C_j. \tag{8.585}$$

The terms $-\ln(Z_j - 1)$ and $-\ln(Z_j + 1)$ have branch cuts that run from $Z_j = 1$ and $Z_j = -1$, respectively, along $Y_j = 0$ to infinity. The function $F(Z_j)$ in (8.581) is multiplied by a factor that is not constant; the branch cut will cause discontinuities in W. We redirect the branch cuts in such a manner that they fall along the boundary of the polygon so that the cuts do not enter the domain outside the polygon. We write the term $\ln(Z_j - 1)$ as:

$$\begin{aligned}
\ln(Z_j - 1) &= \ln(z - z_{j+1}) - \ln\left[\tfrac{1}{2}(z_{j+1} - z_j)\right] \\
&= \sum_{m=j+1}^{n} \ln \frac{z - z_m}{z - z_{m+1}} + \ln(z - z_{n+1}) - \ln\left[\tfrac{1}{2}(z_{j+1} - z_j)\right] \\
&= -\sum_{m=j+1}^{n} \ln \frac{Z_m - 1}{Z_m + 1} + \ln(z - z_{n+1}) - \ln\left[\tfrac{1}{2}(z_{j+1} - z_j)\right].
\end{aligned} \tag{8.586}$$

Splitting the logarithms of the fractions into logarithms of the numerator minus the logarithms of the denominators, we see that the terms cancel one another except the first one ($\ln(z - z_{j+1})$) and the last one, which is cancelled by the term $\ln(z - z_{n+1})$. Thus, the identity holds, but the branch cut of the term $\ln(Z_j - 1)$ is now folded along the polygon and continues from $z_{n+1} = z_1$ to infinity. We apply this approach to the term $\ln(Z_j + 1)$ in (8.585) as well and obtain

$$\begin{aligned}
\ln(Z_j + 1) &= \ln(z - z_j) - \ln\left[\tfrac{1}{2}(z_{j+1} - z_j)\right] \\
&= \sum_{m=j}^{n} \ln \frac{z - z_m}{z - z_{m+1}} + \ln(z - z_{n+1}) - \ln\left[\tfrac{1}{2}(z_{j+1} - z_j)\right] \\
&= -\sum_{m=j}^{n} \ln \frac{Z_m - 1}{Z_m + 1} + \ln(z - z_{n+1}) - \ln\left[\tfrac{1}{2}(z_{j+1} - z_j)\right].
\end{aligned} \tag{8.587}$$

We choose the constant C_j in (8.585) as

$$C_j = -2\ln\left[\tfrac{1}{2}(z_{j+1} - z_j)\right] \tag{8.588}$$

and replace the terms $\ln(Z_j - 1)$ and $\ln(Z_j + 1)$ in (8.585) by expressions (8.586) and (8.587):

$$F(Z_j) = Z_j \ln \frac{Z_j - 1}{Z_j + 1} + \sum_{m=j}^{n} \ln \frac{Z_m - 1}{Z_m + 1} + \sum_{m=j+1}^{n} \ln \frac{Z_m - 1}{Z_m + 1} + 2 - 2\ln(z - z_{n+1}). \tag{8.589}$$

We simplify this:

$$F(Z_j) = (Z_j+1)\ln\frac{Z_j-1}{Z_j+1} + 2\sum_{m=j+1}^{n} \ln\frac{Z_m-1}{Z_m+1} + 2 - 2\ln(z-z_{n+1}). \qquad (8.590)$$

We achieve with this formulation two objectives. First, the branch cuts are all folded along the boundary, and all functions share a term $-2\ln(z-z_{n+1}) = -2\ln(z-z_1)$. This term can therefore be taken out of the functions $F(Z_j)$. We do this by defining a new function H as

$$H(Z_j) = (Z_j+1)\ln\frac{Z_j-1}{Z_j+1} + 2\sum_{m=j+1}^{n} \ln\frac{Z_m-1}{Z_m+1} + 2, \qquad (8.591)$$

so that we may write $F(Z_j)$ as

$$F(Z_j) = H(Z_j) - 2\ln(z-z_1). \qquad (8.592)$$

The logarithmic term is independent of j; we substitute this expression for $F(Z_j)$ in (8.581):

$$\Phi = -\frac{\gamma_0}{32\pi}\sum_{j=1}^{n} Y_j L_j^2\left[H(Z_j)+H(\bar{Z}_j)\right] + \frac{\gamma_0}{16\pi}\sum_{j=1}^{n} Y_j L_j^2[\ln(z-z_1)+\ln(\bar{z}-\bar{z}_1)]. \qquad (8.593)$$

The sum in the second term equals four times the area of the polygon. We demonstrate this by examining a single term $Y_jL_jL_j$. The product $Y_jL_j/2$ is the normal distance between a point z and the line element j, and L_j is the length of the line element. Thus, the product $Y_jL_j^2/2$ is twice the area of the triangle formed by the element and the lines drawn to z from the end points of the element. The triangles formed this way for the entire polygon for some point z are shown in Figure 8.54. It is quite clear that indeed the sum of the triangles equals the area of the polygon when the point z is inside. It is less clear that it is true also for a point z outside the polygon; two of the values of L_jY_j are negative, so that also in this case the algebraic sum of the areas of the triangles is equal to that of the polygon.

We replace the sum in the second term of (8.593) by $4A$, where A is the area of the polygon, and obtain

$$\Phi = -\frac{\gamma_0}{32\pi}\sum_{j=1}^{n} Y_j L_j^2\left[H(Z_j)+H(\bar{Z}_j)\right] + \frac{\gamma_0 A}{4\pi}[\ln(z-z_1)+\ln(\bar{z}-\bar{z}_1)]. \qquad (8.594)$$

8.26.3 The Complex Potential Valid outside the Polygon

The function Φ is harmonic outside the polygon and therefore can be represented as the real part of a complex potential, which is, by definition, a function of z (or $Z_j, j = 1,\ldots,n$) only. Accordingly, we write (8.594) as

$$\Phi = \tfrac{1}{2}[\Omega(z) + \overline{\Omega(z)}]. \tag{8.595}$$

We obtain an expression for Ω valid outside the polygon after rewriting (8.594) as

$$\Phi = -\frac{\gamma_0}{64\pi \mathrm{i}} \sum_{j=1}^{n} (Z_j - \bar{Z}_j) L_j^2 \left[H(Z_j) + H(\bar{Z}_j) \right] + \frac{\gamma_0 A}{4\pi} [\ln(z - z_1) + \ln(\bar{z} - \bar{z}_1)] \tag{8.596}$$

or

$$\Phi = -\frac{\gamma_0}{64\pi \mathrm{i}} \sum_{j=1}^{n} L_j^2 \left[(Z_j - \bar{Z}_j)[H(Z_j) - (\bar{Z}_j - Z_j) H(\bar{Z}_j) \right] + \frac{\gamma_0 A}{4\pi} [\ln(z - z_1) + \ln(\bar{z} - \bar{z}_1)]. \tag{8.597}$$

We are interested in the complex function Ω of which Φ is the real part; this function is

$$\Omega = -\frac{\gamma_0}{32\pi \mathrm{i}} \sum_{j=1}^{n} L_j^2 (Z_j - \bar{Z}_j) H(Z_j) + \frac{\gamma_0 A}{2\pi} \ln(z - z_1). \tag{8.598}$$

The advantage of using the complex potential outside the polygon is that we have access to a stream function. It is not at all clear from (8.598) that this function is indeed holomorphic, as it contains $\bar{z}$. The proof that Ω is indeed holomorphic is given in the next section for the sake of completeness.

8.26.4 The Holomorphic Function Ω Outside the Polygon

The function Ω contains a term $\bar{Z}_j$ and it is not clear that Ω is holomorphic outside the polygon, i.e., has both real and imaginary parts that satisfy Laplace's equation. We verify that this is indeed the case. The partial derivative of Ω with respect to z is

$$\frac{\partial \Omega}{\partial z} = -\frac{\gamma_0}{32\pi \mathrm{i}} \sum_{j=1}^{n} L_j^2 \left\{ H(Z_j) + Z_j \frac{dH(Z_j)}{dZ_j} - \bar{Z}_j \frac{dH(Z_j)}{dZ_j} \right\} \frac{dZ_j}{dz} + \frac{\gamma_0 A}{4\pi} \frac{1}{z - z_1}. \tag{8.599}$$

We differentiate this equation with respect to $\bar{z}$

$$\frac{\partial^2 \Omega}{\partial z \partial \bar{z}} = -\frac{\gamma_0}{32\pi \mathrm{i}} \sum_{j=1}^{n} L_j^2 \left[-\frac{dH(Z_j)}{dZ_j} \right] \frac{dZ_j}{dz} \frac{d\bar{Z}_j}{d\bar{z}}. \tag{8.600}$$

Since $dZ_j/dz d\bar{Z}_j/d\bar{z} = 4/L_j^2$ we can write this as

$$\frac{\partial^2 \Omega}{\partial z \partial \bar{z}} = -\frac{\gamma_0}{8\pi \mathrm{i}} \sum_{j=1}^{n} \frac{dH(Z_j)}{dZ_j}. \tag{8.601}$$

We obtain the derivative of $H(Z_j)$ with respect to z by expressing part of the function in terms of the global variable z

$$\frac{dH(Z_j)}{dZ_j} = \ln\frac{Z_j-1}{Z_j+1} + (Z_j+1)\left[\frac{1}{Z_j-1} - \frac{1}{Z_j+1}\right] - 2\frac{d}{dz}\left\{\sum_{m=j+1}^{n} \ln\frac{z-z_m}{z-z_{m+1}}\right\}\frac{dz}{dZ_j}. \tag{8.602}$$

Simplifying and combining all the logarithms in the sum into the logarithm of a product of terms we see that this can be written as follows, writing dz/dZ_j as $\frac{1}{2}(z_{j+1}-z_j)$

$$\frac{dH(Z_j)}{dZ_j} = \ln\frac{Z_j-1}{Z_j+1} + \frac{Z_j+1}{Z_j-1} - 1 - 2\frac{d}{dz}\ln\frac{z-z_{j+1}}{z-z_{n+1}}\frac{1}{2}(z_{j+1}-z_j) \tag{8.603}$$

or

$$\frac{dH(Z_j)}{dZ_j} = \ln\frac{Z_j-1}{Z_j+1} + \frac{Z_j+1}{Z_j-1} - 1 - \left[\frac{1}{z-z_{j+1}} - \frac{1}{z-z_{n+1}}\right](z_{j+1}-z_j). \tag{8.604}$$

We write $(Z_j+1)/(Z_j-1)$ as $(z-z_j)/(z-z_{j+1})$

$$\frac{dH(Z_j)}{dZ_j} = \ln\frac{Z_j-1}{Z_j+1} + \frac{z-z_j}{z-z_{j+1}} - 1 - \frac{z_{j+1}-z_j}{z-z_{j+1}} + \frac{z_{j+1}-z_j}{z-z_{n+1}}. \tag{8.605}$$

The second, third, and fourth terms cancel and we obtain, with (8.601), and noting that $z_{n+1} = z_1$,

$$\frac{\partial^2\Omega}{\partial z\partial\bar{z}} = -\frac{\gamma_0}{8\pi\mathrm{i}}\sum_{j=1}^{n}\ln\frac{Z_j-1}{Z_j+1} + \sum_{j=1}^{n}\left[z_{j+1}-z_j\right]\frac{1}{z-z_1}. \tag{8.606}$$

The first sum is zero outside the polygon; the second sum is zero because the polygon is closed. The complex potential is therefore indeed holomorphic outside the polygon, i.e.,

$$\frac{\partial^2\Omega}{\partial z\partial\bar{z}} = 0. \tag{8.607}$$

8.26.5 Implementation

Implementation of the solution requires that we use the function (8.552), with $\gamma_0 = 1$, to determine whether the point z is inside or outside the polygon. Storing for each value of z the individual terms $\ln((Z_j-1)/(Z_j+1))$ in memory and feeding the resulting numbers into the functions $H(Z_j)$ renders an efficient system for calculating values of the potential inside the polygon and the complex potential outside it.

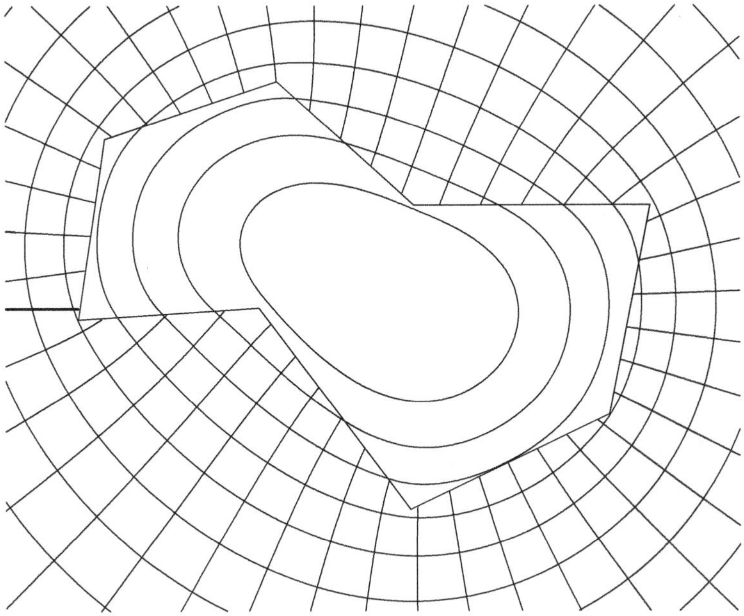

Figure 8.57 Equipotentials for an area-sink of unit strength bounded by a polygon.

Equipotentials are shown in Figure 8.57 for the case of an area-element of unit strength bounded by a polygon; at large distances the element approaches that of a well. The streamlines shown outside the polygon as well as the equipotentials outside the polygon were obtained by contouring the real and imaginary parts of (8.598). We see from the figure that the streamlines and equipotentials form a flow net, and that the equipotentials obtained from (8.594) inside the polygon match perfectly with those outside the polygon.

8.27 The Generating Analytic Element Approach

The generating analytic element approach is a method for determining analytic elements that satisfy differential equations other than the Laplace and Poisson equations. The approach was introduced by Strack [2009b], where the analytic elements for non-singular solutions, wells, line-sinks, and line-doublets are presented. We illustrate the approach for a circular lake with a leaky bottom in a field of uniform flow. The solution can be obtained in terms of expansions. The approach relies on Wirtinger calculus, where z and $\bar{z}$ are used as the independent variables.

The approach is based on first selecting a holomorphic function that has the desired singular behavior; the singularity may be at infinity. The solution for the case under consideration can be written in terms of a series expansion based on the Taylor series, called the generating analytic element, which is singular only at infinity.

We demonstrate in what follows how the generating analytic element approach leads, by a process of successive integration, to a solution of the modified Helmholtz equation. This solution consists of an infinite series of functions; convergence of this series must be determined on a case-by-case basis by considering the form of the solution.

We write the modified Helmholtz equation in the following form

$$\nabla^2\Phi = 4\frac{\partial^2\Phi}{\partial z\partial\bar{z}} = \frac{\Phi}{\Lambda^2}, \tag{8.608}$$

where Λ [L] is the leakage factor. We introduce a dimensionless complex variable Z:

$$Z = \frac{z - z_0}{R}, \tag{8.609}$$

where z_0 is the complex coordinate of some point in the plane, and where R [L] is a chosen parameter with the dimension of length. We write (8.608) in terms of the dimensionless variable Z:

$$\nabla^2\Phi = 4\frac{\partial^2\Phi}{\partial z\partial\bar{z}} = 4\frac{\partial^2\Phi}{\partial Z\partial\bar{Z}}\frac{dZ}{dz}\frac{d\bar{Z}}{d\bar{z}} = \frac{4}{R^2}\frac{\partial^2\Phi}{\partial Z\partial\bar{Z}} = \frac{\Phi}{\Lambda^2}. \tag{8.610}$$

We rewrite (8.610) in terms of a new dimensionless parameter β,

$$\frac{\partial^2\Phi}{\partial Z\partial\bar{Z}} = \beta^2\Phi, \tag{8.611}$$

where

$$\beta = \frac{R}{2\Lambda}. \tag{8.612}$$

We introduce a series of real functions $\underset{n}{H}$ with the property

$$\frac{\partial^2\underset{n}{H}}{\partial Z\partial\bar{Z}} = \underset{n-1}{H} \tag{8.613}$$

and select the first function $\underset{0}{H}$ as the generating analytic element, which we represent as $\underset{0}{\Phi}$, i.e.,

$$\underset{0}{H} = \underset{0}{\Phi}, \tag{8.614}$$

where $\underset{0}{H}$ is holomorphic, i.e.,

$$\frac{\partial^2\underset{0}{H}}{\partial Z\partial\bar{Z}} = 0. \tag{8.615}$$

With the function $\underset{0}{H}$ known, $\underset{1}{H}$ can be found by successive integration of (8.613) with respect to Z and $\bar{Z}$ for $n = 0$. Repeating this procedure will yield expressions for the functions $\underset{n}{H}$. This integration process is possible after writing the Laplacian in the form (8.611), which permits integration in two steps, first with respect to Z and then with respect to $\bar{Z}$.

We write the discharge potential as an infinite series of functions:

$$\Phi = \sum_{n=0}^{\infty} \beta^{2n} \underset{n}{H} \tag{8.616}$$

and substitute this expression for Φ in the differential equation (8.611):

$$\frac{\partial^2 \Phi}{\partial Z \partial \bar{Z}} = \sum_{n=1}^{\infty} \beta^{2n} \underset{n-1}{H} = \beta^2 \sum_{n=1}^{\infty} \beta^{2(n-1)} \underset{n-1}{H} = \beta^2 \sum_{n=0}^{\infty} \beta^{2n} \underset{n}{H} = \beta^2 \Phi. \tag{8.617}$$

We consider only a finite number of terms, N,

$$\frac{\partial^2 \Phi}{\partial Z \partial \bar{Z}} = \beta^2 \sum_{n=1}^{N} \beta^{2(n-1)} \underset{n-1}{H} = \beta^2 \sum_{n=0}^{N-1} \beta^{2n} \underset{n}{H}, \tag{8.618}$$

which results in a difference between $\partial^2 \Phi / (\partial Z \partial \bar{Z})$ and $\beta^2 \Phi$:

$$\frac{\partial^2 \Phi}{\partial Z \partial \bar{Z}} - \beta^2 \Phi = -\beta^{2N+2} \underset{N}{H}. \tag{8.619}$$

If the series of functions converges, then, given an infinite number of terms, the differential equation will be satisfied exactly. The series of functions presented in this section converges absolutely, and we reduce the error due to truncation to within machine accuracy. Thus, for all practical purposes, the solutions satisfy the differential equation exactly.

The procedure can be carried out in terms of complex functions, often an advantage because the generating analytic element is known as a function of a single complex variable. In that case, we use complex functions $\underset{n}{\Xi}$ to replace the real functions $\underset{n}{H}$ so that

$$\frac{\partial^2 \underset{n}{\Xi}}{\partial Z \partial \bar{Z}} = \underset{n-1}{\Xi}. \tag{8.620}$$

The potential is the real part of the sum of functions $\underset{n}{\Xi}$, each multiplied by a term β^{2n}

$$\Phi = \sum_{n=0}^{\infty} \beta^{2n} \underset{n}{H} = \tfrac{1}{2} \sum_{n=0}^{N} \beta^{2n} \left[\underset{n}{\Xi} + \underset{n}{\overline{\Xi}} \right], \tag{8.621}$$

so that

$$\underset{n}{H} = \tfrac{1}{2}\left[\underset{n}{\Xi} + \underset{n}{\bar{\Xi}}\right]. \tag{8.622}$$

8.27.1 The Non-Singular Holomorphic Function

If the generating analytic element does not contain any singularities in the domain of interest, then we can represent the function $\underset{n}{\Xi}$ in general form as

$$\underset{n}{\Xi} = \frac{\bar{Z}^n}{n!}\underset{n}{f}(Z), \tag{8.623}$$

where the functions $\underset{n}{f}(Z)$ have the property:

$$\frac{d\underset{n}{f}}{dZ} = \underset{n-1}{f}. \tag{8.624}$$

The function (8.623) satisfies (8.620). We verify this by first differentiating (8.623) with respect to $\bar{Z}$:

$$\frac{\partial \underset{n}{\Xi}}{\partial \bar{Z}} = \frac{n\bar{Z}^{n-1}}{n!}\underset{n}{f}(Z) \tag{8.625}$$

and next with respect to Z:

$$\frac{\partial^2 \underset{n}{\Xi}}{\partial Z \partial \bar{Z}} = \frac{\bar{Z}^{n-1}}{(n-1)!}\underset{n-1}{f}(Z) = \underset{n-1}{\Xi}, \tag{8.626}$$

which matches (8.620).

8.27.2 The Taylor Series

We apply the generating analytic element approach to generate a function Φ that satisfies (8.608) and has no singularities in the finite domain. We do this by choosing as the generating function the Taylor series:

$$\underset{0}{\Xi} = \sum_{m=0}^{\infty} \underset{0}{\Xi}_m = \sum_{m=0}^{\infty} a_m \frac{Z^m}{m!} = \sum_{m=0}^{\infty} a_m \underset{0}{f}_m(Z). \tag{8.627}$$

We consider a single power Z^m, and construct the function $\underset{n}{f}$ according to the definition (8.624), which yields, denoting the power by the index m,

$$\underset{n}{f}_m(Z) = \frac{Z^{m+n}}{(m+n)!}. \tag{8.628}$$

We apply this to each term in the Taylor series, use (8.623), and obtain

$$\underset{n}{\Xi} = \sum_{m=0}^{\infty} a_m \left\{ \frac{\bar{Z}^n Z^{m+n}}{n!\,(m+n)!} \right\} = \sum_{m=0}^{\infty} a_m \left\{ \frac{(\bar{Z}Z)^n Z^m}{(m+n)!\,n!} \right\}. \tag{8.629}$$

We obtain an expression for the potential Φ by the use of (8.621):

$$\Phi = \tfrac{1}{2} \sum_{n=0}^{\infty} \sum_{m=0}^{\infty} \frac{\beta^{2n}}{(m+n)!\,n!} \left[a_m \bar{Z}^n Z^{m+n} + \bar{a}_m Z^n \bar{Z}^{m+n} \right]. \tag{8.630}$$

Note that each term contains a factor $1/[(m+n)!\,n!]$ that causes a rapid decrease in value of the individual terms beyond a certain term, causing the series to converge to some fixed value. The series converges absolutely; the ratio of the modulus of the $(n+1)$th term divided by that of the nth one vanishes for $n \to \infty$.

The truncation error that results from limiting the series to a finite number of terms consists of two parts: the first part is due to truncation of the Taylor series that represents the generating analytic element; the second part is due to truncation of the series of functions.

8.28 Application: A Leaky Pond

We apply the generating element approach and a method called holomorphic matching (see Strack [2009a]) to the problem of a circular pond with constant head and a leaky bottom in a field of uniform flow. The solution allows us to study the interaction of the pond with the flow field as a function of the leakage factor.

An analytic element for a circular pond was first developed by Miller [2001], who wrote the solution in terms of a general series of powers of z and $\bar{z}$, and then established relationships between the various coefficients; his derivation preceded the publication of the generating analytic element approach and differs from the one presented here. Other than the derivation of the function and the form of the expressions for the coefficients, the solution is equivalent to the one presented by Miller [2001]. Bakker [2002] and [2004] published solutions that involve lakes with leaky bottoms in multi-aquifer systems.

We consider the case of a circular pond with a leaky bottom in a field of uniform flow and recall the equations for semiconfined flow developed in Section 5.1. The head in the pond is ϕ_0 and the expression for the potential in terms of head for this case, valid in the aquifer below the pond, is

$$\Phi = kH(\phi - \phi_0), \tag{8.631}$$

where k is the hydraulic conductivity, and c the resistance of the bottom of the pond,

$$c = \frac{k^*}{H^*}, \tag{8.632}$$

H^* is the thickness of the bottom of the pond, and k^* is the hydraulic conductivity of the material on the bottom of the pond. The leakage factor, Λ, is given by

$$\Lambda = \sqrt{kHc}. \tag{8.633}$$

The aquifer outside the pond also has thickness H and its hydraulic conductivity is k. We define the potential outside the pond also by expression (8.631) for reasons of convenience.

Flow from a pond with a leaky bottom into an aquifer is illustrated in Figure 8.58, where piezometric contours are shown for the case in which $k = 10$ m/d, $H = 10$ m, $\Lambda = 50$ m, $Q_{x0} = 0.6$ m^2/d, and the radius of the pond is $R = 100$ m. The head in the pond is 22 m, and the head nearby and downstream (150 m downstream from the center of the pond) is 19 m. These values were chosen to illustrate a case of significant flow through the bottom of the pond into the aquifer. The maximum head in the aquifer is 20.6 m, at the upstream side it is 21 m, and at the downstream side it is 20.1 m; there is downward leakage throughout the area of the pond. The plot shown in Figure 8.59 refers to the same case, but shows equipotentials rather than piezometric contours, and, outside the pond, streamlines.

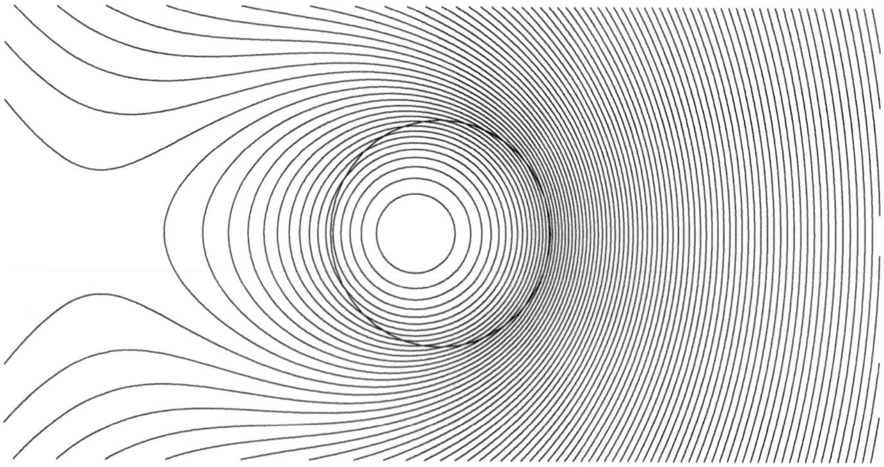

Figure 8.58 Piezometric contours for a pond with a leaky bottom in a field of uniform flow; the leakage is into the aquifer everywhere below the pond.

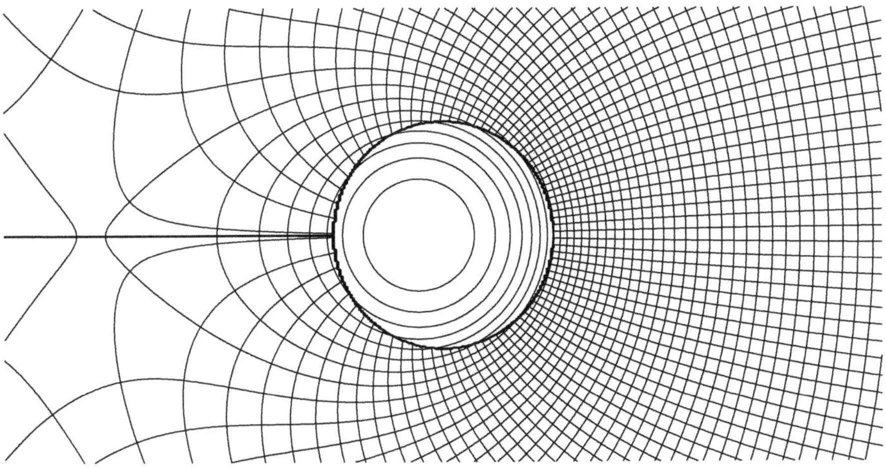

Figure 8.59 Equipotentials and streamlines for a pond with a leaky bottom in a field of uniform flow; the leakage is into the aquifer everywhere below the pond.

8.28.1 The Generating Function

The generating function for uniform flow is the linear function

$$\underset{0}{\Xi} = a_1 Z + a_0, \tag{8.634}$$

where a_0 is a real constant; we define Z as

$$Z = \frac{z - z_0}{R}, \tag{8.635}$$

where R is the radius of the circle that bounds the pond, and z_0 is its center. The Taylor series that serves as the generating analytic element for this case reduces to the linear form (8.634) and the expression for the potential, (8.630), reduces to

$$\Phi = \tfrac{1}{2} \sum_{n=0}^{\infty} \sum_{m=0}^{1} \frac{\beta^{2n}}{(m+n)!\,n!} \left[a_m \bar{Z}^n Z^{m+n} + \bar{a}_m Z^n \bar{Z}^{m+n} \right], \qquad Z\bar{Z} \le 1. \tag{8.636}$$

We separate the term with a_0 from the one with a_1:

$$\begin{aligned} \Phi = & \tfrac{1}{2} a_0 \sum_{n=0}^{\infty} \frac{\beta^{2n}}{(n!)^2} \left[\bar{Z}^n Z^n + Z^n \bar{Z}^n \right] \\ & + \tfrac{1}{2} \sum_{n=0}^{\infty} \frac{\beta^{2n}}{(n+1)!\,n!} \left[a_1 \bar{Z}^n Z^{n+1} + \bar{a}_1 Z^n \bar{Z}^{n+1} \right], \qquad Z\bar{Z} \le 1, \end{aligned} \tag{8.637}$$

and combine terms:

$$\Phi = a_0 \sum_{n=0}^{\infty} \frac{\beta^{2n}}{(n!)^2} (Z\bar{Z})^n + \tfrac{1}{2} \sum_{n=0}^{\infty} \frac{\beta^{2n}}{(n+1)!n!} (Z\bar{Z})^n \big[a_1 Z + \bar{a}_1 \bar{Z}\big], \qquad Z\bar{Z} \le 1. \tag{8.638}$$

8.28.2 The Continuity Condition

We apply the condition of continuity of flow across the boundary of the pond by holomorphic matching (see Strack [2009a]) based on writing the complex discharge function $W = W(Z,\bar{Z})$ along the boundary, in this case defined by $Z\bar{Z} = 1$, as a function of Z only, by replacing $\bar{Z}$ by $1/Z$. The resulting expression is continued analytically, i.e., is used to represent W outside the boundary as a function of Z only, i.e., as a holomorphic function. Since W is continuous, the solution meets the boundary condition, and is the unique solution to the problem, provided that the condition at infinity is matched as well.

We obtain an expression for the discharge function W by differentiating (8.637),

$$W = -2\frac{\partial \Phi}{\partial z} = \Bigg\{ -a_0 \sum_{n=1}^{\infty} n \frac{\beta^{2n}}{(n!)^2} \big[\bar{Z}^n Z^{n-1} + Z^{n-1}\bar{Z}^n\big] - a_1 - \sum_{n=1}^{\infty} \frac{\beta^{2n}}{(n+1)!n!} \big[(n+1)a_1 \bar{Z}^n Z^n + n\bar{a}_1 Z^{n-1}\bar{Z}^{n+1}\big] \Bigg\} \frac{dZ}{dz}, \tag{8.639}$$

where $dZ/dz = 1/R$. We express W along the boundary of the pond, i.e., for $\bar{Z} = 1/Z$ as

$$W = -2\frac{a_0}{R} \sum_{n=1}^{\infty} n \frac{\beta^{2n}}{(n!)^2} \frac{1}{Z} - \frac{a_1}{R} - \frac{1}{R} \sum_{n=1}^{\infty} \frac{\beta^{2n}}{(n+1)!n!} \left[(n+1)a_1 + n\bar{a}_1 \frac{1}{Z^2}\right], \qquad Z\bar{Z} = 1. \tag{8.640}$$

The potential is harmonic outside the pond, and thus can be represented as the real part of a complex potential; the latter is obtained by integration of

$$W = -\frac{d\Omega}{dz}, \tag{8.641}$$

which yields the expression of the complex potential valid outside the pond:

$$\Omega = -\int W dz = -\int W \frac{dz}{dZ} dZ = -R \int W dZ$$
$$= 2a_0 \sum_{n=1}^{\infty} n \frac{\beta^{2n}}{(n!)^2} \ln Z + a_1 Z$$
$$+ \sum_{n=1}^{\infty} \frac{\beta^{2n}}{(n+1)!n!} \left[(n+1) a_1 Z - n \bar{a}_1 \frac{1}{Z} \right] + C, \qquad Z\bar{Z} \geq 1. \tag{8.642}$$

where C is a real constant of integration. We write a_1 as $|a_1| e^{i\alpha_1}$ and determine the real part:

$$\Phi = 2a_0 \sum_{n=1}^{\infty} \frac{\beta^{2n}}{n!(n-1)!} \ln |Z| + |a_1||Z| \cos(\theta + \alpha_1)$$
$$+ |a_1| \sum_{n=1}^{\infty} \frac{\beta^{2n}}{(n+1)!n!} \left[(n+1)|Z| \cos(\theta + \alpha_1) - n \frac{1}{|Z|} \cos(-\alpha_1 - \theta) \right]$$
$$+ C, \qquad Z\bar{Z} \geq 1. \tag{8.643}$$

and combine terms:

$$\Phi = 2a_0 \sum_{n=1}^{\infty} \frac{\beta^{2n}}{n!(n-1)!} \ln |Z| + |a_1||Z| \cos(\theta + \alpha_1)$$
$$+ |a_1| \cos(\theta + \alpha_1) \sum_{n=1}^{\infty} \frac{\beta^{2n}}{(n+1)!n!} \left[(n+1)|Z| - n \frac{1}{|Z|} \right] + C, \qquad Z\bar{Z} \geq 1. \tag{8.644}$$

8.28.3 Continuity of the Potential

The discharge is continuous across the boundary of the pond. The corresponding expressions for the potential can differ only by a constant, which we determine by setting the expressions for the potential approaching the boundary from the outside and from the inside equal to each other. We set $|Z| = 1$ to obtain the expression for Φ along the boundary approaching it from the outside:

$$\Phi = |a_1| \cos(\theta + \alpha_1) \left\{ 1 + \sum_{n=1}^{\infty} \frac{\beta^{2n}}{(n+1)!n!} \right\} + C \tag{8.645}$$

or

$$\Phi = |a_1| \cos(\theta + \alpha_1) \sum_{n=0}^{\infty} \frac{\beta^{2n}}{(n+1)!n!} + C. \tag{8.646}$$

The potential on the inside is given by (8.638); we set $Z\bar{Z} = 1$ and $a_1 Z + \bar{a}_1 \bar{Z}$ equal to $2|a_1|\cos(\theta + \alpha_1)$,

$$\Phi = a_0 \sum_{n=0}^{\infty} \frac{\beta^{2n}}{(n!)^2} + \tfrac{1}{2}|a_1| \sum_{n=0}^{\infty} \frac{\beta^{2n}}{(n+1)!n!} 2\cos(\theta + \alpha_1). \tag{8.647}$$

We set the latter two equations equal to each other and solve for a_0:

$$a_0 = \frac{C}{B}, \tag{8.648}$$

where

$$B = \sum_{n=0}^{\infty} \frac{\beta^{2n}}{(n!)^2}. \tag{8.649}$$

We use this value for a_0 in the expression for the complex potential valid outside the pond, (8.642),

$$\begin{aligned} \Omega = {} & 2\frac{C}{B} \sum_{n=1}^{\infty} \frac{\beta^{2n}}{n!(n-1)!} \ln Z + a_1 Z \\ & + \sum_{n=1}^{\infty} \frac{\beta^{2n}}{(n+1)!n!} \left[(n+1)a_1 Z - n\bar{a}_1 \frac{1}{Z} \right] + C, \qquad Z\bar{Z} \geq 1. \end{aligned} \tag{8.650}$$

If it is given that the flow far away from the pond is uniform, is in the direction α, and has a discharge Q_u, then we may express the constant a_1 in terms of Q_u and α from the equation

$$\Omega \to -Q_u e^{-i\alpha} z \to -Q_u e^{-i\alpha} RZ \tag{8.651}$$

or, with (8.650),

$$-Q_u R e^{-i\alpha} = a_1 \left[1 + \sum_{n=1}^{\infty} \frac{\beta^{2n}}{(n!)^2} \right] = a_1 \sum_{n=0}^{\infty} \frac{\beta^{2n}}{(n!)^2} = a_1 B, \tag{8.652}$$

so that

$$a_1 = -\frac{Q_u R e^{-i\alpha}}{B} = -Q_u R e^{-i\alpha} - Q_u R e^{-i\alpha} \frac{1-B}{B}. \tag{8.653}$$

If the head at some point $z = L$ downstream from the pond is ϕ_1, then the constant C can be determined by applying the condition

$$\Phi = \Phi_1 = kH(\phi_1 - \phi_0), \qquad z = L. \tag{8.654}$$

Figures 8.60, 8.61, and 8.62 are illustrations of a second case of a leaky pond in a field of uniform flow; this case differs from that shown in Figure 8.59 in

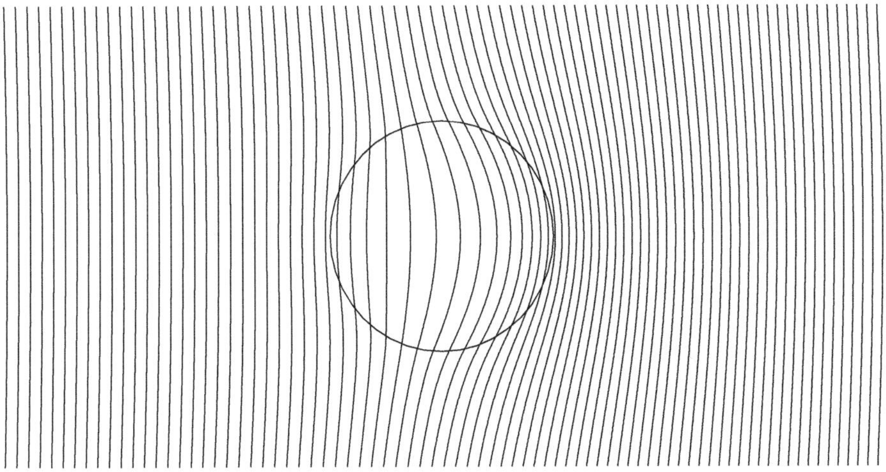

Figure 8.60 A circular pond with leaky bottom in a field of uniform flow; flow enters through the bottom of the pond on the upstream side and leaves the pond on the downstream side. Both upward and downward leakage occur through the bottom of the pond.

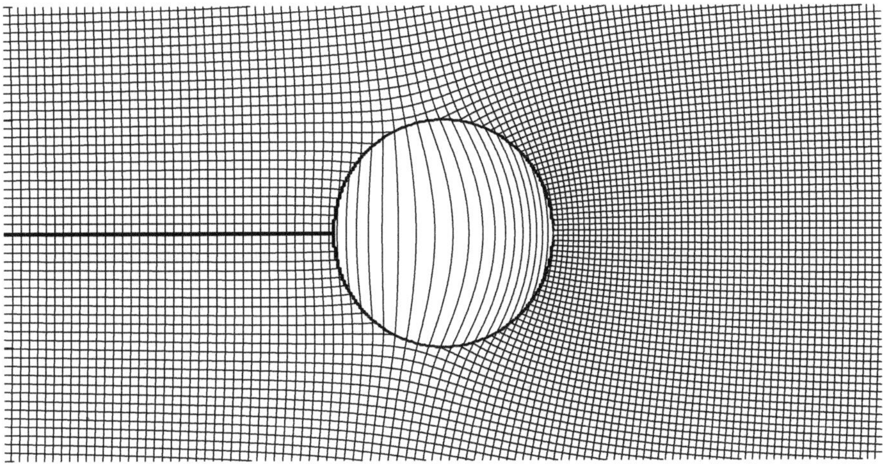

Figure 8.61 A circular pond with leaky bottom in a field of uniform flow; flow enters through the bottom of the pond on the upstream side and leaves the pond on the downstream side. Streamlines are shown outside the pond. Note the branch cut, which shows that there is a net outflow from the pond into the aquifer.

that water enters through the upstream part of the bottom, and leaves it through the downstream part. The hydraulic conductivity is 10 m/d, the aquifer thickness is 10 m, and the leakage factor Λ is $R/2$. The head in the pond is 20 m and a head nearby (at $x = 150R$, $y = 0$) is 19 m. The uniform flow field for all cases is

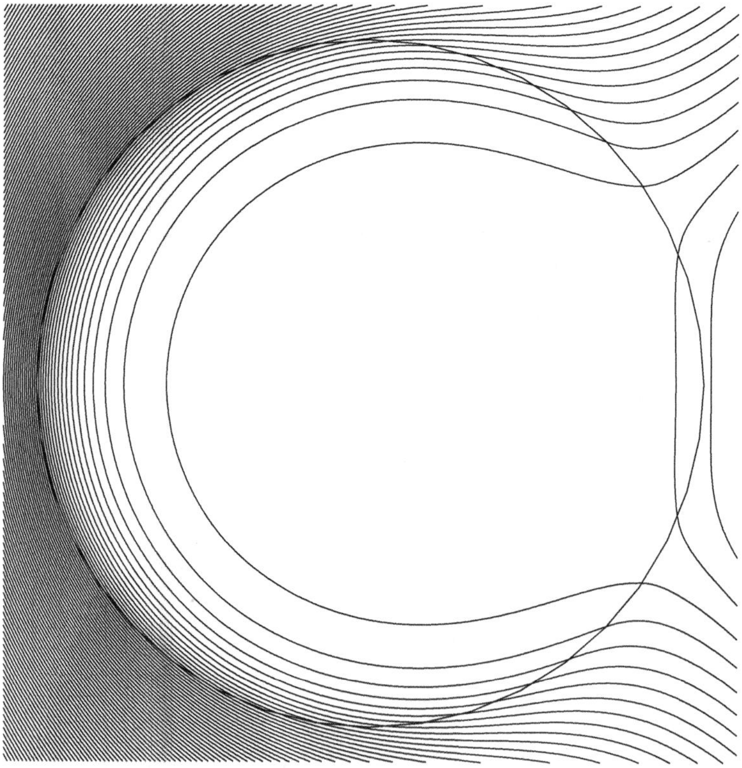

Figure 8.62 Close-up of the flow below the pond; the head varies from 20.2 m to 19.98 m.

$Q_{x0} = 0.6$ m^2/d. Figure 8.60 shows piezometric contours, and Figure 8.61 shows the equipotentials and, outside the pond, the streamlines. The head below the center of the pond is 19.8 m, i.e., below the level of the pond. The head along the fifth piezometric contour to the right of the leftmost point of the pond is 20 m; to the left of this contour the leakage is upward, and to the right it is downward. As a final illustration, a close-up of the area below and near the pond is shown in Figure 8.62. The data for this case are as for Figure 8.60, except that $\Lambda = R/10$ and the head at $z = 150$ m is 19.9 m, i.e., 10 cm below the head in the pond. Note that the heads in the aquifer vary only very little; the range is from 20.2 m on the upstream side to 19.98 m on the downstream side. The small value of the resistance for this case forces the heads below the pond to be nearly equal to that in the pond. The gradient below the pond is highest along the upstream side and is nearly zero in the central portion of the pond.

9

Fluid Particle Paths and Solute Transport

We consider in this chapter (1) the average paths that fluid particles follow on their way through the soil and (2) the transport of solutes. The paths are average; deviations due to the discrete nature of the porous material are ignored. We refer to these average particle paths as path lines.

For steady flow the path lines coincide with the streamlines. For transient flow the pattern of streamlines changes with time; the path lines are different from the streamlines. Closed form expressions for streamlines and path lines exist only for a few cases; we usually resort to numerical methods. We first discuss the computation of points on streamlines and path lines, then cover an approximate technique for determining streamlines and path lines in three dimensions using the Dupuit-Forchheimer approximation. We finish this chapter with a brief discussion of solute transport.

9.1 Numerical Determination of Fluid Particle Paths

The equations for the streamlines usually are implicit in terms of the coordinates x and y. We can determine points on streamlines by the use of the stream function if the potential is harmonic. The application of a contouring procedure to a set of values of Ψ computed at the mesh points of a grid is one way of computing points of a set of streamlines. A disadvantage of this approach is that jumps of the stream function along branch cuts appear in the plot as heavy black lines. This problem can be avoided only by using a special contouring routine that processes the jumps across the branch cuts correctly. An alternative approach is to determine points of a given streamline in such a way that jumps in the stream function can be handled easily; we discuss such a procedure in this section.

If the potential is not harmonic, as in shallow flow with infiltration and in shallow transient flow, then the stream function cannot be used. We cover separately procedures for computing points on streamlines and path lines applicable to such cases.

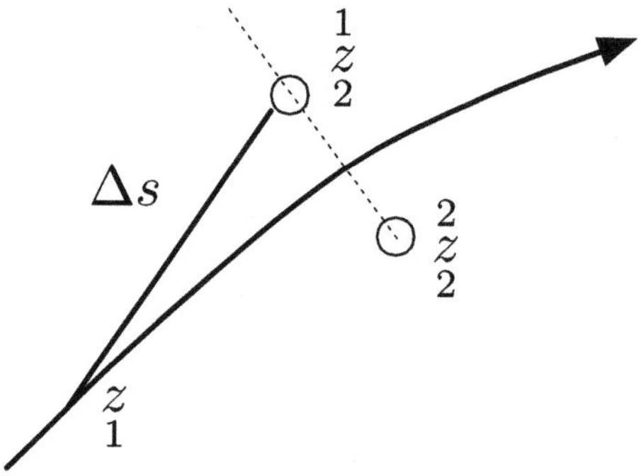

Figure 9.1 Tracing a streamline.

9.1.1 A Procedure for Tracing Streamlines Using the Stream Function

We determine points on a streamline by a process that we refer to as tracing streamlines: we start at any given point of a streamline $\Psi = \Psi_j$ and proceed by calculating complex coordinates z of points on the streamline one by one. Let $z = z_1$ be a point of the streamline $\Psi = \Psi_j$ (see Figure 9.1). We obtain a first estimate of the next point on the streamline, $\overset{1}{z}_2$, as follows:

$$z_2 = z_1 + e^{i\alpha}\Delta s, \tag{9.1}$$

where Δs is some fixed quantity, equal to the distance between z_1 and z_2, and $e^{i\alpha}$ is tangent to the streamline at z_1. The angle α is equal to the argument of $\bar{W} = Q_x + iQ_y$, by definition, and we represent $e^{i\alpha}$ as follows, assuming that $W(z_1) \neq 0$,

$$e^{i\alpha} = \frac{\bar{W}(z_1)}{|W(z_1)|}, \qquad W(z_1) \neq 0. \tag{9.2}$$

We obtain the second approximation $\overset{2}{z}_2$ of z_2 by application of a Newton-Raphson procedure along the equipotential through z_2. We write the following approximation for W, using that Φ is constant along an equipotential:

$$W(z_2) = -\left[\frac{d\Omega}{dz}\right]_{z=z_2} \approx -\frac{\Delta\Phi + i\Delta\Psi}{\overset{2}{z}_2 - z_2} = -i\frac{\Psi_j - \Psi(z_2)}{\overset{2}{z}_2 - z_2}. \tag{9.3}$$

We solve for $\overset{2}{z}_2$:

$$\overset{2}{z}_2 = z_2 + i\frac{\Psi(z_2) - \Psi_j}{W(z_2)}. \tag{9.4}$$

We apply this equation recursively until $|\overset{2}{z}_2 - z_2|$ is less than a certain quantity ε. Tracing streamlines is often done in connection with graphical display; in such cases ε may be taken as the resolution of the graphics device.

9.1.2 Handling the Jumps in Ψ

Where the streamlines intersect a branch cut, the value of the stream function jumps. If such a jump occurs, the magnitude of $\Psi(z_2) - \Psi_j$ in (9.4) will be much larger than the magnitude due to the deviation from the streamline. This magnitude therefore should be checked in the program, and whenever it is larger than some preset value, a procedure should be used that determines which branch cut has been crossed. Assume, for example, that the branch cut is crossed of a well of discharge Q. The constant value Ψ_j then may be updated by subtracting or adding the amount Q. Whether to subtract or add may be decided on the basis of the sign of $\Psi(z_2) - \Psi_j$.

9.1.3 Tracing Streamlines and Path Lines Using the Velocity Vector

The stream function does not exist if the potential is not harmonic, e.g., problems of shallow flow involving infiltration or leakage and problems of shallow transient flow. The procedure outlined above does not apply to such problems; we explain in what follows how points on the streamlines or path lines for such cases may be determined. For transient flow, the streamlines at any time give a graphical representation of the velocity field, whereas the path lines represent the paths followed by the particles. The pattern of streamlines always occupies the entire flow domain and changes with time. A pattern of path lines may be generated by drawing the paths of individual particles that are at given positions at the same time, for example, particles injected through a well. We discuss some elementary procedures for determining streamlines and path lines in what follows.

Steady Flow

Points on streamlines that correspond to the discharge vector (Q_x, Q_y) can be determined by a standard numerical technique. The complex discharge $Q_x + iQ_y$ is given by

$$Q_x + \mathrm{i}Q_y = \overline{W(z)}, \tag{9.5}$$

where

$$W = -2\frac{\partial \Phi}{\partial z}. \tag{9.6}$$

The streamline is tangent to the discharge vector so that

$$\frac{dx}{ds} + \mathrm{i}\frac{dy}{ds} = \frac{dz}{ds} = \frac{\overline{W(z)}}{|W(z)|}, \tag{9.7}$$

where s represents the arc length measured along the streamline. This equation can be integrated numerically. We apply two elementary techniques: the first one is

Euler's method, straightforward, but with errors that increase with the arc length of the streamline. If z_1 represents a point on the streamline, then the next point z_2 is computed by Euler's method as follows:

$$\boxed{z_2 = z_1 + \frac{\overline{W(z_1)}}{|W(z_1)|}\Delta s} \tag{9.8}$$

This approximation amounts to the assumption that the fluid particle moves in the direction of the discharge vector at z_1. An improved technique, the modified Euler's method, is a second-order Runge-Kutta scheme; the fluid particle is assumed to move in the direction of the average discharge over the interval. This yields the approximation:

$$\boxed{\overset{2}{z}_2 = z_1 + \frac{\overline{W(z_1)} + \overline{W(z_2)}}{|W(z_1) + W(z_2)|}\Delta s} \tag{9.9}$$

Further improvement is possible by applying (9.9) again, replacing $\overset{2}{z}_2$ by $\overset{3}{z}_2$ and z_2 by $\overset{2}{z}_2$. This procedure may be repeated until the changes in subsequent approximations of z_2 are less than some specified amount.

The complex variable formulation (9.8) and (9.9) is not restricted to problems where the potential is harmonic, even though the complex potential for such cases does not exist. Note that the stream function is not single-valued for such cases, so that (9.4) cannot be used.

The procedure for tracing streamlines according to (9.8) and (9.9) may be formulated in terms of real variables and generalized to three-dimensional flow. We use a three-dimensional Cartesian coordinate system x_1, x_2, x_3 and represent the velocity vector as v_i $(i = 1,2,3)$. It follows from the definition of a streamline that

$$\frac{dx_i}{ds} = \frac{v_i}{v}, \qquad i = 1,2,3, \tag{9.10}$$

where

$$v = \sqrt{v_1^2 + v_2^2 + v_3^2}. \tag{9.11}$$

We adapt (9.8) to three dimensions:

$$\boxed{\overset{1}{\underset{2}{x}}_i = \underset{1}{x}_i + \frac{v_i(\underset{1}{x}_j)}{v(\underset{1}{x}_j)}\Delta s \qquad (i = 1,2,3)} \tag{9.12}$$

where $v_i(\underset{1}{x_j})$ represents the velocity vector computed at $\underset{1}{x_j}$ $(j = 1,2,3)$. We modify (9.12) by replacing $v_i(\underset{1}{x_j})$ by $\frac{1}{2}[v_i(\underset{1}{x_j}) + v_i(\underset{2}{\overset{1}{x}}_j)]$:

$$\underset{2}{\overset{2}{x}}_i = \underset{1}{x}_i + \frac{\frac{1}{2}[v_i(\underset{1}{x_j}) + v_i(\underset{2}{\overset{1}{x}}_j)]}{v^*}\Delta s, \qquad i = 1,2,3 \tag{9.13}$$

where

$$v^* = \frac{1}{2}\sqrt{\sum_{i=1}^{3}[v_i(\underset{1}{x_j}) + v_i(\underset{2}{\overset{1}{x}}_j)]^2}. \tag{9.14}$$

The time increment Δt corresponds to the time that it takes the fluid to move from $\underset{1}{x}_i$ to $\underset{2}{\overset{2}{x}}_i$ and we estimate it as

$$\Delta t = \frac{\Delta s}{v^*}. \tag{9.15}$$

Because only the ratio of the velocities occurs in (9.10), formulations in terms of velocities and in terms of specific discharges are equivalent: we may use either q_i or $v_i = q_i/n$. The computation of travel times (see (9.15)) must be done in terms of the velocity. It is important to remember that this velocity is an average (the seepage velocity).

9.1.4 Transient Flow

Curves that represent paths of the particles, called *path lines*, and streamlines are not the same for transient flow, in contrast to steady flow. A streamline is tangential to the velocity vector at all of its points. Streamlines form a pattern of non-intersecting curves, unique for each time. Path lines track the particles through time and space; each point on the path line corresponds to a different time.

In the preceding formulation W and the velocity vector are independent of time. The equations can be applied to transient flow, but will yield a pattern of streamlines valid for a particular time if W and the velocity vector are kept constant with time. The same equations can be used to generate path lines, provided that the time dependency of W or v_i is taken into account. We illustrate this for three-dimensional flow and write (9.12) as

$$\underset{2}{\overset{1}{x}}_i = \underset{1}{x}_i + \frac{v_i(\underset{1}{x_j}, \underset{1}{t})}{v(\underset{1}{x_j}, \underset{1}{t})}\Delta s, \qquad i = 1,2,3 \tag{9.16}$$

where $\underset{1}{t}$ corresponds to the time at which the fluid particle is at $\underset{1}{x_i}$. We estimate the time $\overset{1}{\underset{2}{t}}$ at which the particle is at $\overset{1}{\underset{2}{x_j}}$:

$$\overset{1}{\underset{2}{t}} = \underset{1}{t} + \frac{\Delta s}{v(\underset{1}{x_j}, \underset{1}{t})} \tag{9.17}$$

and use this estimate to compute v_i at $\overset{1}{\underset{2}{x_j}}$ and (9.13) becomes

$$\boxed{\overset{2}{\underset{2}{x_i}} = \underset{1}{x_i} + \frac{\frac{1}{2}[v_i(\underset{1}{x_j},\underset{1}{t}) + v_i(\overset{1}{\underset{2}{x_j}},\overset{1}{\underset{2}{t}})]}{v^*}\Delta s, \qquad i=1,2,3} \tag{9.18}$$

where

$$v^* = \frac{1}{2}\sqrt{\sum_{i=1}^{3}[v_i(\underset{1}{x_j},\underset{1}{t}) + v_i(\overset{1}{\underset{2}{x_j}},\overset{1}{\underset{2}{t}})]^2}. \tag{9.19}$$

The next estimate of $\underset{2}{t}$ is obtained from (compare (9.15)):

$$\overset{2}{\underset{2}{t}} = \underset{1}{t} + \frac{\Delta s}{v^*}. \tag{9.20}$$

An alternative procedure for determining path lines is to write the components of the velocity vector as

$$v_i = \frac{dx_i}{dt}, \qquad i = 1,2,3, \tag{9.21}$$

where x_i represents the position of the fluid particle at time t. Numerical integration of (9.21) by Euler's method yields

$$\boxed{\overset{1}{\underset{2}{x_i}} = \underset{1}{x_i} + v_i(\underset{1}{x_j},\underset{1}{t})\Delta t, \qquad i,j = 1,2,3} \tag{9.22}$$

and we obtain a second estimate of $\underset{2}{x_i}$ from

$$\boxed{\overset{2}{\underset{2}{x_i}} = \underset{1}{x_i} + \frac{1}{2}[v_i(\underset{1}{x_j},\underset{1}{t}) + v_i(\overset{1}{\underset{2}{x_j}},\overset{1}{\underset{2}{t}})]\Delta t} \tag{9.23}$$

i.e., we assume that the particle moves from $\underset{1}{x_i}$ to $\underset{2}{x_i}$ with the estimated average velocity. The value of $\overset{1}{\underset{2}{t}}$ is obtained from

$$\overset{1}{\underset{2}{t}} = \underset{1}{t} + \Delta t. \tag{9.24}$$

Equation (9.22) would yield the same answer as (9.16) if Δt in (9.22) were taken as

$$\Delta t = \frac{\Delta s}{v(\underset{1}{x_j}, \underset{1}{t})}. \tag{9.25}$$

By the same token (9.23) yields the same answer as (9.18), if Δt in (9.23) is updated by

$$\Delta t = \frac{\Delta s}{v^*} \tag{9.26}$$

with v^* given by (9.19). The two approaches differ in the step sizes taken along the path line: if a constant step in space is required, then the former approach should be used; if a constant step in time is desired, the latter one should be applied.

We can determine points on path lines by analytic integration of (9.10) only if the function $t = t(s)$ is known beforehand, i.e., if the time associated with each position of the particle on the path line is known, which is true only for a few cases. If we use numerical integration, this function can be determined as the computation proceeds.

Problem

9.1 Consider the (purely imaginary) two-dimensional flow field $v_x = \overset{0}{v}_x e^{-\alpha t}$, $v_y = \overset{0}{v}_y e^{-\beta t}$ where α and β are real constants with the dimension of $[1/s]$.

Questions:

1. Integrate (9.21) to determine the equation for the path line with the initial condition that the particle is at $x = y = 0$ at $t = 0$.
2. Determine t both as a function of x and as a function of y, and use these functions to generate the equation for the path line meant under question 1 by integration of $dy/dx = v_y/v_x$. (The latter equation is obtained by dividing (9.10) for $i = 1$ into (9.10) for $i = 2$.)
3. Determine the equation for the streamline through $x = y = 0$, valid for $t = 0$. Draw both the path line for $0 \leq t \leq \infty$ and the streamline and compare them.

Problem

9.2 Write a computer program to determine streamlines for the case of a well in a field of uniform flow. The well is at the origin, and the discharge of the well is $b * Q_{x0}$. Choose $b = 100$ m, $k = 10$ m/d, and $Q_{x0} = 0.5$ m^2/d. First fill a grid with values of the stream function and obtain a contour plot. Second, start streamlines at points to the extreme left of each streamline produced by contouring. Determine the step size (as a fraction of the width of your window) for which you judge that the errors become unacceptable. Choose a window with the coordinates of the lower left corner at $(-200, -200)$ and the upper right corner at $(200, 200)$.

Problem

9.3 Modify the routine such that iterations are continued until either the error is within a predefined range or a specified maximum number of iterations is surpassed. Repeat the numerical experiment of Problem 9.2.

9.2 3-D Path Lines in Dupuit-Forchheimer Models

Path lines and streamlines can be determined in an approximate manner in Dupuit-Forchheimer models, where the head is constant in the vertical direction as a result of neglecting the vertical resistance to flow; see Section 2.2. The specific discharge does not vary vertically; the horizontal components of the gradient of the hydraulic head do not vary vertically either.

An approach for computing the elevation of points on streamlines in steady Dupuit-Forchheimer models without accretion is considered in Polubarinova-Kochina [1962]. Kirkham [1967] used his vertical slot model to determine points on streamlines for planar Dupuit-Forchheimer models, and Strack [1987] presented an approach for determining streamlines and path lines in general Dupuit-Forchheimer models.

9.2.1 One-Dimensional Flow with Constant Infiltration

We consider a case of shallow one-dimensional flow in a semi-infinite aquifer of constant thickness H, where infiltration occurs at a constant rate N along the horizontal upper boundary of the aquifer (see Figure 9.2). We define a Cartesian (x, x_3)-coordinate coordinate system as shown; we use x_3 to denote the vertical coordinate rather than z, to avoid confusion with the complex variable $z = x + iy$. The potential $\Phi = kH\phi$ fulfills the differential equation

$$\frac{d^2\Phi}{dx^2} = -N \tag{9.27}$$

with the solution

$$\Phi = -\tfrac{1}{2}Nx^2 + Ax + B. \tag{9.28}$$

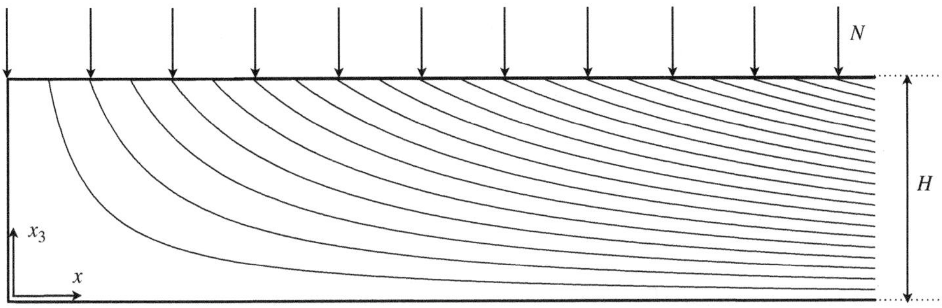

Figure 9.2 Shallow flow with constant infiltration.

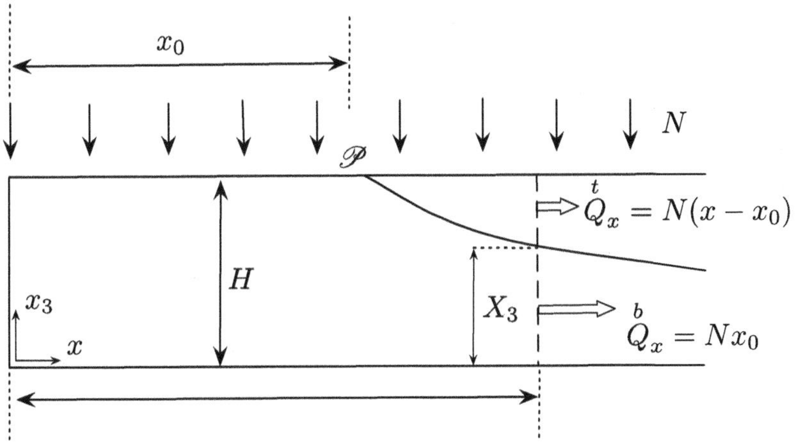

Figure 9.3 The path of a fluid particle.

The discharge vector component Q_x equals

$$Q_x = Hq_x = -\frac{\partial \Phi}{\partial x} = Nx - A. \tag{9.29}$$

The boundary $x = 0$ is impermeable, so that $Q_x = 0$ at $x = 0$; A must be zero and (9.28) becomes

$$\Phi = -\tfrac{1}{2}Nx^2 + \Phi_0, \tag{9.30}$$

where Φ_0 is the potential at $x = 0$.

We define the discharge $\overset{t}{Q}_x$ at any point x as the amount of water flowing between the streamline and the upper boundary. This amount equals the infiltration over the distance $x - x_0$,

$$\overset{t}{Q}_x = N(x - x_0). \tag{9.31}$$

The total discharge Q_x at x equals Nx and the discharge $\overset{b}{Q}_x$ flowing between the base and the streamline equals $Q_x - \overset{t}{Q}_x$,

$$Q_x = Nx, \qquad \overset{b}{Q}_x = Nx_0. \tag{9.32}$$

As a result of the Dupuit-Forchheimer approximation, q_x is constant over the height of the aquifer at x; the elevation X_3 of the streamline above the base is found from continuity of flow:

$$\frac{X_3}{H} = \frac{q_x X_3}{q_x H} = \frac{\overset{b}{Q}_x}{Q_x} = \frac{x_0}{x}. \tag{9.33}$$

This solution corresponds to the exact solution of the problem sketched in Figure 9.2, as we will show. We define the complex variable z as

$$z = x + \mathrm{i}x_3 \tag{9.34}$$

and examine the following complex potential,

$$\Omega = -\frac{NB}{2H}z^2 + \Phi_0, \tag{9.35}$$

where $\Phi = \Re\Omega$ is the potential for two-dimensional flow in the vertical plane,

$$\Phi = kB\phi, \tag{9.36}$$

and B is the width normal to the plane of flow. It follows from (9.35) that

$$W = Q_x - iQ_3 = Bq_x - \mathrm{i}Bq_3 = -\frac{d\Omega}{dz} = \frac{NB}{H}z, \tag{9.37}$$

where the index 3 refers to the vertical coordinate. It follows that

$$q_x = \frac{N}{H}x, \qquad q_3 = -\frac{N}{H}x_3, \tag{9.38}$$

and $q_x = 0$ along $x = 0$, $q_3 = 0$ along $x_3 = 0$, and $q_3 = -N$ along $x_3 = H$, so that the boundary conditions are met. The stream function Ψ equals

$$\Psi = -\frac{NB}{H}xx_3 \tag{9.39}$$

and the value of Ψ along the streamline through $x = x_0$, $x_3 = H$, equals Ψ_0 with

$$\Psi_0 = -NBx_0. \tag{9.40}$$

We obtain the equation for the streamline $\Psi = \Psi_0$ from (9.39) and (9.40) as follows, replacing x_3 by X_3 to indicate that we consider a particular streamline:

$$-NBx_0 = -\frac{NB}{H}xX_3 \tag{9.41}$$

or

$$\frac{X_3}{H} = \frac{x_0}{x}, \tag{9.42}$$

which corresponds exactly to (9.33). It is important to note that the Dupuit-Forchheimer model gives the exact equation for the streamline only if both N and H are constants. The interested reader is referred to Strack [1984] for cases where N varies with position.

The derivation given above for confined Dupuit-Forchheimer flow holds true if the thickness of the aquifer equals $h(x)$, and varies with position such as is the case for shallow unconfined flow. For such cases the thickness H in (9.33) must be replaced by h:

$$\frac{X_3}{h} = \frac{\overset{b}{Q}_x}{Q_x} = \frac{x_0}{x}. \tag{9.43}$$

9.2.2 Streamlines and Path Lines for the General Case

We present approximate equations for the three components of the velocity vector and use them to determine points on streamlines and path lines. We do this by using the complex expression for the divergence and curl of the complex discharge $W = Q_x - \mathrm{i}Q_y$ (see (8.65)):

$$-2\frac{\partial W}{\partial \bar{z}} = \omega = \gamma + \mathrm{i}\delta. \tag{9.44}$$

The function $W(z, \bar{z}, t)$ is, in general, a function of position as well as time and depends on both z and $\bar{z}$, unless the flow is both divergence-free and irrotational. We use the discharge function to determine streamlines in the horizontal plane, the discharge streamlines. In reality, the surfaces in space that are tangent to the specific discharge are not vertical, but in Dupuit-Forchheimer models they are, as a result of the lack of vertical variation of the horizontal specific discharge components. We restrict the flow to irrotational, so that $\delta = 0$, and (9.44) reduces to

$$2\frac{\partial W}{\partial \bar{z}} = \frac{\partial Q_x}{\partial x} + \frac{\partial Q_y}{\partial y} = -\omega = -\gamma. \tag{9.45}$$

If the aquifer is incompressible, the continuity equation in three dimensions in terms of the specific discharge vector components is

$$\frac{\partial q_x}{\partial x} + \frac{\partial q_y}{\partial y} + \frac{\partial q_3}{\partial x_3} = 0. \tag{9.46}$$

If the flow is transient, and the aquifer is compressible, then the continuity equation becomes

$$\frac{\partial q_x}{\partial x} + \frac{\partial q_y}{\partial y} + \frac{\partial q_3}{\partial x_3} = e_0, \tag{9.47}$$

where e_0 is the volume strain rate, taken positive for compression. Note that the compression is distributed uniformly over the thickness of the aquifer. We use that $Q_x = q_x h$ and $Q_y = q_y h$,

$$\frac{\partial q_x}{\partial x} + \frac{\partial q_y}{\partial y} = 2\frac{\partial (W/h)}{\partial \bar{z}} = \frac{2}{h}\frac{\partial W}{\partial \bar{z}} - 2\frac{W}{h^2}\frac{\partial h}{\partial \bar{z}}, \tag{9.48}$$

where

$$-2\frac{\partial W}{\partial \bar{z}} = \gamma = -N - e_0 h + s_p \frac{\partial \phi}{\partial t} \tag{9.49}$$

and s_p is the phreatic storage coefficient, roughly equal to the porosity. We solve (9.47) for $\partial q_3/\partial x_3$ and use (9.48):

$$\frac{\partial q_3}{\partial x_3} = e_0 - \left[\frac{\partial q_x}{\partial x} + \frac{\partial q_y}{\partial y}\right] = e_0 - \frac{2}{h}\frac{\partial W}{\partial \bar{z}} + 2\frac{W}{h^2}\frac{\partial h}{\partial \bar{z}} \tag{9.50}$$

or, with (9.49)

$$\frac{\partial q_3}{\partial x_3} = e_0 - \frac{1}{h}(N + e_0 h) + 2\frac{W}{h^2}\frac{\partial h}{\partial \bar{z}} + \frac{s_p}{h}\frac{\partial \phi}{\partial t}. \tag{9.51}$$

The terms that represent compressibility of the aquifer cancel, and we integrate:

$$q_3 = \frac{x_3}{h}\left[-N + 2\frac{W}{h}\frac{\partial h}{\partial \bar{z}} + s_p \frac{\partial \phi}{\partial t}\right] + q_3(x, y, \underset{0}{x_3}). \tag{9.52}$$

If the flow is confined, the saturated thickness is constant, $h = H$, so that the latter equation reduces to

$$q_3 = -\frac{N}{H} + q_3(x, y, \underset{0}{x_3}). \tag{9.53}$$

If the flow is unconfined, then

$$\Phi = \tfrac{1}{2}kh^2. \tag{9.54}$$

We manipulate the expression for q_3 in the interest of ease in computation by rewriting the terms in (9.52):

$$\frac{\partial \Phi}{\partial \bar{z}} = kh\frac{\partial h}{\partial \bar{z}} \rightarrow \frac{\partial h}{\partial \bar{z}} = \frac{1}{kh}\frac{\partial \Phi}{\partial \bar{z}}. \tag{9.55}$$

Darcy's law in terms of W is

$$W = -2\frac{\partial \Phi}{\partial z} \rightarrow \overline{W} = -2\frac{\partial \Phi}{\partial \bar{z}} \tag{9.56}$$

so that

$$2\frac{W}{h}\frac{\partial h}{\partial \bar{z}} = \frac{2W}{h}\frac{1}{kh}\frac{\partial \Phi}{\partial \bar{z}} = -\frac{2W}{h}\frac{1}{kh}\frac{\bar{W}}{2} = -\frac{W\bar{W}}{kh^2} = -\frac{W\bar{W}}{2\Phi}. \tag{9.57}$$

We use this in (9.52),

$$q_3 = \frac{x_3}{h}\left[-N - \frac{W\overline{W}}{2\Phi} + s_p\frac{\partial \phi}{\partial t}\right] + q_3(x, y, \underset{0}{x_3}). \tag{9.58}$$

If the flow is steady, the transient term is not present, and streamlines and path lines coincide. Note that this equation is the equivalent of equation (9.120) in Strack [1989], which was written in terms of real variables.

9.2.3 Correction of the Phreatic Storage Term

We demonstrate that water stored in pores as a result of a phreatic storage that is less than the porosity may have a significant effect on contaminant transport. To do this, we apply the procedure for determining path lines in three dimensions with v_3 computed by dividing (9.58) by the porosity n. The arc length s refers to a discharge path line rather than to a discharge, streamline and the values of all time-dependent variables (W, ϕ, and h) must be updated as the particle moves. We write the term $W\overline{W}/(2\Phi)$ in terms of the specific discharge components and the saturated thickness h,

$$\frac{W\overline{W}}{2\Phi} = \frac{Q_x^2 + Q_y^2}{k\phi^2} = \frac{h^2(q_x^2 + q_y^2)}{kh^2} = -q_x\frac{\partial h}{\partial x} - q_y\frac{\partial h}{\partial y} \tag{9.59}$$

and use this to rewrite (9.58) after division by n:

$$v_3 = \frac{x_3}{h}\left[-\frac{N}{n} + \frac{s_p}{n}\frac{\partial \phi}{\partial t} + v_x\frac{\partial h}{\partial x} + v_y\frac{\partial h}{\partial y}\right] + v_3(x, y, \underset{0}{x_3}). \tag{9.60}$$

We show in what follows that a special infiltration rate N_p must be added for cases of transient shallow unconfined flow in (9.58) by including a term $(n - S_p)\partial\phi/\partial t$. To focus on the issue of phreatic storage, we take N equal to zero and the base at $x_3 = 0$ to be impermeable, so that the leading and trailing terms in (9.60) vanish.

The material derivative Dh/Dt represents the rate of change of h with time, moving along the discharge streamline with the velocity v, and is

$$\frac{Dh}{Dt} = \frac{\partial h}{\partial t} + \frac{\partial h}{\partial x}\frac{dx}{dt} + \frac{\partial h}{\partial y}\frac{dy}{dt} = \frac{\partial h}{\partial t} + \frac{\partial h}{\partial x}v_x + \frac{\partial h}{\partial y}v_y, \tag{9.61}$$

where x and y are the coordinates of the position of the particle in the horizontal plane. We consider a point of the phreatic surface, $x_3/h = 1$ in (9.60) and replace ϕ by h, according to the Dupuit-Forchheimer approximation,

$$v_3 = \left[\frac{Dh}{Dt} - \frac{n - s_p}{n}\frac{\partial h}{\partial t}\right]. \tag{9.62}$$

Since we trace the path line, the vertical velocity component should be equal to Dh/Dt; (9.62) satisfies this condition only if $s_p = n$. We correct this problem by adding a term γ^* to the expression for v_3:

$$v_3 = \frac{x_3}{nh}\left[-N + n\frac{\partial h}{\partial t} + v_x\frac{\partial h}{\partial x} + v_y\frac{\partial h}{\partial y}\right] + v_3(x, y, \underset{0}{x_3}) + \gamma^*, \tag{9.63}$$

where

$$\gamma^* = -(n - s_p)\frac{\partial h}{\partial t}. \tag{9.64}$$

A physical interpretation of this somewhat surprising result follows. The phreatic storage coefficient S_p will be less than n if the soil above the water table is not entirely dry but contains water. This water is released into the aquifer when the water table rises and the soil becomes fully saturated. By the same reasoning, water is left behind when the water table lowers. This amount of water is represented by the term γ^* given by (9.64). Let there be no infiltration into the aquifer. If $s_p = n$, the path line of any particle at the phreatic surface will follow the rising and lowering of this surface: the particle remains at the phreatic surface. Next, consider that $s_p < n$ and the phreatic surface rises from $h(x,y,t)$ to $h(x,y,t+\Delta t) = h(x,y,t) + \Delta h$ over a time interval Δt at a rate $\partial h/\partial t$. The volume $\Delta V = \Delta x \Delta y \Delta h$ is unsaturated at time t and contains an amount $(n - s_p)\Delta V$ of immobile water. At time $t + \Delta t$, the empty pore space is filled as the volume ΔV becomes saturated. This is in accordance with expression (9.64): there is extraction at a rate γ^*, where $\gamma^* < 0$ because $\partial h/\partial t > 0$ and $s_p < n$. Thus, particles at the phreatic surface will move down with respect to the moving phreatic surface. Similarly, particles at the phreatic surface will move up relative to the phreatic surface if it lowers and $\gamma^* > 0$ because $\partial h/\partial t < 0$. In either case the path lines intersect the locus of points on the phreatic surface on the vertical through the moving particle; this locus represents the transient position of the phreatic surface with respect to the particle.

Although the mechanism described here is a severe simplification of the complicated process at the interface between the saturated and unsaturated zones, it highlights an effect that is important for problems of contaminant transport: if an aquifer is contaminated and the phreatic surface lowers for a period of time, then contaminated water is left behind in the unsaturated zone. This contaminated water will be released again when the water table rises.

9.2.4 Planar Flow

The continuity equation (9.53) for the case of planar flow, i.e., flow where there is no component of flow in the y-direction, reduces to

$$\frac{\partial q_3}{\partial x_3} = -\frac{dq_x}{dx}. \tag{9.65}$$

We integrate this:

$$q_3 = -\frac{dq_x}{dx}x_3 + C, \tag{9.66}$$

where C is a constant of integration. We can use a stream function ψ, because the flow in the vertical plane is divergence-free:

$$\begin{aligned} \frac{\partial \psi}{\partial x_3} &= -q_x \\ \frac{\partial \psi}{\partial x} &= q_3. \end{aligned} \tag{9.67}$$

We combine the second equation of (9.67) with (9.66) and integrate:

$$\psi = -q_x x_3 + Cx + f(x_3), \tag{9.68}$$

where $f(x_3)$ is an arbitrary function of x_3. Integration of the first equation in (9.67) gives

$$\psi = -q_x x_3 + g(x), \tag{9.69}$$

where $g(x)$ is an arbitrary function of x. Equations (9.68) and (9.69) must match:

$$f(x_3) = 0, \qquad g(x) = Cx. \tag{9.70}$$

The resulting equation for the stream function is

$$\psi = -q_x x_3 + Cx. \tag{9.71}$$

Confined and Semiconfined Flows

The expression (9.66) for q_3 simplifies if the aquifer thickness is constant, i.e., $h = H$,

$$q_3 = -\frac{1}{H}\frac{dQ_x}{dx}x_3 \tag{9.72}$$

and expression (9.71) for the stream function becomes

$$\psi = \frac{Q_x}{H}x_3 + Cx. \tag{9.73}$$

Confined Flow with Constant Infiltration The expression for Q_x for flow in a confined aquifer with constant infiltration, shown in Figure 9.2, is

$$Q_x = -Nx \tag{9.74}$$

and the expression for ψ is

$$\psi = -\frac{N}{H}xx_3 + Cx. \tag{9.75}$$

The boundary conditions for ψ are that $x = 0$ and $x_3 = 0$ are streamlines, so that $C = 0$. We can obtain the streamline plot shown in Figure 9.2 simply by contouring the stream function, rather than by determining the equations for the streamlines. The stream function fulfills Laplace's equation for this case, and thus the solution is exact. We can obtain the hydraulic heads using the Cauchy-Riemann equations:

$$\begin{aligned} \frac{\partial \phi}{\partial x} &= \frac{\partial \psi}{\partial x_3} = \frac{N}{H}x \\ \frac{\partial \phi}{\partial x_3} &= -\frac{\partial \psi}{\partial x} = \frac{N}{H}x_3. \end{aligned} \tag{9.76}$$

Integration gives

$$\phi = \frac{N}{2H}\left(x^2 - x_3^2\right). \tag{9.77}$$

Note that the hydraulic head and the stream function combine into a complex potential:

$$\omega = \phi + \mathrm{i}\psi = \frac{N}{2H}z^2 \tag{9.78}$$

so that $w = q_x - \mathrm{i}q_3$ is

$$w = -\frac{d\omega}{dz} = -\frac{N}{H}z. \tag{9.79}$$

Note that the solution is exact only when the stream function is harmonic.

9.2.5 Application: A River with a Leaky Bottom

We considered the case of a river with a leaky bottom in a confined aquifer, illustrated in Figure 5.5. The complete solution to the problem is the sum of the potentials for the symmetrical and the anti-symmetrical cases. The potential for the former case, Φ_s, is

$$\underset{s}{\Phi} = -\underset{s}{Q_{x0}}(x+L) + \Phi_0, \qquad x < -b \tag{9.80}$$

$$\underset{s}{\Phi} = \Lambda \underset{s}{Q_{x0}} \frac{\cosh(x/\Lambda)}{\sinh(b/\Lambda)}, \qquad -b \le x \le b \tag{9.81}$$

$$\underset{s}{\Phi} = \underset{s}{Q_{x0}}(x-L) + \Phi_0, \qquad b < x \tag{9.82}$$

(see (5.42), (5.56), and (5.44)) where

$$\underset{s}{Q_{x0}} = \frac{\frac{kH}{\Lambda}(\phi_0 - \phi^*)}{\frac{L-b}{\Lambda} + \coth(b/\Lambda)} \tag{9.83}$$

(see (5.49)). The expressions for the potential for the anti-symmetrical case are similar:

$$\underset{a}{\Phi} = -\underset{a}{Q_{x0}}(x+L) - \Phi_1, \qquad x < -b \tag{9.84}$$

$$\underset{a}{\Phi} = \Lambda \underset{a}{Q_{x0}} \frac{\sinh(x/\Lambda)}{\cosh(b/\Lambda)}, \qquad -b \le x \le b \tag{9.85}$$

$$\underset{a}{\Phi} = -\underset{a}{Q_{x0}}(x-L) + \Phi_1, \qquad b < x, \tag{9.86}$$

where

$$\underset{a}{Q_{x0}} = \frac{kH(\phi_1 - \phi^*)}{L - b + \Lambda \tanh(b/\Lambda)}. \tag{9.87}$$

The complete solution is the sum of the symmetrical and anti-symmetrical parts:

$$\Phi = -Q_{x0}(x+L) + \Phi_l, \qquad x < -b \tag{9.88}$$

$$\Phi = \Lambda \underset{s}{Q_{x0}} \frac{\cosh(x/\Lambda)}{\sinh(b/\Lambda)} + \Lambda \underset{a}{Q_{x0}} \frac{\sinh(x/\Lambda)}{\cosh(b/\Lambda)}, \qquad -b \le x \le b \tag{9.89}$$

$$\Phi = Q_{x0}(x-L) + \Phi_r, \qquad b < x. \tag{9.90}$$

To determine the stream function, we differentiate the potential in order to obtain expressions for the discharge Q_x:

$$Q_x = -(\underset{s}{Q_{x0}} + \underset{a}{Q_{x0}}) \qquad x < -b \tag{9.91}$$

$$Q_x = -\underset{s}{Q_{x0}} \frac{\sinh(x/\Lambda)}{\sinh(b/\Lambda)} - \underset{a}{Q_{x0}} \frac{\cosh(x/\Lambda)}{\cosh(b/\Lambda)}, \qquad -b \le x \le b \tag{9.92}$$

$$Q_x = (\underset{s}{Q_{x0}} - \underset{a}{Q_{x0}}), \qquad b < x. \tag{9.93}$$

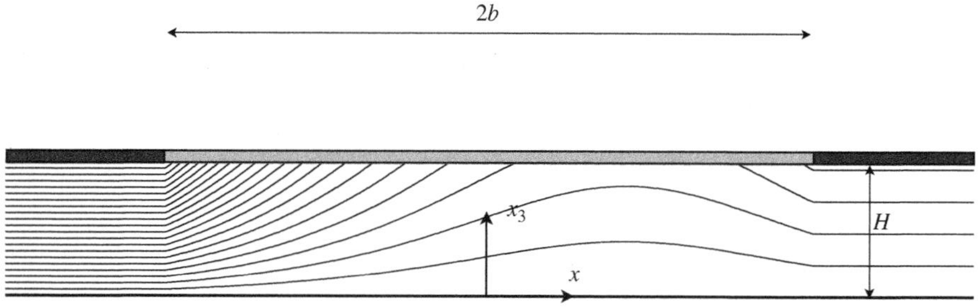

Figure 9.4 Streamlines for flow in a confined aquifer with a river with leaky bottom (the gray rectangle in the upper middle of the figure).

The Streamlines

We determine the stream function for this problem by substituting the expressions for Q_x in (9.73),

$$\psi = -(\underset{s}{Q}_{x0} + \underset{a}{Q}_{x0})\frac{x_3}{H}, \qquad x < -b \tag{9.94}$$

$$\psi = -\left\{\underset{s}{Q}_{x0}\frac{\sinh(x/\Lambda)}{\sinh(b/\Lambda)} + \underset{a}{Q}_{x0}\frac{\cosh(x/\Lambda)}{\cosh(b/\Lambda)}\right\}\frac{x_3}{H}, \qquad -b \leq x \leq b \tag{9.95}$$

$$\psi = (\underset{s}{Q}_{x0} - \underset{a}{Q}_{x0})\frac{x_3}{H}, \qquad b < x. \tag{9.96}$$

The constant C in (9.66) is zero, because the component q_3 is zero where dQ_x/dx is zero. This is true in both zones of horizontal confined flow. A plot of streamlines is shown as an illustration in Figure 9.4, which applies when $b/H = 2.5$, $\Lambda/b = 0.5$, and $\underset{s}{Q}_0/\underset{a}{Q}_0 = 2/3$.

9.2.6 Application: Streamlines in a Stratified Aquifer

We apply the concept of the stream function to plot the streamlines for the aquifer of two strata shown in Figure 9.5. The lower and upper strata are labeled as 1 and 2, respectively, and their hydraulic conductivities and thicknesses are k_1, H_1, k_2, and H_2. The boundary conditions are that the y-axis and the x-axis are streamlines and that there is uniform infiltration through the upper boundary at $x_3 = H = H_1 + H_2$. We choose the (x, x_3)-coordinate system shown in the figure. The infiltration is uniform, and we base our solution on the assumption that the stream function is similar to that for uniform infiltration in a homogeneous aquifer. We define two stream functions, as follows:

$$\begin{aligned} \overset{2}{\psi} &= -axx_3 + Cx \\ \overset{1}{\psi} &= -bxx_3, \end{aligned} \tag{9.97}$$

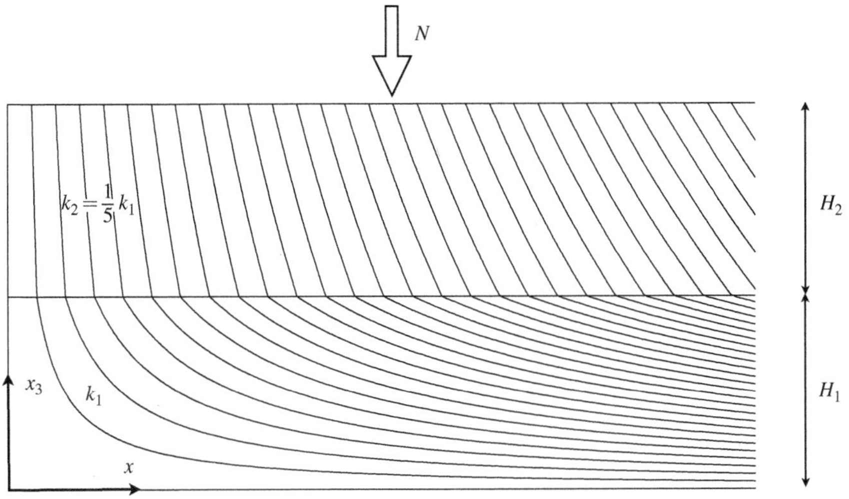

Figure 9.5 Streamlines in a stratified aquifer.

where the superscripts refer to the strata. Note that we kept the general constant C in the expression for the stream function for the upper stratum, but not for the one for the lower one; it would violate the boundary condition along the x-axis. The stream functions contain together three unknowns: a, b, and C. We determine these from the boundary conditions:

$$q_3 = \frac{\partial \overset{2}{\psi}}{\partial x} = -N, \qquad y = H \tag{9.98}$$

$$\overset{2}{\psi}{}^{+} = \overset{2}{\psi}{}^{-}, \qquad y = H_1 \tag{9.99}$$

$$\frac{\overset{2}{q}_x}{\overset{1}{q}_x} = \frac{k_2}{k_1}, \qquad y = H_1. \tag{9.100}$$

The third condition expresses that the head is continuous across the interzonal boundary, so that the gradient is continuous also. We apply the third condition to the stream function:

$$\frac{\partial \overset{2}{\psi}}{\partial x_3} = \frac{k_2}{k_1} \frac{\partial \overset{1}{\psi}}{\partial x_3}, \qquad y = H_1 \tag{9.101}$$

and use (9.98),

$$-aH + C = -N \tag{9.102}$$

and (9.99):

$$-axH_1 + Cx = -bxH_1. \tag{9.103}$$

We divide both sides of the latter equation by x ($x \neq 0$) and subtract the result from (9.102):

$$-a(H-H_1) = -N + bH_1 \tag{9.104}$$

and solve this for b:

$$b = -a\left(\frac{H}{H_1} - 1\right) + \frac{N}{H_1}. \tag{9.105}$$

The final equation to apply is (9.101):

$$ax = \frac{k_2}{k_1} bx. \tag{9.106}$$

We replace b in (9.105) by $k_1/k_2 a$:

$$a\left[\frac{k_1}{k_2} + \frac{H}{H_1} - 1\right] = a\frac{k_1H_1 + k_2H_1 + k_2H_2 - k_2H_1}{k_2H_1} = \frac{N}{H_1} \tag{9.107}$$

or

$$a\frac{k_1H_1 + k_2H_2}{k_2H_1} = a\frac{T}{k_2H_1} = \frac{N}{H_1}, \tag{9.108}$$

where T is the total transmissivity. We solve this for a:

$$a = \frac{Nk_2}{T}. \tag{9.109}$$

We use (9.106) to solve for b:

$$b = \frac{Nk_1}{T} \tag{9.110}$$

and apply (9.102) to find an expression for C:

$$C = aH - N = N\frac{k_2H - T}{T} = N\frac{k_2H_1 + k_2H_2 - k_1H_1 - k_2H_2}{T} = N\frac{(k_2 - k_1)H_1}{T}. \tag{9.111}$$

We substitute the expressions for a, b, and C in the equations for the stream functions in the strata, (9.97):

$$\begin{aligned} \overset{2}{\psi} &= -\frac{Nk_2}{T}xx_3 + N\frac{(k_2 - k_1)H_1}{T}x \\ \overset{1}{\psi} &= -\frac{Nk_1}{T}xx_3. \end{aligned} \tag{9.112}$$

The streamlines shown in Figure 9.5 were produced by programming the stream function in a simple MATLAB program.

Problem

9.4 Determine expressions for the streamlines in a shallow unconfined aquifer with three layers, enclosed between two parallel rivers with heads $\phi_0 > H_1$. The distance between the rivers is L; the hydraulic conductivities of the three layers are k_1, k_2, and k_3; and the thicknesses of the lower two layers are H_1 and H_2. There is a uniform infiltration rate N.

9.2.7 Application: Infiltration from a Circular Pond

We consider the problem of leakage of contaminated water through the bottom of a circular pond above an unconfined aquifer as an application. We assume that the contaminant does not affect the physical properties of the groundwater and that the contamination has occurred over a long time. We refer to the area of the aquifer that is filled with contaminated water as the *plume*. The problem is illustrated in Figure 9.6, where both a plan view and a cross section through the center of the pond are given. We determine equations for points of the streamlines that start at points on the bottom of the pond. The origin of an (x, y, x_3)-coordinate system is at the center of the pond, and there is a uniform flow of discharge Q_{x0} in the x-direction. The infiltration rate through the bottom of the pond is N and the radius of the pond is R. We use the function G_p introduced in Section 2.5, (2.258) and (2.259), and write the potential for this flow problem as:

$$\Phi = -Q_{x0}x + NG_p(x, y; 0, 0, R) + C, \tag{9.113}$$

where

$$2G_p(x, y; 0, 0, R) = -\tfrac{1}{4}(x^2 + y^2 - R^2) = -\tfrac{1}{4}(z\bar{z} - R^2), \qquad x^2 + y^2 \leq R^2 \tag{9.114}$$

$$G_p(x, y; 0, 0, R) = -\frac{R^2}{4}\ln\frac{x^2 + y^2}{R^2} = -\frac{R^2}{4}\ln\frac{z\bar{z}}{R^2}, \qquad x^2 + y^2 > R^2. \tag{9.115}$$

If Φ equals Φ_0 at $x = y = 0$, then $C = -NR^2/4 + \Phi_0$, and we obtain for $z\bar{z} \leq R^2$:

$$\Phi = -Q_{x0}x - \frac{N}{4}z\bar{z} + \Phi_0, \qquad z\bar{z} \leq R^2. \tag{9.116}$$

Outside the pond the potential is harmonic; we write the solution there in terms of a complex potential Ω of z only. It follows from (9.113) and (9.115), with $C = -NR^2/4 + \Phi_0$, that

$$\Omega = -Q_{x0}z - \frac{NR^2}{2}\ln\frac{z}{R} - \frac{NR^2}{4} + \Phi_0, \qquad z\bar{z} > R. \tag{9.117}$$

We determine whether stagnation points exist, and if so, where these are located, as part of our examination of the streamline pattern.

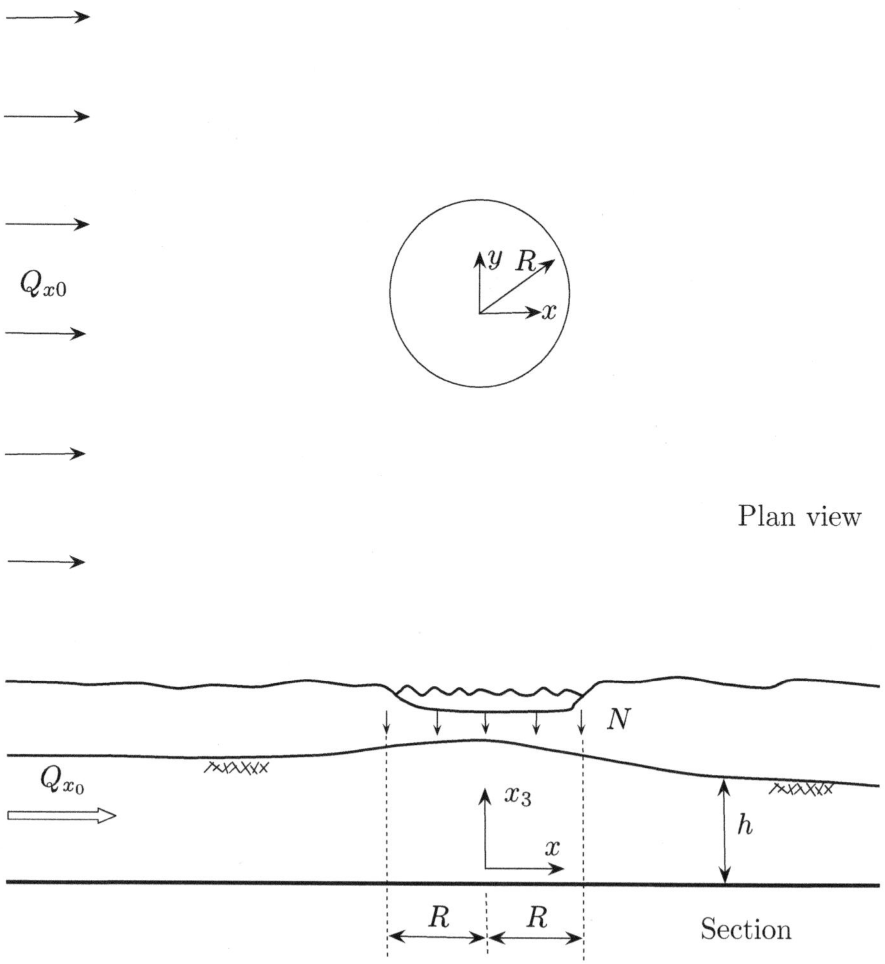

Figure 9.6 Leakage from a circular pond.

The Stagnation Points

The stagnation points may be either below or outside the pond; we first examine whether a stagnation point exists outside the pond. We differentiate (9.117) to obtain an expression for W, and write the location of the stagnation point as $\underset{o}{z_s}$, where the subscript o refers to outside,

$$Q_{x0} + \frac{NR^2}{2}\frac{1}{\underset{o}{z_s}} = 0, \qquad |z|_{\underset{o}{s}} > R \tag{9.118}$$

or

$$\frac{\underset{o}{z_s}}{R} = -\frac{NR}{2Q_{x0}}, \qquad \frac{|\underset{o}{z_s}|}{R} > 1. \tag{9.119}$$

It follows that $\overset{0}{y}_s = 0$ and that the stagnation point exists outside the pond only if

$$\left|\frac{NR}{2Q_{x0}}\right| > 1. \tag{9.120}$$

We obtain the expression for W below the pond from (9.116):

$$W = -2\frac{\partial \Phi}{\partial z} = Q_{x0} + \tfrac{1}{2}N\bar{z}. \tag{9.121}$$

We denote the coordinates of the stagnation point in the aquifer below the pond as $\underset{i}{z_s}$ and obtain, with $W = 0$:

$$\underset{i}{\bar{z}_s} = \underset{i}{z_s} = -2\frac{Q_{x0}}{N}. \tag{9.122}$$

The stagnation point lies below the pond only if

$$\left|\frac{2Q_{x0}}{NR}\right| \leq 1. \tag{9.123}$$

We observe from (9.120) and (9.123) that three cases may occur; for the first case there are two stagnation points, one outside and one below the pond:

$$\left|\frac{2Q_{x0}}{NR}\right| < 1: \quad \frac{\underset{o}{z_s}}{R} = -\frac{NR}{2Q_{x0}}, \quad \frac{\underset{i}{z_s}}{R} = -\frac{2Q_{x0}}{NR}. \tag{9.124}$$

For the second case, the two stagnation points coincide at the boundary of the pond:

$$\left|\frac{2Q_{x0}}{NR}\right| = 1: \quad \frac{\underset{o}{z_s}}{R} = \frac{\underset{i}{z_s}}{R} = -\frac{2Q_{x0}}{NR} = -1. \tag{9.125}$$

For the third case, where $2Q_{x0}/(NR) > 1$, there are no stagnation points at all.

Discharge Streamlines; Vertical Stream Surfaces

Each streamline lies on a single vertical surface, a result of the Dupuit-Forchheimer approximation. The intersection of these vertical stream surfaces with a horizontal plane are the discharge streamlines, i.e., the streamlines for the discharge vector (Q_x, Q_y). We determine the discharge streamlines outside the pond using the stream function Ψ, which equals the imaginary part of (9.117),

$$\Psi = -Q_{x0}y - \frac{NR^2}{2}\theta, \qquad z\bar{z} > R^2, \tag{9.126}$$

where $\theta = \arg(z)$.

The stream function does not exist in the aquifer below the pond, and the discharge function is given by (9.121):

$$W = Q_{x0} + \tfrac{1}{2}N\bar{z} = \tfrac{1}{2}N\left(\bar{z} + \frac{2Q_{x0}}{N}\right) = \tfrac{1}{2}N(\bar{z} - \underset{i}{z_s}), \qquad z\bar{z} \le R^2, \qquad (9.127)$$

or

$$\overline{W} = Q_x + \mathrm{i}Q_y = \tfrac{1}{2}N(z - \underset{i}{z_s}). \qquad (9.128)$$

Since $\overline{W}$ points in the direction of flow, the right-hand side of the equation points in the direction of flow also; this direction is constant and points from the stagnation point to z; the streamlines are straight lines through the stagnation point. If the stagnation point does not exist, i.e., if $\underset{i}{z_s}$ is outside the pond, the streamlines still have this point in common, but have meaning only inside the pond.

Discharge streamlines are shown in Figure 9.7 for the cases that (a) stagnation occurs and (b) no stagnation points exist. The discharge streamlines that bound the plume, the dividing streamlines, are shown in Figure 9.7 as heavy black lines. For case (a) they pass through S_2 and for case (b) they are the curves that touch the pond at two points. The data for the upper plot in Figure 9.7 are $N/k = 0.001$, $Q_{x0}/(NR) = 0.4$, and $\phi_0/R = 0.2$. The data for the lower plot are $N/k = 0.001$, $Q_{x0}/(NR) = 1$, and $\phi_0/R = 0.2$.

Streamlines

We express the elevation of points of the streamlines in the vertical stream surfaces in terms of a parameter s, which is the distance measured along each discharge streamline, with $s = 0$ at the point $\underset{i}{z_s}$. If a stagnation point exists inside the pond, then $\underset{i}{z_s}$ represents the actual stagnation point; otherwise, it is an imaginary stagnation point, as shown in Figure 9.8. We introduce Q_s as the discharge in the s-direction, i.e.,

$$W\overline{W} = Q_s^2, \qquad Q_s = -\frac{d\Phi}{ds}, \qquad Q_s = \frac{N}{2}s. \qquad (9.129)$$

There is infiltration at rate N below the pond, so that

$$\gamma = -N. \qquad (9.130)$$

We use these quantities in (9.58):

$$q_3 = \frac{x_3}{h}\left[-N - \frac{Q_s^2}{2\Phi}\right] + q_3(z_0,(x_3)_0). \qquad (9.131)$$

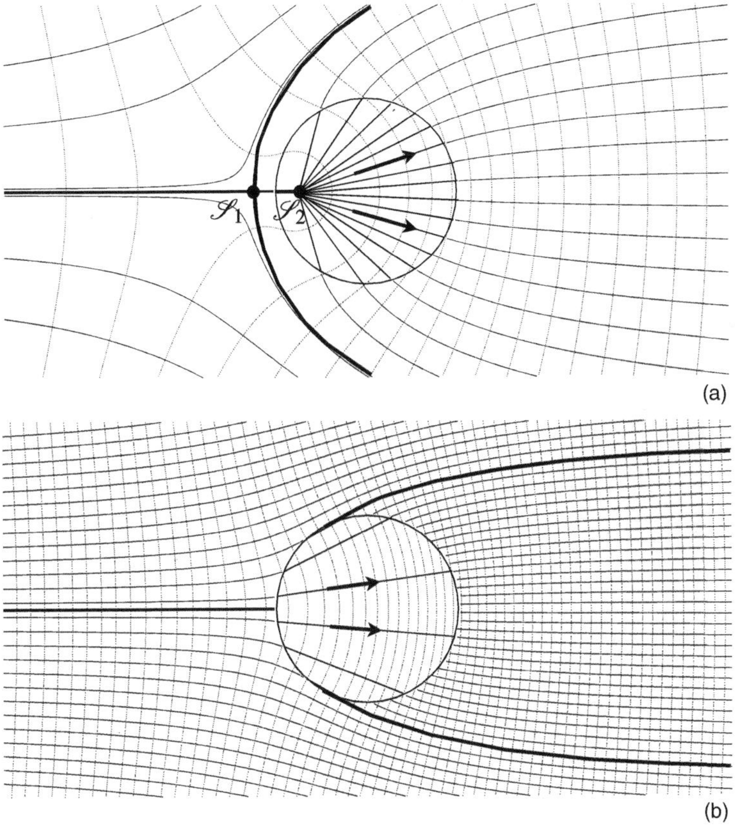

Figure 9.7 Discharge streamlines; the upper figure is for case (a) with stagnation points, and the lower figure is for case (b), without stagnation points. (Source: adapted from Strack [1984].)

The streamline is tangent to the flow direction, so that

$$\frac{q_3}{q_s} = \frac{hq_3}{Q_s} = \frac{dx_3}{ds} = x_3\left[-\frac{N}{Q_s} + \frac{1}{2\Phi}\frac{d\Phi}{ds}\right], \tag{9.132}$$

where we applied Darcy's law to the second term in the brackets to write Q_s as $-d\Phi/ds$. We divide by x_3 and integrate:

$$\int \frac{dx_3}{x_3} = \int \left[-\frac{N}{\frac{1}{2}Ns}ds + \frac{1}{2\Phi}d\Phi\right] \tag{9.133}$$

or

$$\ln\frac{x_3}{(x_3)_0} = -2\ln\frac{s}{s_0} + \tfrac{1}{2}\ln\frac{\Phi}{\Phi_0} = \ln\left[\frac{s_0^2\Phi^{1/2}}{s^2\Phi_0^{1/2}}\right] = \ln\left[\frac{s_0^2 h}{s^2 h_0}\right] \tag{9.134}$$

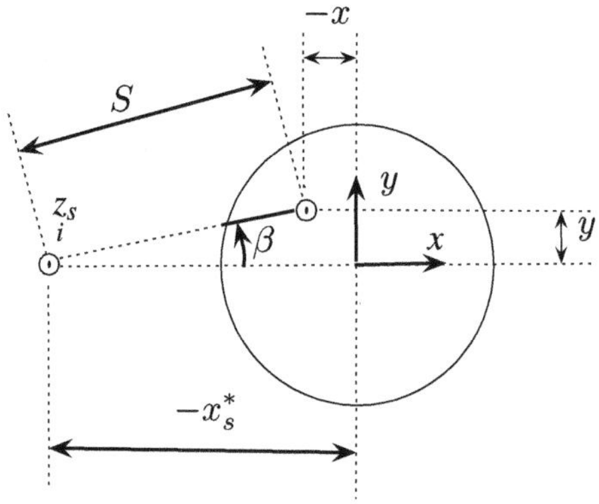

Figure 9.8 The definition of s as the distance from the imaginary stagnation point $\underset{i}{z_s}$ to a point on the streamline.

so that

$$\frac{x_3}{(x_3)_0} = \left[\frac{s_0}{s}\right]^2 \frac{h}{h_0}, \qquad z\bar{z} \leq R^2. \tag{9.135}$$

We use this to calculate x_3 at any point of a streamline starting at s_0 with elevation $(x_3)_0$ in the aquifer below the pond. The infiltration rate is zero outside the pond; the first term in (9.133) is not present; the expression for x_3 reduces to

$$\frac{x_3}{(x_3)_b} = \frac{h}{h_b}, \tag{9.136}$$

where $(x_3)_b$ and h_b are the values of x_3 and h at the intersection of the discharge streamline with the boundary of the pond.

For the case of Figure 9.7(a), all discharge streamlines contained between the dividing discharge streamlines emanate from point S_1 at $z_s = -2Q_{x0}/N$ inside the pond. The vertical surfaces through these discharge streamlines therefore each contain one streamline that starts at S_1. It follows from (9.129) that s is zero at S_1, where $Q_s = 0$. All streamlines starting at $s = s_0 = 0$ are on the aquifer base, since x_3/h in (9.135) vanishes for $s_0 = 0$; all infiltrated water fills the entire space contained between the dividing discharge streamlines, the phreatic surface, and the aquifer base.

The plot of streamlines in a vertical plane through the x-axis reproduced in Figure 9.9(a) corresponds to Figure 9.7(a). The plot extends from $x/R = -1.25$ to $x/R = 1.25$ and contains the two stagnation points, which lie at $x/R = -1.25$ and $x/R = -0.8$. The streamlines start at equal intervals between $x/R = -1$ and $x/R = 1$; there are 21 streamlines. A similar drawing is shown in Figure 9.9(b) and corresponds to Figure 9.7(b). For this case, stagnation does not occur, $x_s^*/R = -2$, and the bottom

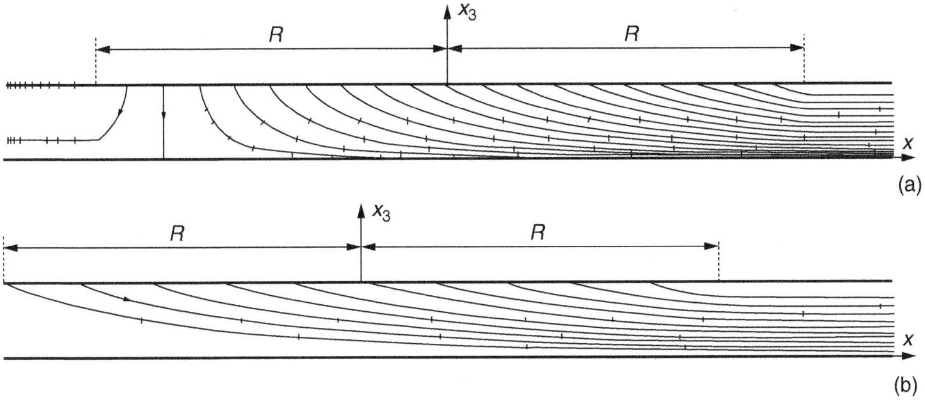

Figure 9.9 Streamlines in the (x,x_3)-plane. (Source: Strack [1984].)

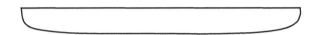

Figure 9.10 Cross section through the plume at $x/R = 2$. (Source: Strack [1984].)

of the plume is the lowest curved streamline. The plot extends between $x/R = -1$ and $x/R = 1.5$. For both cases, the value of Φ_0 equals 0.02 kR^2. A discontinuity in slope of the streamlines occurs below the boundary of the pond. This discontinuity results from coupling the component q_3 directly to the infiltration rate, which is discontinuous across the boundary of the pond.

The plume extends over the full height of the aquifer only if a stagnation point exists outside the pond; otherwise, the bottom of the plume is a curved surface above the base. A section through the plume parallel to the (y,x_3)-plane is shown in Figure 9.10. The cross-section is at $x/R = 2$. Points of the curved bottom of the plume are determined as follows. First, the width of the plume at the given location is determined by intersecting the dividing discharge streamlines with the plane $x = x_1$. Second, each discharge streamline $\Psi = \Psi_j$ inside the plume is intersected with the downstream side of the circle, which yields a value for $\overset{p}{s}_0 = \overset{p}{s}_{0j}$. Third, the discharge streamline is traced below the pond until it intersects the circle on the upstream side, giving s_{0j}. Fourth, s_{0j} and $\overset{p}{s}_{0j}$ are substituted for s_0 and s, respectively, in (9.136), which yields the elevation x_{3j}/h_j of the bottom of the plume at $\overset{p}{s}_{0j}$. The elevation of the plume at the desired cross section is found, finally, as a function of the saturated thickness h by the use of (9.135). The above procedure is implemented in an elementary computer program, written merely for the purpose of illustration. This program is used to show, in Figure 9.11, how the cross section of the plume varies as a function of x_s^*/R. When

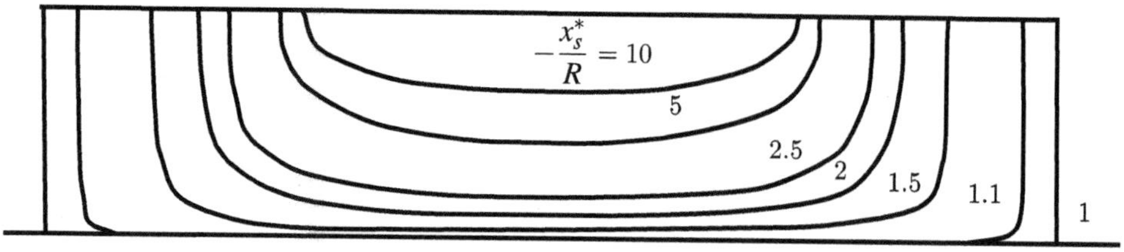

Figure 9.11 Cross sections through the plume for various values of $-x_s^*/R = 2Q_{x0}/NR$ at $X/R = 2$. (Source: Strack [1984].)

x_s^*/R equals -1, the plume is rectangular and fills the entire space between the dividing discharge streamlines. As the ratio $2Q_{x0}/(NR) = -x_s^*/R$ increases, the area of the plume decreases; the plume becomes both shallower and narrower. The data used for this case are the same as those for Figure 9.7(b), except that the saturated aquifer thickness was increased ($\phi_0/R = 1$) in order to make the curves more visible.

The plume attains the geometry described above after the pollution has occurred for a long time; we determine the front of the plume as it moves through the aquifer as a function of the time recorded since the pollution began. The marks shown on the streamlines in Figure 9.9 correspond to constant intervals of the dimensionless time $t^* = Nt/(nR)$, where n is the porosity. The front of the plume at any time is obtained by connecting corresponding marks.

It should be noted that the technique is approximate and gives a simplified picture of the real streamline pattern. For example, the vertical faces of the plume shown in Figure 9.11 for $x_s^*/R = -1$ do not occur in reality, because both horizontal components of the specific discharge vector vary over the thickness of the aquifer near the bottom of the pond. Haitjema [1987] compares the results presented here with those obtained from a true three-dimensional analysis and reports that the errors are small for ratios of R/H greater than 5. For cases where three-dimensional effects are too important to be neglected, a truly three-dimensional solution may be imbedded inside the Dupuit-Forchheimer model (see Haitjema [1985]).

Problem

9.5 For the case of Figure 9.7(b), determine:

1. The value of Ψ at point T_1
2. The elevation at $x = R$, $y = 0$, of the streamline that starts at the center of the pond.

9.2.8 Application: An Elliptical Pond in Uniform Flow

When examining the plume of contaminated water that entered the aquifer through the bottom of a circular pond, we noted that the plume occupies the entire saturated thickness of the aquifer, if stagnation points exist. The occurrence of a stagnation point on the upstream boundary of the pond is of particular interest, as the entire plume is contained in an area that just encloses the pond. This is important when using a well to capture the contaminated water; a well with a discharge equal to all infiltrated water of the pond removes all contaminant, under ideal conditions, as shown in Figure 2.35.

We consider a pond with an elliptical boundary rather than a circular one. The analyses for the two cases are similar, but the results are quite different: the oblong shape of the elliptical pond affects both the width and the depth of the contaminant plume. The orientation of the major axis of the pond relative to the direction of the uniform flow has a major influence on the shape of the plume. In terms of contaminant removal, the case of a single stagnation point on the upstream boundary of the pond is of special interest; we examine whether the streamlines through the stagnation point, the dividing streamlines, contain the entire plume. We consider two extreme cases, first where the major axis of the ellipse is aligned with the flow, and second where the major axis is at 90 degrees to the flow. We use the solution and the results for an elliptical pond in a field of uniform flow, presented in Section 8.22.

Major Axis of the Pond Aligned with the Flow

There is uniform flow in the x-direction, and the major principal axis of the pond is in the x-direction as well. Stagnation points exist if the following condition is satisfied (derived on page 314 (8.467)):

$$\frac{Q\nu}{\pi Q_{x0}}\frac{2}{L} = \frac{\pi abN\nu}{\pi Q_{x0}}\frac{2}{L} = \frac{2ab\nu N}{Q_{x0}L} \geq 1. \tag{9.137}$$

The case of a stagnation point on the boundary of the pond on the upstream side is of interest, and for this orientation of the pond the dividing discharge streamlines through the stagnation point contain the entire plume, which occupies the saturated thickness of the aquifer, as we demonstrate in what follows. When (9.137) applies, the stagnation point is at $x = -a, y = 0$; all streamlines that begin at the stagnation point are inside the boundary of the pond and reach the base of the aquifer. The stagnation point is at the boundary for the case of Figure 9.12; the data are $L =$ 100 m, $k = 10$ m/d, $a = 50.2$ m, $b = 5.02$ m, $Q_{x0} = 0.1$ m^2/d, and $N = 0.0198$ m/d.

In contrast to the case of infiltration through the bottom of a circular pond, the discharge streamlines below the pond are not straight, but curved. We obtain equations for the discharge streamlines below the pond by differentiating the applicable expression for the potential, (8.461). We replace Z by $zL/2$ for the present case,

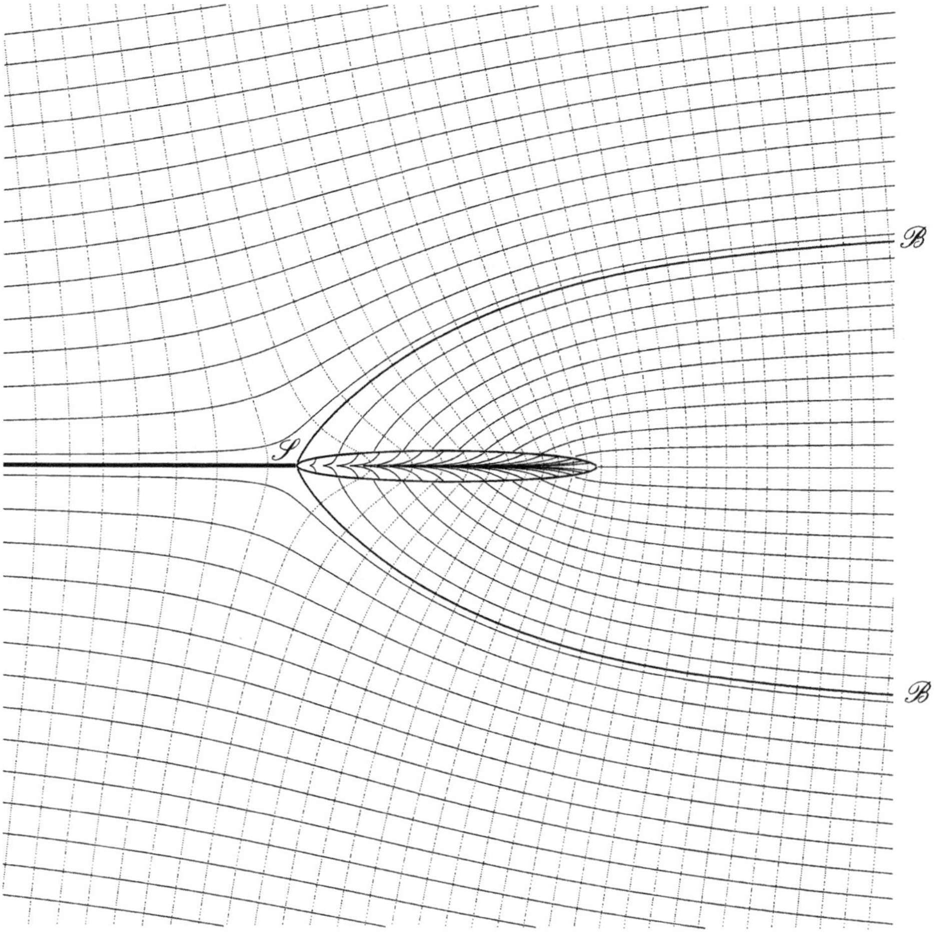

Figure 9.12 A plume caused by infiltration through the bottom of an elliptical pond; the stagnation point is at the leftmost point of the pond. The plume occupies the space bounded by the heavy black lines; the cross section ℬ-ℬ is occupied over the entire saturated thickness by the plume.

$$\Phi = -\tfrac{1}{2}Q_{x0}(z+\bar{z}) - \frac{N}{4}\left\{z\bar{z} - \frac{\nu^2}{2}\left[z^2+\bar{z}^2\right]\right\} + \frac{NL^2}{16}\left(\frac{1}{\nu^2} - \nu^2\right), \qquad \chi\bar{\chi} < 1, \tag{9.138}$$

and differentiate with respect to z:

$$W = -2\frac{\partial\Phi}{\partial z} = Q_x - \mathrm{i}Q_y = Q_{x0} + \frac{N}{2}(\bar{z} - \nu^2 z). \tag{9.139}$$

We obtain the components of the discharge vector by separating real and imaginary parts,

$$Q_x = Q_{x0} + \frac{N}{2}(1-\nu^2)x$$
$$Q_y = \frac{N}{2}(1+\nu^2)y \tag{9.140}$$

and obtain the equation for the streamlines from the definition, $dy/dx = Q_y/Q_x$:

$$\frac{dy}{dx} = \frac{N/2(1+\nu^2)y}{Q_{x0} + N/2(1-\nu^2)x} \tag{9.141}$$

or

$$\int \frac{dy}{y} = \frac{1+\nu^2}{1-\nu^2} \int \frac{dx}{2Q_{x0}/[N(1-\nu^2)] + x}. \tag{9.142}$$

Recall that

$$a = \frac{L}{2}\frac{1+\nu^2}{2\nu}, \qquad b = \frac{L}{2}\frac{1-\nu^2}{2\nu}, \qquad \frac{b}{a} = \frac{1-\nu^2}{1+\nu^2}, \tag{9.143}$$

which we use to rewrite (9.142):

$$\int \frac{dy}{y} = \frac{a}{b}\int \frac{dx}{x+\beta} + C^*, \tag{9.144}$$

where

$$\beta = \frac{2Q_{x0}}{N(1-\nu^2)} = \frac{2Q_{x0}}{N4\nu b/L} = \frac{Q_{x0}L}{2b\nu N}. \tag{9.145}$$

We integrate (9.144),

$$\ln y = \frac{a}{b}\ln(x+\beta) + \ln C, \tag{9.146}$$

where we wrote the constant C^* as $\ln C$. We combine the logarithms and take the exponential of both sides:

$$y = C(x+\beta)^{a/b}. \tag{9.147}$$

This equation represents curves; if $b = a$, i.e., in the case of a circular pond, this equation becomes linear. The stagnation point is on the boundary of the pond if the constant β equals a.

We wish to establish whether the discharge streamlines leave the area below the pond at the stagnation point, $\mathscr{S}$, and examine the angle that discharge streamlines make with the x-axis at $\mathscr{S}$, where $x = -a$, $y = 0$. Equation (9.147) yields $y = 0$ for $x = -a$; we use it to examine the discharge streamlines through $\mathscr{S}$. We obtain an expression for the tangent to the curve on differentiation:

$$\frac{dy}{dx} = \frac{a}{b}C(x+a)^{a/b-1}. \tag{9.148}$$

This derivative is zero for $x = -a$ since the exponent is larger than 0; the tangents to all discharge streamlines through $\mathscr{S}$ are parallel to the x-axis there; these discharge streamlines are inside the area of the pond until they leave through its boundary; for the case of Figure 9.12, the exponent a/b is 10. The streamlines through the stagnation point lie in the vertical planes through the stagnation point; all reach the base of the aquifer, as shown for the case of a circular pond in Figure 9.9.

Discharge streamlines are plotted inside the ellipse as shown in Figure 9.12; the plots of the streamlines are made independently of the flow net outside the ellipse and have a different meaning; the stream function, used to plot the flow net, does not exist as a single-valued function inside the pond area.

Major Axis of the Pond at Right Angles to the Flow

The plot shown in Figure 9.13 corresponds to the same data as Figure 9.12, except that now the major principal axis of the pond is perpendicular to the flow. For this case, discharge streamlines pass underneath the pond on its upstream side; the plume does not extend everywhere all the way to the base of the aquifer, but it does so at the stagnation point $\mathscr{S}$. The discharge streamlines below the pond are curved, as for the previous case, but their tangents are normal to the x-axis at the stagnation

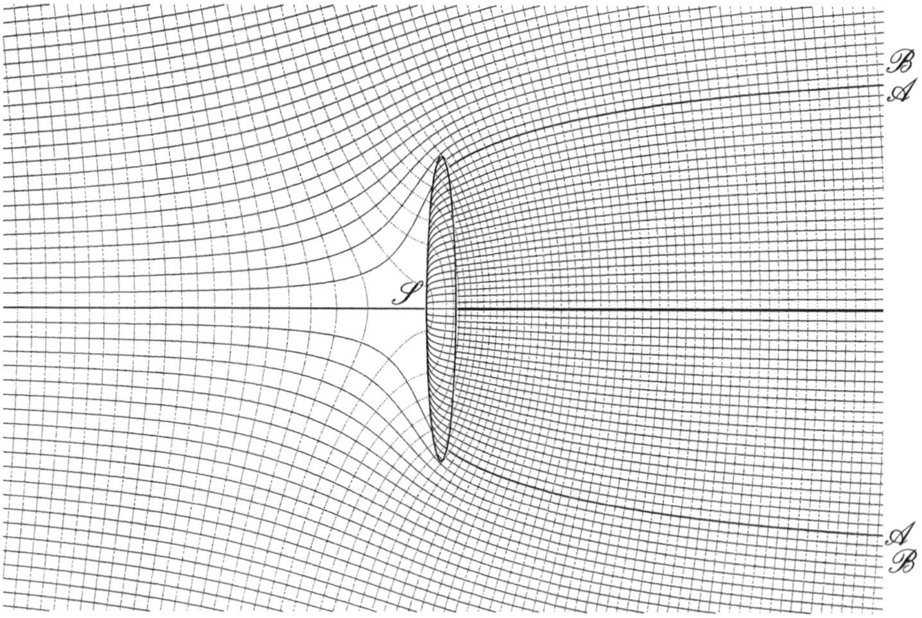

Figure 9.13 A plume caused by infiltration through the bottom of an elliptical pond; the stagnation point, $\mathscr{S}$, is at the leftmost point of the pond. The plume occupies the space bounded by the heavy black lines; the cross section $\mathscr{B}$–$\mathscr{B}$ is not occupied over the entire saturated thickness by the plume.

point. We apply an analysis as for a pond with its major axis aligned with the flow. The angle α that the major axis of the ellipse makes with the x-axis is $\pi/2$, and we replace Z by $-2\mathrm{i}z/L$ in the expression for the potential below the pond, (8.461), so that the point $z = \mathrm{i}L/2$ maps onto $Z = 1$:

$$\Phi = -\tfrac{1}{2}Q_{x0}(z+\bar{z}) - \frac{N}{4}\left\{z\bar{z} + \frac{\nu^2}{2}[z^2+\bar{z}^2]\right\} + \frac{\mathrm{N}L^2}{16}\left(\frac{1}{\nu^2} - \nu^2\right), \qquad \chi\bar{\chi} < 1. \tag{9.149}$$

We differentiate this to obtain W,

$$W = -2\frac{\partial\Phi}{\partial z} = Q_{x0} + \frac{N}{2}\left(\bar{z} + \nu^2 z\right), \tag{9.150}$$

so that

$$\begin{aligned} Q_x &= Q_{x0} + \frac{N}{2}x(1+\nu^2)x \\ Q_y &= \frac{N}{2}(1-\nu^2)y. \end{aligned} \tag{9.151}$$

We express dy/dx as Q_y/Q_x to obtain the equation for the streamline:

$$\frac{dy}{dx} = \frac{N/2(1-\nu^2)y}{Q_{x0} + N/2(1+\nu^2)x} = \frac{1-\nu^2}{1+\nu^2}\frac{y}{2Q_{x0}/(N(1+\nu^2)) + x}. \tag{9.152}$$

We again use (9.143) to express this in terms of a and b:

$$\frac{dy}{dx} = \frac{b}{a}\frac{y}{x+\mu}, \tag{9.153}$$

where

$$\mu = \frac{2Q_{x0}}{N(1+\nu^2)} = \frac{Q_{x0}L}{2a\nu N}. \tag{9.154}$$

We integrate (9.153) to obtain the equations for the streamlines,

$$y = D[x+\mu]^{b/a}. \tag{9.155}$$

The latter equation is similar to the one that applies when the major principal axis of the ellipse is parallel to the flow. One difference is that μ replaces β; the constant μ equals b for the case in which the stagnation point is on the boundary, i.e, at $x = -b$. The second, and most important, difference is that the exponent is now b/a, rather than a/b; this is less than one, so that the tangents to the streamlines through the stagnation point $\mathscr{S}$ at $x = -b$ are normal to the x-axis; see Figure 9.13.

The discharge streamlines that start at points on the x-axis fall along the x-axis; $y = 0$ is the solution to (9.155) for $x = -\mu$. The three-dimensional streamlines through the stagnation point that are contained in the vertical surface of the discharge streamline reach the base of the aquifer.

We examine the angles between the discharge streamlines and the x-axis at $\mathscr{S}$ by differentiating (9.155) with respect to x. The derivative dy/dx is

$$\frac{dy}{dx} = \frac{b}{a} D(x+b)^{b/a-1}. \tag{9.156}$$

We wish to determine whether the discharge streamlines leave the area below the pond at $\mathscr{S}$, or not, and compare the derivative (9.156) with the derivative dy/dx obtained from the equation of the ellipse. The equation for the ellipse is, for this case,

$$\left(\frac{x}{b}\right)^2 + \left(\frac{y}{a}\right)^2 = 1. \tag{9.157}$$

We solve for y,

$$y = a\sqrt{1-\left(\frac{x}{b}\right)^2} \tag{9.158}$$

and differentiate:

$$\frac{dy}{dx} = -\frac{a}{2}\frac{2x/b}{\sqrt{1-(x/b^2)^2}} = -\frac{a}{b}\frac{x}{\sqrt{(b-x)(b+x)}}. \tag{9.159}$$

The ratio of the derivatives in the neighborhood of $\mathscr{S}$, where $x \approx -b$, is

$$\frac{(dy/dx)_s}{(dy/dx)_e} \approx \frac{\frac{b}{a}D(x+b)^{b/a-1}}{\frac{a}{\sqrt{(b-x)(b+x)}}} \approx \frac{b}{a^2} D\sqrt{2b}(x+b)^{b/a-1}(x+b)^{1/2} \tag{9.160}$$

or

$$\frac{(dy/dx)_s}{(dy/dx)_e} \approx \frac{b}{a^2} D\sqrt{2b}(x+b)^{b/a-1/2}. \tag{9.161}$$

When $b/a > 1/2$, the ratio of the slopes is zero at the stagnation point, $x = -b$; i.e., the discharge streamline will fall inside the ellipse near the stagnation point and the plume extends all the way down to the base of the aquifer. The opposite is the case when $b/a < 1/2$; the streamlines will fall outside the area below the pond. Note that the plume in Figure 9.13 does not extend all the way down to the base of the aquifer. The boundary of the plume, the section $\mathscr{B}-\mathscr{B}$, is wider than the streamlines that define section $\mathscr{A}-\mathscr{A}$, which delineate an area with flow equal to the total infiltrated amount. The actual plume is wider, because clean water flows underneath the pond and mixes with the infiltrated water. The case shown in Figure 9.14 corresponds to $b/a = 1/2$ and the stagnation point is on the boundary at $x = -b$. The bounding streamlines indeed exactly contain the infiltrated water;

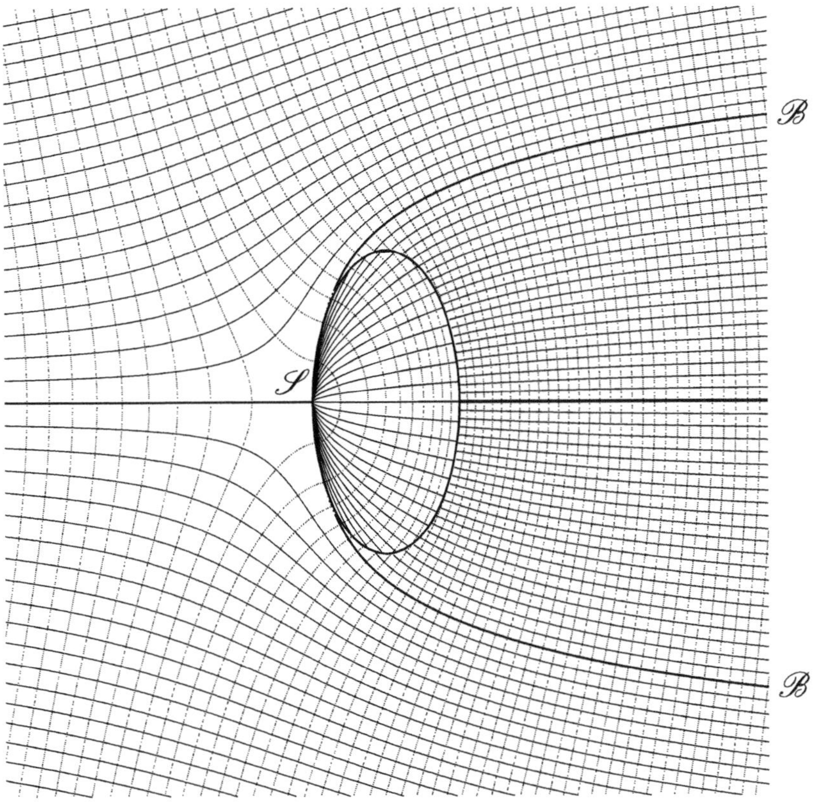

Figure 9.14 An elliptical pond with the stagnation point $\mathscr{S}$ at $x = -b$. The plume fills the entire saturated aquifer between the two bounding streamlines, $\mathscr{S} - \mathscr{B}$.

the plume occupies the aquifer over its entire saturated thickness. Streamlines are shown inside the plume, using the equations for the streamlines derived above. All discharge streamlines that emanate from the stagnation point $\mathscr{S}$ lie entirely inside the ellipse.

Comparison of the Plumes

The shape of the plume is affected by the orientation of the ellipse relative to the direction of flow. Aligning the pond with the flow results in a narrower plume than placing the pond at right angles to the flow direction. The cause of this phenomenon is that the plume reaches deeper into the aquifer when the major axis of the ellipse is aligned with the flow. This effect becomes much more pronounced if there are no stagnation points. This is illustrated in Figures 9.15 and 9.16; the data used to generate the figures are $Q_{x0} = 0.1$ m^2/d, $N = 0.002$ m/d, $a = 50.2$ m, and $b = 5.02$ m. If the major axis of the pond is normal to the flow, the plume is much wider than if the major axis is aligned with the flow, as shown in Figure 9.16. We

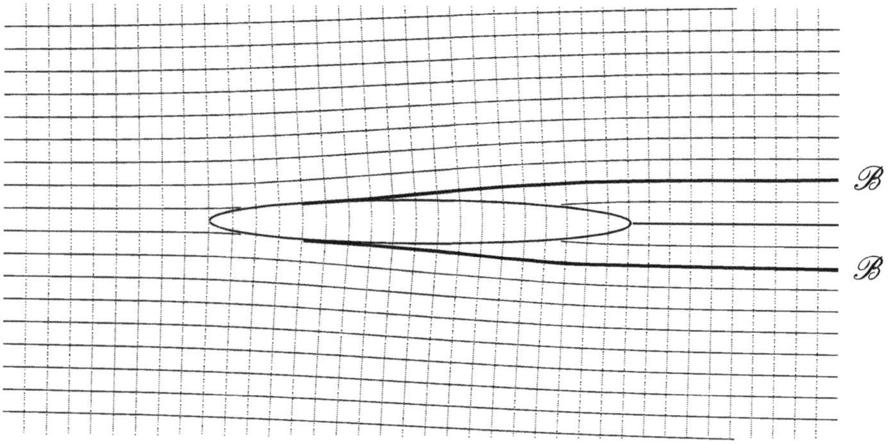

Figure 9.15 Plume for a pond in a field of uniform flow; there are no stagnation points. The plume does not reach the base of the aquifer anywhere.

observe that if we wish to limit the plume in width, rather than in depth, we should place the pond such that the major axis is in the direction of flow. If we desire to limit the depth of the plume, we should place the major axis of the pond at right angles to the flow. This effect is most noticeable when no stagnation points exist. As the total infiltration increases, the plume increasingly extends to the aquifer base and the plumes become similar. This is illustrated in Figure 9.17; the two plumes downstream from the pond are much more similar than for the case where no stagnation points exist.

9.3 Solute Transport

Solute transport is governed by a number of different phenomena: advection, chemical reactions, diffusion, and dispersion. We divide transport of solutes into different components: an advective component, a chemical component, a diffusive component, and a dispersive component. We focus here on the first two: advective transport and the chemical interaction between water and solid particles.

9.3.1 Simplifying Assumptions

We base the following analysis on three simplifying assumptions:

1. The solutes are present in the flow only in small concentrations and do not influence the physical properties of the groundwater.
2. The porosity and hydraulic conductivity are not affected by the solutes.
3. The chemical interaction between solutes and solid particles is described fully by instantaneous linear sorption.

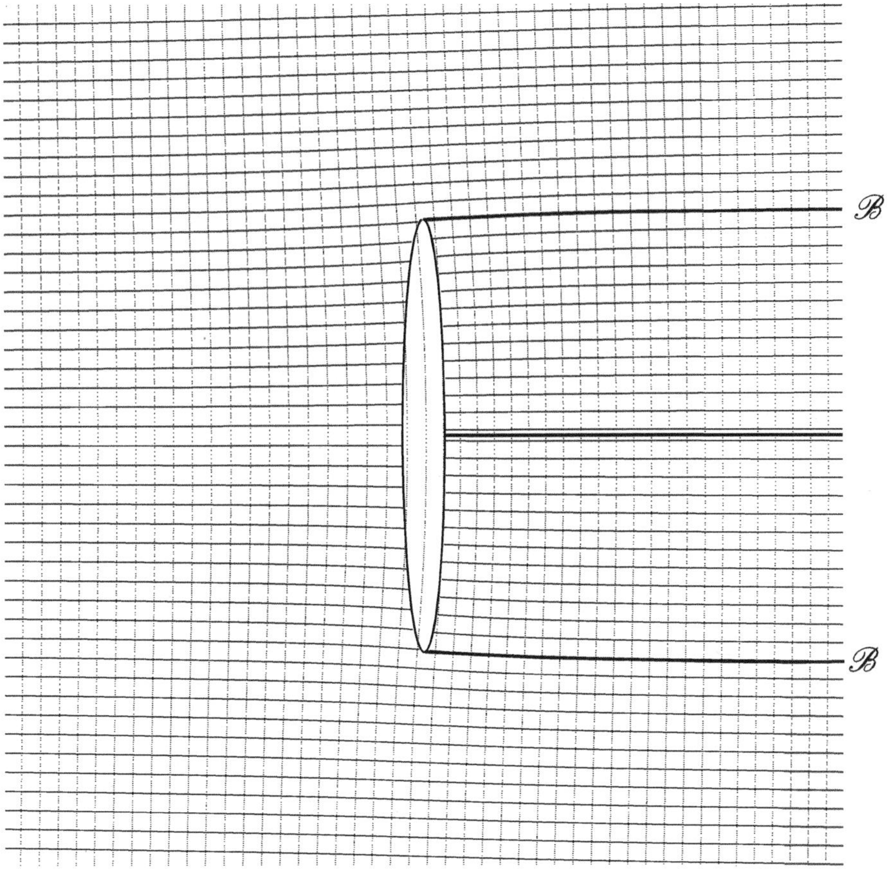

Figure 9.16 Plume for a pond in a field of uniform flow; there are no stagnation points. The plume does not reach the base of the aquifer anywhere. The major axis of the pond is normal to the flow.

Flow and the transport are independent of one another as a result of assumptions (1) and (2); the transport of solutes can be solved *after* determining the flow field.

9.3.2 Solute Movement

Advective transport is the main component of solute transport in most applications. Advection is the movement of solutes with the fluid, with the dissolved particles behaving as water particles.

We consider contaminated water that occupies an area $\mathscr{A}$ in an aquifer. If the only form of transport is advective, then the contaminated water moves as a plug through the aquifer. This plug is bounded at all sides by path lines, and its front moves with the velocity of the groundwater, as illustrated in Figure 9.18. In reality,

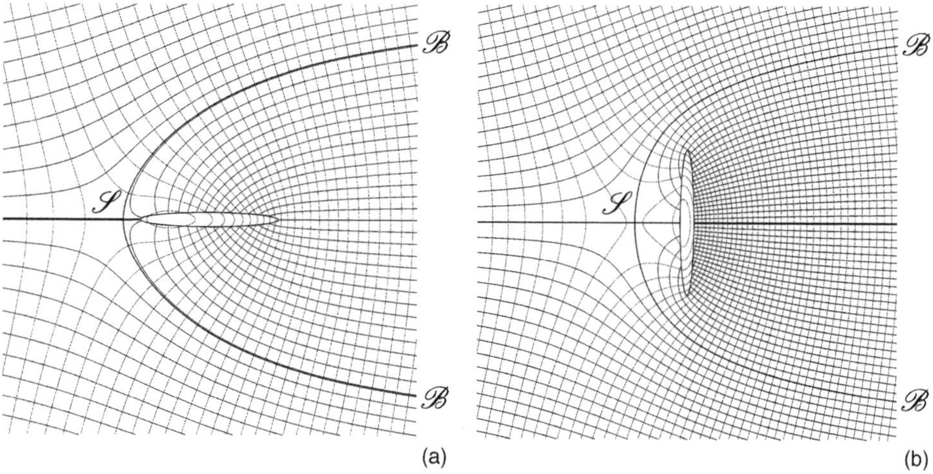

Figure 9.17 A pond with its major axis parallel to the flow (a) and a pond with its major axis normal to the flow (b). The total infiltration is 1.81 times that for the case in which point $\mathscr{S}$ is on the boundary.

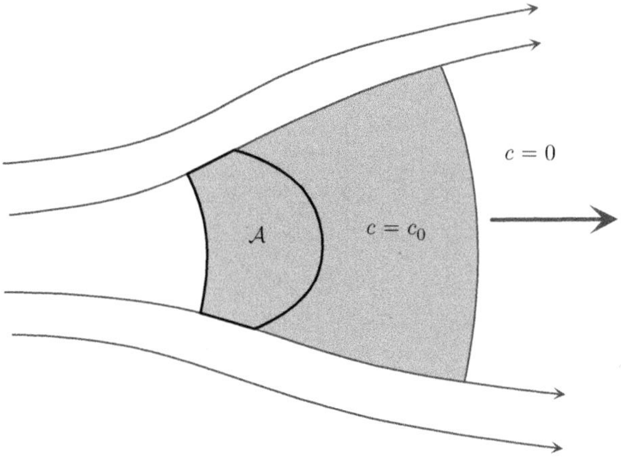

Figure 9.18 Plug flow.

the front moves slower than the groundwater. This is the result of chemical reactions that take place between solutes and solid particles; this reduction in velocity is called retardation.

The front is usually not sharp: there is a gradual transition in concentration from $c = c_0$ to $c = 0$. This may be caused by decay such as in radioactive materials or by another effect, called *hydrodynamic dispersion*.

Hydrodynamic dispersion is caused by inhomogeneities in the porous material and results from differences in velocity of neighboring water particles. Dispersion caused by the inherent inhomogeneous nature of the porous medium on the particle scale is called *microscopic dispersion*; the effect of microscopic dispersion is negligible in most porous media, but is important in fractured rock. Dispersion caused by inhomogeneities on a larger scale, such as gravel or clay pockets, and, in particular, stratification, is much more important and is called *macroscopic dispersion*. Dispersion has a smoothing effect on both the front and the sides of the plug that would exist if there were no dispersion. It is important to note that the concentration outside this plug will now be non-zero, since some particles will move faster than average. As stated above, the effect of hydrodynamic dispersion will not be included here.

9.3.3 Transport Equation

We define the mass flux (F_x, F_y, F_3) as the mass of a contaminant c carried by the specific discharge vector as

$$(F_x, F_y, F_3) = (cq_x.cq_y, cq_3). \tag{9.162}$$

The concentration c is the concentration of the mobile phase of the contaminant, as opposed to the concentration, c_s, of the sorbed phase, i.e., the mass of contaminant that is bonded with the solid particles. We define both concentrations as mass per unit volume of fluid.

We use the concept of divergence, introduced for the specific discharge vector in equation (1.36). The divergence of the mass flux is the mass of contaminant carried away at a point per unit time and unit volume, which equals the rate of decrease in mass at a point. This rate of decrease is composed of the decrease in mass of both mobile and sorbed phases. We include decay, with a rate factor λ:

$$\frac{\partial F_x}{\partial x} + \frac{\partial F_y}{\partial y} + \frac{\partial F_3}{\partial x_2} = -n\lambda(c + c_s) - n\frac{\partial}{\partial t}(c + c_s). \tag{9.163}$$

The porosity enters in this equation because we defined concentration as mass per unit volume of fluid. We use (9.162) and (9.163) and rearrange:

$$-\left[\frac{\partial}{\partial x}(cq_x) + \frac{\partial}{\partial y}(cq_y) + \frac{\partial}{\partial x_3}(cq_3)\right] - n\lambda(c + c_s) = n\frac{\partial}{\partial t}(c + c_s). \tag{9.164}$$

The chemical interaction between solutes and solid particles is described fully by instantaneous linear sorption. The assumption of instantaneous linear sorption (see, e.g. Javandel et al. [1984]) implies that

$$c = \frac{1}{R}(c + c_s), \tag{9.165}$$

where R is a dimensionless constant called the retardation factor. If R is independent of time, then

$$\frac{\partial}{\partial t}(c+c_s) = R\frac{\partial c}{\partial t}, \tag{9.166}$$

so that (9.164) becomes

$$nR\frac{\partial c}{\partial t} + \left[\frac{\partial}{\partial x}(cq_x) + \frac{\partial}{\partial y}(cq_y) + \frac{\partial}{\partial x_3}(cq_3)\right] = -n\lambda Rc. \tag{9.167}$$

We differentiate the terms within the brackets by parts and divide both sides by a factor nR,

$$\frac{\partial c}{\partial t} + \frac{1}{nR}\left[q_x\frac{\partial c}{\partial x} + q_y\frac{\partial c}{\partial y} + q_3\frac{\partial c}{\partial x_3} + c\left(\frac{\partial q_x}{\partial x} + \frac{\partial q_y}{\partial y} + \frac{\partial q_3}{\partial x_3}\right)\right] = -\lambda c. \tag{9.168}$$

For the sake of simplicity we treat the flow as incompressible, so that the divergence is zero and the term in parentheses vanishes. We introduce velocity components v_x, v_y, and v_3:

$$\frac{\partial c}{\partial t} + \frac{1}{R}\left[v_x\frac{\partial c}{\partial x} + v_y\frac{\partial c}{\partial y} + v_3\frac{\partial c}{\partial x_3}\right] = -\lambda c. \tag{9.169}$$

This equation is the transport equation for solute transport with retardation and decay in an incompressible porous medium, with the effect of hydrodynamic dispersion neglected. This transport equation is a first-order partial differential equation, which can be integrated along its characteristics.

9.3.4 Integration along Characteristics

We introduce the apparent velocity with components

$$v_x^* = \frac{v_x}{R}, \qquad v_y^* = \frac{v_y}{R}, \qquad v_3^* = \frac{v_3}{R}. \tag{9.170}$$

If x, y, and x_3 represent the coordinates of a point moving with the apparent velocity, then

$$v_x^* = \frac{dx}{dt}, \qquad v_y^* = \frac{dy}{dt}, \qquad v_3^* = \frac{dx_3}{dt}. \tag{9.171}$$

Integration of these differential equations yields a curve that represents the path of a point moving with the apparent velocity; we call this curve the *apparent path line*. We combine (9.170) and (9.171) with (9.169):

$$\frac{\partial c}{\partial t} + \frac{\partial c}{\partial x}\frac{dx}{dt} + \frac{\partial c}{\partial y}\frac{dy}{dt} + \frac{\partial c}{\partial x_3}\frac{dx_3}{dt} = -\lambda c. \tag{9.172}$$

The left-hand side of this equation is the total derivative of c with respect to time along the apparent path line:

$$\frac{dc}{dt} = -\lambda c. \tag{9.173}$$

The term dc/dt represents the rate of change in concentration at a point that moves with the apparent velocity. If there is no sorption, i.e., if R equals unity, then dc/dt is equal to the material-time derivative Dc/Dt, which represents the rate of change in concentration of a material point, which moves with the actual average velocity of the fluid. The apparent path line is a special curve; the differential equation gives information only regarding the change of the concentration along this curve. Such a curve is called the *characteristic* of the differential equation.

To illustrate the meaning of the derivative dc/dt, we introduce the distance s traveled along the apparent path line and represent the magnitude of the apparent velocity as v^*. The differential equation

$$v^* = \frac{ds}{dt} \tag{9.174}$$

defines a curve in the (s, t)-plane that represents the characteristic of the differential equation. This characteristic passes through the point (s_0, t_0) corresponding to the position s_0 that the point moving along the apparent path line occupies at time t_0. The slope of the curve represents the apparent velocity v^*.

The total derivative dc/dt represents the change in concentration with respect to time, moving along the apparent path line with the apparent velocity v^*. Thus, dc/dt is the limit, for $\Delta t \to 0$, of $\Delta c/\Delta t$ where Δc is the difference in concentration between points $\mathscr{A}$ and $\mathscr{B}$ on the path divided by the projection Δt of the segment $\mathscr{A}\mathscr{B}$ on the t-axis (see Figure 9.19). Note that the values of s at points $\mathscr{A}$ and $\mathscr{B}$ are not the same; Δs represents the distance moved with velocity v^* along the path over the time interval Δt. The differential equation (9.173) can be integrated:

$$c = c_0 e^{-\lambda(t-t_0)}, \tag{9.175}$$

where c_0 is the concentration at time $t = t_0$ and position $s = s_0$:

$$c_0 = c(t_0), \qquad s = s(t_0). \tag{9.176}$$

In the absence of decay, i.e., if $\lambda = 0$, (9.175) reduces to

$$c = c_0, \tag{9.177}$$

which expresses that the concentration remains unchanged when moving along the apparent path line with the apparent velocity v^*: the front moves with this velocity and there is plug flow.

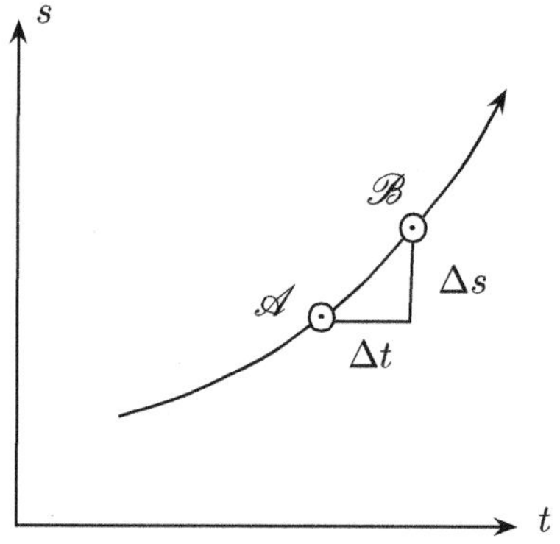

Figure 9.19 The characteristic.

The retardation factor R is usually expressed in terms of yet another factor, called the distribution factor K_d.

9.3.5 The Distribution Factor

The distribution factor K_d is defined as

$$K_d = \frac{S}{c}, \tag{9.178}$$

where S is defined as the ratio of the mass of sorbed material to the mass of solid material, both contained in some volume V of porous material. If ρ_b [kg/m^3] is the bulk density of solid material, then

$$S = \frac{c_s n V}{\rho_b V} = n\frac{c_s}{\rho_b}. \tag{9.179}$$

The factor n occurs in the numerator of the fraction because c_s is defined as mass per unit volume of fluid. We combine (9.178) with (9.179):

$$K_d = \frac{n}{\rho_b}\frac{c_s}{c}. \tag{9.180}$$

We use (9.165) to express R in terms of K_d as

$$R = 1 + \frac{c_s}{c} = 1 + \frac{\rho_b K_d}{n}. \tag{9.181}$$

The units of the distribution factor are [L^3/M]. The distribution factor is the quantity associated with retardation that is measured in the laboratory.

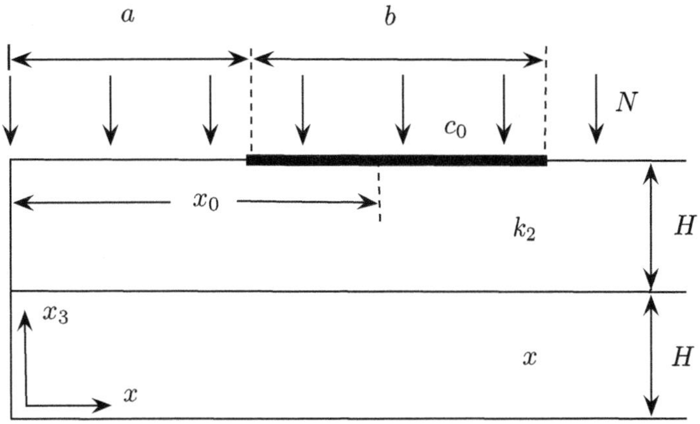

Figure 9.20 Contaminant transport in a stratified aquifer.

9.3.6 Contaminant Transport in a Stratified Aquifer

As an example, consider the case of steady flow in a stratified aquifer solved in Section 9.2 and shown in Figure 9.5. We assume that contaminated water of concentration c_0 enters the aquifer over a section $a \le x \le a+b$, $y = H_1 + H_2$, of the upper boundary, from time $t = t_0$ onward (see Figure 9.20). The retardation factor is R. We wish to determine the time it takes for a point $\mathscr{P}$ to reach the plane $x = L$. We identify the path of point $\mathscr{P}$ by the initial condition:

$$x = x_0, \qquad x_3 = H_1 + H_2, \qquad t = t_0. \tag{9.182}$$

The apparent path lines coincide with the streamlines, because the flow is steady. We set the thicknesses of the strata equal:

$$H_1 = H_2 = H. \tag{9.183}$$

We derived the equations for the stream functions in the two strata, (9.112),

$$\begin{aligned} \overset{2}{\psi} &= -\frac{Nk_2}{T} x x_3 + N\frac{(k_2-k_1)H_1}{T} x \\ \overset{1}{\psi} &= -\frac{Nk_1}{T} x x_3. \end{aligned} \tag{9.184}$$

We set $T = (k_1+k_2)H$ and replace H_1 and H_2 by H:

$$\begin{aligned} \overset{2}{\psi} &= -\frac{N}{H}\frac{k_2}{k_1+k_2} x x_3 + N\frac{(k_2-k_1)}{k_1+k_2} x \\ \overset{1}{\psi} &= -\frac{N}{H}\frac{k_1}{k_1+k_2} x x_3. \end{aligned} \tag{9.185}$$

We obtain an equation for the streamline through a point $x = x_0$, $x_3 = 2H$, in the upper stratum from the first equation in (9.184) by setting $\overset{2}{\psi}$ equal to its value at the starting point:

$$-Nx_0 = -N\frac{k_2}{k_1+k_2}x\frac{x_3}{H} + N\frac{k_2-k_1}{k_1+k_2}x \rightarrow \frac{k_2}{k_1+k_2}\frac{x_3}{H}x = x_0 + \frac{k_2-k_1}{k_1+k_2}x, \tag{9.186}$$

so that the equation for the streamline is

$$\frac{x_3}{H} = \frac{k_1+k_2}{k_2}\frac{x_0}{x} + \frac{k_2-k_1}{k_2}, \tag{9.187}$$

which is a hyperbola. The streamline intersects the interzonal boundary, $x_3 = H$, at x_I, with

$$1 - \frac{k_2-k_1}{k_2} = \frac{k_1}{k_2} = \frac{k_1+k_2}{k_2}\frac{x_0}{x_I}, \tag{9.188}$$

so that

$$x_I = \frac{k_1+k_2}{k_1}x_0. \tag{9.189}$$

We obtain the equation for the streamline in the lower stratum from the second equation in (9.184) by setting $\overset{1}{\psi}$ equal to $-Nx_0$, which gives

$$-Nx_0 = -N\frac{k_1}{k_1+k_2}\frac{x_3}{H}x \rightarrow \frac{x_3}{H} = \frac{k_1+k_2}{k_1}\frac{x_0}{x}. \tag{9.190}$$

The specific discharges $\overset{1}{q}_x$ and $\overset{2}{q}_x$ in strata 1 and 2 are

$$\begin{aligned} \overset{2}{q}_x &= -\frac{\partial\overset{2}{\psi}}{\partial x_3} = \frac{N}{H}\frac{k_2}{k_1+k_2}x \\ \overset{1}{q}_x &= -\frac{\partial\psi}{\partial x_3} = \frac{N}{H}\frac{k_1}{k_1+k_2}x. \end{aligned} \tag{9.191}$$

We obtain expressions for the apparent velocities $\overset{1}{v}_x{}^*$ and $\overset{2}{v}_x{}^*$ in the two strata by dividing $\overset{1}{q}_x$ and $\overset{2}{q}_x$ by a factor nR:

$$\overset{2}{v}_x{}^* = \frac{1}{nR}\frac{k_2}{k_1+k_2}\frac{N}{H}x \tag{9.192}$$

$$\overset{1}{v}_x{}^* = \frac{1}{nR}\frac{k_1}{k_1+k_2}\frac{N}{H}x. \tag{9.193}$$

We have, by definition,

$$v^* = \frac{ds}{dt}, \tag{9.194}$$

where s represents the arc length measured along the apparent streamline. The streamline is tangent to the velocity vector,

$$v_x^* = \frac{dx}{ds} v^* = \frac{dx}{ds}\frac{ds}{dt} = \frac{dx}{dt}, \tag{9.195}$$

where v_x^* is the projection of v^* onto the x-axis. We compute the component v_x^* at each point of the streamline; v_x^* is independent of x_3 in each stratum; we compute the travel time of point $\mathscr{P}$ by evaluating the following integrals:

$$\int_{t_0}^{t} dt = \int_{x_0}^{x} \frac{dx}{\overset{2}{v_x}{}^*}, \qquad H \leq x_3 \leq 2H, \quad x_0 \leq x \leq x_I \tag{9.196}$$

$$\int_{t_I}^{t} dt = \int_{x_I}^{x} \frac{dx}{\overset{1}{v_x}{}^*}, \qquad 0 \leq x_3 \leq H, \quad x_I \leq x \leq \infty, \tag{9.197}$$

where x_I is given by (9.189); we use (9.192) with (9.196) to evaluate $t - t_0$ in the upper aquifer:

$$t - t_0 = \frac{nRH}{N}\frac{k_1+k_2}{k_2} \ln \frac{x}{x_0}, \qquad x_0 \leq x \leq \frac{k_1+k_2}{k_1} x_0. \tag{9.198}$$

Point $\mathscr{P}$ reaches the interface between the strata when $t = t_I$ and $x = x_I = (k_1+k_2)x_0/k_1$:

$$t_I - t_0 = \frac{nRH}{N}\frac{k_1+k_2}{k_2} \ln \frac{k_1+k_2}{k_1}. \tag{9.199}$$

We assume that this interface is reached before arriving at the boundary at $x = L$, and integrate (9.197) between x_I and L, using (9.193):

$$t_L - t_I = \frac{nRH}{N}\frac{k_1+k_2}{k_1} \ln \left[\frac{k_1}{k_1+k_2}\frac{L}{x_0}\right]. \tag{9.200}$$

We determine the arrival time of point $\mathscr{P}$ of the plume, defined by the initial condition (9.182), by adding (9.199) to (9.200).

Problem

9.6 Consider the foregoing example. Introduce the dimensionless time $t^* = tN/(nRH)$ and compute the dimensionless time $t_L^* - t_0^*$ at which point $\mathscr{P}$ reaches the boundary at $x = L$ if

$$x = x_0 = a, \qquad k_1 = k, \qquad k_2 = 5k, \qquad a = b = H, \qquad L = 10H,$$

Let Δt^* be $(t_L^* - t_0^*)/5$ and consider five points $\mathscr{P}$ of the front on apparent path lines defined by the initial conditions $t = t_0$, $y = 2H$, with $x = H$, $x = 5H/4$, $x = 3H/2$, $x = 7H/4$, and $x = 2H$ for the five path lines, respectively.

Questions:

1. Plot the five apparent path lines and mark the points that correspond to $t_0^* + j\Delta t^*$ $(j = 1,2,3,4,5)$. Sketch the position of the front of the plume for each of these values of t^*.
2. Express the concentration c in terms of the parameters of the problem for all marked points if there is a decay rate λ.

10

Finite Differences and Finite Elements

Finite difference and finite element methods are common numerical techniques for modeling groundwater flow problems. These methods differ from those discussed thus far in that the hydraulic head is approximated by discretization throughout the flow domain. As a result, not only are the boundary conditions approximated, but also the differential equation itself. Characteristic for both methods is that the flow domain is bounded. This is true also for applications to regional flow in contrast to the analytic element method, where the aquifer system is modeled as being infinite in extent; boundary conditions are applied only along internal boundaries that are physically present in the aquifer system.

Advantages of both the finite difference and finite element methods are that the hydraulic conductivity can be easily varied throughout the aquifer system, the formulations are well suited for modeling transient flow, and they are comparatively straightforward. Both methods are discussed in detail in the literature on groundwater mechanics, e.g., Verruijt [1982], Wang and Anderson [1982], and Bear and Verruijt [1987].

A number of computer implementations of finite difference and finite element techniques exist. The most popular finite difference program is MODFLOW, e.g., Mehl and Hill [2001], [2003], [2005], and MODFLOW-2005. A popular finite element program is FEFLOW; manuals for this program are available online.

There is an extensive literature on groundwater flow modeling using numerical methods. We limit the treatment in this text to a discussion of the basic principles of such models; it is important, when applying numerical software, to understand the basic principles involved, as well as the strengths and weaknesses of the methods; we focus on these aspects.

10.1 Finite Difference Methods

Finite difference methods are based on a discretization of the flow domain into a mesh that is usually rectangular, where the potential or head is computed at the gridpoints by solving the differential equation in finite difference form throughout the mesh. There exist a variety of numerical techniques for solving the resulting system of linear equations, and the method may be applied with comparative ease to problems where the hydraulic conductivity varies from node to node. Our purpose is to explain the basic principles involved in the method for a square mesh and constant hydraulic conductivity.

10.1.1 Steady Flow

We formulate the problem in terms of the discharge potential for the sake of consistency within the text; most applications of finite differences, such as MODFLOW, are formulated in terms of the hydraulic head. The differential equation for steady flow in an aquifer with homogeneous hydraulic conductivity is

$$\frac{\partial^2\Phi}{\partial x^2}+\frac{\partial^2\Phi}{\partial y^2}=-N(x,y), \tag{10.1}$$

where N represents a given rate of infiltration. The flow region is discretized by a square mesh as shown in Figure 10.1(a). The nodes are identified as (i,j) where i and j are integers such that the coordinates of node (i,j) are given by

$$x=(i-1)a, \qquad y=(j-1)a, \tag{10.2}$$

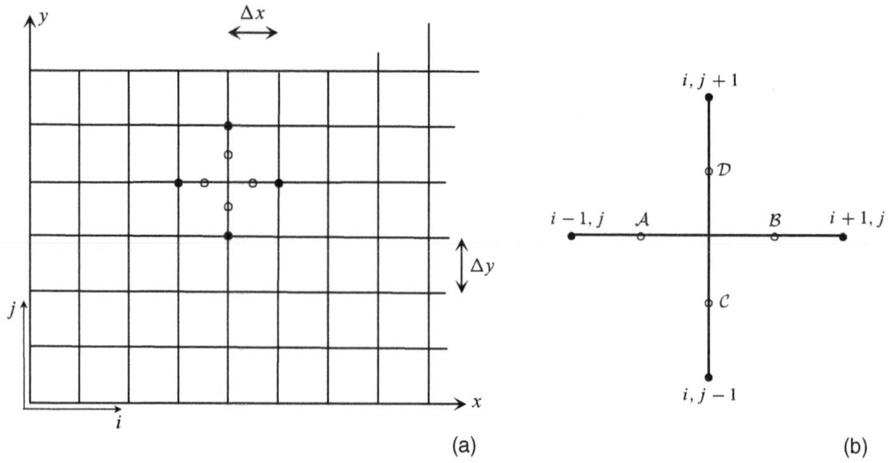

Figure 10.1 Discretization by a rectangular mesh.

where a is the distance between grid points,

$$a = \Delta x = \Delta y. \tag{10.3}$$

The node (i,j) with its four surrounding grid points is shown in Figure 10.1(b). Points $\mathscr{A}$, $\mathscr{B}$, $\mathscr{C}$, and $\mathscr{D}$ lie midway between nodes, and we approximate the derivative $\partial\Phi/\partial x$ at points $\mathscr{A}$ and $\mathscr{B}$ as

$$\left[\frac{\partial \Phi}{\partial x}\right]_{\mathscr{A}} = \frac{\Phi_{i,j} - \Phi_{i-1,j}}{a}, \qquad \left[\frac{\partial \Phi}{\partial x}\right]_{\mathscr{B}} = \frac{\Phi_{i+1,j} - \Phi_{i,j}}{a}. \tag{10.4}$$

We express the second derivative of Φ with respect to x at node (i,j) in the same way:

$$\left[\frac{\partial^2 \Phi}{\partial x^2}\right]_{i,j} = \frac{\Phi_{i-1,j} + \Phi_{i+1,j} - 2\Phi_{i,j}}{a^2}. \tag{10.5}$$

The derivative $\partial^2\Phi/\partial y^2$ at node (i,j) has the same form, and we obtain the following expression for the Laplacian:

$$\left[\nabla^2 \Phi\right]_{i,j} = \frac{\Phi_{i-1,j} + \Phi_{i+1,j} + \Phi_{i,j-1} + \Phi_{i,j+1} - 4\Phi_{i,j}}{a^2}. \tag{10.6}$$

We use this in the differential equation (10.1) and represent N at node i,j as $N_{i,j}$,

$$\Phi_{i-1,j} + \Phi_{i+1,j} + \Phi_{i,j-1} + \Phi_{i,j+1} - 4\Phi_{i,j} = -a^2 N_{i,j}, \tag{10.7}$$

and solve for $\Phi_{i,j}$:

$$\Phi_{i,j} = \tfrac{1}{4}\left[\Phi_{i-1,j} + \Phi_{i+1,j} + \Phi_{i,j-1} + \Phi_{i,j+1} + a^2 N_{i,j}\right]. \tag{10.8}$$

We apply this equation at each node and use it together with the boundary conditions to obtain values of the potential at all nodes as explained below.

Boundary Conditions in Terms of Φ

In the Dirichlet problem the potential is known along the entire boundary, and therefore at each boundary node. Equation (10.8) need not be applied at the boundary nodes but only at the interior ones. A common method for solving the system of equations (10.8) is by the Jacobi method, where each value for $\Phi_{i,j}$ is first estimated at all interior nodes, and values for successive iterations are obtained by applying (10.8) at each node using the values computed in the previous iteration:

$$\Phi_{i,j}^{(m+1)} = \frac{1}{4}\left[\Phi_{i-1,j}^{(m)} + \Phi_{i+1,j}^{(m)} + \Phi_{i,j-1}^{(m)} + \Phi_{i,j+1}^{(m)} + a^2 N_{i,j}\right], \tag{10.9}$$

where m is the iteration index: $m = 1$ for the first iteration.

A method that converges more rapidly is the Gauss-Seidel iteration. In this method we apply (10.8) at nodes $(2,2), (3,2), \ldots, (n_i - 1, 2)$ successively, where n_i

is the number of nodes in the row $j = 2$. We continue with row $j = 3$ in the same way, and proceed until all nodes have been used. In this way, two newly computed values are used in the expression for $\Phi_{i,j}^{(m+1)}$:

$$\boxed{\Phi_{i,j}^{(m+1)} = \tfrac{1}{4}\left[\Phi_{i-1,j}^{(m+1)} + \Phi_{i+1,j}^{(m)} + \Phi_{i,j-1}^{(m+1)} + \Phi_{i,j+1}^{(m)} + a^2 N_{i,j}\right]} \tag{10.10}$$

Boundary Conditions in Terms of $\partial\Phi/\partial n$; Mixed Boundary Conditions

In the Neumann problem the boundary conditions are specified in terms of $\partial\Phi/\partial n$. In order to approximate $\partial\Phi/\partial n$ at boundary nodes, the mesh is extended one row of nodal points beyond the boundary, as illustrated in Figure 10.2. The values for either i or j are zero for nodes (i,j) outside the boundary, which are called imaginary nodes. For the case illustrated in Figure 10.2, $\partial\Phi/\partial n$ is given along the boundary $x = 0$, $i = 1$, and $\partial\Phi/\partial n$ is approximated for node $1,j$ as

$$\left[\frac{\partial\Phi}{\partial n}\right]_{1,j} = \left[\frac{\partial\Phi}{\partial x}\right]_{1,j} = \frac{\Phi_{2,j} - \Phi_{0,j}}{2a}, \tag{10.11}$$

so that

$$\Phi_{0,j} = \Phi_{2,j} - 2a\left[\frac{\partial\Phi}{\partial n}\right]_{1,j}, \tag{10.12}$$

which is used to compute the value of $\nabla^2\Phi$ at the boundary nodes; expression (10.10) becomes

$$\Phi_{1,j}^{(m+1)} = \tfrac{1}{4}\left\{2\Phi_{2,j}^{(m)} + \Phi_{1,j-1}^{(m+1)} + \Phi_{1,j+1}^{(m)} + a^2 N_{1,j} - 2a\left[\frac{\partial\Phi}{\partial n}\right]_{1,j}\right\}. \tag{10.13}$$

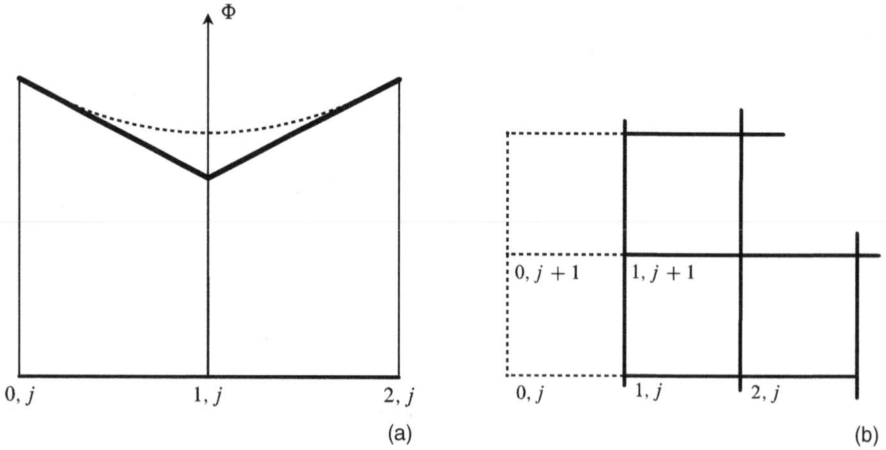

Figure 10.2 A boundary with $\partial\Phi/\partial n$ given.

Boundary values of $\partial\Phi/\partial n$ are applied by using (10.13) instead of (10.10) at the boundary nodes.

Mixed boundary value problems are solved by extending the mesh beyond those boundaries where $\partial\Phi/\partial n$ is given, using equations of the form (10.13), while setting Φ equal to prescribed values at those nodes where the head is prescribed.

As a result of the discretization, it is a relatively simple matter to adapt the finite difference scheme to include such phenomena as leakage. The right-hand side of the equation will involve a leakage term which is dependent on the potential Φ, which merely results in a modification of equation (10.10) for the potential at node i,j. Varying values of the hydraulic conductivity can be taken into account by formulating the problem in terms of a function F,

$$F = \Phi/k, \tag{10.14}$$

that is independent of the hydraulic conductivity. The differential equation to be discretized then becomes

$$\frac{\partial}{\partial x}\left[k\frac{\partial F}{\partial x}\right] + \frac{\partial}{\partial y}\left[k\frac{\partial F}{\partial y}\right] = -N(x,y). \tag{10.15}$$

Problems of transient flow may be solved by discretizing the applicable differential equation both in space and in time. The resulting scheme may be written either explicitly or implicitly in terms of the potential, as outlined below.

Problem

10.1 Write a computer program using Gauss-Seidel iteration for solving the problem of flow through a strip between two impermeable walls, bounded by two equipotentials that are a distance $2L$ apart; the equipotentials are normal to the impermeable walls. The head at the upstream equipotential is ϕ_1 and at the downstream one it is ϕ_2.

10.1.2 Transient Flow

We discuss application of the finite difference method to problems of transient flow in unconfined aquifers. The formulation for problems of transient flow in confined aquifers is simpler, and it is left to the reader to adapt the scheme discussed below to problems of confined transient flow.

The differential equation for shallow transient flow in an unconfined aquifer is given by (7.15),

$$\nabla^2\Phi = S_p\frac{\partial\phi}{\partial t} - N. \tag{10.16}$$

There are two different methods of discretization of this equation with respect to time: the explicit formulation and the implicit one.

Explicit Finite Difference Formulation

In an explicit formulation, the discretization in time is:

$$\left[\nabla^2\Phi\right]_t = S_p \frac{\phi_{t+\Delta t} - \phi_t}{\Delta t} - N_t. \tag{10.17}$$

This discretization is called explicit because the Laplacian is computed at time t, so that we obtain an explicit expression for the head at time $t+\Delta t$:

$$\boxed{\phi_{t+\Delta t} = \phi_t + \frac{\Delta t}{S_p}\left[\nabla^2\Phi + N\right]_t} \tag{10.18}$$

A benefit of the scheme is that linearization of the differential equation is not necessary. The head at each node at time $t+\Delta t$ is obtained by computing $\nabla^2\Phi$ by application of (10.6), in which the values of Φ at time t are used.

The explicit scheme has the advantages of flexibility and simplicity, but gives meaningful answers only if Δt is less than some critical value, Δt_{crit}:

$$\Delta t < \Delta t_{\text{crit}}. \tag{10.19}$$

The value of the critical time step may be found by trial and error: if Δt exceeds Δt_{crit}, the solution becomes unstable, giving results that are clearly in error. Alternatively, we estimate the value of Δt_{crit} by using the expression derived below for the case in which $N = 0$.

The Critical Time Step

Stability of the numerical solution requires that a small disturbance, caused, for example, by rounding, damps out rather than amplifies. We consider that all values of Φ in the mesh are equal to some constant value Φ_0, except for node i,j where $\Phi = \Phi_0 - \delta\Phi$, and for the four neighboring nodes where $\Phi = \Phi_0 + \delta\Phi$; $\delta\Phi$ represents a disturbance. The expression (10.6) for the Laplacian then becomes

$$[\nabla^2\Phi]_{i,j} = \frac{4\Phi_0 + 4\delta\Phi - 4(\Phi_0 - \delta\Phi)}{a^2} = 8\frac{\delta\Phi}{a^2}. \tag{10.20}$$

Substitution in (10.18) yields, with $N = 0$,

$$\phi_{t+\Delta t} = \phi_t + \frac{8\Delta t}{S_p}\frac{\delta\Phi}{a^2}, \tag{10.21}$$

where $\phi_{t+\Delta t}$ and ϕ_t represent the head at node (i,j). The disturbance $\delta\Phi$ corresponds to a disturbance $\delta\phi$ in the head; $\phi_{t+\Delta t} - \phi_t$ must be less than $2\delta\phi$ in order that the disturbance decreases with time. Hence, the condition for stability is

$$\frac{8\Delta t}{S_p a^2}\delta\Phi < 2\delta\phi. \tag{10.22}$$

With

$$\delta\Phi = \tfrac{1}{2}k[\phi_0 + 2\phi_0\delta\phi + (\delta\phi)^2] - \tfrac{1}{2}k\phi_0^2 = k\phi_0\delta\phi + \tfrac{1}{2}k(\delta\phi)^2 \tag{10.23}$$

(10.22) becomes

$$\frac{8\Delta t}{S_p a^2}[k\phi_0\delta\phi + \tfrac{1}{2}k(\delta\phi)^2] < 2\delta\phi, \tag{10.24}$$

so that

$$\Delta t < \frac{S_p a^2}{4k(\phi_0 + \frac{1}{2}|\delta\phi|)}. \tag{10.25}$$

We obtain a lower bound for the critical time step by taking for $\phi_0 + \frac{1}{2}|\delta\phi|$ the maximum value of ϕ that is expected to occur in the grid. Denoting this value as $\phi_{\max}$, we obtain

$$\Delta t_{\text{crit}} = \frac{S_p a^2}{4k\phi_{\max}}, \tag{10.26}$$

where

$$a = \Delta x = \Delta y. \tag{10.27}$$

The reader may verify that for problems of one-dimensional flow the expression for the critical time step becomes

$$\Delta_{\text{crit}} = \frac{S_p(\Delta x)^2}{2k\phi_{\max}}. \tag{10.28}$$

It is good practice to keep the time step well below the critical value, e.g., $\Delta t = \frac{1}{2}\Delta t_{\text{crit}}$.

Problem

10.2 Adapt (10.26) to the case of shallow transient confined flow.

Problem

10.3 Write an explicit finite difference program to solve a problem of transient one-dimensional flow in a confined aquifer of width L with the following initial and boundary conditions:

$$t=0, \qquad 0<x<L, \qquad \Phi=\Phi_0$$

$$t>0, \qquad \left\{ \begin{array}{ll} x=0, & \Phi=\Phi_1<\Phi_0 \\ x=L, & \dfrac{\partial\Phi}{\partial x}=0 \end{array} \right\}. \tag{10.29}$$

Write a computer program to evaluate the following exact solution to this problem (see Carslaw and Jaeger [1959]):

$$\Phi=\Phi_1+\frac{4}{\pi}(\Phi_0-\Phi_1)\sum_{m=0}^{\infty}\frac{1}{2m+1}e^{-[\pi^2(2m+1)^2/4]\tau}\sin\left[\frac{\pi}{2}(2m+1)\frac{x}{L}\right], \tag{10.30}$$

where

$$\tau=\frac{k}{S_s}\frac{t}{L^2}. \tag{10.31}$$

Choose $L=100$ m, $H=10$ m, $S_s=2*10^{-4}$ m^{-1}, $k=10^{-6}$ m/s, $\phi_0=15$ m, and $\phi_1=12$ m, and compare the explicit finite difference solution with the exact one.

Implicit Finite Difference Formulation

In the implicit finite difference formulation, we take the value of $\nabla^2\Phi$ at some time between t and $t+\Delta t$. The resulting system of equations is implicit in terms of the unknowns and must be solved at each time step, for example, by Gauss-Seidel iteration. Because of the implicit nature of the formulation, the differential equation for shallow transient unconfined flow must be linearized. The linearized form of the differential equation is given by (7.15):

$$\nabla^2\Phi=\frac{S_p}{k\bar{\phi}}\frac{\partial\Phi}{\partial t}-N. \tag{10.32}$$

We represent the Laplacian as a weighted average between the approximations at times $t_0+n\Delta t$ and $t_0+(n+1)\Delta t$, where t_0 is the onset of transient flow:

$$\nabla^2\Phi=\nu\nabla^2\overset{n+1}{\Phi}+(1-\nu)\nabla^2\overset{n}{\Phi}, \qquad 0\le\nu\le 1, \tag{10.33}$$

where the superscripts refer to the time level. We introduce a function $\overset{n}{F}_{i,j}$ as

$$\overset{n}{F}_{i,j}=\tfrac{1}{4}\left[\overset{n}{\Phi}_{i-1,j}+\overset{n}{\Phi}_{i+1,j}+\overset{n}{\Phi}_{i,j-1}+\overset{n}{\Phi}_{i,j+1}\right], \tag{10.34}$$

so that $\nabla^2\Phi$ may be written as follows (see (10.33) and (10.6)),

$$\nabla^2\Phi=\frac{4}{a^2}\left[\nu(\overset{n+1}{F}_{i,j}-\overset{n+1}{\Phi}_{i,j})+(1-\nu)(\overset{n}{F}_{i,j}-\overset{n}{\Phi}_{i,j})\right]. \tag{10.35}$$

We discretize the differential equation (10.32):

$$\nu(\overset{n+1}{F}_{i,j} - \overset{n+1}{\Phi}_{i,j}) + (1-\nu)(\overset{n}{F}_{i,j} - \overset{n}{\Phi}_{i,j}) = \frac{a^2 S_p}{4k\overset{n}{\phi}_{i,j}} \frac{(\overset{n+1}{\Phi}_{i,j} - \overset{n}{\Phi}_{i,j})}{\Delta t} - \frac{a^2}{4}\overset{n}{N}_{i,j}, \tag{10.36}$$

where $\overline{\phi}$ is taken as $\overset{n}{\phi}_{i,j}$. This equation is solved for the unknown value $\overset{n+1}{\Phi}_{i,j}$:

$$\boxed{\begin{aligned}\overset{n+1}{\Phi}_{i,j} = \frac{4k\overset{n}{\phi}_{i,j}\Delta t}{a^2 S_p + 4k\overset{n}{\phi}_{i,j}\nu\Delta t}\Bigg[&\nu\overset{n+1}{F}_{i,j} + (1-\nu)(\overset{n}{F}_{i,j} - \overset{n}{\Phi}_{i,j}) \\ &+ \frac{a^2 S_p}{4k\overset{n}{\phi}_{i,j}\Delta t}\overset{n}{\Phi}_{i,j} + \frac{a^2}{4}\overset{n}{N}_{i,j}\Bigg]\end{aligned}} \tag{10.37}$$

Further improvement may be achieved by replacing the term $\overset{n}{N}_{ij}$ by a weighted average over the time step, using a formula of the form (10.33). Equation (10.37) may be solved by the use of Gauss-Seidel iteration. The implicit scheme reduces to an explicit one for $\nu = 0$ (compare (10.33)). For $\nu \neq 0$ the scheme is *implicit*; the finite difference formulation then is called a *backward* one, in contrast to the *forward* one for $\nu = 0$. For $\nu = 1$ the scheme is *fully implicit*; the method is called a *central finite difference scheme* for $\nu = \frac{1}{2}$.

The advantage of the implicit formulation is that the time step may be chosen much larger than for the explicit one. Disadvantages are a greater complexity, and that the Gauss-Seidel iteration must be applied at each time step. The explicit formulation is particularly attractive for three-dimensional problems; the formulation may be adapted to problems of three-dimensional flow simply by generating a three-dimensional mesh and changing the expression for $\nabla^2\Phi$ in (10.18).

10.2 The Finite Element Method

The finite element method differs from finite difference methods in two respects. In the first place the domain is discretized into a mesh of finite elements of any shape, such as triangles. In the second place the differential equation is not solved directly but replaced by a variational formulation. There exist various variational formulations (see Desai and Abel [1972] or Zienkiewicz [1977]; we follow the approach on the basis of a minimum principle here; the discussion closely follows Verruijt [1982]).

We first discuss the variational principle, apply the finite element method to an elementary flow problem, and finally discuss the finite element formulation for transient flow.

10.2.1 The Variational Principle

We write the differential equation for two-dimensional groundwater flow in a form that is commonly used in finite element formulations. Darcy's law may be written in terms of the discharge vector (Q_x, Q_y) as

$$Q_x = -T\frac{\partial \phi}{\partial x}$$
$$Q_y = -T\frac{\partial \phi}{\partial y}, \tag{10.38}$$

where the transmissivity T may be a function of position. For shallow confined flow T equals kH, for shallow unconfined flow T equals $k\phi$, and for two-dimensional flow in the vertical plane T equals kB. We write the continuity equation in the form

$$\frac{\partial Q_x}{\partial x} + \frac{\partial Q_y}{\partial y} = N - \frac{\phi - \phi^*}{c}, \tag{10.39}$$

where the term $(\phi - \phi^*)/c$ is added to cover cases of flow in leaky aquifers (see Section 5.1). We combine (10.38) with (10.39):

$$\frac{\partial}{\partial x}\left[T\frac{\partial \phi}{\partial x}\right] + \frac{\partial}{\partial y}\left[T\frac{\partial \phi}{\partial y}\right] + N - \frac{\phi - \phi^*}{c} = 0. \tag{10.40}$$

The differential equation is applied in a domain $\mathscr{D}$ with boundary $\mathscr{B}$, which is subdivided in segments $\mathscr{B}_1$ and $\mathscr{B}_2$ with boundary conditions of the form

$$\phi = f, \qquad (x,y) \in \mathscr{B}_1 \tag{10.41}$$

and

$$-T\frac{\partial \phi}{\partial n} = g \qquad (x,y) \in \mathscr{B}_2, \tag{10.42}$$

where $\partial/\partial n$ represents differentiation in the direction of the outward normal to $\mathscr{B}_2$.

The variational principle that replaces the differential equation (10.40) is based on minimizing the following functional U (see Verruijt [1982]):

$$U(\phi) = \tfrac{1}{2}\iint\limits_{\mathscr{D}} \left\{T\left[\frac{\partial \phi}{\partial x}\right]^2 + T\left[\frac{\partial \phi}{\partial y}\right]^2 - 2N\phi + \frac{\phi^2 - 2\phi\phi^*}{c}\right\} dxdy + \int\limits_{\mathscr{B}_2} g\phi ds. \tag{10.43}$$

We demonstrate that the functional U has an absolute minimum if ϕ satisfies (10.40) with the boundary conditions (10.41) and (10.42). Let χ be a function that differs from ϕ by a variation λ,

$$\chi = \phi + \lambda, \tag{10.44}$$

where λ is an arbitrary function of x and y that vanishes on $\mathscr{B}_1$,

$$\lambda = 0, \qquad (x,y) \in \mathscr{B}_1. \tag{10.45}$$

For the sake of brevity, we perform the analysis in indicial notation, replacing the coordinates x,y by x_i $(i = 1,2)$ and adopting the Einstein summation convention. The expression for $U(\chi)$ then becomes

$$U(\chi) = \frac{1}{2}\iint\limits_{\mathscr{D}} \left[T\partial_i\chi\partial_i\chi - 2N\chi + \frac{\chi^2 - 2\chi\phi^*}{c}\right]dx_1dx_2 + \int\limits_{\mathscr{B}_2} g\chi ds, \tag{10.46}$$

where $\partial_i\chi = \partial\chi/\partial x_i$ represents the derivative of χ with respect to x_i, and the Einstein summation convention implies that

$$\partial_i\chi\partial_i\chi = \partial_1\chi\partial_1\chi + \partial_2\chi\partial_2\chi = \left[\frac{\partial\chi}{\partial x_1}\right]^2 + \left[\frac{\partial\chi}{\partial x_2}\right]^2. \tag{10.47}$$

Substitution of $\phi + \lambda$ for χ in (10.46) gives, with

$$\partial_i\chi\partial_i\chi = \partial_i(\phi+\lambda)\partial_i(\phi+\lambda) = \partial_i\phi\partial_i\phi + 2\partial_i\phi\partial_i\lambda + \partial_i\lambda\partial_i\lambda, \tag{10.48}$$

the following expression for $U(\chi)$:

$$\begin{aligned} U(\chi) = {} & \frac{1}{2}\iint\limits_{\mathscr{D}} \left[T\partial_i\phi\partial_i\phi - 2N\phi + \frac{\phi^2 - 2\phi\phi^*}{c}\right]dx_1dx_2 + \int\limits_{\mathscr{B}_2} g\phi ds \\ & + \iint\limits_{\mathscr{D}} \left[T\partial_i\phi\partial_i\lambda - N\lambda + \lambda\frac{\phi - \phi^*}{c}\right]dx_1dx_2 + \int\limits_{\mathscr{B}_2} g\lambda ds \\ & + \frac{1}{2}\iint\limits_{\mathscr{D}} \left[T\partial_i\lambda\partial_i\lambda + \frac{\lambda^2}{c}\right]dx_1dx_2. \end{aligned} \tag{10.49}$$

The first two terms represent $U(\phi)$; we use that

$$\begin{aligned} & \partial_i(T\lambda\partial_i\phi) - \partial_i(T\partial_i\phi)\lambda \\ & \quad = \lambda\partial_iT\partial_i\phi + T\partial_i\phi\partial_i\lambda + \lambda T\partial_i\partial_i\phi - \lambda\partial_iT\partial_i\phi - \lambda T\partial_i\partial_i\phi = T\partial_i\phi\partial_i\lambda \end{aligned} \tag{10.50}$$

to write (10.49) as

$$\begin{aligned} U(\chi) = U(\phi) - & \iint\limits_{\mathscr{D}} \lambda\left[\partial_i(T\partial_i\phi) + N - \frac{\phi - \phi^*}{c}\right]dx_1dx_2 \\ & + \frac{1}{2}\iint\limits_{\mathscr{D}} \partial_i(T\partial_i\phi\lambda)dx_1dx_2 + \int\limits_{\mathscr{B}_2} g\lambda ds \\ & + \frac{1}{2}\iint\limits_{\mathscr{D}} \left[T\partial_i\lambda\partial_i\lambda + \frac{\lambda^2}{c}\right]dx_1dx_2. \end{aligned} \tag{10.51}$$

The first integral vanishes since ϕ satisfies the differential equation (10.40). The integrand in the last integral is positive since $T > 0$ and $1/c \geq 0$, so that this integral equals some positive number $|M(\lambda)|$,

$$|M(\lambda)| = \tfrac{1}{2} \iint_{\mathscr{D}} \left[T\partial_i\lambda\partial_i\lambda + \frac{\lambda^2}{c} \right] dx_1 dx_2. \tag{10.52}$$

We apply the divergence theorem to the second integral in (10.51):

$$U(\chi) = U(\phi) + \int_{\mathscr{B}} T\lambda \frac{\partial \phi}{\partial n} ds + \int_{\mathscr{B}_2} g\lambda ds + |M(\lambda)|. \tag{10.53}$$

The first integral vanishes along $\mathscr{B}_1$, by (10.45), and cancels with the second one along $\mathscr{B}_2$, by (10.42). What remains is

$$U(\chi) = U(\phi) + |M(\lambda)|. \tag{10.54}$$

We observe from (10.52) that $M(\lambda)$ vanishes only if $\lambda = 0$, i.e., if $\chi = \phi$: the functional $U(\chi)$ reaches an absolute minimum when the function χ equals the solution to the differential equation with boundary conditions, as asserted.

The application of the variational principle consists of minimizing the functional U for a chosen class of functions that meet the boundary condition along $\mathscr{B}_1$ and contain a number of parameters. For example, if χ is a function of x and y and a set of n parameters ϕ_j,

$$\chi = \chi(x, y, \phi_1, \phi_2, \ldots, \phi_n), \tag{10.55}$$

then $U(\chi)$ will be a minimum if

$$\frac{\partial U}{\partial \phi_j} = 0, \qquad j = 1, 2, \ldots, n, \tag{10.56}$$

which represents n equations for the n unknowns ϕ_j. The accuracy of the solution is affected both by the choice of the function χ and by the number of parameters; that (10.56) is fulfilled does not mean that the solution obtained is exact; it means that the best solution is found within the class of solutions considered.

10.2.2 The Finite Element Method for Steady Flow

A common choice for the function χ is obtained by discretization of the flow region in a mesh of n triangular elements, approximating the head as varying linearly inside each triangle, and taking the transmissivity T as being constant in each triangle. The integral $U(\chi)$ can then be written as a sum of integrals U_m over the n elements $\mathscr{D}_m$ $(m = 1, 2, \ldots, n)$:

$$U = \sum_{m=1}^{n} U_m. \tag{10.57}$$

We represent the head inside the triangle $\mathscr{D}_m$, $\overset{m}{\phi}$, as

$$\overset{m}{\phi} = \overset{m}{a}_1 x + \overset{m}{a}_2 y + \overset{m}{a}_3. \tag{10.58}$$

We derive expressions for the coefficients $\overset{m}{a}_j$ from the three linear equations, (10.57), obtained by letting m be 1, 2, and 3,

$$\begin{aligned}
\overset{m}{a}_1 &= (\overset{m}{b}_1 \overset{m}{\phi}_1 + \overset{m}{b}_2 \overset{m}{\phi}_2 + \overset{m}{b}_3 \overset{m}{\phi}_3)/2A_m \\
\overset{m}{a}_2 &= (\overset{m}{c}_1 \overset{m}{\phi}_1 + \overset{m}{c}_2 \overset{m}{\phi}_2 + \overset{m}{c}_3 \overset{m}{\phi}_3)/2A_m \\
\overset{m}{a}_3 &= (\overset{m}{d}_1 \overset{m}{\phi}_1 + \overset{m}{d}_2 \overset{m}{\phi}_2 + \overset{m}{d}_3 \overset{m}{\phi}_3)/2A_m,
\end{aligned} \tag{10.59}$$

where A_m is the area of the triangle and where

$$\begin{aligned}
&\overset{m}{b}_1 = \overset{m}{y}_2 - \overset{m}{y}_3, && \overset{m}{b}_2 = \overset{m}{y}_3 - \overset{m}{y}_1, && \overset{m}{b}_3 = \overset{m}{y}_1 - \overset{m}{y}_2 \\
&\overset{m}{c}_1 = \overset{m}{x}_3 - \overset{m}{x}_2, && \overset{m}{c}_2 = \overset{m}{x}_1 - \overset{m}{x}_3, && \overset{m}{c}_3 = \overset{m}{x}_2 - \overset{m}{x}_1 \\
&\overset{m}{d}_1 = \overset{m}{x}_2 \overset{m}{y}_3 - \overset{m}{x}_3 \overset{m}{y}_2, && \overset{m}{d}_2 = \overset{m}{x}_3 \overset{m}{y}_1 - \overset{m}{x}_1 \overset{m}{y}_3, && \overset{m}{d}_3 = \overset{m}{x}_1 \overset{m}{y}_2 - \overset{m}{x}_2 \overset{m}{y}_1.
\end{aligned} \tag{10.60}$$

We consider that the boundary $\mathscr{B}_2$ is a streamline so that $g = 0$ (see (10.42)) and the integral along $\mathscr{B}_2$ vanishes. We further set N and $1/c$ equal to zero so that the functional (10.43) becomes, representing the approximate solution as ϕ,

$$U = \tfrac{1}{2} \iint_{\mathscr{D}} \left\{ T \left[\frac{\partial \phi}{\partial x} \right]^2 + T \left[\frac{\partial \phi}{\partial y} \right]^2 \right\} dxdy. \tag{10.61}$$

The contribution of the mth element to U is

$$U_m = \tfrac{1}{2} T_m \iint_{\mathscr{D}} \left\{ \left[\frac{\partial \phi}{\partial x} \right]^2 + \left[\frac{\partial \phi}{\partial y} \right]^2 \right\} dxdy, \tag{10.62}$$

where T_m represents the transmissivity of the mth element, which is taken to be constant. Using (10.58) in (10.62) we obtain

$$U_m = \tfrac{1}{2} T_m \iint_{\mathscr{D}} (\overset{m}{a}_1{}^2 + \overset{m}{a}_2{}^2) dxdy = \tfrac{1}{2} T_m A_m (\overset{m}{a}_1{}^2 + \overset{m}{a}_2{}^2). \tag{10.63}$$

Using (10.59), this may be written as

$$U_m = \sum_{i=1}^{3}\sum_{j=1}^{3} \frac{\frac{1}{2}T_m A_m}{4A_m^2}\left[\overset{m}{b}_i\overset{m}{\phi}_i\overset{m}{b}_j\overset{m}{\phi}_j + \overset{m}{c}_i\overset{m}{\phi}_i\overset{m}{c}_j\overset{m}{\phi}_j\right] = \tfrac{1}{2}\sum_{i=1}^{3}\sum_{j=1}^{3}\overset{m}{P}_{ij}\overset{m}{\phi}_i\overset{m}{\phi}_j, \qquad (10.64)$$

where

$$\overset{m}{P}_{ij} = \frac{T_m}{4A_m}(\overset{m}{b}_i\overset{m}{b}_j + \overset{m}{c}_i\overset{m}{c}_j). \qquad (10.65)$$

An expression for the functional U in terms of the heads at the nodes is obtained by summation over all contributions U_m:

$$U = \tfrac{1}{2}\sum_{m=1}^{n}\sum_{i=1}^{3}\sum_{j=1}^{3}\overset{m}{P}_{ij}\overset{m}{\phi}_i\overset{m}{\phi}_j = \tfrac{1}{2}\sum_{i=1}^{M}\sum_{j=1}^{M}P_{ij}\phi_i\phi_j, \qquad (10.66)$$

where M represents the number of nodal points and ϕ_i the head at node i. Application of (10.56) gives the following system of linear equations:

$$\boxed{\sum_{j=1}^{M} P_{ij}\phi_j = 0, \qquad i = 1,2,3,\ldots,n} \qquad (10.67)$$

where use is made of the symmetry of P_{ij} (compare (10.65)).

The finite element method is quite suitable for numerical applications; the coefficients of the matrix P_{ij} can be generated from the geometry of the triangular mesh. The system of equations (10.67) may be solved by Gauss-Seidel iteration as discussed for the finite difference method in Section 10.1. For listings of finite element programs see, for example, Verruijt [1982], Wang and Anderson [1982], and Bear and Verruijt [1987].

Example 10.1: Flow through a Strip

We consider the trivial case of flow through a strip of uniform transmissivity T as an illustration of the method. The flow region is the rectangle with sides of 2 m and 1 m shown in Figure 10.3.

The heads along sides 1–2 and 5–6 are ϕ_a and ϕ_b, and sides 2–4–6 and 1–3–5 are streamlines. The flow domain is subdivided in four triangles as shown in Figure 10.4. The first step is to express the constants $\overset{m}{a}_1$ and $\overset{m}{a}_2$ in terms of the heads at the nodal points. Application of (10.59) and (10.60) to the four triangles $\mathscr{D}_m$ gives, representing the head ϕ at node j as ϕ_j $(j = 1,2,\ldots,6)$,

$$\overset{1}{a}_1 = \phi_4 - \phi_2, \qquad \overset{1}{a}_2 = \phi_1 - \phi_2, \qquad \overset{2}{a}_1 = \phi_3 - \phi_1, \qquad \overset{2}{a}_2 = \phi_3 - \phi_4$$

$$(10.68)$$

$$\overset{3}{a}_1 = \phi_6 - \phi_4, \qquad \overset{3}{a}_2 = \phi_3 - \phi_4, \qquad \overset{4}{a}_1 = \phi_5 - \phi_3, \qquad \overset{4}{a}_2 = \phi_5 - \phi_6.$$

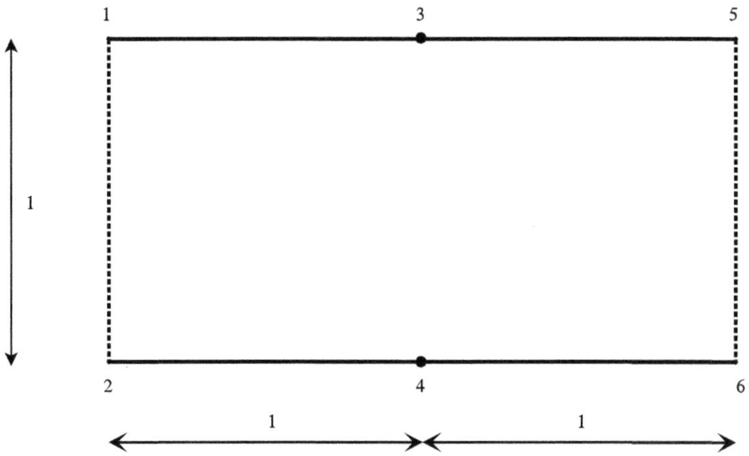

Figure 10.3 Flow in a strip.

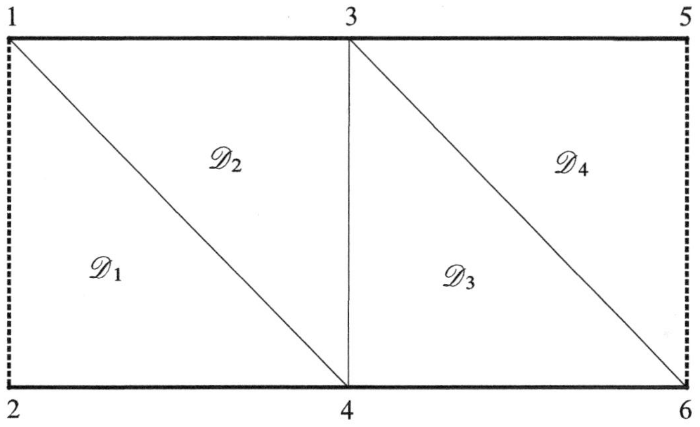

Figure 10.4 Finite element mesh for flow in a strip.

The heads ϕ_1, ϕ_2, ϕ_5, and ϕ_6 are given,

$$\phi_1 = \phi_2 = \phi_a, \qquad \phi_5 = \phi_6 = \phi_b, \tag{10.69}$$

so that the only two unknowns in the problem are ϕ_3 and ϕ_4. Differentiation of the functional U obtained from (10.57) and (10.63) with respect to ϕ_3 and ϕ_4 yields

$$\begin{aligned} T[(\phi_3-\phi_1)+2(\phi_3-\phi_4)-(\phi_5-\phi_3)] &= T[4\phi_3-\phi_1-2\phi_4-\phi_5]=0 \\ T[(\phi_4-\phi_2)-2(\phi_3-\phi_4)-(\phi_6-\phi_4)] &= T[4\phi_4-\phi_2-2\phi_3-\phi_6]=0 \end{aligned} \tag{10.70}$$

with the solution

$$\phi_3 = \phi_4 = \tfrac{1}{2}(\phi_a + \phi_b), \tag{10.71}$$

which corresponds to the exact solution for this case.

Problems with Leakage and Infiltration

For problems with leakage and infiltration, i.e., $N \neq 0$ and $1/c \neq 0$, the integrands in the expressions for U_m will contain a term $-N\phi + \frac{1}{2}(\phi^2 - 2\phi\phi^*)/c$. In the finite element method it is straightforward to account for variation with position of the infiltration rate N, the resistance c, and the head ϕ^* in the adjoining aquifer. Both the infiltration rate N and the resistance c are taken constant inside the elements, and the head ϕ^* is interpolated linearly inside the triangle $\mathscr{D}_m$ between the values $\overset{m}{\phi}{}^*_j$ $(j = 1,2,3)$ at the three corner points as follows:

$$\overset{m}{\phi}{}^* = \overset{m}{p}_1 x + \overset{m}{p}_2 y + \overset{m}{p}_3, \tag{10.72}$$

where $\overset{m}{p}_j$ $(j = 1,2,3)$ are obtained from 61.21 by replacing $\overset{m}{a}_j$ by $\overset{m}{p}_j$ and $\overset{m}{\phi}_j$ by $\overset{m}{\phi}{}^*_j$. Following Verruijt [1982], we introduce local coordinate systems $(\overset{m}{\xi}, \overset{m}{\eta})$ for each triangle, with the origin at the centroid,

$$\begin{aligned} \overset{m}{\xi} &= x - \tfrac{1}{3}(\overset{m}{x}_1 + \overset{m}{x}_2 + \overset{m}{x}_3) \\ \overset{m}{\eta} &= y - \tfrac{1}{3}(\overset{m}{y}_1 + \overset{m}{y}_2 + \overset{m}{y}_3). \end{aligned} \tag{10.73}$$

We obtain an expression for U_m following a procedure similar to the one used to derive (10.64). This expression is presented here without derivation:

$$U_m = \sum_{i=1}^{3}\sum_{j=1}^{3}\left[\tfrac{1}{2}\overset{m}{P}_{ij}\overset{m}{\phi}_i\overset{m}{\phi}_j - \frac{1}{c_m}\overset{m}{R}_{ij}\overset{m}{\phi}_i\overset{m}{\phi}{}^*_j\right] - \sum_{i=1}^{3}\overset{m}{Q}_i\overset{m}{\phi}_i, \tag{10.74}$$

where c_m is the resistance in element m, and

$$\overset{m}{P}_{ij} = \frac{T_m}{4A_m}\left[\overset{m}{b}_i\overset{m}{b}_j + \overset{m}{c}_i\overset{m}{c}_j\right] + \gamma_m \overset{m}{R}_{ij} \tag{10.75}$$

$$\overset{m}{R}_{ij} = \frac{1}{4A_m}\left[\overset{m}{d}_i\overset{m}{d}_j + \overset{m}{Z}_{xx}\overset{m}{b}_i\overset{m}{b}_j + \overset{m}{Z}_{yy}\overset{m}{c}_i\overset{m}{c}_j + \overset{m}{Z}_{xy}(\overset{m}{b}_i\overset{m}{c}_j + \overset{m}{b}_j\overset{m}{c}_i)\right] \tag{10.76}$$

$$\overset{m}{Q}_i = \tfrac{1}{3}A_m N_m, \tag{10.77}$$

where N_m represents the infiltration rate in element m and the coefficients γ_m are expressed in terms of the resistance c_m as

$$\gamma_m = \frac{1}{c_m}. \tag{10.78}$$

The constants Z_{xx}, Z_{yy}, and Z_{xy} are given by

$$\overset{m}{Z}_{xx} = \frac{1}{A_m} \iint_{\mathscr{D}_m} \overset{m}{\xi}{}^2 d\overset{m}{\xi} d\overset{m}{\eta} = \tfrac{1}{12}(\overset{m}{\xi}{}_1^2 + \overset{m}{\xi}{}_2^2 + \overset{m}{\xi}{}_3^2) \tag{10.79}$$

$$\overset{m}{Z}_{yy} = \frac{1}{A_m} \iint_{\mathscr{D}_m} \overset{m}{\eta}{}^2 d\overset{m}{\xi} d\overset{m}{\eta} = \tfrac{1}{12}(\overset{m}{\eta}{}_1^2 + \overset{m}{\eta}{}_2^2 + \overset{m}{\eta}{}_3^2) \tag{10.80}$$

$$\overset{m}{Z}_{xy} = \frac{1}{A_m} \iint_{\mathscr{D}_m} \overset{m}{\xi}\overset{m}{\eta} d\overset{m}{\xi} d\overset{m}{\eta} = \tfrac{1}{12}(\overset{m}{\xi}_1\overset{m}{\eta}_1 + \overset{m}{\xi}_2\overset{m}{\eta}_2 + \overset{m}{\xi}_3\overset{m}{\eta}_3), \tag{10.81}$$

where $\overset{m}{\xi}_j$ and $\overset{m}{\eta}_j$ $(j = 1,2,3)$ denote the local coordinates of the corner points of the triangle.

We minimize the functional U obtained from (10.57) and (10.74) and obtain a system of n linear equations in terms of the n heads ϕ_m, which may be written in the form (compare (10.67))

$$\boxed{\sum_{j=1}^{n} P_{ij}\phi_j = Q_i + \frac{1}{c_m}\sum_{j=1}^{n} R_{ij}\phi_j^*, \qquad i = 1,2,3,\ldots,n} \tag{10.82}$$

where ϕ_j^* represents the head in the adjoining aquifer at node j.

10.2.3 The Finite Element Method for Transient Flow

The above finite element formulation may be readily modified to cover problems of transient flow. We adapt the differential equation (10.16) to the case of variable transmissivity and obtain, including the leakage term,

$$\frac{\partial}{\partial x}\left[T\frac{\partial \phi}{\partial x}\right] + \frac{\partial}{\partial y}\left[T\frac{\partial \phi}{\partial y}\right] + N - \frac{\phi - \phi^*}{c} = S_p\frac{\partial \phi}{\partial t}. \tag{10.83}$$

Verruijt [1982] integrates this differential equation over a short time interval Δt and obtains

$$\frac{\partial}{\partial x}\left[T\frac{\partial \overline{\phi}}{\partial x}\right] + \frac{\partial}{\partial y}\left[T\frac{\partial \overline{\phi}}{\partial y}\right] + N - \frac{\overline{\phi} - \phi^*}{c} - \frac{S_p}{\Delta t}(\phi_+ - \phi_-) = 0, \tag{10.84}$$

where $\overline{\phi}$ represents the average value of ϕ over the time interval and ϕ_+ and ϕ_- are the values of ϕ at the end and the beginning of the time interval, respectively. We use an implicit formulation, as in Section 10.1 (compare (10.33)), so that $\overline{\phi}$ may be written in terms of ϕ_-, ϕ_+, and a coefficient ν as

$$\overline{\phi} = \nu\phi_+ + (1-\nu)\phi_-. \tag{10.85}$$

The interpolation corresponds to a forward difference for $\nu = 0$, to a backward difference for $\nu = 1$, and to a central difference for $\nu = 1/2$. Verruijt [1982] suggests a value somewhat smaller than 0.5. Using (10.85), (10.84) becomes

$$\frac{\partial}{\partial x}\left[T\frac{\partial\overline{\phi}}{\partial x}\right] + \frac{\partial}{\partial y}\left[T\frac{\partial\overline{\phi}}{\partial y}\right] + N - \frac{\overline{\phi}-\phi^*}{c} - \frac{S_p}{\nu\Delta t}\left[\overline{\phi}-\phi_-\right] = 0. \tag{10.86}$$

It follows that the transient effects are represented in (10.86) by a term that is similar in form to the leakage term. The equations derived for flow with leakage may be adapted to cases of transient flow by replacing ϕ by $\overline{\phi}$, ϕ^* by ϕ_-, and $1/c$ by β where

$$\beta = \frac{S_p}{\nu\Delta t}. \tag{10.87}$$

If the flow is both leaky and transient, (10.74) becomes

$$U_m = \sum_{i=1}^{3}\sum_{j=1}^{3}\left[\tfrac{1}{2}\overset{m}{P}_{ij}\overset{m}{\overline{\phi}}_i\overset{m}{\overline{\phi}}_j - \frac{1}{c_m}\overset{m}{R}_{ij}\overset{m}{\overline{\phi}}_i\overset{m}{\phi}{}^*_j - \beta_m\overset{m}{R}_{ij}\overset{m}{\overline{\phi}}_i\overset{m}{\phi}_{j-}\right] - \sum_{i=1}^{3}Q_i\overset{m}{\overline{\phi}}_i, \tag{10.88}$$

where β_m is the value of β in element m, and $\overset{m}{\overline{\phi}}_j$ and $\overset{m}{\phi}_{j-}$ represent the average head and the head at the beginning of the time interval at corner point j of element m, respectively. The expressions (10.75) through (10.77) and (10.79) through (10.81) for the various constants in (10.88) remain valid, but (10.78) is replaced by

$$\gamma_m = \frac{1}{c_m} + \beta_m. \tag{10.89}$$

The system of equations (10.82) is modified to account for the additional term in (10.88) as follows:

$$\boxed{\sum_{j=1}^{n} P_{ij}\overline{\phi}_j = Q_i + \sum_{j=1}^{n} R_{ij}(\phi^*_j + \beta^*\phi_{j-})} \tag{10.90}$$

The head ϕ_+ valid at the end of the time interval is obtained from (10.85) as

$$\phi_+ = \phi_- + \frac{1}{\nu}(\overline{\phi} - \phi_-). \tag{10.91}$$

The above developments were carried out in terms of heads in order to facilitate the understanding of computer programs on the market today, most of which are written with the head as the dependent variable. A formulation in terms of potentials may have advantages over the one in terms of heads. In the first place, problems where both confined and unconfined flows occur can be handled elegantly; in the second place the programs can be readily adapted to various types of flow, e.g., interface flow. A formulation in terms of potentials may therefore be preferred and will be discussed below.

10.2.4 Formulation in Terms of Potentials

We introduce a potential F for a vector with components U_x, U_y so that

$$\begin{aligned} U_x &= -\frac{\partial F}{\partial x} \\ U_y &= -\frac{\partial F}{\partial y}, \end{aligned} \tag{10.92}$$

where

$$\begin{aligned} U_x &= \frac{Q_x}{k} \\ U_y &= \frac{Q_y}{k}. \end{aligned} \tag{10.93}$$

The function F, defined as

$$F = \Phi/k, \tag{10.94}$$

is independent of the hydraulic conductivity. Expressions for F applicable to the various types of flow discussed in Chapter 2 are obtained from the definitions of Φ by division by k; for example, F equals $\frac{1}{2}\phi^2$ for unconfined flow and $H\phi - \frac{1}{2}H^2$ for confined flow. Constants, such as $-\frac{1}{2}H^2$, may be added to F without affecting the governing differential equations as follows from (10.92): the derivatives with respect to x and y of these constants vanish. It is important to note, however, that the aquifer thickness H now must be kept constant. Discontinuities in H may be allowed, provided that the corresponding jumps in potential are properly accounted for. Note also that the varying thickness of the flow domain, as occurring in shallow unconfined flows and interface flows, is automatically accounted for in the definition of the potential F. With

$$\frac{\partial Q_x}{\partial x} + \frac{\partial Q_y}{\partial y} = \frac{\partial [kU_x]}{\partial x} + \frac{\partial [kU_y]}{\partial y} = -\frac{\partial}{\partial x}\left[k\frac{\partial F}{\partial x}\right] - \frac{\partial}{\partial y}\left[k\frac{\partial F}{\partial y}\right], \tag{10.95}$$

the differential equation (10.83) becomes

$$\frac{\partial}{\partial x}\left[k\frac{\partial F}{\partial x}\right] + \frac{\partial}{\partial y}\left[k\frac{\partial F}{\partial y}\right] + N - \frac{\phi - \phi^*}{c} = S_p\frac{\partial \phi}{\partial t}. \tag{10.96}$$

For problems of horizontal confined flow the head ϕ is a linear function of F. In all other cases the relation between F and ϕ may be written as

$$\phi = \frac{F}{h}, \tag{10.97}$$

where h is the thickness of the flow domain. This thickness may vary with position, but depends on the solution and is to be determined iteratively. We write the differential equation (10.96) in terms of F and h as

$$\frac{\partial}{\partial x}\left[k\frac{\partial F}{\partial x}\right] + \frac{\partial}{\partial y}\left[k\frac{\partial F}{\partial y}\right] + N - \frac{F - F^*}{c^*} = S_p\frac{\partial (F/h)}{\partial t}, \tag{10.98}$$

where

$$F^* = h\phi^*, \qquad c^* = hc. \tag{10.99}$$

If the flow is steady and $1/c = 0$, the finite element formulation in terms of potentials is obtained from the one given above in terms of head simply by replacing T everywhere by k and Φ by F. For cases of leakage, the formulation remains valid, and F^* and c^* may either be kept constant using some average value for h or be updated in the iterative process using newly computed values of h.

Transient flow in confined aquifers may be covered by replacing S_p in the equations by S_s with

$$S_s = \frac{S_p}{h} = \frac{S_p}{H}. \tag{10.100}$$

For problems of shallow unconfined transient flow, integration of the differential equation over a time interval Δt yields, taking h to be constant over the time step and equal to the value h_-, obtained at the end of the previous time step,

$$\frac{\partial}{\partial x}\left[k\frac{\partial \overline{F}}{\partial x}\right] + \frac{\partial}{\partial y}\left[k\frac{\partial \overline{F}}{\partial y}\right] + N - \frac{\overline{F} - F^*_-}{c^*_-} - \frac{S_{s-}}{\Delta t}[F_+ - F_-] = 0, \tag{10.101}$$

where

$$F^*_- = \phi^* h_-, \qquad c^*_- = ch_-, \qquad S_{s-} = S_p/h_-. \tag{10.102}$$

It is seen by comparison of (10.101) with (10.84) that the finite element scheme in terms of potentials is obtained from the one in terms of heads by replacing T by k, c by c^*_-, ϕ^* by F^*_-, S_p by S_{s-}, $\overline{\phi}$ by $\overline{F}$, ϕ_+ by F_+, and ϕ_- by F_-. The difference between the two approaches, for the case of unconfined flow, is that in the formulation in terms of heads the transmissivity T is taken constant over each element, whereas in the present formulation the aquifer thickness varies as $\sqrt{2F}$ (F is a linear function) over the elements, but is kept constant over a time interval. Thus, the error is shifted from an error in aquifer thickness to an error in the computation of the terms for leakage and storage.

Appendix A

Sinusoidal Tidal Fluctuation

We apply the method of separation of variables to the governing differential equation for transient flow

$$\frac{\partial^2 \Phi}{\partial x^2} = \frac{s_p}{k\bar{h}} \frac{\partial \Phi}{\partial t}. \tag{A.1}$$

We assume that the solution has a period $2\pi/\omega$; we will verify this assumption after determining the solution. We pose a solution of the form

$$\Phi(x,t) = \Re\{F(x)e^{i\omega t}\}. \tag{A.2}$$

This technique is known as separation of variables; we assume that the solution can be written as the product of a function of x only, $F(x)$, and a function of t only, $e^{i\omega t}$, which has a period of $2\pi/\omega$. The function $e^{i\omega t}$ is complex and $F(x)$ is also complex. Since $\Phi(x,t)$ is a real function, the solution will be the real part of $F(x)e^{i\omega t}$.

We determine the function $F(x)$ by requiring that the complex function

$$f(x,t) = F(x)e^{i\omega t} \tag{A.3}$$

fulfills the differential equation (A.1). We substitute expression (A.3) for $f(x,t)$ in the differential equation and obtain

$$e^{i\omega t}\frac{\partial^2 F}{\partial x^2} = \frac{s_p}{k\bar{h}}(i\omega)Fe^{i\omega t}. \tag{A.4}$$

Division by $e^{i\omega t}$ gives

$$\frac{\partial^2 F}{\partial x^2} = i\frac{s_p}{k\bar{h}}\omega F \tag{A.5}$$

or

$$\frac{\partial^2 F}{\partial x^2} - \frac{F}{\lambda^2} = 0, \tag{A.6}$$

where λ is a complex number with

$$\frac{1}{\lambda}=\sqrt{i\frac{S_p\omega}{k\bar{h}}}=\sqrt{\frac{S_p\omega}{k\bar{h}}e^{i\pi/2}}=\sqrt{\frac{S_p\omega}{k\bar{h}}}e^{i\pi/4}=\sqrt{\frac{S_p\omega}{k\bar{h}}}\left[\frac{1}{\sqrt{2}}+\frac{i}{\sqrt{2}}\right]. \tag{A.7}$$

The differential equation for the function $F(x)$ has the same form as the differential equation for two-dimensional semiconfined flow; the solution of (A.6) is

$$F(x)=c_1e^{x/\lambda}+c_2e^{-x/\lambda}. \tag{A.8}$$

Note that λ and F are complex, as the coefficients c_1 and c_2 also are. The complex number $1/\lambda$ may be written in the form

$$\frac{1}{\lambda}=\sqrt{\frac{S_p\omega}{2k\bar{h}}}(1+i)=\mu(1+i), \tag{A.9}$$

where

$$\mu=\sqrt{\frac{S_p\omega}{2k\bar{h}}}. \tag{A.10}$$

Substitution of (A.9) for $1/\lambda$ in (A.8) yields

$$F=c_1e^{\mu(1+i)x}+c_2e^{-\mu(1+i)x}. \tag{A.11}$$

Expression (A.2) for Φ becomes

$$\Phi=\Re\{c_1e^{\mu(1+i)x}e^{i\omega t}\}+\Re\{c_2e^{-\mu(1+i)x}e^{i\omega t}\} \tag{A.12}$$

or

$$\Phi=\Re\{c_1e^{\mu x}e^{i(\omega t+\mu x)}\}+\Re\{c_2e^{-\mu x}e^{i(\omega t-\mu x)}\}. \tag{A.13}$$

We determine the real part and obtain

$$\begin{aligned}\Phi=&\,e^{\mu x}[\Re(c_1)\cos(\omega t+\mu x)-\Im(c_1)\sin(\omega t+\mu x)]\\&+e^{-\mu x}[\Re(c_2)\cos(\omega t-\mu x)-\Im(c_2)\sin(\omega t-\mu x).\end{aligned} \tag{A.14}$$

This equation represents the general solution for the class of problems defined by a periodic boundary condition applied at some given value for x in a semi-infinite aquifer. The next step is the determination of the constants in the solution. The boundary condition is applied at $x=0$ and the aquifer is defined by $0\le x<\infty$; the boundary condition is given by

$$\Phi(0,t)=\Delta\Phi\sin(\omega t). \tag{A.15}$$

We observe from (A.14) that (A.15) can be satisfied only if c_1 is zero, i.e.,

$$c_1=0. \tag{A.16}$$

Application of boundary condition (A.15) to (A.14) yields, noting that $c_1 = 0$,

$$\Delta\Phi \sin(\omega t) = \Re(c_2)\cos(\omega t) - \Im(c_2)\sin(\omega t). \tag{A.17}$$

Hence

$$\Im(c_2) = -\Delta\Phi, \qquad \Re(c_2) = 0, \tag{A.18}$$

so that

$$c_2 = -i\Delta\Phi. \tag{A.19}$$

We substitute zero for c_1 and $-i\Delta\Phi$ for c_2 in expression (A.14) for Φ and obtain

$$\Phi = \Delta\Phi e^{-\mu x}\sin(\omega t - \mu x). \tag{A.20}$$

The assumption that Φ can be represented by (A.2) appears to be correct; both the differential equation and the boundary conditions are fulfilled.

Appendix B

Numerical Integration of the Cauchy Integral

We consider numerical integration of Cauchy integrals. We begin with the Cauchy integral of the real part of Ω, given by (8.119). Note that this integral can be used only if $\Omega(z(\delta))$ are the boundary values of a function that is known to be holomorphic inside the unit circle.

B.1 Complex Integrand

$$\underset{j}{a_n} = \frac{1}{\pi}\oint_{C} \Phi(z(\delta))\mathrm{e}^{-in\theta}d\theta. \tag{B.1}$$

We evaluate this integral first by dividing the circle $\mathscr{C}$ into N segments, each of length $\Delta\theta$, and multiply the arc length of each of these segments by the value of the potential Φ at the center of each segment of arc. The radius of the circle in the $\underset{j}{Z}$-plane is 1 and the coordinate of the center of the arc number k is

$$\delta_k = \mathrm{e}^{\mathrm{i}\theta_k}. \tag{B.2}$$

We choose the first interval to span the arc between $\theta = -\frac{1}{2}\Delta\theta$ and $\theta = \frac{1}{2}\Delta\theta$, where

$$\Delta\theta = \frac{2\pi}{N}. \tag{B.3}$$

The value of θ_k thus becomes

$$\theta_k = k\Delta\theta, \qquad k = 1,2,\ldots,N. \tag{B.4}$$

The potential Φ requires a global value of its complex argument, i.e., of $z(\delta)$, which for the kth point is

$$z_k = z(\delta_k) = z(\mathrm{e}^{\mathrm{i}k\Delta\theta}) = \underset{j}{z} + \underset{j}{Z}\underset{j}{R} = \underset{j}{z} + \mathrm{e}^{\mathrm{i}k\Delta\theta}\underset{j}{R}. \tag{B.5}$$

We are now in a position to evaluate the integral, which reduces to the following sum

$$\underset{j}{a_n} = \frac{1}{\pi}\sum_{k=1}^{N}\Phi(z_k)\mathrm{e}^{-\mathrm{i}n\theta_k}\Delta\theta = \frac{\Delta\theta}{\pi}\sum_{k=1}^{N}\Phi(z_k)\mathrm{e}^{-\mathrm{i}n\theta_k}, \quad n > 1, \tag{B.6}$$

or, with (B.3)

$$\underset{j}{a_n} = \frac{2}{N}\sum_{k=1}^{N}\Phi(z_k)\mathrm{e}^{-\mathrm{i}n\theta_k}, \quad n > 1, \tag{B.7}$$

Efficiency of evaluation can be improved significantly by first computing and storing the values of Φ at the N points, and retrieving them as needed for each value of n, i.e., for each coefficient $\underset{j}{a_n}$. The coefficient a_0 must be computed separately, and is purely real:

$$a_0 = \frac{1}{N}\sum_{k=1}^{N}\underset{\text{other}}{\Phi}(z_k). \tag{B.8}$$

B.2 Cauchy Integral of the Stream Function, Ψ

If the boundary value problem is given in terms of the stream function, we must apply the boundary values of the stream function along the circle. The expression for the Cauchy integral is now in terms of Ψ, (8.120):

$$a_m = \frac{\mathrm{i}}{\pi}\int_0^{2\pi}\Psi(\delta)\mathrm{e}^{-\mathrm{i}m\theta}d\theta, \qquad m = 1,\ldots, \qquad \delta \in \mathscr{C}. \tag{B.9}$$

Numerical evaluation leads to

$$\underset{j}{a_n} = \frac{2\mathrm{i}}{N}\sum_{k=1}^{N}\Psi(z_k)\mathrm{e}^{-\mathrm{i}n\theta_k}, \qquad j = 1,2,\ldots. \tag{B.10}$$

The constant a_0 must be evaluated separately and is purely imaginary:

$$a_0 = \frac{\mathrm{i}}{N}\sum_{k=1}^{N}\Psi(z_k). \tag{B.11}$$

Note that this result is valid if Ψ is continuous; if branch cuts intersect the boundary, the integral must be divided into sections where Ψ is continuous.

List of Problems with Page Numbers

References

M. Abramowitz and I. A. Stegun. *Handbook of Mathematical Functions*. Dover Publications, New York, 1965.

E. E. Allen. Note 169. *MTAC*, 8:240, 1954.

V. I. Aravin and S. N. Numerov. *Theory of Fluid Flow in Undeformable Porous Media*. Daniel Davey, New York, 1965.

W. Badon-Ghyben and J. Drabbe. Nota in verband met de voorgenomen put boring nabij Amsterdam. *Tijdschr. Kon. Inst. Ing.*, pages 8–22, 1888–1889.

M. Bakker. Two exact solutions for a cylindrical inhomogeneity in a multi-aquifer system. *Adv. Water Resour.*, 25:9–18, 2002.

Modeling groundwater flow to elliptical lakes and through multi-aquifer elliptical inhomogeneities. *Adv. Water Resour.*, 27:497–506, 2004.

K. W. Bandilla, I. Janković, and A. J. Rabideau. A new algorithm for analytic element modeling of large-scale groundwater flow. *Adv. Water Resour.*, 30:446–454, 2007.

J. Bear and A. Verruijt. *Modeling Groundwater Flow and Pollution*. Reidel, Dordrecht, The Netherlands, 1987.

M. A. Biot. General theory of three-dimensional consolidation. *J. Appl. Phys.*, 5:339–404, 1941.

J. Boussinesq. Recherches théoriques sur l'écoulement des nappes d'eau infiltreés dans le sol. *Journal de Math. Pures et Appl.*, 10:363–394, 1904.

C. W. Carlston. An early American statement of the Badon-Ghyben-Herzberg principle of static fresh-water-salt-water balance. *Am Jour. Sci.*, 261(1):89–91, 1963.

H. S. Carslaw and J. C. Jaeger. *Conduction of Heat in Solids*, 2nd ed. Oxford University Press, London, 1959.

I. A. Charny. A rigorous derivation of Dupuit's formula for unconfined seepage with a seepage surface. *Dokl. Akad. Nauk SSSR*, 79:937–940, 1951.

H. H. Cooper and C. E. Jacob. A generalized graphical method for evaluating formation constants and summarizing well field history. *Trans. Am. Geophys. Un.*, 27:526–534, 1946.

H. Darcy. *Les Fountaines Publiques de la Ville de Dijon*. Dalmont, Paris, 1856.

S. N. Davis and R. J. M. de Wiest. *Hydrogeology*. Wiley, New York, 1966.

C. S. Desai and J. F. Abel. *Introduction to the Finite Element Method*. Van Nostrand Reinholt, New York, 1972.

J. DuCommun. On the cause of fresh water springs, fountains and c. *Am. Jour. Sci.*, 1st. ser. (44):174–176, 1828.

J. Dupuit. *Études Théoriques et Pratiques sur le Mouvement des Eaux dans les Canaux Decouverts et à Travers les Terrains Perméables*, 2nd ed. Dunod, Paris, 1863.

P. Forchheimer. *Hydraulik*. B. G. Teubner, Leipzig, 1914.

R. A. Freeze and J. A. Cherry. *Groundwater*. Prentice Hall, Englewood Cliffs, NJ, 1979.

N. K. Girinskii. Complex potential of flow with free surface in a stratum of relatively small thickness and $k = f(z)$ (in Russian). *Dokl. Akad. Nauk SSSR*, 51(5):337–338, 1946a.

Complex potential of fresh groundwater flow in contact with brackish water (in Russian). *Dokl. Akad. Nauk SSSR*, 58(4):559–561, 1946b.

H. M. Haitjema. Modeling three-dimensional flow in confined aquifers using distributed singularities. PhD thesis, University of Minnesota, 1982.

Modeling three-dimensional flow in confined aquifers by superposition of both two- and three-dimensional analytic functions. *Water Resour. Res.*, 21(10):1557–1566, 1985.

Modeling three-dimensional flow near a partially penetrating well in a stratified aquifer. In *Proceedings of the NWWA Conference on Solving Groundwater Problems with Models*, page 9, 1987.

Analytic Element Modeling of Groundwater Flow. Academic Press, San Diego, CA, 1995.

C. Hastings. *Approximations for Digital Computers*. Princeton University Press, Princeton, NJ, 1955.

H. R. Henry. Saltwater intrusion into fresh-water aquifers. *J. Geophys. Res.*, 64(11), 1959.

A. Herzberg. Die Wasserversorgung einiger Nordseebaden. *Z. Gasbeleucht. Wasserverzorg.*, 44:815–819, 824–844, 1901.

S. Irmay. Calcul du rabattement des nappes aquiferes. *VIemes Journées de Hydraulique*, 7 (Question I, Nancy, France), 1960.

C. E. Jacob. The flow of water in an elastic artesian aquifer. *Trans. Am. Geophys. Un.*, 21: 574–586, 1940.

I. Javandel, C. Doughty, and C.-F. Tsang. *Groundwater Transport: Handbook of Mathematical Models*. American Geophysical Union, Washington, DC, 1984.

D. Kirkham. Explanation of paradoxes in Dupuit-Forchheimer seepage theory. *Water Resour. Res.*, 3(2):609–622, 1967.

J. A. Liggett. Location of free surface in porous media. *J. Hyd. Div. ASCE*, HY4(102): 353–365, 1977.

S. W. Lohman. Ground-water hydraulics. *U.S. Geol. Survey Professional Paper 708*, 1972.

S. Mehl and M. C. Hill. Development and evaluation of a local grid refinement method for block-centered finite-difference groundwater models using shared nodes. *Adv. Water Resour.*, 25(5):497–511, 2001.

S. W. Mehl and M. C. Hill. Local grid refinement for MODFLOW-2000, the good, the bad, and the ugly. In *Proceedings, MODFLOW and More 2003, Understanding through Modeling*, pages 55–59, International Ground Water Modeling Center, School of Mines, Golden, CO, 2003.

S. W. Mehl and M. C. Hill. MODFLOW-2005, the U.S. Geological Survey Modular Ground-Water Model – Documentation of Shared Node Local Grid Refinement (LGR) and the Boundary Flow and Head (BFH) Package. In *Modeling Techniques, Section A, Ground Water*, number 6 in Techniques and Methods 6-A12, chapter 12. U.S. Department of the Interior, 2005.

R. D. Miller. Circular area-sinks for infiltration and leakage. Master's thesis, University of Minnesota, 2001.

L. M. Milne-Thomson. *Theoretical Aerodynamics*, 4th ed. Dover Publications, New York, 1958.

MODFLOW-2005, The U.S. Geological Survey Modular Ground-Water Model – Documentation of Shared Node Local Grid Refinement (LGR) and the Boundary Flow and Head (BFH) Package.

M. Muskat. *The Flow of Homogeneous Fluids through Porous Media*. McGraw-Hill, Ann Arbor, 1937.

P. Y. Polubarinova-Kochina. *Theory of Groundwater Movement*. Princeton University Press, Princeton, NJ, 1962.

O. D. L. Strack. A single potential solution for regional interface problems in coastal aquifers. *Water Resour. Res.*, 12(6):1165–1174, 1976.

Three-dimensional streamlines in Dupuit-Forchheimer models. *Water Resourc. Res.*, 20 (7):812–822, 1984.

The analytic element method for regional groundwater modeling. In *Solving Groundwater Problems with Models*, pages 929–941, Denver, CO, February 1987.

Groundwater Mechanics. Prentice-Hall, Englewood Cliffs, NJ, 1989.

Principles of the analytic element method. *J. Hydrol.*, 226:128–138, 1999.

Theory and applications of the analytic element method. *Reviews of Geophysics*, 41(2): 1–16, 2003.

Comment on "Steady two-dimensional groundwater flow through many elliptical inhomogeneities" by Raghavendra Suribhatla, Mark Bakker, Karl Bandilla, and Igor Janković. *Water Resour. Res.*, 41(W11601), 2005.

Using Wirtinger calculus and holomorphic matching to obtain the discharge potential for an elliptical pond. *Water Resour. Res.*, 45(W01409), doi:10.1029WR00Y 128, 2009a.

The generating analytic element approach with application to the modified Helmholtz equation. *J. Eng. Math.*, 64:163–191, March 2009b.

Vertically integrated flow in stratified aquifers. *J. Hydrol.*, doi: http://dx.doi.org/10.1016/j.jhydrol.2017.01.039, 2017. (in press).

O. D. L. Strack and B. K. Ausk. A formulation of groundwater flow in a stratified coastal aquifer using the Dupuit-Forchheimer approximation. *Water Resour. Res.*, 51:1–20, 2015.

O. D. L. Strack and H. M. Haitjema. Modeling double aquifer flow using a comprehensive potential and distributed singularities: 1. Solution for homogeneous permeability. *Water Resour. Res.*, 17(5):1535–1549, 1981.

O. D. L. Strack and T. Namazi. A new formulation for steady multiaquifer flow; an analytic element for piecewise constant infiltration. *Water Resour. Res.*, 50(10):7939–7956, October 2014.

O. D. L. Strack, R. J. Barnes, and A. Verruijt. Vertically integrated flows and the Dupuit-Forchheimer approximation. *Ground Water*, 44(1):72–75, 2005.

R. Suribhadla, M. Bakker, K. Bandilla, and I. Janković. Steady two-dimensional groundwater flow through many elliptical inhomogeneities. *Water Resour. Res.*, 40(W04202): 1–10, 2004.

C. V. Theis. The relation between lowering of the piezometric surface and the rate and duration of the discharge of a well using ground water storage. In *Trans. Am. Geophys. Un., 16th meeting*, volume 2, pages 519–524, 1935.

G. Thiem. *Hydrologische Methoden*. J. M. Gebhardt, Leipzig, Germany, 1906.

P. van der Veer. Calculation methods for two-dimensional groundwater flow. PhD thesis, Delft University of Technology, the Netherlands, 1978.

A. Verruijt. *Theory of Groundwater Flow*. Macmillan, London, 1970.

Theory of Groundwater Flow, 2nd. ed. Macmillan, London, 1982.

K. von Terzaghi. *Theoretical Soil Mechanics*. Chapman & Hall, London, 1943.

H. Wang and M. P. Anderson. *Introduction to Groundwater Modeling*. Freeman, San Francisco, 1982.

W. Wirtinger. Zur formalen Theorie der Funktionen von mehrenen komplexen Veranderlichen. *Mathematischen Annalen*, 97:357–375, 1927.

E. G. Youngs. Exact analysis of certain problems of ground-water flow with free surface conditions. *J. Hydrol.*, 4:277–281, 1966.

Seepage through unconfined aquifers with lower boundaries of any shape. *Water Resour. Res.*, 7(3):624–631, 1971.

O. C. Zienkiewicz. *The Finite Element Methods*. McGraw-Hill, London, 1977.

Index